MW01503110

SPAS T. RAIKIN

Emeritus Professor of History
East Stroudsburg University of Pennsylvania

REBEL WITH A JUST CAUSE

A POLITICAL JOURNEY
AGAINST THE WINDS OF THE 20TH CENTURY

Volume II

Sofia–Moscow
2001

© Spas T. Raikin

PENSOFT Publishers
Acad. G. Bonchev Str., Bl. 6, 1113 Sofia, Bulgaria
Tel.: (02) 979 34 60, 979 34 06
Fax: (02) 70 45 08
E-mail: pensoft@mbox.infotel.bg
Pensoft Online Bookshop: *www.pensoft.net*

© **PENSOFT Publishers**
First published 2001
ISBN 954-642-130-8

A catalogue record for this book is available from the British Library

Printed in Bulgaria, April 2001

To my wife Ruby,
To my children Ted and Fia,
And to my Grand children
Stephanie and Gregory.

To Norman & Judith

J Rollin

5/20/03

BY THE SAME AUTHOR:

Political Problems Confronting the Bulgarian Public Abroad (In Bulgarian)

Vol. 1. *Problems of Bulgarian foreign policy.*

Vol. 2. *Problems of the Bulgarian Agrarian National Union.*

Vol. 3. *Problems of the Bulgarian Orthodox Church.*

Vol. 4. *Communism, Socialism, Fashism.*

Vol. 5. *Human rights*

"Nationalism and the Bulgarian Orthodox Church," in <u>Religion and Nationalism in Soviet and East European Politics</u>, Pedro Ramet, (ed.) Duke Press Studies, Durham, N. C, 1984.

"The Bulgarian Orthodox Church," in <u>Eastern Christianity and Politics in the Twentieth Century</u>, Pedro Ramet (ed.), Duke Press Studies, Durham and London, 1988.

"Bulgaria," in <u>Eastern Europe, Politics, Culture and Society since 1939</u>, Sabrina P. Ramet, (ed.) Indiana University Press, Bloomington and Indianapolis, 1998.

"The Schism in the Bulgarian Orthodox Church," in <u>Bulgaria in Transition, Politics, Economics, Society and Culture After Communism</u>, John D. Bell, (Ed.) West View Press, Boulder, Colorado, 1998.

"The Communists and the Bulgarian Orthodox Church, 1944-48. The Rise and Fall of Ex-arch Stefan," in <u>Religion in Communist Lands</u>, Keston College, Winter, 1984

"Successes and Failures of Atheism in Bulgaria," in <u>Religion in Communist Dominated Areas</u>, New York, Spring and Summer 1988, Parts I and II.

<u>Encyclopedia of Eastern Europe</u>, Richard Frrucht (ed.), (New York and London: Garland Publishing, 2000) articles: "Bulgarian Emigres," "Todor Alexandrov," "Ivan Bagrianov," "Panayot Hitov," "Dimo Kazassov," "Traicho Kostov," "Andrey Liapchev," "Alexander Malinov," "Ivan Mikhailov," "Nikola Mushanov."

CONTENTS

INTRODUCTION

This autobiography was written for the benefit of my wife, my children and grand-children who have been urging me for many years to leave an account of my life experiences in Bulgaria. During the years I have given them little tid bits of my childhood and my escape from the country of my birth. They have been intrigued by my past and I could no longer resist their pleas.

Re-reading my Memoire, I could not escape the conclusion that I have poured my soul in it in expressing my love for the people closest to me – my father, my mother and my brothers, and my bitter apprehension of the conditions, the circumstances and the individuals which forced me to seek relief beyond the border, overcoming enormous difficulties and risks, and succeeding at the end under the protective hand of God. My life was an endless struggle of the Good versus the Evil and I am grateful that I lived long enough to see my cause vindicated by history. My dedication to the cause of Good, and my struggles against the Evil were sustained by my deep faith and hope for an ultimate liberation of Bulgaria.

The Bulgarian version of this Memoire is planned as a five volume collection of reminiscences and documents of my life experiences and activities since about the 1930s. This first volume is the only one to appear in both languages – English and Bulgarian. After finishing the manuscript I tried to figure out under what title I should publish it. It occurred to me that all my life has been an uninterrupted Journey Against the Winds of the XX Century. I felt that this should be the general title of all volumes. For the English version of this volume I chose to use the title *Rebel with a Just Cause* – not to counter James Dean's classic, but because it best expresses the essence of my private and public endeavors. For its Bulgarian version I had chosen the title *Trudovak*, my status in the construction troops where I did my military service. The word literary means "laborer" but is used exclusively to mean "labor soldier." The service, in my times known as "Trudova Povinnost," later changed to "Construction Troops," was in effect a penal institution for political opponents to the communist regime. As it will be seen in the text, the word "soldier" is a misnomer, totally inappropriate for the type of duties a *Trudovak* had to perform. It was more a "peculiar institution," as slavery was known in the United States in ante-bellum times, rather than a military service.

I had chosen the term "Trudovak," even though it covered a brief period of my past, 1950-1951, because it proved to be a turning point in my life. It was a climactic point of

1

my struggles as a peasant boy battling abject poverty, theology student persecuted by the Communist regime and a political activist challenging the most brutal oppression of the century. It led me to the risky decision to run away from the Trudovak camp and seek safety in the free world – to continue my struggles and build a personal career in the academic world. However, I realized that by stressing this brief period of my life, I was diminishing the value of my other experiences in those difficult years. This is how I came to choose the title <u>Blue and Red Bulgaria</u> 1922-1951 for the Bulgarian version. This volume, <u>Rebel with a Just Cause</u>, being English version of <u>Blue and Red Bulgaria</u>, ergo, English version of volume I of the series <u>A Political Journey Against the Winds of the XX Century</u>, is being designated as volume II of these series.

Having reviewed and copied my Dossier in the archives of <u>Durzavna Sigurnost</u> (State Security), the Bulgarian "KGB", containing hundreds of documents, relative to the events described in chapter XI of this volume, I dcided that it would be the natural sequence and companion to this Memoire, to be designated as Volume 3 of my <u>A Political Journey Agaist the Winds of the XX Century</u>, though it may be published after Volume 4. The said collection of documents will be published, appropriately, under the title <u>Desertiory</u> (Deserters), which is the title of the Dossier of our group. The State Security had titled my personal dossier by the name of "Iskariot" (Judas) and has referred to me in all documents as "Renegade of the Motherland", "Bandit", et al. The volume will be published with an introduction, to be folowed by the text of all documents in the archives which I have copied.

In the fourth volume of the project, entitled "Bezotechestvenitsi" (Stateless persons, literally meaning "people without a country") I intend to concentrate on my peregrinations as a stateless person in a refugee camp in Greece, and my studies in Switzerland and England, until my arrival in America, sponsored by Metropolitan Andrey. In the same volume I will publish my diary of the first month and a half after arriving in the United states, my problems with Metropolitan Andrey, excerpts from my correspondence with Bulgarian and foreign friends, my services in resettling Bulgarian refugees in the United States and my personal and professional settling down in America. In the fifth volume, entitled "Bulgarian National Front" I will cover the activities of that organization, as reflected in my articles published in "Borba" – its organ – to which I was the editor. In this, and in the following volumes, I will publish a vast collection of documentary material concerning the BNF and other Bulgarian Associations – the American Bulgarian League, The Bulgarian National Council and the Anti-Bolshevik Blok of Nations, to which I was Secretary General – 1959-1961. It will all focus on the Bulgarian Liberation Movement abroad.

If future historians become interested in this page of our past, I hope to save much material for their consideration, which, otherwise, may be lost. Should I, for some reason, fail to finish this ambitious plan, there is an enormous amount of material in my archives, which, I hope, will be somehow preserved. When I, and all Bulgarian emigrants in exile, were engaged in these activities, we felt that we were serving a noble cause for which destiny had chosen us to be its messengers. I never missed a moment in my life when I did not have an eye on history, as insignificant as we otherwise were. I have always looked to history as a human drama, and to us as actors playing on the big stage, small men in the crowd or leading political figures, with all our sins and charms, with all of our weaknesses and strengths, illusions and hopes, but always conscious that we were on a mission to help our nation which had given us our lives and for God, who had ordained this mission to us. How well we have fulfilled our duty to our fellow men of our times, and to those who have come after us, is not for us to judge. But one thing which I would not admit to anyone to question is our sincerity and dedication to our sacred cause. We fought the winds in our journey, and we defeated them. What follows in these volumes is history, nothing less, and nothing more.

Sometimes in October 1965, as a member of the faculty of the History Department at State University of New York, the Potsdam Campus, I had to deliver my first lecture in a team-teaching experiment of History of Western Civilization. As I took a place at the lecturn in the auditorium, ready to begin my discussion, I looked at the audience of students and caught the hundreds of pairs of eyes fixed on me. The front row was occupied by my colleagues in the department. In one fleeting moment, I closed my eyes and in my mind flashed the long path I have travelled from the muddy streets of the small Bulgarian village of Zelenikovo, through trials and tribulations, to this temple of the American Academia where I was called to celebrate the triumph of my dreams. I have defied all curses of the fates and have reached the mountaintop of my salvation and redemption. If the reader has the patience to reach the end of these pages, he or she will understand my feelings of victory over the forces of evil pursuing me in this age of fascism and communism. I do not claim to be a hero and this is not a heroic story. This is an ordinary story of an ordinary man, who lived in an extraordinary age.

Even before putting the finishing touches to this first volume, I was struck by the thought whether my experiences had any value at all. From the very beginning, to the very end, they are not exclusively my destiny. Hundreds, thousands of young men and women have trudged the same paths. Perhaps many others have written their vitae, or never written at all, and my contribution is simply an useless addition to other people's similar experiences. To that, many of my fellow fighters have suffered much worse things, have ended their lives in tragedies, as I have indicated on many occasions such instances in my memories. Burdened by such thoughts, at one point I was ready to abandon this project, my reminiscences of my childhood, my studies, my service as *trudovak*, in that penal institution, and my odyssey in the Rodopy Mountains, followed by my twilight years in Europe and America. But while going through all this examination of my path, I realized that I owed this confession to myself. During the writing I relived the feelings which had troubled me in the times when the events recalled here, had occurred. I could not so easily throw them in the basket of oblivion. I was deeply impressed by my long path of struggles against all kinds of adversities and injustices, having overcome them with patience and tenacity. It could not but please me. Who knows, these notes may fall in the hands of some young men, doomed to the same fate, on the verge of giving to desperation and loss of confidence in themselves and in life. If they read these pages they may be inspired to confront the misfortunes in life and overcome them with the same perseverance and determination with which I have confronted them. All my life I have pushed the boulder of Sisyphus towards the peak, it had always slipped from my grasp, I have again and again placed my shoulder under it, until throwing it over the challenges in my path. If all this helps at least one of these unhappy brothers in destiny, then all my labors have not been in vain.

All along this line of thinking, contemplating the myriad, trivial at first glance events and processes described on these pages (the reader may omit them if they appear to be boring him) I, having, taught Cultural Anthropology for some years, and archeology as introduction to my classes in History of Civilization, have come to realize that such trivia in appearance, are an important aspect of the achievements of humanity, and may interest future, as well as present students of our not so distant past, which, for the young, may look like ancient history. In my life time things have changed astronomically. Our ways of doing things some six or seven decades ago are remnants of the Neolithic age culture. In this respect our ways of life, our political, social and personal experiences, and historical events as well, reflected in my reminiscencies, described in their truthful features, may interest ethnographers, sociologists, political scientists, psychologists and historians. The picture I have presented is an authentic photograph of the past, without any attempt to change things as they have happened for considerations of current prejudices and attitudes, and will help to understand better our now ending XXth century.

The reader will notice that in many instances I have made big leaps to events having occurred decades earlier, or decades later, to place my experiences in the context of my entire life, in the conditions created after 1989. I have made these leaps in order to put them in proper perspective. I have felt satisfaction that such juxtapositions have come to vindicate me, confirming my visions of some five to seven decades ago. It is for this reason that I have deemed it appropriate to include at the end of this first volume my *Impressions of Bulgaria* which I visited in 1991, forty years almost to the day when I left it.

I am including these *Impressions* in order to contrast pre-communist and communist to post-communist Bulgaria, or rather to review the challenges facing the new democratic governments in the light of the past as seen by me. I must add here, that things have changed since 1991, the country has been led from totalitarianism to democracy, the old political and economic system has been dismantled – root and branch – but the incompetence and the amateurism of the new statesmen have made things worse and the country is going now through an economic crisis closely comparable to the years of transition from private ownership to socialism in 1950-1951.

I am also including in the text of my Reminiscence two essays written two weeks after crossing the border – "Trudovak II, the Peculiar Institution. Balchik," Chapter X, and "The Bulgarian Orthodox Church in the Conditions of the Red Terror," Chapter VIII (4). They complete the picture of my experiances in Bulgaria. All these reports, written in the Reception Center for Refugees, in the city of Drama, Northern Greece in July, 1951, have the value of documents of contemporary events, and, it is fair to say, reflect the spirit of the wounded animals, as we felt at the times. Still they are an important source for judging the times and the events subject to this tale. After all, I see this volume as a book of history. My role, from the beginning to the end is only incidental.

In these essays, and in this book, I have mentioned the names of many individuals – most of them by now departed from this world – for whom I have retained good or bad memories. To those whom I have mentioned in good standing, and are still alive, I could only express my gratitude for the happiness which we have shared in those difficult times. This same goes for those who have fallen asleep in the Lord, but my memories still hold feelings of pleasant remembrances. But also there are those of whom I have spoken ill, for their behaviour in the days of my travails. Again, most of them may have departed this world, and if I have mentioned them in unpleasant events in our past, it is not because of some grudge that I still have against them – God forbid! The times for that have long gone – but simply to characterize the times we lived in. And who knows, may be during the years of my exile they have atoned for their behaviour, and may have died as martyrs of the same evil, which had afflicted me. For those who have never expiated their sins, let God, not me, be their judge. And if there are still some of these people alive, and find my remarks reprehensible, or untrue, or unfair, or may be that they have long corrected themselves, to them I extend my apologies. I have not made my remarks with any intent for revenge, in some instances I myself may have been wrong, not knowing the exact circumstances of the events under consideration. All I have tried to do is to leave a true picture of the days we lived in. My memories cover an age which has no equal with its cruelty, its injustices, its moral, social, economic and political imperfections. There is no way to praise it for anything, there is no way of justifying it on account of its total lack of nobility. For some it may have been a blessing, but for me and myriads like me, at least until 1951, it was a time for suffering, agony and torments. Many of us were dragged, or pushed into the field of political struggles not because of a choice, but because of irrestitible circumstances, always inspired by ideals for a better society, more just than what we have been destined to live in. It may be that in seeking this ideal many of us have followed paths of illusions and have paid for their deception with their lives, but their sincerety and deep beliefs in the ideals pursued by them can never be questioned and will always move in our memories with the halo of self-sacrifice. Those with whom I have shared similar visions, I take this opportunity to honor them. Those, with whom I have not shared the same views, and in my story I have ventured to expose as evil doers, at this time I beg them for forgiveness, as I forgive all those who have caused me harm and injustice. Not too many years will pass, or to quote Alexander Pushkin's *Stancy*, "I chei ni bud o blizoi chas," there will be no one left of our generation. But let the memory of our times be preserved, and if any one could draw some lessons from our experiences and correct the things which have caused the evil and our suffering, this work of ours have not been in vain.

I have titled the English version of this volume *Rebel with a just Cause*, a variation of James Dean's film because, in effect, all my life has been a rebellion in the name of a cause – to expose injustice, or what I considered injustice in this world. My personal satisfaction of my successful response to all challenges which I have met is my priceless reward for all my efforts, and all my experiences, most of them not of very pleasant nature, and I would not exchange it for anything in this world, least of all to condemn it to oblivion. I have fought invincible *sub-sole* forces and have never surrendered to them in my life of seventy seven years plus as at the time of this writing. I cannot do that in the golden autumn of my life. All this may sound as a Faustian flame in me, but it is gratifying to view my life in such terms. So, I decided to leave this manuscript intact, and if possible, to see it published. Let the public know, now and in the future, my struggles and be my judge. History is written in more objective terms decades after some events have taken place. And if someone discovers these pages and uses them, his readers may have a better understanding of our problems in the 20th century.

Finally I have to apologize for the twists and turns of my English language. My style of writing may not be easy to follow sometimes, but it reflects the structure of my thinking and I rather leave it as it is. I have never taken even one hour of formal instruction in the subject and whatever marks this may have left on the text, let it be for my judgment and my condemnation. Because of the distance between Stroudsburg and Sofia, where the final copy is edited, it is possible that here and there the proverbial comma may appear or disappear and change the meaning of the text. This English version of the first volume is expanded here and there and at variance with the Bulgarian text of the first volume. I should also mention that here and there the reader may notice repeat references to some events and experiences which I have had. These repetitions have crept in inadvertently or have been inevitable. I have tried to eliminate most of them, but I may have missed some.

It is appropriate that I should express here my gratitude to those who have helped me in preparing this account of my turbulent life. I owe much to my family for their moral support which has sustained me in carrying out my crusading activities for the past four and a half decades and have encouraged me to make this account known to the younger generation of my Bulgarian people and American friends, free from the challenges and the ordeals which we, the older generation of Europeans, had gone through. I am very much obliged to those, without whose technical support this series of writings and their publication would not have been possible. In this respect the help of Marincho Nachev Marinov, student at the University of Sofia, with his painstaking efforts to decypher my handwritten notes of 1951, and preparing the text for publication, was invaluable for me. I am very much obliged to the staff of East Stroudsburg University, especially to Patt A. Newhart and Carolyn Gallagher of the Multilith Department for their assistance throughout the years in reproducing much of the materials which I have needed for my work, and especially to Luis Vidal of Instructional Resources for his services in producing the text on University facilities from the computer disks printed in Bulgaria.

It may be out of order, but I have to mention here all those individuals, most of whom have by now fallen asleep in the Lord, but at some critical moments in my life have, knowingly or unknowingly, intervened in my destiny and have effected a turning point, for the good, in my path of victories and successes against all adversities which have plagued me. I am forever grateful for their help. The reader will see these names as he plows through these pages, but I want to honor them as I begin telling my story.

In the first place I will mention my teacher in the elementary school, Boniu A. Bonev, who first lighted the spark of intellectual curiosity and the quest for a new moral order of things based on the principles of justice and truth. In the course of my growing to adulthood and all through my life, that spark loomed into an inextinguishable fire in my search for greater knowledge and greater justice. So many times friends and acquaintances have asked me to tell them about my family, seeking to explain my rise from ashes to honorable stations in life with heredity. No! There is no hereditary link to my life experiences. My family, parents and brothers, and as far as I have been able to establish, are and were virtually illiterate peasants. I owe all my good fortunes to that dedicated village teacher, a missionary of good in that God forsaken land.

Next I have to mentioned the village "Professor", Koicho Manevsky, a farmer with broad interests in matters social and political – secretary of the local Social Democratic party. He was instrumental in arranging and helping me enter the Plovdiv Theological Seminary. When a leading Bulgarian politician, Georgi Govedarov, in his campaign for election, visited the villages and asked if they wanted him to do something for them, Koicho told him to help a boy who could not afford it to go to high school. The reader will find the story told later in greater detail. The "Boy" was me. Without this chance event, I would have never gone further. Koicho was executed by the Communists when they come to power.

Dimitar Poptomov, a teacher in the Plovdiv Seminary, who stopped me from giving up my Seminary studies, in a one minute conversation, before I had even registered, saved my future. Yovcho D. Filchev, a young communist, partisan, saved me from the wrath of the red cannibals. Without his intervention, I might have gone to St. Peter in

1944. Then there was a man whom I had never met, Petar Peltekov, from the village of Brani Pole, but whom I had come to trust on account of his son, a fellow Trudovak, who misled us to defect from the Labor Military Service, but we successfully crossed the border on 19 June, 1951. Christopher King, an Englishman, Director of the Refugee Office of the World Council of Churches in Athens, saved me from disappearing into the jungles of Brazil by arranging a scholarship for me in Switzerland, until my departure for the United States in 1954. It would not have been possible without the sponsorship of Metropolitan Andrey of the Bulgarian Orthodox Church in New York. Dr. John Worinen, a Finn, professor at Columbia University, closed his eyes for my poor English language at the time and let me have a graduate degree, to be able to apply for a teaching job in the American universities. Dr. Carl Wilson, Dean of Social Studies at Rio Grande College in Ohio, took the chance and offered me a position as Assistant Professor of History, which opened the doors for me to better academic institutions – the State University of New York where Dr. Walter Wakefield sought my help for his studies of Medieval Heresies, specifically Bogomilism, and the East Straudsburg University of Pennsylvania, where I was warmly received by Dr. Kurt Wimer and settled down for good. It is with gratitude and humble thoughts that I am mentioning all these men along the paths of my turbulent life. Last, but not less important was the help I received from Drazha Parlapanska, my close friend, sister of Georgi Parlapansky and my aunt. I confided to her that I was going to study in the university in Sofia and needed a note that I have been stricken from the list for ration cards. She told me to meet her next day in her office in the municipal building. She was a secretary there. When I went to see her, she met me in the hallway, shook hands with me and I felt the crumpled paper intended for me between our palms. She had forged the signature of her brother, put an official seal, and surreptitiously passed it to me. Without that note I could not go to Sofia, my life would have taken an unknown course and I most probably would not be writing these Reminiscences now.

While struggling to raise a family and build a career as History Professor, I became deeply involved in Bulgarian exile politics. It was the late Dr. Kalin Koichev who introduced me to the A B C of this thorny field by way of the Bulgarian National Front. Over thirty years we worked together and struggled together , sacrificing all that we could to the cause of liberation of Bulgaria from communism. I am ever grateful for his friendship and inspiration to work for the public good.

May this book be a lesson for all those who lead the Bulgarian nation in the XXI century, if they venture to read it, so that all the miseries which we, the older generation, have passed through in Blue and Red Bulgaria, should never be suffered again.

Spas T. Raikin
Stroudsburg, Pennsylvania
11 November, 2000

I

BACK TO THE FUTURE

Sallus populi suprema lex esto!
Marcus Tullius Cicero
De Legis.

1. "Why Did You Escape from Bulgaria?"

On June 19 1951, about 4.00 p.m., seven fugitives pierced the Iron Curtain, crossing the border between Bulgaria and Greece, south-west of the town of Smolyan, at the corner where it turns south-east. Raggedy, starved, exhausted of their forty four days long ordeal in the Rodopy mountains, clutching carbines in their hands, they flew over the virgin meadow at the crest of the mountain between two distant military posts. No barb wires, no electrical charges, no mine fields, no frontier guards and dogs, least of all some Iron Curtain did they see running through the two feet high grass covering the long stretch of cleared of trees frontier, as far as they could see on the left and the right. There was only one well travelled path, from East to the West, where the guards walked on their rounds in patrolling the frontier. But this day, and at this hour, they were nowhere to be seen, not a living soul, and the scared fugitives rapidly passed from the land of tyranny, to the world of freedom. One of these seven was the author of this autobiography.

Many a time I have been asked abroad:"Why did you escape from Bulgaria?" My response has always been short and to the point:"Because I did not like communism! I could not live under this regime.!" The next question I had to answer to satisfy the curiosity of my friends was:"How did you escape?" Again my answer was short:"I crossed the border into Greece!." The conversation would then end with: "Was it difficult?" and "Very difficult, indeed!" Sometimes, answering my students in the University, I allowed myself to give more details and could read in their eyes sorrow and amusement. When I told the story of our escape, the forty four days in the Mountains, and the crossing of the border to a CIA agent who was investigating the case, and finished my monologue, he stared into the space, kept quiet a few minutes briefly – we were sitting outside, in the garden of the reception center – looked me in the eyes and said: "How in hell did you succeed?" I looked up and told him: "May be He, upstairs, had stretched His hand over us and led us out of the Hell in Bulgaria."

I do not know whether they, in the CIA, believed us, or suspected that we might have been sent by the Communists to infiltrate their services. Who knows! Such accusations were raised much later, but it never amounted to anything. Such accusations were made when I got involved in the political struggles among the exiles, but it was normal to invent derogatory accusations for all those playing on the political stage. But I have never come into possession of any credible evidence that the official services had ever doubted my story. For some reason, I very soon came out against these services, for their activities very often revealed their incompetence, and even hostility against genuine political exiles.

Months after I had written the above mentioned lines, I received my dossier from CIA, under the Law of Freedom of Information. I had requested this dossier in 1993. Reading through it, I was pleasantly surprized to see references to numerous reports of CIA informers or interviewers where all their sources had given glowing evaluations of my integrity and my political reliability, during the critical years of 1955-63. I withdrew from the political field in 1963, even before I had received my first appointment as a professor. There was only one report where doubts were expressed, and then only as a supposition, that everybody who comes over the border from the Communist world, might be a plant of the red intelligence services, which put me in a limbo, in the framework of this report. This information had come from a Church World Service official in Athens, Greece, a Ms. Mollie Rule. Ms Rule had arrived in Athens after I had departed for Switzerland, so I had never met her. But as it turned out one of my friends from Athens, a refugee, had become her helper. He had been for some time involved with the American organized subversive movement in Bulgaria known as "Goriani," but for some reason had been dropped by the CIA and sent to the refugee camp in Lavrion where we met. We became close friends and met on numerous occasions in Athens where I was given a scholarship by the World Council of Churches to study Theology. He enjoyed the privilege of frequent permits to visit Athens, where we met and shared our hopes and anxieties. He openly confessed that he had been a Legionair in Bulgaria. Our friendship continued after I left Greece and we exchanged a long correspondence, which may find its place in my subsequent volumes. I remember one occasion when he became vitriolic in his protestations at me, about some innocent, but inconsequential indiscretion in my letter to him. I ignored the whole thing, and eventually I arranged for his immigration to the United States. Now, reading my dossier from CIA, and the comments of Ms. Rule, I reluctantly thought of him as having been her mentor against me. Later on he became a grand persona in the Bulgarian community in Washington, D.C. In the meantime, after his arrival in the U.S. our paths crossed and we became bitter political enemies. It is after I read my CIA dossier that I thought that he might have been my black angel in Athens, Greece, whispering in the ears of Ms. Rule behind my back.

For whatever it is worth, I want to mention here an incident which took place in Greece while I was a refugee there. Sometimes, in 1952 I believe, the Chief of the Church World Service Mr. Christopher King asked me to accompany them to show me some monastery or something like that in Athens. There, a gentleman, armed with photocameras, asked me through King to give him permission to take pictures of me.

I had no objection and he placed me against a wall and started with his cameras. He was not making pictures for his family album. He was taking close shots, front and profile. Much later, when the affair of the notorious Philby came out, I remembered that the man in the newspapers looked very much like the photographer in Athens. I learned that in those days he had been in Athens, 1951 or 1952, that he had been Chief British Intelligence officer for the Balkans, that he was at the same time agent of KGB, and advisor of the CIA for the Balkans. I often thought if his reports had reached Sofia, and if Sofia had not tried, using his services, to compromise me and remove me from consideration for a job in their agencies. If anything like that had taken place, they had done me a favor, because I could not have tolerated their methods and their policies. I would have left them very fast and would have involved myself in Bulgarian exile affairs. After re-reading my remembrances in this autobiography, I was able to comprehend better, or rather to recognize, that in all circumstances, from the very beginning of my life, until defecting to Greece, and during all decades that followed, to this very day, I have been, in the strictest sense, independent in my judgments, and never, and to nobody a slave to grave for ideas, individuals and organizations, which had appeared to be false in my judgement.

My doubts about the interest of Philby in me after I crossed the border were further strengthened by another bit of information in the CIA dossier. I read there that in the middle of July, 1951, one month after my arrival in Drama, the British Intelligence Service, the Scotland Yard, during the tenure of the same Philby as a very high official there, I think chief of their Balkan department or higher, had taken energetic steps to have me transferred immediately to Athens. The transfer did take place and it appeared to me that it was done under some sort of an ultimatum. I have never been able to explain this and in my mind there reminded the conviction that the reception authorities may have had some bad information about me and were ridding themselves of me. Mr. Tsokalidis, the Bulgarian under a Greek name who acted as some sort of a confidant to the Greek or American police, explained to me in confusion that order had come from the United Nations offices to dispatch me immediately to them to be interviewed. I was dispatched, not in a very pleasant manner – I will discuss this case in the next volume – which puzzled me. I did not even stay in Salonica, as usually was the case. I arrived in Athens and was sent immediately to the refugee camp in Lavrion. Never did I see any United Nations representatives, to be interviewed by them. But one day, weeks later, I was taken to a British Intelligence officer in Athens, a very crude looking and unpleasant man, who after some formalities, made a proposal to me to work for them. I gave him the same answer which I had given to the CIA agent in Drama, that I was not interested in this kind of work, that I was looking forward to continuing my education and serving the Bulgarian Orthodox Church in exile. This was the end of the story. Was Philby working on this line? I will never know.

Just to complete these speculations, I have to mention that my Bulgarian dossier in State Security, a summary of which I have received a few years ago, contains a notation, somewhere in the 1960s, what they had established: I was not working for any Intelligence Service.

But the story of my escape was not simple, it could not be explained in a few words, or tainted by some sinister conspiracy of the Communist Secret Services. The reasons for my escape, are rooted in my personal life experiences, in many events which I had lived through, going far back before September 9 1944. What happened in 1951 was just the inevitable end of a drama which begins with my birth and developed in conditions and influences which cannot be explained just as "I did not like communism." This, of course, is true, but behind it lies a personal tragedy which led me to the Rodopy Mountains and the happy ending with the successful crossing of the border. This personal tragedy I want to tell here, in the first volume of my projected collection. In this volume I will offer my explanations as to why and how I became a political refugee.

It is not going to be an exaggeration if I say that no less than a hundred times I have thought of writing the story of our escape from Bulgaria, and probably as many times I had really started it, but I never went very far. During our odyssey in the Rodopy Mountains I kept a little diary, but sometimes, somewhere and somehow I lost it or inadvertently destroyed it. Recently I read an account of these events written by my friend Stefan Peltekov. He was one of the trio with whom we escaped from the *Trudovak* service in Balchik. It prompted me to return to my intentions of the past and record my own memories of what really happened. What I read in Stefan's book is a romantic adventure, perhaps for the purposes of making his tale interesting, perhaps out of desire to expose the Communists for what they were, but it all is done in such a primitive way and style, that it defeats all his intentions, and gives a distorted picture of the events. He has not attempted to play down anyone of his companions, nor to make himself and his father and uncle big heroes, though he appears to have appropriated to himself the role of a leader. Our story is big enough to need his distortions, and our difficulties serious enough, in order to be maximized by invented events. It would have been better if he had changed nothing, rather than try to add sensationalism in our tragic situation.

Some of the inventions of Stefan diminish the dignity of our risky adventure and make a mockery of our challenge to communism. His explanations about the immediate cause for our running away from the military camp is very much an infantile naiveté and in contradiction to every logic of the situation. I will discuss it further down. He has made up a story of an incident between me and Zdravko which never took place and, considering all circumstances, impossible to happen. In this story Zdravko, in exasperation from the setback in our initial planning, which exposed us to certain death, had decided to surrender himself to the police authorities when we were already in the Rodopy. The dialogue invented by Stefan, is a silly fabrication. In this dialogue I had threatened Zdravko that if he moved to leave us, I was going to kill him with my knife. First, there never was such a contingency. Second, I never had a knife to perform this threat. And third, Zdravko was twice as strong as I was and it would have been enough for him to give me a kick, so I would go flying onto the ground. I, the theologian, was going to kill Zdravko with a knife! What is more interesting in this episode is, that Stefan never mentioned in his book, that we were armed with carbines, he and Zdravko had the good ones and I had a small Turkish thing outfitted by some village smith. Why should I have threatened Zdravko with a knife, when he had the best gun in hand, Stefan had the other, and mine was a joke? When I was reading his book I wondered why he had never mentioned our guns. But at the end, where he tells at length his spiritual peregrinations in the United States – having become a high level functionary of a society of Theosophs, it became clear to me that such an admission, in writing, would have exposed him to his Theosophic following, where fire arms, an instrument for killing, are a tabu. So, the knife was more appropriate for me to kill Zdravko. Another tale revealing his naivete is a story of Zdravko, having gone to look for food supplies, gets involved in a wedding celebration of some relative of his, where he drinks, and dances and enjoys himself. And all that when the Communist Militia was looking all over for us.

But there are events in Stefan's book which were unknown to me, and he might have learned about them after 1989 from his relatives, as to how our case was handled by the police authorities and the military, the investigations, the arrests, etc., I was ready to complement my story by quoting in extenso from his book, even though, in the light of what I already observed, one could not be sure what is fantasy and what is truth. But even with such reservations, I was prepared to add something to my story. For this purpose I contacted his widow and asked permission, under the copyright laws, to do that. She categorically refused, and, therefore, I will go without his information. In fact, I am glad that I would not have to use it, for having to repeat stories which may be blatantly untrue. Stefan died sometimes in the spring of 1999, soon after we resumed relations, and he sent me his book. He was a good soul, and may be he had found a solace in his turbulent life with the Theosophs.

But let me return to the subject of this book and my story. The reader who ventures to learn this story should know from the very beginning that he will not encounter tragic events where people are subjected to horrible tortures in prisons and concentration camps, where common humanity seems to have yielded to satanic rituals, to bestial cruelty and unimaginable miseries. Had not this author taken the risk of escaping from the country where he was born and raised, he would not have survived the hellish conditions so ably described by Petko Ogoisky and Nedelko Geshev in their books *Notes of the Bulgarian suffering – 1944-1989*, and *Belene, the Island of the Forgotten*. We have not seen other descriptions of the bestialities practiced in Bulgaria over political prisoners, besides the reports in the press of the inhumanities committed at Skravena and Slunchev Briag, where human bodies were fed to the pigs, but all that we have read is enough for anybody to curse these long gone days.

2. Conspiracy Behind the Wagonette

On May 6, 1951, St. George's Day, Stefan, Zdravko and I escaped from the military labor camp in Balchik, on the Black Sea, near the Romanian border, where we were serving our regular military service as common laborers. We had served the previous year, 1950, in Bezmer, in South-East Bulgaria, in the district of Yambol. From Balchik we walked to Varna, some fifty kilometers, then took the train and the next morning, May 7 1951, about 10.00 a.m. arrived in Plovdiv, Filipovo railway station. We left the train to avoid arriving at the Central Station, fearing that we might be intercepted by the Police, or Militia, as they called it at that time. From there we were led by Stefan's uncle, Boris, and his co-conspirators to the Rodopy Mountains. This is how our odyssey and our adventures began. I use the words odyssey and adventure, because when the entire story is told, it will clearly fall in these categories, though the second word, adventure, in fact, expresses a fearful and frustrating experience. Yet, I knew why I took this full of dangers road. Right or wrong, I had come to the conclusion, while still serving in Bezmer, that I would not have survived, would not have come out alive from the military labor camp. My conviction was further confirmed after I was transferred, the next March, to the Balchik camp, after a three months lay off. Sometimes after our arrival there, Stefan, Zdravko and I started discussing the possibility of an escape – to the border and abroad. Stefan was in touch with his father and his brother-in-law, Vassil, a corporal in the regular army. They assured him that it all could be arranged, that they were already working on the plan., and eventually by the second half of April, they had it all set up. Stefan was giving us their assurances in such a convincing way, that we found ourselves committed to go.

I never entertained any doubt or hesitation that I had to try anything that would take me out of Bulgaria. While I was in Sofia, theology student, I planned an escape with Radko P. Todorov, through Yugoslavia. His village, Gradets, was close to the border and he had somebody who would take us across the frontier. In fact, Radko assured me, he was doing these things. We confided our plan to Arckhimandrit Gorazd. He had been abroad, in Constantinople, but for some stupid mistake had returned to Bulgaria. We could never understand why he had done that. We wanted from him information about the Church abroad and whether we could rely on Bishop Andrey to help us. He did not give us any assurances on that, he was evasive, but he did not try to disuade us from running away. To his credit, he never revealed our plans to anybody. These plans fell through anyway.The man who was supposed to take us over the border vanished from the scene and we continued to brood over the failure.

Now things looked better. The assurances given to us by Stefan that it was all arranged and in secure hands impressed us. He corresponded with his people in coded language and was receiving encouraging instructions. Many a time, after it was all over, especially after I read the book of Stefan, I could not help wondering how these people – Peter (Stefan's father), Boris, his uncle, Vassil, his brother-in-law, Maria – his sister, and Velika – his mother, had ventured into this affair, with all the risks involved, and all the dangers which were hanging over the members of the family who would be left behind, to encourage Stefan to take his chances and to drag us with him. Peter and Boris were simple peasants. Boris was a good soul, a friendly and unassuming elderly man, widower and with no children on the side. But Peter was very firm, full of himself, and convincing in his courageous presentation of the facts. Perhaps, the explanation for the whole affair, of this adventureous course which his son, and we with him, were to undertake, should be sought in him. He could stop it and could have told his son not to entertain such ideas. He should have thought of his wife, who, if things did not work out, would have to carry the burden of living alone while the rest of them could either be killed, or rotting in jails and concentration camps. Who knows, it may be that he himself had been pushed to the brink of the precipice, and was unable to rationalize his actions. Vassil, the brother-in-law, was serving as a corporal in the Plovdiv 9th Regiment, may be 22 or 23 years old, with a pregnant wife, was another mystery. He, we were told by Stefan, had been a Legionair, member of the Bulgarian pro-nazi youth movement, and may be was still living with the fantasies of the past, unable to recognize the realities, and place his family first. Had he thought it over whether his pregnant wife could go through all that? Or may be it was an infatuation with the idea of fighting communism?

Zdravko and I were bachelors, with no family ties to freeze us in our tracks – no wives, no children to think about. If we were taking the risk, we were risking ourselves, with no fear that it all would affect the rest of our relatives. While the preparations were going on, I was not asking myself the questions which the Peltekovs should have asked themselves. Were they led into this adventure by somebody else, somebody who was encouraging them, and if not genuinely involved in the struggle against communism, may be deceiving them?

The only thing that I was interested in and concerned about was how feasable and reliable the plans for proceeding from Plovdiv to the border were. From Balchik to Plovdiv it was our responsibility, which I was prepared to take, but what about after arriving in Plovdiv? After many discussions and communications between Stefan and his family, I was persuaded that the plans were sound, and firm like iron. Stefan persuaded us, Stefan convinced us of that. His father and Vassil had given him absolute guarantees that a man was going to take us to and across the border.

How and when this idea to escape from Balchik and head for the border was first broached, I do not clearly remember. It seems to me it was one afternoon when we were working in the field. We were digging dirt from high mounds, loading it on wagonettes and then pushing them to fill a ravine. A small railtrack cracked and swung along, often breaking, then repairing it, and pushing, and pushing. One afternoon, exhausted, disillusioned, hungry and disgusted, resting over the shovels, the three of us, Zdravko blurted it out: "I wish there was some way for me to run away from here, as far as I can." Stefan then started speaking of the connections which his father had and that some people were operating some canal to Greece. This is how it started. We asked questions, we delved into it, we fantasized about life abroad. Zdravko and Stefan wanted to go to school and become some specialists. I would look for opportunities in the Church and also continue with my studies of history. We did not much talk about politics in exile, or joining some liberation army to come back to Bulgaria. We had no information of that kind and were little concerned about contingencies of returning back. The power of communism was so overwhelming, and so firmly rooted in Bulgaria, the international situation was so discouraging, that we did not spend much time with illusions for return to a liberated Bulgaria.

Like in Bezmer, where we served the year before, we were working on a project to build a military airfield. Rumor had it that a chain of such fields was stretching from Bezmer – the last, or the first such project – to Balchik and then on all the way to Moscow. We were leveling the ground for laying of a runway for the Soviet Migs. There were too many mounds that had to be removed and we were called to perform this task by hand. There was no machinery to do the excavating and no trucks to move the dirt around. It was like the ancient Egyptians building their pyramids – the primitive way. It was a heavy, exhausting work – digging the ground with picks, loading the wagonettes with one cubic meter or more of dirt, and then pushing it hundreds of metters away. It was not much different from Bezmer.

We had become used to it, always cursing and castigating to hell the people who had sent us here – when there was no traitor among us – and were passing the days, counting the remaining months for this service. It was to last twenty four months, but somehow a rumor had it that the service was going to be extended to thirty six months. This broke our hearts and made us indifferent to the dangers which expressing our indignation from time to time publicly held for us. We were supposed to have free speech and if we noticed something wrong with the service, to speak our minds. Somehow I took seriously this allowance, and tried to exercize my right now and then. Also, after having served in Bezmer, I felt something like a veteran, and entitled to be heard if something was going wrong. I tried to phrase my observations in the language of the establishment, but still to make my point. The problem is, that no matter how you say it, if it is true it is easily recognized as a hostile agitation. At one of the company meetings, after the long tyrade of the political officer, we were given a chance to express our observations. I got up and started speaking about the food. I argued that to be able to perform better in our work and raise the production percentages, we needed physical energy. Physical energy was generated by more and better food. I observed that in comparison with Bezmer, the bean soup served in Balchik was poorer in quantity of beans, as compared to that in Bezmer, and asked if the portion was reduced. The political officer explained that the General Staff of the Construction Troups (Trudova povinnost) had appointed a special commission of nutritionists, that they had calculated the calories supplied by the beans, and have found that for the amount of work we do, a smaller portion of beans was adequate. I did not sit quietly, but continued my observations: "I have not counted the calories in my can, but I have noticed that two hours after lunch, I am hungry and hardly capable to continue working at the same pace.

The commanding officers continuously pressed for higher and higher percentages of production and were searching for methods where such percentages could be increased. There was much talk about stakhanovism, about dedication, about competition with corresponding units at other camps and all sorts of devices to make us work harder. At one point the political officer revealed a new plan for raising the work percentages. He told us that from now on, we will try to work by assigning one, instead of two men to every wagonette. I could not restrain myself and murmured, loud enough to be heard: "I hardly managed to work with another man, and now I have to do it alone? It is preposterous." But I was answered by another trudovak, I do not remember his name, who got up and shouted: "Of course, we could do it. Let us try it!" He was a former university student, one of those who were trying to curry favor with the officers, after being expelled from the law school, so they could assign him to a better job. It was decided to try it, any way. And it did not work. A few days later I found myself alone, face to face with that *trudovak* in the fields, looked in his eyes and with all the suppressed fury in my soul, clenching fists and gnawing of teeth, screamed at him: "Traitor!" He looked at me in disbelief, turned yellow, and walked away. He did not betray me. He did not report it to the officers. Because he was here, and God knows where else, for the same reason that I was – political activism in opposition to communism.

3. To Take Your Head in Your Hands ...

Things were proceeding uneventfully the first month. I was recognized by the officers and the *trudovaks* as "*Daskala*" (The teacher). This is how I had represented myself. After all, I was a teacher in the Sofia Theological Seminary before being called to serve. One day the political officer met me in the yard of the camp, stopped me, looked at me sideway and said: "You are no *daskal* at all. You are a pop (a priest)." He was saying more than there was to it. I was a theologian, but I was not an ordained priest. But for him the worst thing was to be a pop, rather than a Theologian. He gave me to understand that he had received the report from my village Communist authorities, and they had blackened me as enemy of the people.

Things were getting hot for me. One evening, I was just about to fall asleep, the Corporal of our group, standing at my feet, tapped me slightly on the thigh and asked me to follow him. He led me to an empty barrack, where, in the darkness, I soon recognized the political and a few low level officers. One of these low level officers, a man by the name of Popchev, I believe from the village of Goliamo Konare, now Suedinenie, under whose command I was, asked me: "Daskale, what are you planning to do after you finish your service?" I was a little flabergasted, but answered: "I think I will go teaching." He continued: "It is good! But the way you are going, you will hardly be able to get a job." He went on citing my occasional remarks critical of the order in the service. He stressed that I am sort of a role model for the other trudovaks, young men with less education, who looked at me as an example and listened to me and my opinions. I was not helping them to control this mass of people. In a way, I was subverting the service and their authority. He called my questions at the meetings of the company provocations and observed that this had to stop. "You are a university graduate," continued he, "you are a *daskal*, you are an authority for them. The majority of the trudovaks here are illiterate boys, and you, instead of being an example for them, encourage hostile attitudes towards our government." I realized the insecurity of my position and kept quiet. What could I say? I tried to play the penitent sinner, that I may have made mistakes, but that I, honestly, beginning from now, would be very careful in my behaviour, and would try to help them in any way I can.

By that time our preparations for running away were in quite advanced stage, and I would not risk messing them up. They asked me about Zdravko. I told them that we were friends, that he had little patience and sometimes misspoke out of frustration, but otherwise he was nothing more. They asked me of another young man, Koliu, who, incidentally, belonged to the group of the trudovak who had contradicted me at the meeting and I had called him "traitor". I did not know anything about Koliu, and left it at that. They asked me to tell Zdravko, "if he was not asleep yet," to come to them. Zdravko was sleeping next to me. When I snuggled myself under the blanket, I whispered to him to pretend that he is sleeping. They did not come to look for him. A few days later I saw Koliu going to the water fountains to wash his can. I went there, took the last of the eight spigots on the opposite end, and, without looking at him, whispered: "Be careful! They asked me about you." He understood, said "Thank you," and hurriedly went away.

While all these things were happening, we were already well into finalizing our plans for escape. I repeatedly asked Stefan for assurances from his people, that it all had been arranged and that there was no room for failure. Our plan was simple: we would run away from the camp, walk to Varna, take the train at about 6.30 in the evening and next day would be disembarking at Filipovo station, in the Karshiaka section of Plovdiv. We would be met there by the people who would help us and lead us to the border. When it was all arranged and I was fully convinced in the reliabilty of Stefan's people, I most carefully looked at every link of the chain of events to happen. At one point I almost gave out our plans. I wrote a letter to Radko p. Todorov in Sofia, to find me a compass. I mailed the letter in town, to avoid the controls of the political officer. Radko replied to me, that he could not find the book for which I was asking him. I received his letter unopened by the camp controls, so, uncensored. If they had seen it, and asked me what book I had asked for, I do not know what may have happened. But, nothing came out of it and we continued our planning. This went on for the whole month of April. We had arranged, and planned, to escape from the camp on May 1, anticipating that it was not going to be a working day. But, since we had lost one day of work, because of rain, May day was declared a work day, and we had to make up for lost time. So, our plan fizzled. They had been waiting in Plovdiv for us to show up. And they had given up on us. So, we postponed the escape for the next Sunday, which was also St. George's day.

Saturday evening, before the big day, we had our last conference. We retreated in the ravine, some two hundred feet far from the front of the barracks, lied down there, away from anybody, not to be seen, and went over our plans again. We had to make a decision to go, or not to go. We drew lots. On a piece of paper I wrote: "We will go!" On another piece – "We will not go!" I do not remember who pulled the crumpled piece, but it was for "Go!". Then we had to vocally express each individual's decision. Everyone said: "We will go!" The decision was made, definitively and unconditionally: We were going to start our long journey the next day, Sunday morning, St. George's day, as soon as the Company Assembly was over.

Many a time, ever since, I have thought of that evening, trying to remember my feelings on the eve of this turning point in my life. How, and why I decided to take this jump into the unknown? I could never recover any of my thoughts. It seems to me I was looking at the next day not much differently than any other day in service: no excitement, no fear, as if some sort of numbness had seized my whole being. What better, and what worse I could expect from that day ahead of us? Only a vague idea that we might succeed, was wandering in my mind. So many times I had told my co-conspirators:"we are taking our heads in our hands, we may succeed, we may fail." Never did I think as to what may happen if we failed. The worst that could happen was death. But it did not seem to worry me. We could be shot dead somewhere, most probably when crossing the border. I did not think of prisons or jails. Now the hour was near, and I did not feel anything! Only a wish that this hour came as soon as possible, and then, let it be, whatever comes. What was I looking for in the future, and why was I throwing myself into this risky undertaking? There was nothing clearly defined in my mind, except the thought that I was to free myself of this slave, penal servitude. Sometimes it crossed my mind that I might be considered for work in the free Radio Stations in the west, which, in my estimation, were doing a very poor job. But I did not know anybody to help me get in. What mostly tempted me was to find myself in an academic atmosphere, surrounded by books, and books, and books, as I was in the School of Theology. But, facing the new day of the unknown, I could not concentrate on anything. Now, forty five years later, after reading the account of Stefan, I am asking myself again some of these questions: why did I risk to jump in this dangerous, mortally dangerous adventure, where I could lose my life?

Rethinking all that now, I asked myself the same question in a different way which I had never thought before: under what circumstances I would not have taken this road? The answer clearly crystalized in my mind: Without the certainty that we would be met in Plovdiv by Stefan's father, and the people who were going to take us to and across the border, I would not have moved. This certainty was provided by Stefan. He had no doubts whatsoever, his trust in his father was so overwhelming that at the end, I also put my trust in him. I did not know these people. They were playing now a key role in our decision to go. But I did not put my trust in Stefan. He was a good fellow, good soul, friendly, intellectually quite sophisticated, but did not seem to be very well suited for practical life. He was more of a dreamer and not too well prepared for political games. I sort of saw him to be quite shallow, with lack of understanding of more complicated situations, and rarely showing some initiative. Now I realized that he was dominated by his father, who was manipulating him as an addendum to his own strong will. I did not rationalize these father-son relations at that time because I did not know his father, and, perhaps, because I was seeking a way out of the slavery of the trudovak service and was ready to accept any man who would have extended his hand to help me. Behind Stefan I saw the hand of his father, and this hand seemed strong and certain in its grasping of the situation. This is how I was dragged into this web of conspiracy, and by inertia, was pulled and pushed into it.

4. Zdravko.

Zdravko was a different story. He simply resented the work, he resented the officers who were always after him for one thing or another, mainly for refusing to deliver the quota of assigned work. But he never followed their admonitions and continued as before. He had graduated from High School, and had some vague ideas about continuing to study, but never explained what really he wanted to be. Before being drafted as trudovak, he was involved in youth organizations of the Bulgarian Agrarian movement on the opposition side. He was not allowed to register for studies in the university. He was convinced, that he had no future in Bulgaria, that his life would be wasted in occupations not much different from what the trudovak service was offering him. So, he did not see why he should overexert himself on account of the threats of the officers. Otherwise he was physically strong and healthy. If he worked, as much as he did, it was to kill time, not himself. He did not think much of the officers and their orders. But he hated from the bottom of his heart all that was around him under the sign of trudovak service. During our preparatory discussions he, like myself, made it clear that he did not want his family involved. I do not remember him speaking of a father, but a mother and sister he had. Whatever he was doing and was going to do, concerned only himself and nobody else.

It is only recently, last month, November 1999, that I learned the true story behind Zdravko's escape. I received a letter from his brother, dated October 23, 1999, in response to questions asked by me in our correspondence. Here is the text of this letter:

Jordan K. Damianov
K. Fotinov 2
Haskovo, Bulgaria

October 23, 1999.

Dear Mr. Raikin,

Thank you, from the bottom of my heart, for the letter and the enclosed recollections of your escaping from the camp of trudovaks to the free world in 1951.

In your letter you have written very little about yourself and nothing as to how you live now and what are your intentions of visiting your motherland Bulgaria, to see it, so that we could see each other here, in our long-suffering country. If you come to Bulgaria, even for a short time, please let me know, so we could meet and converse as brothers. You do not write anything about my brother, Zdravko – is he alive? Have you seen his name in some paper if he has passed away? I still believe that he is alive. I was only 14 years old when he left us in 1951 in our enslaved motherland.

In your letter you are asking me what kind of a young man was my brother Zdravko, and what is his family backgroiund.

I have been told, while I was still a kid, that my great grand father, Damian, had been a well to do man, with much land, and had been continuously elected mayor of the village of Belashtitsa. I remember nothing about my grand father Georgi, father of my father – Konstantin.

I remember that prior to 9 September 1944 my father had his own butcher's shop, and his brother, my uncle, had a small bar. I remember that after 9 September it all was nationalized and turned us into poverty. They had pointed a gun at the temple of his head, but did not kill him. He had become dejected, developed some bad illness, had turned black like a charcoal and had died. My mother had three brothers who were shot dead after September 9. Now we even do not know their graves. I remember how they were paraded in the village with handcuffs.

The age difference between me and Zdravko was considerable, but as much as I remember he was very strong and handsome young man. He was being expelled from the school. The State Security was following him, he was arrested on many occasions and beaten to a pulp. Our people were covering his black and blue body with fresh skins to save him. He used to write poems, and recite them. Long time after he was lost, the young men in the village composed songs about him and cried for him.

There was no Zdravko. The people in Belashtitsa mourned him and created many legends. Tormented after him, my mother died very early. I and my other four brothers and sisters were left orphans. Now I remember that when you have escaped, a military general came to our house, asking my father where was Zdravko. My father confronted him with accusations, asking them to give him back his son. "I gave him to serve you, I want him from you." This is how this horrible visit of the General in our house ended in Belashtitsa. I remember how our house was surrounded by armed men with cocked guns and grenades, so that a bird could not fly in or out. This continued for long time. The parents of my mother lived in the village of Novosel, some six kilometers from Belashtitsa. When you have been in the Mountain, my brother Zdravko had gone to grand mother Todorka for food. She had given it to him, but on his way back he had run into an ambush and saved himself by a miracle. The next morning grand-mother Todorka had been arrested and had been beaten to death. Likewise, my grand-father died from beatings... So, this is how they finished our family.

In 1951 I was in the first year of the Business School in Plovdiv. During the summer vacations I worked in the brick factory. There I pushed a wagonette of half a ton for sixteen hours a day, in order to make money for the school and to help my mother with a few levs. She was working in the Kolkhose farm, paid 20 pennies (Bulgarian) a day. This is how my childhood and teenage years were spent. After I finished the school I was also drafted as a trudovak. Days in and days out we were digging canals with picks and shovels. Many young men from our platoon jumped from the building to the ground and killed themselves there. Every evening about 1.00 a.m. I was being awaken and taken to the political officer for questioning. He would hold me there until 4.00 a.m., one hour before the whistle would blow for getting up. Our sufferings as trudovaks were no less severe than your's in Bezmer and Balchik. After your escape the borders were solidly fortified, with signaling technology, so that a bird could not fly over. At once all escapes from Communist Bulgaria ceased.

Having graduated from Business school with perfect score (A), I was not allowed to continue my education. My desire to study further evaporated. I was subjected to persecutions similar to those of Zdravko. This forced me to run away from Plovdiv and establish myself as a stranger in Haskovo, where I am to this day. After a long wait, twenty years later, I entered the Institute of Varna and graduated from there. After that, until now, I worked as a financial official.

You ask in your letter what has happened to us after your escape. Nothing good. We were persecuted everywhere. We were not allowed to take appropriate jobs, except in the quarries to break stone into gravel or some black and poorly paid work. We were declared "enemies of the people." They tormented our children too. Your escape from the Communist tyranny brought to you salvation, and to us oppression. On account of that that you had slipped out of their hands, they poured out their wrath on us. It is not possible to describe it in words. A man who has not experienced it on his back cannot fathom it all.

After 45 years, on November 10, 1989, a palace revolution took place. Since then our revolutionaries, who in reality are people of the Bulgarian Communist Party, never remembered you, the refugees from the Communist paradise, nor us, who suffered as guilty for your becoming traitors of the people, for having left Bulgaria. The red nomenclature overnight turned into capitalists, and took out of the country all what they could. They left the people to starve and live in misery. Ten years already they are stealing, and there is not even one to have been condemned by the courts. You are far away from us, so how could you help us?

As for us, only the suffering was left, now as before. If my brother Zdravko did not escape, we who were left behind were not going to be branded as "relative to emigrant in the West." My study would not have been postponed for twenty years and we would not have been left orphans. But there are things that man alone cannot solve. It is FATE which decides these things. This is what happened to you and to us.

I think that in this letter I have been exhaustive and satisfied your wish to write you in a common language for the events we lived through, afflicting us after your escape.

I am glad that now you are well. You have become professor. Most probably you live better and a more peaceful life than we do here. I am sad that we know nothing about Zdravko, in substance. I cannot accept it that he is not among us, the living, that he has passed away so young from this world – away from relatives and friends. We all here loved him tremendously and continue loving him, even though we have lived in separation and darkness for many decades. Please, write me as much in details as possible about Zdravko. I have the address of his son Karlos in Sao Paolo – Brazil, so that if you wish, you could write him and share with him the fact that we are exchanging letters – with me, his uncle Jordan. His address is ... Write him also in my name and give him my address... I will eagerly wait for your reply. Best regards from me and my family.

Jordan Damianov.

I wrote back to Mr. Damianov that I will publish his letter in my Memoires. It is a priceless document of the events I am discussing in this book and testimony for the times we have lived through. I have written to Zdravko's son and I am expecting a reply, if he is alive in Brazil. I will continue collecting information on Zdravko in Brazil and try to aleviate the pain in his brother's heart.[*]

5. Another Story of Our Escape

So, our decision of that Saturday evening was definitive and irrevocable. For over a month we have been preparing for our escape, we had discussed it, we had gathered and obtained all the information which we were interested in for its successful execution. It was not caused by some incident or an accident. It was not a reaction to a sudden event. My friend, Stefan, in his book "Forty Years in Exile", in Bulgarian, published in Sofia, 1993, has concocted the following story, which distorts and cheapens our risky undertaking. He recites the speech of the platoon commander, Lieutenant Draganov at the meeting immediately preceding our defection. Draganov told the trudovaks under his command that on May 3rd a group of them had moved some big pipes from a runway. One of the trudovaks, Nikola, from the village of Brezovo (Nikola was from my village) had played hookie and was ordered to move them during the night. He had moved four pipes, but left one on the runway. Next morning a Soviet Mig was taking off, hit the pipe and exploded. And the Lieutenant, in Stefan's version, continued:

This, comrades, is a sabotage, and your platoon has committed this sabotage deliberately. People's Republic of Bulgaria lost a plane. Our People's republic lost over three million levs. Well, tell me, how is it possible that this comrade of yours commit this sabotage, if he was not ordered by someone to do it? You all know that Nikola is a simple boy, and he could not think of such a horrible act. Comrades here is the finger of a group of conspirators, sons of fascists and opposition members who are attempting to sabotage our national defense, our national economy and to destroy socialism. These enemies of the people should be punished. They deserve the gallows. These enemies of the people, tools of the American imperialists, are here and are now gloating for the enormous damage which they have caused to our country. I will tell you their names. They are Slavi Radoikin (This is me, the name he uses for me in his book), Stefan Petrov (That is him, his correct name, without his surname, Peltekov) and Zaprian Danov (this is Zdravko Damianov). The Company Commander has

---

[*]Jordan died on December 23, 2000, after taking me to Belashtitsa, Markovo and Brani Pole - of heart attack.

ordered the Officer on Duty to arrest Slavi, Stefan and Zaprian at once, after this meeting is adjourned.

And Stefan continues his story:

"When the meeting was over … Stefan touched Slavi Radoikin and Zaprian Danov who were in front of him, and whispered to them:

"Did you hear the accusation? What are we going to do now?"

"I do not know", both of them replied.

"What a slander", Stefan continued. "Nobody of us has not said a word to Nikola to commit sabotage, but they are accusing us."

"But how could we tell them that we are innocent?, added Slavi, What justice exists today in Bulgaria?"

"No, there is no justice, this is true" said Zaprian. Everybody knows that."

"There is no time to think it over," said Stefan."Let us defect, immediately, before the meeting is over."

"Where are we going to defect?" asked Zaprian.

"I do not know", replied Stefan. This is not the time to think of that. But I am deciding to run away. Are you going to come with me?

"We have no other choice," replied simultaneously his two friends.

"Let us go, and let it come what may"

The three trudovaks, unnoticed, stepped aside under the cover of the fog..."

This is a concoction from the beginning to the end. As far as I am concerned, the reader will discover the real reason for my escape, in the following pages. My motives were much more complex. My life, from the day I was born, to this very day, led me to this act of desperation. The climax of my martyrdom in childhood and youth was reached in Balchik. If it was some imagined sabotage, committed by a young gypsy, inspired by us, we would have been arrested instantly, before any announcements at a general trudovak meeting. There was much, much more that had transpired in my personal life before that fateful day. I had seen in the extended hand of Stefan's father the lifesaving rope, which I grabbed and held firmly to it. It was the accumulation of all the torments of my spirit, of the griefs of my soul, all the heartaches suffered for so many years, all the tears and pains silently shed and suppressed during all these years since I was born. It is a chain of dreadful events which led me to this fateful day that was going to begin with the sunrise on May 6th, 1951. So, let me go back and chart the course of my life which was to be crowned with a risky, but successful challenge to my destiny.

II

GROWING UP IN THE "GOOD OLD TIMES"

"I am no prophet, nor a prophet's son; but I am herdsman and a dresser of sycamore trees, and the Lord took me from following flock, and the Lord said to me: Go, prophesy to my people Israel." Amos: 7:14.

1. From Babunovo to Zelenikovo. At the Foot of Sakar Tepe

On October 26 1922, the day when Alexander Stamboliysky had begun delivering his Swan Song in the National Assembly, I have opened my eyes to look at this unhappy world. I have read his speech so many times, and have found myself in it, if I had delivered it. I was born in Zelenikovo, Plovdiv district. At that time it still had its Turkish name, Hamzalare, after the name of some Hamza Bey, who may have owned it as a fief from the Turkish sultans. Zelenikovo is located 45 kilometers North-East of Plovdiv, at the foot of Sredna Gora, between the then villages of Rahmanly, now Rosovets, and Abrashlare, now Brezovo. These names were changed to their present Bulgarian nomenclature in 1934, but all the surrounding fields and mountains still have their Turkish names.

A local, self styled historian, Drazha Parlapanska – she will be mentioned again – has established that the village was in existence in the eleventh century, known under the name of Babunovo. When she told me recently of her finding, I told her that at that time the Bulgarian Bogomils, a dualistic heresy, were sometimes called Babuni, which may very well indicate that the population of the village, so far back, had fallen in the heresy of Bogomilism. Drazha spoke to me of interesting archeological sites near the village, indicating that some great battles had been fought there, supposedly by Crusaders against Bulgarians. She also told me of some remnants of Thracian culture in the vicinity of the village. I remember that when I was still a little kid, my grandmother was covering one of her crocks where she kept the cheese with a large marble plate – some 12 by 12 inches, and one inch thick with a Thracian horseman on one side. This plate must have come from the vicinities of the village, which would indicate that some sort of a settlement was there some five hundred years B.C., and from there, God knows how long before that people have been inhabiting these areas, and the village itself. And who knows what had happened to these people and races which have lived there. It is quite possible that the present generation may have some of the blood of those coming from ancient times.

I do not know how the name Zelenikovo has originated and who thought of replacing Hamzalare with this pretty name. May be because, looked at from the nearby long low hill on the east, rising from the banks of the river, it was, and still is, covered with greenery – Zelenikovo may be freely translated as Greentown. It is surrounded by high and sharply pointing hills, which are easy to recognize as mountains, extensions of Sredna Gora. North-West of it is the Yunchal, a Turkish word meaning Yun's Rock, with a high massive rock on one side which had always fascinated me. The Yunchal and the surrounding forests were used by half of the village for pasture during the summer months. Its sad glory is, that in October 1944 the Communist leaders had taken there six villagers, including the priest, and had killed them in the cruelest possible manner. This will be discussed later. One could not mention the Yunchal without thinking of the victims of the political storm that swept the country in the darkest days of our recent history. Behind the Yunchal and a number of mountainous hills, was the highest peak of Sredna Gora, Bratan. South of it was rising a conic spire-like peak, piercing the skies, with a Turkish name, hill Kissevriy. It was in the limits of the village of Rozovets. Next to Kissevriy, extending to the South, was a thick forest of yoke-elm trees, Dumlaliy, whose crest stretched about a mile down to a small elevation, known as the Bear's Hollow. Next to the Bear's Hollow was rising high in the skies, steep and magnificient, ending with sharp peak like a needle, my beloved Sakar Tepe. This is how it looked from the village, and from my window facing East. It looked like a replica of Vesuvius, without a volcano. The Thracian plain, reaching all the way to the Rodopy Mountains, perhaps forty miles or so, begins at the foot of this last extension of Sredna Gora. On the right side of Sakar Tepe, to the South-West, was a smaller hill, Kutchukia, as if it was standing as a guard to its big brother. All along, below Dumlaliy and the Bear's Hollow, in parallel with the river Giul Dere, running from North to South was Ruta (u pronounced as in much, and A as in dam). A bare hillock, a mile and an half long and perhaps three quarters of a mile wide, was slowly sloping towards the river and it was visible from every point of the village, used as a pasture for flocks of sheep. Standing there, one could see all of Zelenikovo with its green beauty and red tiles covering the roofs of the houses. It ended with a steep slope at the river bed. Hundreds of tracks, used by the animals, led upward to the plain meadows, which we, as children, used to climb up during spring and summer, and play on the wide grassy fields. Every morning and every evening one could see the hundreds of sheep, or goats and cow herds climbing and descending from all parts of the village. At the Southern end this hillock was a deep ravine, with water running towards the river, if and when there was water, but forested with much shrubbery, leading to the foot of Sakar Tepe and the Bear's Hollow. As little kids we spent much of our time in the spring playing in this ravine. It was the shortest way to Sakar Tepe.

During my childhood, when I was studying or dreaming at the window, I was looking at Sakar Tepe. The sun would rise around it, and depending on the seasons, it would gradually move from North to South and back. I would stare at it, lifting my eyes from the books for a short rest. It has left a deep imprint on my mind. This moment, writing these lines, I have pictures of it, taken in my recent visits there, taped opposite me on the panelled wall. And I still look at these pictures as I had stared at it for two decades in my early life. When I feel tired, and stop writing or typing, I lean back on my chair and look at Sakar Tepe, and with it I feel that I am back in my native land and my native village. After my first return to Bulgaria in 1991, after forty years of exile, in my *Impression*, Sakar Tepe was given an honorable mention. One of my classmates from the Seminary, to whom I had sent a copy, wrote to me back, that when they had their class reunion, he had read to them this portion of my *Impressions* where I mentioned Sakar Tepe. Here is this portion of the text:

Only Sakar Tepe was rising over the village as a silent witness of the sufferings and the misery through which the villagers had gone through. Sakar Tepe! It was embedded within me. I could not have enough looking at it. Two decades I had stared at its sharp peak high in the skies through the window of my crumbling down century old house, from early morning to late in the evening where I studied my lessons, learned to write and dreamed my youthful dreams. How many times I had climbed to reach this peak and from there to embrace with my eyes the entire Thracian Plain North of Plovdiv? How many times I have sought to penetrate through the farthest reaches of my vision, in the blue skies, its outline on the far, far away horizon? I would climb, on the Plovdiv hills, and from there seek to see Sakar Tepe. I would press my face to the glass of the highest round window in the Seminary, and seek the peak of Sakar Tepe. Rarely was I successful, and it filled my heart with joy and inner satisfaction. And how many times I have remembered it abroad?

God willed that I should see it again – so magnificient, so proud and so unchanged. What a feeling of victory had I experienced, and what inner satisfaction had I achieved when, in olden days, I ran, and ran to reach its peak? Now, standing in the middle of this desert, after forty years of wandering around the world, climbing and reaching other peaks in life, I was contemplating Sakar Tepe and a thought flashed in my mind: perhaps it was this peak, perhaps the inspiration and the pleasure to climb it in bygone times gave me the strength and the energy which have sustained me all these years in my exile misfortunes! Who knows!

For anyone who has never experienced such a long separation from his native land, all these emotional reactions to a mountain peak may seem naive and ludicrous, but for me Sakar Tepe has always been and will always remain, a Holy Land. At the very peak of it, in a rock, the centuries of erosion had dug a hole, more or less resembling a giant man's foot step. The tradition had it that Krali Marko, King Marco, a medieval legendary leader, who sided in fact with the Turks, but in the ethnic folklore has been portrayed as an invincible hero, fighter for the Orthodox faith, had stepped there with one foot, and has stretched his other foot to step on a hill, some three miles away, Gyol Tepe. So we all, when reaching the peak, tried our foot in this hole.

South of the Yunchala and west of the village, as from the foot of Sakar Tepe, there began the Thracian plain, reaching the Rodopy Mountains. It was a hilly land until it reached Brezovo, but was all cultivated for agricultural use. First came the *Voditsite*, then *Mal Tepe, Ala Bair, Babunar, Kudjouolare, Gyol Tepe, Kushuolan*, and then turn to the East to reach the Gul Dere River, embracing, *Chervinatsite, Chomlek Dere, Baramliytsa* and *Yurtishtata*. It should be noted that all these names were Turkish, leftover from Otto-man times. They were never changed. In some of these places there have been Turkish villages, like in *Kurdjalare*, but with the advance of the Russian armies in 1878, the entire Turkish population had fled. Their lands have been bought by our people. I heard the story, that my Grandfather had bought from a fleeing Turk a vast acreage, South of Zelenikovo, for the price of a "banitsa", a baked pita in a large pan made up of dough, cheese, plenty of butter, plus a pair of *navushta*, special strips of woollen clothe to cover the legs up to the knee in winter time. I was familiar with all these places, but few of them were my favorite hangouts. I will speak of them later.

2. Gyul Dere (River of Roses)

On the East side of the village was our small river – Gyul Dere – River of Roses. Indeed our region – Zelenikovo and Rosovets (Village of Roses), had had a highly developed Rose industry. Between the two villages, along the River, there were three rose oil factories – that of Botev, of Batsurov and Shipkov. The river itself – as a river – was not much of a water trek. In the late fall, winter and spring it had enough water, to need a bridge here and there for crossing. Only one of them had more or less permanence, because there must have been ten-fifteen houses on its left bank – a dillapidated structure which was shaking when walking over it. In the summer it was dry all the way. In May-June the Rose factories were pouring out the refuse, djibri, of the boiled rose petals, the little water that there was would take it to a point – one or so kilometer down the river and as the water would dry out, the rose petals formed a black crust which would stay there until a big rain in Rosovets would wash it out.

During the spring, summer and fall, Giul Dere was as busy as the Times Square. Here and there, there were natural springs where the water was running out of the ground, for a few hundred yards, until it would dry out. Since its space from one embankment to the other was wide from between fifty and a hundred yards, while the water bed was no more than two to five yards, there was much space that was free, and sometimes formed wide meadows for pastures – as were the segments between the village and the Vrazbova Vurba, then the Palamorski briag, next *Sriadnata Vodenitsa* and *Mochuraka*, and below the *Poundiovata Vodenitsa*, down to Brezovo's frontier – known as *Teniov Gyol*. And as the river bed was swinging left and right between the two embankments, it was leaving wide fields with lucious green grass for pasture of sheep and cattle on both sides. Here and there, clusters of willow trees, poplar trees and elm trees bordered the river valley. Under the gigantic elm trees, along the Mochurak, the sheep flocks rested during the hot summer, under their cool shade. They were milked here by the attendants. The cattlemen would bring their cows, oxen and calves from their pastures in the Oak Forests for watering and rest – keeping them there from around 12 o'clock to two o'clock. The field at the *Sredna Vodenitsa*, one kilometer from the village, would accomodate hundreds of cattle for watering. We, the kids, would be eating our lunch and then jumping in the big pool of water... Big? Some fifteen feet by ten feet, and about six feet deep. It was a muddy, brown water, but this was the best that we could have. Rarely I ventured in, mostly watching from the side, because I never learned how to swim. But that was the biggest excitement and pleasure that could be had. Still, from one end of the river to the other, there was life, people going up and down to their preoccupations with sheep, with cattle, and working on their small patches of vegetable gardens. It looked like it was the busiest place in the entire village territory.

The Gyul Dere offered the only entertainment during the summer. After a big rainfall in the Rozovets Mountains, the water streams would run down the river, and by the time it would reach our village it would be "Goliama reka," a "big river." It was a fascinating performance of nature, fearfully threatening for us children. The whole village – women, men and children – would gather along the embankments and watch the wild play of the muddy waters – splashing against the embankments, from one end in the West to the other end on the East, running over the green pastures and open spaces, and dragging all that was movable along the way. Very often some donkey left on the pasture, with tied front legs in the river and in the absence of its owners, who might be away in the fields, helpless against the elements of nature, would be drowned. Sometimes an occasional sheep, or a pig, would be watched with sympathy by the peasants on the embankment struggling to survive. The wild water would undermine the embankments and big chunks of land would fall with a splash in the water, thus widening its berth. Too often, in this process of erosion, big willow trees and poplar trees would be undermined and toppled over in the waters and dragged away. This was a spectacle which attracted all the attention – for the sorrow of those who owned them, and amusement for those who were watching. The wild streams would rage for an hour or two, with irresistible force, forming big muddy waves which hurled themselves against the banks then turn back to meet other giant waves starting from the middle. It was a show of nature, which we enjoyed watching. After it would be over, we would run to our places along the river outside the village, down to Tenyov Giol, to see what damages our properties had suffered. I could not count the times when I found myself on Tenyov Giol, where our summer garden was and every year my father and I had planted hundreds of willow trees, and watch them topple one after the other. The ravages of the Giul Dere were remembered as the hurricanes in the United States. It was a tempest in a tea pot, but this was the biggest drama that we could watch as demonstration of the powers of nature.

3. Tenyov Gyol and Baidachka

Let me say more about Tenyov Gyol. This is where our Bakhchia (garden) was. The name was derived from some mentally retarded man by the name of Tenyo, who had drowned in one of the pools of water which was there. In fact it was the most damp place in the whole river, water bursting from under the ground and running for hundreds of yards, through our garden. It was sort of a swamp. Our place was about an acre. For many years my father planted there all kinds of vegetables – tomatoes (he usually had the biggest tomatoes in the village), peppers, eggplants, pumpkins – especially the white kind – which grew to gigantic sizes and in the fall we baked them, to delight in their sweet taste, or make out of them the only delicacy which we knew – *rachel*, cooked in the preparatory stage of sugar-beets mollasses, called *petmez*. Later on my father planted the whole place with apple and pear trees and wild plums, which he grafted to make huge delicious red apples, pears and plums. He taught me how to do the grafting. When I recently visited Bulgaria, my younger brother told me that he is the official graft man in the village. I asked him:"Who taught you to do grafting." "You did," said he. I have forgotten all about it. And since my father was very much involved in planting trees, I also got involved in it, and enjoyed it. I would be watching during the summer the willow trees and poplar trees which I had planted early in the spring – from branches of old trees – how they would grow buds and I was exhilarated with joy of my success.

When I visited there in 1992, I found one single apple tree still standing there. I recalled that I had grafted it. And of course it had a few apples, ripe in September, and I had the pleasure of renewing my acquaintence by eating again this fruit. All else had gone. Only a few bushes of quince trees were growing in the now dry swamp. I asked my brother: "Who planted these quince trees here?" He answered, again: "You did it!." I had completely forgotten about them. But I had not forgotten the willow trees which I had planted along the river. They had been spared by the raging waters of Giul Dere and had survived. They had grown, and grown, and reached unusual heights. They were not the kind of weeping willows, were a much sturdier trees. But time had taken its toll on them. They were dead trees. Branches were sticking high in the sky, their stumps were all covered with brush, but they loomed high, very high above all that surrounded them. I felt humbled, and I felt shaken, looking at these giant dead willow trees. I told them:"So, you old friends, you waited forty years for me to come and see you, to welcome me in my lost paradise." I could not have enough of them. I had my brother take pictures of me standing in front of them. What a spiritual treat was all that for me! This was a reminder of my long gone past, my childhood, my youth...

But in 1991-1992, Giul Dere was not the same as I had left it some forty years ago. Now it was a desert. Better say, it was a jungle. It was a wasteland. I could not recognize most of its sections. The old landmarks were gone. Gone were the sheep and cattle. They were now in the farm. Gone were the private gardens all along the river. Giul Dere apparently was abandoned. Gone were the sheep, gone was the cattle, gone were the pastures, gone were the elm trees, gone were the willow trees, gone was the river. They had built dams above the village, had made lakes west of the village for watering the crops and no water was allowed to go down through and below the village. The entire berth of the river was let go and all kinds of brush had taken over – thorns, wild blackberries, wild roses, wild grass, here and there some uncared for small trees – nothing was left of the past. There was the time this was a living paradise, the living garden of Eden. Now it was a jungle, it was a wild, and only small paths were criss-crossing it. As I was walking with my brother, he often asked me if I recognize this place or that place. I could not recognize any. I remember every foot of land along Giul Dere as it was, but there was nothing left there to remind me of the old life. It was a wild jungle. What they had lost!

But they had found something to preserve. When they were looking for more water for the village – the Yunchal reservoir was running dry during the summer, they discovered Tenyov Giol. When I arrived in Zelenikovo I could not have enough of drinking water there. My sister-in-law asked me if I knew where this water was coming from. I had no idea. "From Tenyov Giol" explained Vessa. Sometimes in the 1960s the village authorities were considering the possibility of bringing more water to the village, but could not figure where to find it. Then someone suggested, "Go to Tenyov Gyol, drop a pump, there is always water there." This is what they did, and found plenty of water. Then they took it with pipes to the hill fronting Sakar Tepe, Kutchukia, and then with pipes down to the village. The water which I could not have enough of, was the water I had grown on while I was young, in my childhood.

When my brother was taking me in the fields to show me my favourite places, we reached Baidachka, a hilly land, some 300 feet above the river level, rising from the very embankments of the Giul Dere and perhaps one kilometer, North of Tenyov Giol. The pipes leading to the Kutchukia were passing through our property. So, let me say a few words about Baidachka. After Tenyov Giol, this was my second favourite place. It was a poor piece of land, probably an acre, some one hundred yards left of Giul Dere, rising probably some two hundred meters above the river, sort of a platoeau, because on the other side of it the land was inclining into a deep ravine, and on the South was steeply descending to a road. We owned it from the road up, some one hundred meter, and then the plateau, but between the plateau and the incline towards the road, there was a Thracian funerary mound which was good for nothing, except to stand on the top and look down the valley of the river – all the way to Tenyov Giol, and the surrounding areas. There was a little forest, bilima, some fifty meters below the mound where we always had a hut, built of branches and covered by straw. This hut was my "summer house" and my fall cabin.

After August, when all the crops of the fields East of it, up to the big forest, oak forest Dabravata – oak forest – were collected, and up to November, for as long as I could take the animals – sheep or cattle – to pasture, I would leave them alone in the forest and in the fields. I would make myself a fire at the entrance of the hut, and in the long rainy days, lounge around in the dirt and read, and read, and read, while listening to the bells of the animals, so they would not go away. During the week, after about 1937, about 11.00 o'clock, I would run down and across the river, run through the fields on the other side, reach the main road from Brezovo to Zelenikovo and wait for the postman – an old man, diado Petko – who travelled by a horse buggy up to Rozovets and Babek, and buy myself a newspaper from him. He carried the papers of the previous day to the villages. Then I would run back to my hut, some one kilometer or so, and then read the paper, in fact, devour it from end to end, with its international news, its national news, feature stories, serial novels, etc. Perhaps for some it would be a lonely life, but for me it was life filled with satisfaction. Every day I carried with me books, from the "Public Library" in the village, the Chitalishte, from one to three books in my bag, and read, the whole day. Outside, the drizzle and mist were filling the air. The bells of the animals grazing in the forest were ringing and I could hear them. The fire was looming, and I would turn one side or the other of my body to keep warm. The days were passing by and I enjoyed my reading. This is how I best remember Baidachka. Most of our work, which was the domain of my father, was around Tenyov Giol and Baidachka. Sometimes he was at the one place and I at the other. He would put his hands around his mouth and shout with all his force to me to come to him with the animals. It was about a kilometer distance but I could hear him.

When I visited Zelenikovo in 1992, I wanted my brother to take me there. We started on the main road to Brezovo, then switched to the river which had turned into a jungle. Nothing was recognizable. It all had changed so much, the old visible marks were gone, and only after he would tell me where we were, I would see the general outline – the Vrazbova Vurba, the Palamorski Briag, the Sredna Vodenitsa, the paths towards the Oak Forest, etc. We climbed on the opposite side of the river and followed the crest of the plateau, leading down to Baidachka. It was a memorable walk for me. I saw the Thracian mound some twenty feet high, from afar. As we were approaching it, I saw how it was covered with thorns, wild growth, brush mostly. Once we had planted it with grape vines, but now they were all gone and their place taken by the wild. We could climb it from the south, which was not covered by brush, but even there we had to criss cross among the big, five six feet tall thorns. I have not seen this type of thorn in America. When I reached the top, I looked down to Tenyov Giol, but the old Tenyov Giol was gone and what I saw was a jungle spreading in front of me. On the top of the mound was an old, dry, dead, pear tree. I asked my brother, "How this pear tree got here?" He explained to me that it had been growing there as a wild pear, but that I had grafted it and it had been a "Mehmedka". I had forgotten all about it. I looked at this scroony remnant of the pear tree, its dead stem and few dead branches, and something was rising in my throat, like a sorrow and pain. I was ready to start crying. Apparently some treasure hunters had tried to dig through the mound seeking old things, but had abandoned it. My brother told me that when he saw it, there was a formation of a clay pot indicating that some small vessel had been there. On the South side, in a small ravine, a fig bush was luciously green, in this dried corner. There were two figs, and they were ripe. I picked them and again asked my brother how this fig tree happened to be here. He again told me, that I had planted it there. I remembered nothing.

4. A Visit with the Old Oak Tree

I cannot finish this story of my attachment to various places around Zelenikovo. I have to say a few words about the "Chervinatsite" – the lands south of the village, known as the place of the red pears. One kilometer or so from the village was our biggest piece of land – some five acres. I wanted to see it. One afternoon, we walked with Ruby out of the village just to get some fresh air, and ended at the "Mogilkite" – two Thracian burrial mounds, some fifteen feet high, adjacent to the village. They were so poor now, with overgrown nasty thorny plants, the whole section where in old times life was bubbling in every corner – sheep, cows, pigs, donkeys, kids... It was our favorite playground from where we looked to Sakar Tepe, down the river, the village behind us, the "Ruta" – the long stretch of the grass covered elevation, some fifty feet above the river bed, all along the length of the village – now covered with ugly brush .... The mogilki area was covered with high dry grass, thorns, a large dump of garbage, and lots of abandoned rusting machinery of the collective farm, which barns, storage places, crumbling walls, in all directions, were nearby. To the Southwest I noticed a lake, which was not there, in my time. They had built a dam, had captured the river water from above the village, and were keeping it for use at the farm, for watering the animals, and for growing fish. It was at its lowest level, following a long dry season. We decided to go down and take a look at it. Then we climbed to the cultivated lands to the left of it. It was much sliding and skidding on the dry grass, and what was left of the wheat stems after the summer harvest.

But I was not ready to turn back. Something was pulling me further away from the village. From the distance, of one kilometer or so, I noticed in the mist of dry air, some sort of a black enormous ball like a shadow, in the form of a big cloud sitting on the ground. Somehow, in my mind, this was the place where our old piece of land was, and it looked like this big foggy looking ball was the place where once an enormous oak tree stood in the middle of our property. I pleaded with Ruby, to take a little more, a little more, and then we would turn back. I was not walking, I was almost running and Ruby was following me grunting and sliding. Something was pulling me in that direction. I could not help it. And the closer we went, the clearer the outline of the foggy ball became, until we were close enough to recognize it, that that was what was left of the old oak tree. In a few minutes I was standing before this big ball.

It stood there, in the middle of our once biggest land property. My grandfather had bought this land from its former Turkish owner, for a pumpkin banitsa – a sort of a pumpkin danish, the size of a large pizza, where the grated pieces of pumpkin were wrapped into home made fillo and baked in the oven – a Bulgarian delicacy. Half of the land was owned by my aunt, wife of my father's brother Stephen and my cousin, Penka. Together the two pieces were approximately ten acres. But the oak tree was in the middle of our property. For many years, since I remember myself about the age of seven or eight, I have spent long days and long hours under this oak tree, mostly with my father, especially when I was old enough to work there –cultivating the spring vegetation, plowing, sowing, weeding, digging the soil around each plant (corn, beans, sunflower, watermelons, melons, etc.) harvesting the fruits of them, of wheat, barley and rye.

As I remember, I was always there with my father, and every day at noon, we ate our lunch under this big oak tree. Under its thick shade, we took a short nap, or rather, while my father was taking his nap, I was reading newspapers and books, which were all the time with me. Sometimes, while rolling his cigarette in a piece of paper cut to size from a newspaper, he would look at me in amazement and would lovingly reprimand me: "I do not understand you! What is so much interesting in these books and papers, that you spend all your time over them? Take a nap, relax a little, it is a long day and we have much to do." I have never forgotten one of these reprimands. But I continued my reading. And I enjoyed it. The shade of this tree was my home away from home.

Approaching from afar I felt like I was going to a long lost friend. In my mind I was recalling all my life under it – in long hot summers and in long days and nights in the fall where I would tend the sheep and seek protection from the rain. And, as I was coming closer and closer, able to discern its shape, form and surrounding, I felt like I was stabbed in the heart from the present state it was in. Its long and lucious branches were cut about five feet above the stem crown, small growth of weaker branches under the cuts testified that it was still alive, but what was surrounding the stem was an abomination. It was all thorns, wild roses (shipki), long weaving vines of thorny wild blackberry bushes and all kinds of dry high grass and a thicket of brushwood. The stem of the oak tree was hardly accessible, but I pushed my way through and embraced it. My eyes were tearing. A painful clump was pushing in my throat and I felt like crying. It would take three or four men to spread their hands around the tree trunk. I walked around it. I felt that I was embracing an old dying friend. As if he was waiting for me to return from faraway, before expiring. It looked, this old oak tree, as a martyr who had seen it all around it, witnessing the death of the old world. I contemplated its injured body, I remembered its magnificence in times long past. In its present lamentable state I recognized a martyr who had suffered physical and moral abuse in the long four decades that I was away from it. Evil hands had disfigured it, no men have sought its shade and its protective powers, it was surrounded by a bunch of parasites sucking its blood in a little patch of a wasteland, of a desert, of desolation. For an instant my mind shifted to Bulgaria and for all that I had seen, this remnant of the oak tree illustrated its state - devastated, dying, turned into a social and physical ruin, a jungle of thorny vines, of thorny wild thicket of wild brushwood, infested with detestable parasites, helplessly gathering its last breath before expiring. Embracing this tree I felt its might of the past, and looking at its present state up, I recognized its helplessness and its sorrow, its suffering from the cuts it had endured as if its hands and legs were being amputated. And I thought that those same people who chased me out of the country for forty years, are the same people who had hurt this old friend of mine.

While I was going through this agony, Ruby patiently and understandingly stayed about fifty feet away, giving me time to live through my long forgotten past, to relive my torturous memories over the lost greatness of this old martyr in this vast field of the dead socialist world. Leaving it, I turned around to take again and again another look, Ruby picked several marble stones from the property. I still have them on my bookshelves where I keep some of my most precious mementoes. I grabbed a handful of dust from the ground and put it in my pocket. I am keeping it at home in a small box, and I wish that when I fall asleep in the Lord, it be tucked with me in my resting place.

5. Roses and Thorns

Ala-Bair was not on the list of my favorite places. That is where our garden of roses was. "Rose Garden" and "Rozober" (picking the rose petals) sounds very romantically but for those who had trudged the furrows between bushes before sun rise, after a rain, it is neither romantic nor smelling of roses. It was torture, especially for a young kid of ten or so years. To begin with, the planting of roses required digging straight trenches of fifty or so yards, one foot deep and foot and half wide. Old rose bushes were cut to the ground, and the branches layed in the trench, to be covered with dirt. All this was done in the winter or early spring when the ground would be thawed. In April-May, young shoots would grow from the ground, to be cultivated for protection from weeds. They would be ready to produce blooms in two years or three. The rows were six feet apart. In the fall the roots had to be covered with soil, to protect them from freezing, and in the spring they had to be uncovered, so when it would rain, the water would not run away. Since the branches would incline on both sides, one had to push them aside to reach the roots. It was a miserable struggle with thorns, weeds, dirt, and whatever was in the way. During the summer it was necessary to plow among the rows with the wooden plow pulled by the cows. In the morning all blooms would be picked, and the next morning the entire garden would be loaded with blooms as if it had not been touched. I was recruited for all these chores probably at ten. Early in the morning, about four o'clock, my father would awake me, having already prepared the cart and the cows. I would drag myself half asleep and continue my snoozing while the cart was trudging along the fields, following a path full of holes, and in half an hour or so we would be there, but I managed to grab some sleep, a minute here, a minute there, until we started the work. One would see on propaganda posters and pictures young girls, roses above their ears, baskets in their hands, smiles on their faces, and one would think that this whole thing was a celebration of some sort and immeasurable pleasure. Well, it was not. Nobody picked roses in baskets. Instead, a home made bag, long and big enough to take as much flowers as possible, hanging around one's neck, so one could pick with both hands, was used. And there you start. My father at one side of the row, I on the other. The plucking of the blooms was easy. Under the flower was its seed pod, it was tender and only a light twisting on the side, would terminate its life. The flowers are pushed in the bag, and so on, and so on ... for hours. The full bags are emptied in sacks lined up at the end of the rows. The picking was to be finished by nine or ten o'clock, because it was loosing its value for the rose factories. By noon all blooms had to be delivered at the factory, where they were weighed and recorded.

The "Rosober," the "picking of the roses," was a tedious and miserable work. It was made even worse when it was cold, or a rainy morning. Your hands would freeze from the morning dew, your feet, in *tsarvuli*, would be collecting clay mud which, as you feel it, weighed a ton. So, you have to stop every 20 feet to shake it off. After the sun rises bees attack the flowers in swarms and if you do not see the lovable otherwise insect, it would sting your fingers. But you go, and go, and are bored. I used to kill boredom by concentrating on some issue of my interests, mostly on social issues, and figure how they could be resolved, issues of religion, of science, and whatever would come to my mind, to take me away from the repetitious picking the flowers, tugging them in the bag and dragging my mud covered feet. When it is all over, we would load the sacks on the cart and head for the village. There, at the factory office, the sacks were weighed, the kilos were recorded, and then ... waiting for the next morning to go through the same procedure. The roses delivered by all peasants were sent by trucks to the factories along the river – the Factories of Botyov, of Batsurov and Shipkov, and later to Raicho Grozev of our village. The rose petals would be dumped in enormous bins, similar to what a cement mixer in this country is, even bigger A fire would be lighted under them. The thing would be boiled for hours. The steam would be passed through a cooler, where it would form driplets of water. This water would then be boiled in smaller bins, again turning it into vapors, again passed through coolers, and at the end the oil of roses would separate from the water. It would take some three tons of rose petals – three thousand kilograms – to produce one kilogram, 2.20 pounds, of rose oil. When the process is completed, the oil is sealed in metal bidons of ten kilograms each and dispatched to the factory owners for sale to interested customers. For some reason it is the Bulgarian oil of roses which is of the highest quality and sought after on the international perfume market. It was the gold of Bulgaria in those days. In May, 1939, I was appointed as assistant to the book-keeper of the village consumer cooperative, which at that time had taken over the trade of rose production and the bidons with the oil of roses were often stored in our office. The containers were sealed, but somehow the smell of it was so strong that it was causing nausia for those in the room. I was sick of the nauseating smell as long as this season lasted.

After the season of picking the roses is over, there would come the day when the factory owners had to pay the peasants. The price of kilogram roses was set well in advance. What profits they realized in their trade, was not known to us. Neither did we care. We expected to be paid for so many kilograms, so many levs. Apparently it was a profitable business, because year after year they were there to collect our roses. Our three villages – Rosovets, Babek and Zelenikovo – were able to provide three, later four rose oil producing factories. Payment day was a big event for the village. On that day it was demonstrated what was the worth of the peasant labor. For most of the peasants in those years there were two ways to see cash in their hand – when they are paid for the roses, and when they are paid for their tobacco in January. In the appointed day which never came about too soon, the agent of the businessman would come to the village, sit in the office of the village tax collector, guarded by armed men and a bag full of money. Next to him would sit the tax collector, and when the peasant is called to sit in front of them, he would be told how many kilograms he had delivered, how much it was worth, and also, by the tax collector, how much taxes he owed to the state and to the municipality. The agent would pay first to the tax collector the stated amount, and if something was left for the peasant, he would be handed the amount. If nothing was left for him, he would walk away empty handed and visibly shaken and cursing them all. Some peasants were well off and received their pay. Most of the peasants received very little, or nothing.

As it turned out, my father could never satisfy the tax collector, and walked away empty handed. But he was a little appeased that at least his taxes were paid, more or less. But things sometimes did not work as they should have. In 1933 the company of Botyov, its owner, had gone bankrupt, and after collecting the roses, never came to pay the peasants. My father was one of them. For many years the village was to talk about this bankruptcy. So my father's taxes remained unpaid. There followed one warning after another by the Tax Collector – he was the second most important official in the municipality after the mayor. I think his name was Petar Manev. He was threatening that if we did not pay the taxes, he would come and search the house for valuables. How could we pay taxes when we rarely saw one lev in the house? We would be waiting for the chickens to lay one or two eggs, to take them to the local grocery store and buy salt or other most essential necessities. Most of the time my father was buying things from the store of Tsaniu Hadjimitev, on credit, and pay in goods or work when he could.

And so, the long waited day for a visit by the tax collector came. He was accompanied by two heywards, armed with their carbines. There they stood in front of the house. My father, with squinting eye and utter displeasure, listened to the new threats. At end he sourly told the tax collector: "Take it, whatever you find in the house, take it. I have no money to pay." What were they going to take? Our house was, in the language of the day: *"Trun da zavlachish, niama kakvo da zakachish,"* – "to drag a thorn through it, there is nothing to pick up." So, the heywards stormed the house looking for valuables. There were none. Our silverware was wooden spoons, our dishes were wooden bowls (*kopani*), our blankets were raggedy remnants, our clothes – rags. But in a few minutes the heywards emerged from the house carrying the two copper buckets used by my mother to bring water from the *cheshma* (water fount), and my father's *bukul* (u pronounced as in much) – a wooden container to take drinking water to the fields. In the house there was not one grain of wheat or barely – it was before threshing times. There was hardly a few pounds of flouer for bread... So the two copper buckets and the bukul were going to pay our taxes? My God! I never forgot this scene. To the last iota it happened as I describe it. I did not exaggerate anything. I have not forgotten anything. I was eleven years old. My father was looking at the official in bewilderment, but said nothing. My mother was crying standing on one side. I was clutching her hand and crying... Two weeks later they called my father to recover the buckets and his bukul. Well, what were they going to do with them? They could not sell them. No one would buy them. How much they would have fetched, one percent of the taxes? Hardly. In all my life I have never seen such stupidity of the authorities, nor could I ever imagine such poverty of judgment ... So, this is how things went on in the "good old times."

6. Happy Childhood

My memories of early childhood are probably the only shining page in my troubled life. They begin with the gang of rascals in the neighborhood when we were between four and five, to seven-eight years old, with endless games along the river, at the Thracian mounds outside the village, on the street square in our section of the village, and in the meadows surrounding the lower part of Zelenikovo. As spring would descend upon us, we fled to the river, climbed the ravine leading to the Bear's Hollow and the Kitchukya, picked crocusses, primroses, violetes, hellebore – it was not much of a flower, but was one of the earliest arrivals and refreshingly beautiful – and anything that was prying the soil open to meet the warm sun. We struggled along the deep paths dug out by domestic animals being taken for gazing, leading to Sakar Tepe, squeezing through brush, bushes and thickets, meeting here and there some slow moving turtle with irresistible delight, and from some point turning back to the river. Lined along the water stream, we would throw in our stalks of hellebore and watch to see which one will move ahead of the others. It meant good luck for those winning the contest. Then we would start collecting stones and building a bridge, a row of stone piles, then jump from one to the other to cross the water. We all would miss the stone sometimes, and found ourselves in the water, to run home to change.

But it was fun and pleasure, in camaraderie and in play. When we did not go to the river and beyond, this was actually in the fall, we would play on the street. One of the most popular games was "Rokan." We would draw on the ground a number of squares, of different sizes and forms, then throw a shard or a piece of glass from square to square, jump on one foot from square to square, avoiding the one where the piece was and on the way back, when reaching the square where the shard was, we stood on one foot, leaned, grab it, and finished to the front. Errors and failures were scored as losses, and whoever made the least errors, would win the game. But the most popular game was the game of "Topka", "ball". In structure it followed almost in everything the American baseball, only that instead of a bat we would use our hand to hit the ball when pitched by the opposite team, or something similar to the ping-pong paddle. The ball was the size of an average apple, made of rags sewn together in that shape The rest was the same – three bases, catching the ball, touching the runner, etc. the game was won by the team which scored the highest number of players reaching the home base. Both, boys and girls, played it – divided into two mixed teams. There was a game played by boys only called "Pile" (Chicken). A three-prong branch of wood, standing on three legs, was being placed some twenty feet from the starting line. The boys, with sticks of two to tree feet long, would throw them to hit the "pile". If they missed, they would be out of the game. If they hit the "pile", the guardian from the other team would run to put it back on its place and then run after the player who would try to reach from base to base, until returning home. The guardian of the "Pile" would have a short stick, one and a half foot long, called "Petel", (rooster), would run to hit the successful thrower, or any one running from base to base, and if he succeeded to touch him with the "petel" the runner was out. The team which would succeed to have more players reaching the home base was winner of the game. A similar game for boys was the "Liuf" (the *iu* pronounced as in "putre"). There the server would throw to the player at the start line a stick, three quarters of a foot long and one inch thick; the player would intercept and hit the liuf with his stick, and then run to the bases. The opposing team would pick the liuf, return it to the server, who, having received it, would run to touch the player. The principles of these games – "Topka", "Pile" and "Liuf" were the same, all similar to baseball. All this time everybody was screaming for one thing or another, disputing many incidences in the game, there was uproar all around, sky-high, running and screaming till full exhaustion.

Other games, very popular, were "gonenitsa" and "Krishku-Mishku" – "run to catch" and "Hide and Seek." For both of these games the group would designate one of the players, boy or girl, who would, in the first game, run after the entire group, to catch whomever he could. The looser will then run, in the next game to catch another of the players. In the "Hide and Seek," the designated boy or girl would close his eyes, count to ten, during which time all players would hide wherever they could, splitting usually. The "seeker" would look for them, then if discovering someone, would try to touch him and run to the base. At the same time all players would watch for the seeker and if he or she passed them they would run to the base to save themselves from being touched. In both games the designation of the runner or the seeker was done by a magic meaningless formula, in fact there were three formulas: "Ala bala tuta kala ista vitsa dupa tittsa, talion, talion ..." I have forgotten the end of it. Another formula was: "San dan di limone, san dan kuzakus, sipi done kuzakus, nai posle pod limon, sin gos." The third formula: "Ini mini du du mini drano pa drano pa sandi to pa, karolino marolino haru pa." Before the game would start all players would form a circle, each one would extend a fist. One of the players would receive the last syllable, would pull out and the formula is repeated. Until it comes to two players and one of them looses, to serve as runner or as seeker.

When I was seven or may be six, one of the games of "krishku-mishku" led to a serious injury to me. I ended up with a broken leg. The whole group was trying to hide behind a bunch of logs resting upward on my uncle's house. I was reaching the group when a heavy log slided sideway over me. When it was moved out I could not get up, with excruciating pain in my left leg under the knee. I was taken home, put to lie down and screaming with pain. It went on for two or three days. Someone told my father to take me to a Turkish hodja, imam or something, some twenty kilometers across forests and mountains. All along I screamed from the pain while the cart pulled by the oxen, was suddenly passing a big or small pot hole on the country dirt road. Finally, we found the hodja. He washed my leg, twisted it this way and that way, oiled it with whatever oil it was, wrapped it in strips of cloth, then placed two pieces of flat thin wood planks, wrapped it again and sent us home with instruction not to unwrap it for thirty days. All the time I was screaming from the pain. Very soon it subsided, but I could not walk. After a week I started dragging myself on my behind to move from place to place. Who knew in those days about crutches which would have come in handy? There was one benefit to all that. I was supposed to start school in September that year, but for reasons I will discuss later, I did not go. Two teachers, Dimitrina Vuleva and Mara Angelieva often came to pester my mother and my father for not sending me to school. After I broke my leg, they stopped coming. In any event, after a month all strips and planks put by the hodja were removed and my leg was as good as new.

But our greatest pleasure was dancing our folks "horos". Every evening, spring and fall, after dinner, we would fly out our homes to the little square in our section of the village, where we played the games which I described, and we danced. We did not have music, but we had voices and as we were holding hands, twenty or so kids, beginners' five or six years old, up to thirteen or so, forming a circle where one would lead, boy or a girl, we sang the popular songs, half of the kids singing a strophe, then the other half repeating it. Most popular was the "narodno horo," four steps forward advancing to the right, four steps back, always advancing. There was the "Paidushko horo" – I will not describe them – there was the "Kassapsko horo" (Butcher's dance), there was "Eleno mome," there was the "forward-backward" horo, and the most popular dance with its "only-in-Bulgaria" type of music structure, the "Ruchenitsa." The "Ruchenitsa" was danced individually, not holding hands and going in circle. There boy and girl, or boy and boy or girl and girl, sometimes just one individual, sometimes three in the center, and most of the time the whole group, would sing, dance in all forms of configurations, until exhaustion. It was a peppy dance, where one would raise his right hand waiving a handkerchief, with left hand placed on the back. All the time there was singing, there was screaming, there was havoc, there was laughing and exhuberance.

As the evening progressed, the women of the neighborhood would start a low fire, all bringing some wood from home, would sit around it spinning cotton, or wool, picking clusters of wool apart for sending it to a carder, knitting socks or pullovers, and every possible small kind of work. The older girls, fifteen or so, would be there, with the women. Young men, "Ergeni" (Bachelors), looking for wives, already having found their girl-friends, would come and sit among the girls under the watchful eyes of the mother and chit-chat the whole evening. This was what was called *"Sedianka,"* literally meaning "sitting together" in a group. While we kids would be dancing our hearts out, the sedianka would be going to late in the night, or until someone had to go and sleep. At the same time, on the opposite side of the street, the men of the neighborhood would make their own "sedianka", around a slow burning fire. They would be lying in the dusty street or sitting on small stones. All they did was talk about their daily work, about all happenings in the village, conditions of work in the fields, their domestic animals, … and politics, and news as they understood them. Some of them would go uptown, to the center square, where a loudspeaker attached to the public radio in the chitalishte was broadcasting, and then return to the neighborhood and report it to the rest. They would be rolling their cigarettes in paper cut from newspapers, home produced tobacco, and sometimes tell locker room jokes. There were times that I loved to snuggle in there and listen to their adult conversations. Later on, when I started reading newspapers and listening to the radio, I would join them and bring to them all national and international news. But the main thing was the dancing which we all enjoyed to our hearts content. Our songs, our screaming, our racket overcame everything that was going on around and sometimes we would be asked to tone it down.

One of our favorite occasions was what in the west passes as *All Souls Day*, Halloween in America, and in England, the Guy Fawkes Day. This was the Saturday evening before Meat Sunday. We all would be prepared for that. Our fathers or brothers would take a wood branch, thick enough to be split in one side, or naturally branching in two like a fork. They would place between these two branches a large amount of "ruzhenitsa," straw, which had not been cut, but left in long straws, usually from barley, and tightened together. This thing we called "Kurukulnik", a meaningless word, but somehow relating to a weapon to chase evil spirits. After sun set, with the advance of darkness, we would light the straw and waive the kurukulniks in circles over our heads until the straw would burn out. All this was taking place outside of the village, around the Thracian mounds, to avoid starting fire in the village. It didn't matter if that February day, it always fell in February, it was snowing, or raining. But we enjoyed it and would not miss it for anything in the world.

Christmas was a very special holiday for us. We celebrated it on January seventh. The day before was "little Christmas." On that day we, below ten years old, would be outfitted with new cloths, wherever and however possible, with a specially made stick, a replica of Joseph's stick for tending his sheep. For this purpose our fathers would cut a fresh wood, half or three quarters inch thick, three feet long, would remove the bark and wrap it in spiral circles with leek strips. Our leek was two feet or more long, not the short one-foot plant that is sold in our supermarkets. Then a small metal container filled with kerosene, with a wick protruding from a narrow pipe on the top, the "gaze", is lighted, (sometimes used to provide light in the night – it was our flashlight) and if burning, it emitted a stream of black smoke. Placing the stick, wrapped spirally with strips of leek, would have the open spaces blackened. Here we have the candy miniature batons made of sugar. Early the day before Christmas, dressed in warm clothes, with a colourful miniature plaid of red, white, green and black across our shoulders bag, would trudge in the deep snow if this was the case, walk from door to door in the neighborhood in groups of three to five, knock at the door and yell to the people in the house: "Slavite li Boga?" – "Do you glorify God?" The man or the woman would open the door and would say, "Yes, we glorify him," and give everyone of us dry prunes, or dry pears, some apple, and inevitably a special Christmas "kravaiche," a home-made baked in the form of a jumbo pretzel, or simply a small home-baked donut, sometimes some wallnuts, but the most appreciated gift was a five-cents piece which we would use to buy candies.

The same way we celebrated "New Year's Day," which for us fell on January 14th, St. Basil's Day. Instead of the special sticks, now we would carry a branch of dogwood, or more often than not a branch of a hornbeam – I have not seen this tree in America – but the majority of our woods near Zelenikovo were hornbeams. We would hit the people of the house with our branches, often quite big, "for health" during the year. If someone was still sleeping we would be admitted in the bedroom and while they would cover their heads we would strike them. We would be taken to the barns to strike also the animals, all "for good health" and "bereket" – a Turkish word meaning abundance of fruits. It all revolved around the idea of a blessing for the new year with good health and prosperity. As to our other winter pleasure, I just have to mention that we, kids, enjoyed the snow. With home made sleds, two pieces of wood, held together with a plank on the top and horns in the front, we would go to the Thracian mounds outside the village, climb to the top and slide downward some fifty feet, roaring with laughs and screams …and freezing, until someone would come from home and take us back to warm up.

7. When I Was a Little Shepherd … and a Little Cowboy … and a Little Farm Worker

So, this is where I was growing – in the fields around the village. When I was four or five, I outgrew my only summer outfit – a bell shaped shirt, down to the middle of my calfs, with no underwear, barefooted, and always dirty. Once in a while the shirt would be washed in the river by my mother. That was where all wash was done. After I outgrew my childhood outfit, I gained my trousers, which were always dirty, and always ripped at the knees and at the back, so that my mother would mend them by attaching patches from discarded clothes – never matching colors or designs, whatever, but still good for running after lambs, or sheep, or cows, doing small chores around the house, and playing on the street with the kids, until I was old enough to be included in the daily preoccupations of the family.

My first assignment to work was to join the other kids, older than me, to take the oxen in the morning to graze them in the woods, in the Oak Woods (Dubravata). It was not a very successful first attempt. At some point, around eleven o'clock, my two oxen, Sariu and Pencho, disappeared, I could not find them in the woods, they vanished. I broke down crying and ran to find my family, harvesting barley in the Chervinatsite – place where our big old oak tree was, which I described above. I was all tears when I arrived there. I was not scolded, I was consoled and my older brother, Petar, was sent to look after my lost charges, some two miles away to take care of missing Sariu and Pencho. He found them and stayed there for the day. But I had to go there the next day again. I got used to the operation. The worse calamity in this daily assignment with the oxen was when, in the fields, or resting after watering at the river, at noon, they were attacked by a bug, the *buzhgalka* (gadfly). The bug would sting the cattle on the back and deposit its eggs under their skin. The next spring the eggs would turn into cocoons under the skin and eventually break out as bugs. The cattle apparently recognized the buzz of this bug (nature is a mysterious thing) and when they heard it, they would raise their tails up and run as wild in whatever direction they could. One could not stop them, and one could not control them. It was a terribly frustrating experience, but it was part of life which one could not avoid.

As I was growing up and gaining experience, and as I did not very much enjoy the mundane company of the other kids – I was more interested in reading while grazing the animals, I started going alone, and with the few animals – which I had, I could graze them in small corners around the river, where the grass was better. Our cattle population started increasing – my father exchanged Sariu for a cow with a calf, which was white like a goose, and I named it *Patka*, goose. It became and remained my favorite pet for as long as I was in the village and she was still alive. Then came other calfs, to which at end was added a baby buffalo. My task was getting more complicated. But I managed marvelously by training my animals to behave. My favorite place was a water-mill-canal. The water of the river was captured at the Mochurak section, then ran between the planted fields – about one, one and a half feet deep and two-two and a half feet wide; then the embankments, some three to five feet on both sides, covered with lucious grass.

I trained my cattle to line up in the water, positioning them in a pecking order – the one who would peck them all first, so the next one would not dare to push him or her, and so to the end. Sometimes I would have seven or eight animals in line. If anyone ventured to misbehave, or attempt to get out to go into the planted fields for dessert – corn, or sunflower, or wheat, whatever, one yell from me and a raised stick would bring it back in the water. The grass was virgin and by the time the end of the canal was reached, near our Tenyov Giol property, they were well fed and would go to rest. All along I would be reading while standing on the side, and watching with one eye for violations of the order. It turned out that my cattle was the best fed and putting on weight during the summer.

So did I do with our sheep when this was my assignment. I especially enjoyed grazing the sheep in the spring. There were times when I was tending the sheep of our whole neighborhood, sometimes from fifty to seventy animals. Spring is the time, especially March and April, when pregnant sheep would give birth to their lambs. In warm sunny days I would watch for animals which would go into labor. I would then keep the whole flock graze at the same field, until the happy mother comes out of labor, often with my help as a midwife, lick the newly born and clean it up. There were times when I had two or three births a day. In the evening, returning to the village, I would lead the flock, lambs in my hands and the worried mothers flanking me, looking at their babies and uninterruptedly bleating. In the village I would hand over the lambs to the neighbors whose sheep had the baby. I knew all sheep, to whom they belonged, and I knew the character of each sheep - which were the trouble makers, which were the leaders and which were the slowpokes.

Most of the times I had a donkey with me, which would be the leader, and which I would ride on, on bare back, when going home or when needed to cross the river. Sometimes I would race it to test my skills as a horseman, and quite often I would lose my balance and fall, or the donkey, annoyed at me, would throw me on the ground. I still remember one of these falls, when I almost had a broken shoulder.

During the summer tending the sheep was not very pleasant. The flocks were not brought home during the night for months – from June to September. They would be kept on land left for fallow, fenced in small circles, *chalmars*, to fertilize the soil. The fence would be moved every week, so that the entire property is fertilized. It was an effective, organic fertilization. Early in the morning they would be taken out to pasture and about ten or eleven brought under the heavy shades of gigantic elm trees along the river. There the owners would come for the milking. The shepherd would herd the flock into a fenced circle, will sit on a rock at the opening with a wooden bucket between his feet (vedro), and the owners would let the sheep one by one to the shepherd to milk them. The owners took their turns, as per the number of sheep they had. After milking forty to sixty sheep, you feel like vomiting. Someone would bring me food for the next 24 hours – bread and cheese, onions and garlic, or whatever was left over from the home cooking. Often, enough exhausted, you cared less for food than for sleep. And you go and huddle in some comfortable corner nearby, stretch in your night cloths, and fall asleep. About six p.m., it is still hot, but you let the sheep out to pasture again. And if it is hot, the cursed animals would not open their mouths, they huddle to each other, still sleeping. You try to spread them, and they run again to line up in a pile back to back. It would take one or two hours to start grazing. Then you keep them until eleven in the evening, then take them to the field for fertilization, in the fenced circle (the *chalmar*), you grab a nap with them and about four o'clock let them to graze again. After the summer is over, sometimes in September or October, when it gets cold to spend the nights in the fields, or because during the days it is cooler, they graze longer, so taking them home was a respite for the shepherds. I believe I spent one summer this way. I did not like it at all. Many years later, in exile, when I was writing complicated analysis of events of the peasant life in Bulgaria, I would remember a little song, a poem really, written by Ivan Vazov, our national poet, or somebody else of prominence in the literary world. The song went like this:

> When I was a little shephard
> and I grazed my sheep
> I was very pleased and thankful
> Even though I was a poor kid.

I often remembered this little song, as I sat in exile, and thought that its author had never experienced the misery of the peasant life and by this song was deceiving the poor kids of the villages to accept their lot as God given paradise. It was a romantic dream, but it was a savagery, I thought in later years. I have lived this life. I have not come to describe it, as yet, but the miseries and my crushed hopes and dreams experienced when I was fifteen or sixteen, and many years thereafter, did not quite square with this little song. Now I look back and see that there were two sides of the story. One side is the simplicity of life and the many moments of inner peace, while the other side is the sense of injustice and unsatisfied dreams.

During these times, at 14 and 15, my father was teaching me to do all kinds of work around the house, the yard's work, and in the fields. My earliest responsibility during the winter was to cut wood for the stove and take them in on a daily basis. I struggled with the heavy axe, managed it so and so, and then in three or four trips to the house, finished the job. It seems to me that all this time I wanted to run to the street and play with the kids. I must have been ten years old or so, when I protested and did not want to do the work. My father grabbed a *zhugul*, a stick used to harness the cows, gave me one good hit across my thighs, and I ran to do my chores. This was the one time that I remember my father hitting me. At that age my other chores during spring and fall were to follow my father when he was plowing with the wooden, neolithic plow, pick up rocks and throw them to the end of the field, or nasty weed roots.

It was tedious and killing chore, but I had to do it so many times and die of boredom, walking in the furows behind my father. And as I was growing, he put me behind the wooden plow, while he was walking alongside with me, teaching me the profession of agriculture. Weak, skinny, hardly keeping myself walking, falling left and right and holding the plow handle, I struggled to follow, grasping with my last ounce of strength the primitive tool. The important thing was to keep it straight and press down the *paleshnik*, a metal plate with sharp front end, to penetrate some four-five inches under the surface. I felt very proud when I would finish one single furrow straight. Little by little I got used to the work and mastered the skill. Then my father would let me do it alone, and when he was confident enough, he would go to the other fields to do other work. But it often happened that the cows would sense that my father was not there, or I confused them with my wrong actions – pulling the wrong ropes tied to their horns to direct them, or hitting them and punching them with the special long stick, the *Osten* or *Kopralya*, with a filed nail at end, and they would deviate from the course, mad at me, and drag the plow and me in any direction, left and right as they chose. This would irritate me, I would get mad, sometimes in helplessness, with tears in my eyes, I would try to get them back to the furrows. But this too, in time, was mastered and I became a specialist. So much so, that my father went around to announce in pride: "Ah, I taught Pankata (this was the diminutive of my name) to plow. He will be able to make his living." I heard this from Drazha Parlapanska in my recent visits to Bulgaria. Alas! It had not been in the cards for Pankata to plow with the wooden monstrosity. He was to become Professor in the United States.

As soon as the snow would have melted and the spring had begun, the work in the fields would start. But since in the morning I would be in school, after it was over, about four o'clock, my first chore was to take the lambs for grazing it was actually in our rose garden, next to our house. Sometimes they would not graze, and I was killed by frustration. How do you make a lamb to start grazing? In March I would run to the fields outside the village, seek the shepherd with our sheep, separate them and take them to graze in the wheat or rye fields which we owned. It was not so bad. In May we would start planting the summer crops. I would follow my father behind the plow and drop corn seeds in the furrow. He would return and cover them with the next furrow, and so for hours. So was also the planting of sunflower. Planting of beans was different. There you drop the beans in the ground and kick dirt over them. Then would come planting of tobacco. Since early March, the beds were prepared to sow the seeds for seedlings. By late May they would be ready, and as soon as it would rain, we would run to the fields to transplant them. One day my father left me home to plant tobacco next to our house, where formerly was the rose garden. I did the planting in rows in the morning. Then felt sleepy and went to take a nap. When I awoke the sun had passed noon, it had been very hot, and all the seedlings I had planted, without watering them, were all dead. I got a good lecture that evening, but the next day I had to repeat the job, watering the seedlings after planting them. And then, there always was some work to be done in our garden at Tenyov Gyol.

I was overwhelmed with work, chores of all kinds, since most of the time it was only me and my father. I was running from one place to another, from school to the fields, for one thing or for another. Sometimes I would have enough of it, and like all kids, I wanted to play in the street. Most of the kids did not have these chores. I wanted to join them to play *liuf*, or *pile*, or *chelik* (I have not mentioned this game. The *chelik*, Turkish for steel, was a small stick one foot long, with a sharp end which was stuck upward – inclined in the ground, and we had to hit it from distance with our sticks), or "*gonenitsa*," or "*Krushku-Mushku*" or dance the *horo* and *rutchenitsa*.

After I graduated from 7th grade, I spent two years out of school. The summer began about June 5th when the barley would be ready for harvesting, to be followed by the rye, and then – the wheat. Every living soul of the village was going to the fields. As I was growing, some 13 or 14 years of age, I was trained to wield the sickle, while the left hand was fitted in the *palamarka* (wooden glove, three fingers in holes, front side with sharp angle ending four inches in a sharp prick and a plate covering the back of the hand – to enable the harvester to accumulate as big a chunk of harvested straws, cut about ten inches above the ground, and to protect the hand from being cut with the sickle. We would arrive early morning, with the rise of the sun, line up, bend down and start harvesting – one, or two, or as many harvesters as the family had – men and women. Sometimes two families would join together and work on each other fields. The sun would be rising, travelling its diapazon, throwing its hot fire above the fields, the harvesters push, and push, and push – cutting, laying behind them the cut straws into *rakoiki* – three or four hands, until they reach the end of the field, and then start another *chakum* – some five to six feet wide for every harvester. We had no harvesting machines.

As the *rakoiki* increase, a man, assigned this task, will pull the longest bunches of straw, cut for this special purpose low to the roots, or sometimes pulled with roots. He would divide them in two, tie them together at the ear's end in a special knot, spread it on the ground and pile across it eight to ten *rakoiki*, bring around the two ends of the tied stuff, press in the middle with his knee and with a special *kolche*, a wooden stick, and in a masterly manner, tie the two ends together. Thus, formed into *snops*, some two feet in diameter, the snops were dragged to the center of the field and as there would be enough of them, they were arranged in formes, two by two in a cross figure, one above the other, with the ears in the middle, the whole figure called *krustets* up to four snops high and others added next to it as the work would progress.

With many harvesters, this task was done by one man. Under the burning rays of the sun, the harvesting would go until sun set. The fields were filled with people. Women and young girls would break the monotonous and harduous work singing their folk songs, chanting antiphonally where one group would sing one set of lines, and the other would repeat it. Sometimes neighboring groups would compete in the singing of these songs. While the mothers would be singing and harvesting, behind them were the young babies – from a month to two years – in a *liulka* (cradle suspended on four poles tied together, on the top, covered around with a sheet) under a tree or near behind the harvesters, always having a head of garlic hanging somewhere to keep the snakes away. I had mastered these arts – of harvesting and tying the *snops* to perfection and sometimes was proud to beat the group in speed and skill. This went on for years, until I left forever the village in 1946, twenty four years old.

The worst calamity for the peasants in the harvest days were the hail storms, which quite often hit us – sometimes light, sometimes devastating. The frozen balls, sometimes as big as a walnut, and even bigger, destroyed the crops, and wherever it hit, who ever was hit, that meant no wheat for the owner, and no bread. The situation was most of the times saved by the fact that properties planted with grain crops were scattered all around

46

the village, miles away from each other, so if one field was hit, others would be spared. I have lived quite a few times through these calamities.

It often happened, that following an oppressing heat wave, we would see rising black clouds from the west, crawling towards us, covering the sun and a deep shadow over-whelms the sky, causing some hellish quietness, making us to run for cover. We would rush to collect all *rakoiki* into snops and form *krustsi* before it hits us. And then all hell would break loose. Vicious winds would swarm around us, blowing all over and making the trees twist and turn like small toys of silk materials, branches cut out, flying all over, trees uprooted. The skies were split by deadly lightening, discharged over high trees, and if people and animals had sought refuge there, would end up being killed. Deafening thunders would scare any living soul. It was a horror that few could stand. If there were any dogs among us, they would huddle in our feet, while the icy scourge, the ice balls would be beating mercilessly the open fields. We would watch how the ripe unharvested wheat or rye or barley was being reduced to straw on the ground, after being twisted by wind in every direction in the vast fields in front of us. The murderous vehemence would rage as a uncontrollable beast for twenty minutes, for half an hour, and then move to the East, to leave us behind bewildered, despondent, helpless over the sorry remnants what only half an hour had been a rich soul-filling generosity of nature, reducing to smitherines our hopes, our rewards for long and arduous labor. Now it was all gone. This fearful event in the life of the peasant has been best described in our Yavorov poem "*Gradushka*" (Hail Storm). I do not know if there is anything like that in world literature. Recited by a good actor, it will make one's hair stand up. All that weighed heavily on the minds of the peasants, and as harvest would begin, it was racing with the time to finish it as fast as possible, to avoid such disasters.

8. The Threshing Field (The Kharman)

After harvest was over, we prepared for the threshing. It was a primitive operation, inherited from neolithic times. While the men would be bringing in carts all the snops from the fields, the women and the kids would prepare the threshing field, the *kharman*. Bringing the snops was not an easy work. The oxcarts were fitted with special side-supports, *ritly* – long and high on both sides of the extended vehicle wooden frames with thin bars all along the length. The *snopy* were lined up from bottom to top in an orderly fashion, as many as possible, then the oxcarts would be driven to the threshing field and built in long, wide and high up to fifteen feet stacks, twenty by ten, by fifteen, to wait for the threshing. Quite often the overloaded oxcarts would break down, sometimes on un-even roads, with steep inclines on the one side, would overturn, and then redoing the whole thing would take a long time and frustration. I had to go through all that many times. But, this was life, *what do you do*?

Preparing the threshing field (*kharman*) was in itself a complex procedure. Every peasant, for many years, had his threshing field at the same spot, on public grounds, west of the village, on the *poliana* (open field, pastures outside the village). As soon as har-vest is finished women and kids grab hoes and brooms, and remove the grass, making the surface even. The field was a round circle, some forty feet in diameter. After the grass is removed and the field is evened, big kegs of water are brought, and the harman is soaked, then, as it dries, it is sealed with diluted cow's excretions. It is then ready for use.

As soon as all the *snopy* are brought from the fields and lined up in *kamary* around the threshing field, the work begins. Every morning, at sun rise, the snops are pulled down from the stacks, are untied and the straw spread at even depth on the field. When it is all done, a *dikanya* is brought out – three planks of willow wood, two inches thick, with flintstones encrusted at the bottom in a geometric pattern diagonally – the cows are brought in, harnessed to pull the *dikanya* over the field, an "operator" will take its place over it, holding the ropes tied to the bottom of the horns of the cow – to "drive" them in the threshing field, armed with an *osten* – the five-feet long stick with an iron pin at the end, and the threshing begins – going around, and around, for a thousand times at least, or more – from seven o'clock until after two or three o'clock. The "operators" change – men, women and children, taking their turns. Every half an hour or less, when it appears that the surface of the crop is cut to some degree, the whole household lines up with wooden forks to turn over the straw and expose it to the flintstones the unaffected part of it. Then again on the *dikanya* where sometimes a small chair is brought in for the operator to sit down when he tires of standing up.

As the day progresses, the sun throwing hot air to burning degrees, the crop gets chopped to smaller and smaller pieces. At some time the forks are no longer effective and are replaced by a *graba* – a wooden instrument similar to the fork, but with four or five "teeth" – one foot long, and the turning over continues, until the *graba* is no longer of any use, and is replaced by a *lopata* – a wooden shovel. Both the fork (vila) and the graba are wooden tools masterfully fashioned. Finally the straw is reduced to small pieces, to one inch long, and the grain has fallen under it. It was a long and boring procedure. You stand there behind, on the *dikanya*, you sit on the small chair, often feeling like falling asleep, but you urge and urge and urge. Sometimes the cows had to do their natural things in execrating, and you will catch it with a special *lopata*, (shovel) and throw it out of the *kharman*. Sometimes you would miss, but it is in order of the work involved. The cows from time to time would reach to the bottom of the stuff and fill their mouths with grain, but, as the Bible says, you do not tie the mouth of the ox when it is doing the threshing. Who had read the Bible in those days and follow its admonitions, but the poor animals also had to eat. All in all, this whole procedure was boring, our bodies were burning from the fire coming down from the sun, we fought falling asleep, the monotony was killing, but the work had to go on and on.

After it is over, there comes the time to separate the grain from the straw. The cows are relieved of their duty and are taken to pasture. In early years it was me to go out. But as my younger brother grew up, he took this chore. I joined the troop at the threshing field. We would first grab the *greblo* – a contraption with five feet long handle, at bottom spreading in triangle made up of thin planks, and we, the men, would push the crop to the center of the field. The women behind us would sweep the grain to the center, until it is all collected. Then we start with the *grabba* again: throwing the collected stuff as high as we could, five-six feet, and watching how the wind would blow the straw to one side, while the grain is falling down. If there was wind, we would rush to take advantage of it. If the wind stopped, we also had to stop and wait until it started blowing again. All the time a woman with a broom would go over the grain pile and sweep to the side the heavy particles which the wind would not separate from the straw. When the procedure is over, we would remove the straw to the end, outside of the field, where it would accumulate until taken to the house and stored in the special hayloft, to feed the cows in the winter. The work with the grain will continue. It is then passed through a sieve where all foreign materials are removed, and it is taken in big bags home.

Quite often my father would have "rented" land from other people, had plowed it one summer into fallow, and planted the grains in the fall. Then we would harvest it, thresh it, and when the grain was ready for taking home, the owners of the land would wait on the side and take their "rent" for the land – half of the grain. I was young, I had no say there, but I watched and my heart was sinking to see these men take away half of the hard labor done by my father, me and my family. It was a glaring injustice in my eyes but there was nothing I could do. Our labor was exploited by people who had not moved a finger to produce the grain. The memory of these events and injustices never left my mind. When I first read that the land should belong to those who cultivate it, I could not but accept as gospel truth this economic mathematics. Quite often, at the verge of harvest, we ran short of grain and flour, and my father would go back to these same people to beg for advance of grain, to be returned from the new harvest, together with the "rent." I still remember the day when we harvested some barley, even before it was ready, how we threshed the stalks in the house, how we dried the grain in the sun and rushed it to the mill. Still, it was not ready for the day when we were to finish the last bite of bread. That evening we sat around the table, with no bread, only some soup of *lobuda* (goose foot), a wild weed, and tried to put up with the bad turn of events. I remember how my older brother, Petar, a teenager, may be 18 or 19 years old, exploded crying for the misfortune which had visited us. I do not remember, or given to understand, how my father felt at this point. It was not easy for him, I guess, to see his three children going to bed hungry. So, that is how it was in those "good old days."

In my early childhood our *kharman* was next to the Thracian mounds outside the village, but when I was thirteen or so, my father could squeeze it in our backyard, which made things much easier for us. At that time technology was reaching us, and somehow we managed to get a *viyatchka*, a contraption which separated the grain from the shaft with mechanically operated panel wings, so we did no longer depend on the wind. This is why we could move the threshing into our backyard.

It had a drum, four wings inside, two by four feet, made of thin lumber, attached to an axel, which was run by a hand mechanism outside the structure – a foot and a half in diameter cogwheel – which at the same time moved a number of other parts, with different functions: the straw was deposited in a holding trunk on top with an opening to let the contents slide into the machine, where a grade of teeth was moving left and right, causing the stuff to fall upon a sieve which also was moving left and right, while the wind caused by the rotating wings in the drum blew the straw out front in open space. The grain fell through the cieve to another sieve, moving left and right, where all sand and dirt would fall on the ground, while the grain would slide to the back, to fall on the ground under the drum. It was removed periodically. Bad straw, made up of connecting knuckles of the plants, which could not be blown out, because of their weight, known as *kotsaly*, would be falling in a groove and discharged to the ground separately. So, when the whole procedure was finished, we had four products – the grain, the straw, the kotsaly and the dirt. Each one would be appropriately taken care of.

This operation required a minimum of two people to run it – one to turn the wheel and one to fill the trunk on top with a *graba*. They would stop from time to time to remove the respective accumulations. Things turned in such a way, that my father and I ran the whole thing most of the times – my mother was sick, my younger brother out with the cows and my father filling the trunk. When he would be running the wheel, I would be filling, and when I was turning the wheel, he would be filling. Both functions were quite heavy for me. I would turn the wheel holding the handle with two hands, and would exert all my force, to move the heavy contraption. Soon I would get tired and my father would take over. The filling of the trunk was also heavy work, lifting the stuff over the top was not quite for my strength. But it had to be done. With pain and exertion it was all finished by the end of the day, only to start again the next morning. And so on for two weeks at least.

In those days I was very weak, very slim, skin and bones. And I exerted every ounce of energy I had.Once I overheard my father murmuring to himself, looking at me, at my bulging ribs and reed thin legs: "This one, also will not make it." He was refering to my two brothers and one sister who had died in childhood, younger than me. One of my little brothers died when I must have been six or seven years old. I suppose, from pneumonia. In those days there were no antibiotics, and there was no doctor in the village. After a week or so, little Totiu was dead. I still remember how my father took the little coffin and carried it in his hands to the cemetery. I loved the kid. I cried my heart out. Going to church with my mother and watching the Holy Mother of God holding baby Jesus I prayed for a miracle, to return my little brother to us. I prayed Sunday after Sunday. Well, in a year or two, in 1932, my prayers were answered. I had a new brother, Stoiu, still alive in Bulgaria for whom I prayed all the forty years in exile, not knowing what was happening to him.

But in 1937 or 1938, my father was seeing me, in the midst of the work, as one who was going to be dead too. This scared the wits out of me. No! I would not die, I swore to myself, clinching my fists and gnawing my teeth. I had to survive! I had to live! In those days it was believed that exposing oneself long to the sun was good for the health and I started burning myself, my skinny body, under the sun. It was not the sun that saved me, but I was dead-set to survive. I did. I am still here, just passing the seventy eight. No, there was not much food to improve my diet, just dry bread and sometimes a piece of feta cheese, or raw bacon if it was left from the winter months. But I never, never thought bad of my father uttering his sad premonition. He had lost three children, and watching me grow like a bean stalk, what else could have he thought? Poor man!

After the threshing season, it was the turn of the corn. In the beginning of August the upper part of the plant, above the ears, was cut, left to dry up and then stored for forage in the winter. The cows loved it. About the end of August the ears are cut and carted to the threshing place in big piles. Then the entire neighborhood would come some evening to help peel the ears, which then would be dried in the sun and then passed through a contraption, *ronkachka*, run by motor, where they would be passed through a three teethed wheels and all grain separated from the ear. Next would come the harvest of the sunflower. The large round heads would be hit repeatedly with a stick until all grains are out. With this the summer work is finished, until at the end of September would start planting the winter crops – wheat, rye and barley. By the time it is finished, sometimes in October, there would be a short interval of rest, and then, when the fogs start rolling across the village, the time of preparing the tobacco for sale would start.

During the summer, sometimes in July and August, the tobacco plants, two to three feet high, with lucious sticky leaves spread all out, are ready for picking when the leaves

50

turn yellow. Brought to the village, they are pierced at the stem on a flat metal needle, *guberka*, some one foot or a little more long. At one of the ends in a hole is passed through a thin strong thread, *kanap*, and as the *guberka* is filled, the leaves are pushed back to the kanap, a five-six feet long string to make a *niza*. This niza is then attached to a *saruk* – a long thin hard wood, which would support it, being tied every two feet with a piece of dry covers of the corn ears. It is then taken to a specially built sun-exposed scaffolding and kept there ten days until it is dried, and then taken off the wooden *saruk* and hanged in a dry safe place.

In November the fog softens the dry leaves and the time for the final operation has arrived. Again the whole neighborhood would gather – one evening in our house, another evening in somebody else's house, for many, many weeks until it is all done. The leaves are pushed out of the string, then taken one by one, matched for size and quality, and superimposed – some fifteen twenty leaves. They are tied at the top where the stems are crossed into a *lepeze* and ordered in a stack tied tighter in a *denk*, which is appropriately wrapped and stored for sale. The agents of the Tobacco companies would turn in the village sometimes in January and go from house to house to bargain for the goods. They would pull at random *lepezes* smell it, examine it and make their offers: so much for first quality, which is estimated to be, say, 10 percent; so much for second quality – 20 percent; so much for third quality – 30 percent; the rest discard. They would take it for nothing just "to free" the peasants from it. The peasants would stare at the agents, mull over the offers, and most of the time decide to accept it. What could they do? The deals were made, and the agents would come to collect the goods and pay. The village tax collector is again there, and again, as in the payment of the roses, pick from the top his share. If something is left for the peasants, good – if not, too bad. The peasants would walk out, practically empty-handed, and would curse and swear the agents, the tax collectors, the government and all those who stole their hard labor. It was not so for those who did not have unpaid taxes.

I was involved much in the cultivation of the tobacco, from end to end. When I was in the refugee camp in Greece, and a Brazilian agency recruiting laborers visited us in Lavrion, I was asked what was my profession. Theologian would not go very far, so I answered: farming. They asked me to describe the procedures for producing tobacco. I explained the whole process and they listened very intrigued to my narrative. A few weeks later, the representative of Church World Service told me that he did not know how much I knew about farming tobacco. The comission was impressed and signed me up. I was almost ready to depart for Brazil, when, on the eve of it, a scholarship had been arranged for me to go to Geneva, and so I missed the boat. Fate had determined that I would practice agriculture in my garden patches in the United States, but never raising tobacco.

There was a way out of this injustice. Tobacco was government controlled substance. There was tax to be paid for it. So, it could be sold only to authorized tobacco company agents. It turned out that it was a valuable good to be sold on the black markets. In the darkness of nights, some mysterious men would cross over the Balkan, the Sredna Gora, with their mules and horses, would be led to our house by their contacts, would buy at twice the price the agents would pay, would load their animals and disappear in the night. They would take it to far away corners and resell it at profit to other peasants where tobacco was not grown. They would roll their cigarettes in pieces cut from newspaper, and so the government would be cheated of its taxes. So, this was life and I learned very early to keep my mouth shut when such a transaction would take place in my presence.

51

9. Zelenikovo in Bygone Times

But let me go back to Zelenikovo and say a few words about the village itself. It was a big village, some five hundred or so houses and families, an estimated population of approximately 2500 people, all farmers, except for a few craftsmen - one or two tailors, one or two shoemakers, smiths, ox-cart makers, less than a dozen teachers, a priest, a few bars, a few clerks in the municipality, few store owners, and a tinsmith. Some forty four kilometers North-East of Plovdiv, it was situated between Brezovo, former name Abrashlare, and Rahmanliy – now Rozovets. Abrashlare was twice as big as Zelenikovo and was sort of a county seat. High government offices – regional police headquarters, the District Court of Peace Justices, the local bank, the county post office and telephone-telegraph office, biggest mill in the region, and the regional doctor. The people there were all farmers like in Zelenikovo. The Rahmanliy-Rozovets people – were more sophisticated, with more people scattered around the country and the government, and always much better informed on national and international affairs, than Zelenikovo and Brezovo peasants were. My mother was from Rosovets.

The people in my village, all of them farmers, as I mentioned it, were less sophisticated, hard working, mostly middle class land owners, with many of them of under the acreage needed land to produce enough for their survival and renting from absentee owners to make the difference. In my childhood most of them dressed in Turkish type of trousers (potury) and jackets – abas. They all wore a cape, made of heavy goat wool when going to the fields, called *Yamurluk*, with a hood. By the time I left the village, about 1940, most of them had discarded this attire and dressed western style. My father never changed, even though the time came when my mother was no longer able to renew his wardrobe and he was unable to buy new clothes. He walked around in the same cloths that he slept with, reduced to rags after a long use. Around the waist this Turkish outfit was completed with a long turban-like piece (poyas) seven feet long, colorful, but strong to keep the back of the man together in the heavy work which he was doing. This outfit was completed with a kapa – sort of a hat, fur-like finish, which from the long use always appeared greasy and crumpled. The women were dressed in the winter in their woolen skirts, *sukman*, woolen hand-made pullovers, jackets and in the summer cotton made skirts and blouses. There was a lady in our neighborhood, Kaka Dona, (Kaka is a polite addressing of a women older than you, as Bati was the name by which younger people would address their older brothers, and also older men). who was the village seamstress. Only the teachers, and people who had lived outside the village, and had returned, were dressed western style.

But it was rapidly changing in the 1930s. We, the children, until we would reach the age of five, were dressed during the summer in a bell-shaped overall, without underwear, while the girls were dressed better. My biggest thrill was to show off when my mother would knit me a new sweater. From early spring, to late fall, we, the kids, sloped around in mud, in dust, in brushes and all that wild beastly behaviour which is difficult to describe. When we started going to school, we put on our feet wooden slippers, held by a piece of leather across the top, a "shoe" called *nalumi*. In the winter we would replace them with a cow-skin adapted to the feet – *tsarvuly* (A pronounced as in Much). The tsarvul was made of a piece of cow or pig skin, from ten to sixteen inches long, and about ten inches wide, according to the size of the foot to wear it. Two corners of the narrow side were pulled together and threaded with thin skin stripes bound together, to form the front end. The same thin skin thread is then passed through holes on the sides, half a dozen or so, to reach to the back. When pulled to shorten the "shoe", they form the tsurvul, which is then fitted to the foot, wrapped in *navushta*, (rags of old clothes) or specially made cloth of hard wool, masterfully wrapped around the calves and tied together with *kadunki*, black strings of goat's wool, in geometric figures. Should add that this type of "shoe" is no longer used, except in ethnographic museums or folk dance performances to show ethnic culture. But in my days this was the prevailing use for daily activities type of a shoe. When they would dry up, they become most uncomfortable and caused blisters on the feet. I hated the thing and when my father insisted that I put them on before going to the fields with the cattle or the sheep, I would put them on, and once out of the village, take them off, hide them in some bush, and look for them when returning home in the evening, having gone during the day barefooted. But in fall, winter and spring, one could not avoid them. They were very easy to let the water in though. Since there was no street electricity in those days, returning home late at night, one could not see where to step. One would follow known tracks on the street, sometimes, after rain, water would fill potholes, they would look like dry spots, one steps in and finds himself in water to the ankles, the tsarvul filled with water...

All, or almost all Zelenikovo men, had nicknames and referred to each other by these names. When I returned to my village, I met many people whom I could not recognize. How do you recognize someone you have not seen for forty of fifty years? People were introduced to me by their nicknames, so I could remember them. These nicknames are a curious assortment of qualifications. I will use only the English translations here: Taniu the Tar, Encho the devil, Grozu the turnip, Deniu the donut, Ivan the Spider, Ivan the Rabbit, The White Hat, Ivan the male Turkey, Matiu the Sardine, Ivan the goose, Todor the goldsmith, Mincho the sour weed, Koliu the beetle, Todor the Beardless, Denue the mad man, Doncho the lizard, Tsanue the gun-shell, Mussolini, Peniu the pig, Koliu the little barrel, Koliu Komunata etc., etc. When a woman was explaining to me that she was the daughter of Encho the Devil – he was manager of the Consumers Cooperative Store in old times, I could not control myself exploding in hearty laughter. In those days nothing was so funny, but now every name that they mentioned with the nickname made me laugh.

I should say something about the public accommodation places in the village. There were three or four grocery stores where one could buy anything that was not produced on the farms and was imported from the city of Plovdiv. The Cooperative Consumer store was quickly overtaking the big store in the center of the village of Tsanu Hadjimitev. There were some four or five *kruchmi* – bars where men crowded in the winter evenings, but where we, the young, had no access until seventeen or nineteen. There was no legal age set up, but by tradition a man had to be a little grown up to get in there. Small and skinny men like me could not go near the door. Neither did I care for it. But we were readily admitted in the *kafene* houses, coffee houses. They were small, narrow, crowded places where the adults sipped Turkish coffee, and inevitably playing cards on the few tables. That's where most of the people met in the fall and winter evenings, and so also in spring and summer times.

Closer to our neighborhood was Kafene "Suglasie" (Concord), held by Pencho Radicala. The nickname of the whole family was "Radikala," having belonged to the old Radical Party in Bulgaria. Day and night people played cards there, or backgammon, surrounded by kibbitsers, and all commenting on the games, arguing in loud voices and having fun.

Pencho, who on Sundays played violin for the village horo, was serving coffee. The loosers in any game paid for the coffee to all players. All the time there were discussions of politics, Bulgarian and international. I loved to spend my time there, when I was of age permissible to enter, after Middle School. I loved to listen to the discussions, which were quite sophisticated and informed. Center of the discussions was Pencho's brother Dimitar. He was paralyzed, could not turn in any direction without turning his whole body, walked with difficulty with crutches, was tall, slim and always seemed to be in pain. But his mind was sharp like a razor. He was informed on all matters, in politics and any field of knowledge, considering the limitations of the village life. I was fascinated by his discussions. Politically he was a rightist, was a nationalist, and eagerly discussed any issue brought up before him. Sometimes I would join in the discussion, make my contribution, and was listened to by the entire congregation. I had read enough, I was informed enough for their milieu, I was reading the newspapers and loved to share my ideas with theirs. Their father, Diado Tsaniu Radikala, a short sturdy man of considerable intelligence, sometimes in the past a mayor of the village, to whose credit was the bringing the water system into the village, was regularly there and we became close friends. He was bitter anti-communist. I remember my last conversation with him, in 1945, following the trial by the "People's Court" when it appeared that Ivan Bagrianov was going to be executed and expressed my regrets about it, because I had a good opinion of him as a man who stood firmly for the peasants. Diado Tsniu spared no words condemning him. "Let them hang him" said he. "If he had brains he would have fled from Bulgaria before they (the communists) came to power." So I enjoyed the atmosphere of Kafene "Suglasie" – as poor and filled with cigarette smoke as it was. When I came back to Bulgaria, I was anxious to take a look at it. It was a pathetic sight. The then new brick building was eroding, the window was dirty and covered with dust, the door was dilapidated and inside it was a pitiful remnant of the once popular hangout. The walls had never been painted, the plaster here and there was falling, and the place had been turned into some sort of a dirty store – milk collection center. It was dirty and filthy and ... looked so small that I could not figure how in old times it could accommodate so many people and be such a cozy corner, during the long, rainy and snowy nights.

10. The Map of Zelenikovo

As a village, Zelenikovo was well planned. The main street was, actuality the *shosse*, the road (I hesitate to call it highway) connecting it with Plovdiv, by way of Brezovo, through our village and reaching Rozovets. East of the Main street was a section extending from end to end, North to South, divided to blocks by streets reaching to the river, and the West side, the larger section of the village, extending also North to South and cut into blocks by perpendicular streets only that two parallel streets, stretched from the chitalishte to the South, and from the central square to the North. On the west side of the village, a glade, the Polyana, stretching from end to end, north to south, then turning East to reach the river, was flanking the village.

In spring and fall the entire polianna (glade) was a pasture for sheep, and in July and August it would be all turned into threshing fields. It is there that we, as kids, did our "Mardi Grass" rituals which I have already mentioned. It was one of those things that we eagerly anticipated in February and enjoyed it. In the spring we would take the lambs after school to pasture there, playing to exhaustion, and felt happy. In those days the poliana was clean shaven – by the sheep grazing – and there was always something going on there. In the winter, when snow would cover it, we would take our sleds there, mount the mogilki and slide down as far as we could go, to return home frozen, red, and satisfied of the pleasures. On Christmas Eve we would go to the Mogilki, the Thracian Mounds, and leave there some hay for the donkeys of Diado Koleda (our Santa Claus).

When I visited this place in 1991, I was heart broken. Gone was the old poliana. It was covered by overgrown nasty weeds, what we called *Magareshki truni* – Donkey's Thorns – dry, high to the knees grass, at one end near the village, near our parallel street, a garbage dump, and around the mogilki scatterred all over abandoned and broken agricultural machinery and parts, adding to the desolation of the place. Thorns were growing all over. Here and there peasants had tied their mules to graze. I climbed to the top of the mogilka, I stared at Sakar Tepe, at the village, and all over, and left depressed, regretting the present sorry state of the once beautiful paradise enjoyed by the kids.

There were three main squares in the village. The biggest one was in front of the Church – close to 200 by 150 feet, right on the Main Street, ending in the east with the Church, which, on the back side, bordered on the river. It was a low structure, half of it below ground level. This is how churches were built in Turkish times, in the 18th and 19th century, hidden behind high walls, so that they could not be seen from outside. The roof was covered with ceramic tiles, who knows how old. Inside, it consisted of two sections – the front nave was three feet higher than the main nave. All baptisms and weddings would begin in the front nave, and then moved to the main one. The altar was at the eastern side. It was separated from the main nave with the traditionall iconostas, a wooden wall, decorated with the icons of Jesus Christ, the Holy Mother of God, St. John the Baptist on the side of Jesus, and St. Petka, patron of the church, left of the Holy Theotokos. The third icon on each side was an Angel. The Royal doors were an interesting wood carving portal. Above the major icons, just mentioned, was a long frame from end to end, with numerous small icons, depicting the main events in the life of Christ. On each side of the altar, in the nave, was the Kliros – the place where the chanters stood and responded to the priest's invocations. When I was seminarian, and was on vacation, I practiced my Byzantine music there. I think, I was pretty good at it. The regular chanter was Diado Milcho. For decades he murmured his old tunes. Here was also a young man, Boyu, I believe. In my recent visits to Zelenikovo I learned that his son was the local police officer, Militia man. At one point, when I was surrounded by many people, he came very close to the group with his motorcycle, but did not come to greet me.

The ceiling of the old Church was in three sections in longitude. The central one was some four feet higher than the lateral ones. As a kid I was amused and read a hundred times a Biblical verse, Mat. 7:12, somehow in reverse rendition: "Do not do to the others what you don't want them to do to you", written from end to end in big letters. On the opposite side was the text: "Love thy neighbor, as you love yourself." Nobody who attended the church even once, could miss this message and could have any doubts about the substance of Christian social philosophy. Recently, June 1996, when I was in Zelenikovo, I saw a picture of a wedding, with over a hundred people attending in a long line in front of the Church. It happened to be the wedding of my favorite aunt, Plina, in 1919. Behind them was the bell tower – a wooden dilapidated structure, sort of a well, two feet high, made up of lumber and over it the bell, suspended on two beams over the wall.

Sometimes in the 1930s a new bell tower was constructed, probably fifty feet high, a modern addition to an old building, which still stands there. I remember a funeral of a Communist partisan in October 1944, held as a civil ceremony on the square in front of the church. At a given time, in memory of the dead man, some ten militiamen, partisans, shot their guns aiming at the tin cross on the top of the bell tower. They virtually destroyed the tin ornamentation and only the iron skeleton was left. When I visited the village in 1991, I looked for this remnant of the old dark days. It had been rebuilt. I also learned from Drazha Parlapanska that archeological evidence, found around the holy table of the altar, has indicated the antiquity of this place for worship, who knows, perhaps in ancient times it had been a pagan religious shrine. What impressed me in 1991 about the square in the village in front of the church was that while all streets (except the one where my brother's house is) were paved, the church square was not and in rainy days was turning into a pool of mud. Near the gate of the church I saw a large stock of paving cement blocks. I do not know how long they have been there, but apparently the Communist authorities had left this square unpaved purposely. What a foolishness! It was the largest square in the village.

When I returned to Zelenikovo in 1991 and visited the Church, it looked to me exactly the way I had left it four decades earlier. The icons were the same, the flag stone floor was the same, and the service books were the same, just somehow falling apart. Diado Milcho, the old psalt (the man responding to the priest from the antiphonal kliros, with whom I had shared so many services, was no longer alive. The new Priest, Father Kolentsov, was assisted by his daughter and the congregation of some ten or so old ladies, singing in a choir, as good as they could. I saw the Gospel book read by Father Kolentsov. It was falling apart. When I went to Sofia I purchased a new one from the Synodal bookstore and took it to America, intending to have it bound with golden-like covers, but when I was told that it would cost $600.00, I gave up. The book itself, silver-covers, cost me only $2.00. The ceiling was repaired, probably damaged by leaks, covered with lumber planks. Gone was the wisdom of the old priests in large letters which had impressed me so much in my childhood, my first lessons in moral law. On the north wall were the same old frescoes, which I had not forgotten. One was St. Elija, on a chariot over the clouds. My grandmother had told me that when it was thundering it was St. Elija racing his horses. She would take me to the church and show me this icon. I never forgot this story and in my classes in the future never missed the chance to mention it when I was discussing the origins of religion. These frescoes were big – some three-foot by three foot. One of them, I believe, was an icon of St. Cyril and Methodius. I do not recall now whether it was on this or another Icon that the eye of one of the saints was gouged, one inch deep in the wall, in times past, when some Turks had entered the Church. There was an interesting big carving over the Iconostas: two beasts facing each other, a cross between them, but in mild mood, as if testifying to the magnificence of the faith. Next to a column was an icon of Theotokos, very old, silver-plated. Next to it was a rickety candle stand. Years later Father Kolentsev told me that they had purchased a new stand there, a solid brass piece, with the money which I and my wife had donated to the Church.

Further up the main street, some 3-400 feet, was the central square, sort of triangular enclosure, probably about 100 feet each side, out of which some five streets were branching

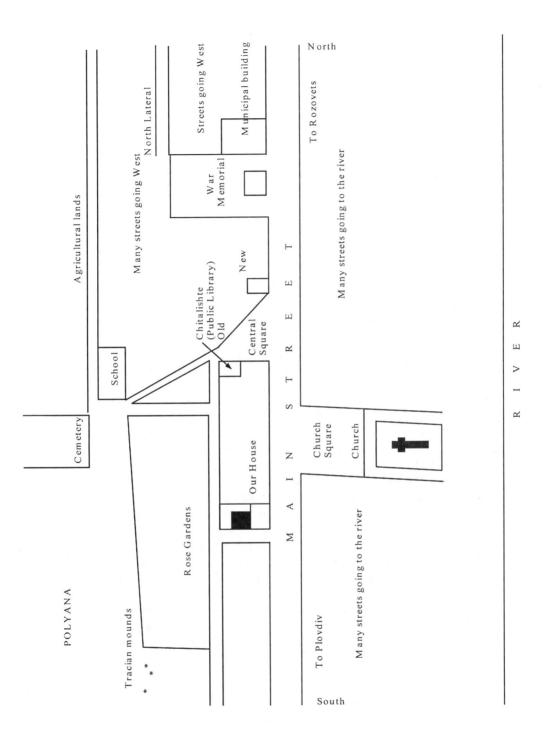

POLYANA

– two of them, the main street, going down to the church, and the other to Rosovets. In my time there were three taverns around the square, two coffee houses, a barber's shop, and the big and impressive grocery store of Tsaniu Hadjimitev, supposedly the wealthiest man in the village. When the communists had come to power, the story goes, he had been tied and tossed half dead on the square in front of his store, as a capitalist exploiter.

Sometimes this square was used for holidays' horo. It would start here and when the square could not accomodate more people, the music would lead the entire crowd to the next square, some five hundred feet north. During the summer, the entire village, better say those of the men who were interested of news, would gather in Central Square, practically would pack it, and listened to the loudspeaker connected to the radio in the *chitalishte*, the public library. It is here that one evening, in July 1940, I heard a German speaking orator who seemed very excited. I asked, arriving in the square, who was speaking, and someone told me that it was Hitler, delivering his famous speech offering England to negotiate peace. Nobody understood German, but the National Radio was carrying it. Now, all these buildings were either razed, or dilapidated signs of the past. The grocery store of Diado Tsaniu, preserved in its renaissance structure, was closed, the plaster on all sides falling and the former *chitalishte* in a miserable shape. Two of the big houses, with stores on the ground floor, the *Kurtovite kushti*, were no longer there. In their place was a small park dedicated to the village partisans, with a monument and a memorial marble plate with their names carved on. Further up was the modern new *chitalishte*.

The most important place for me here was the old *Chitalishte* "Georgi Sava Rakovsky," named after one of the leading figures of the Bulgarian movement for liberation in the 19th century. The word, "Chitalishte" literally means "Reading room." The Church and the Chitalishte were the two most important non-governmental institutions in all villages of Bulgaria. The *Chitalishte* institutions appeared back in the time when Bulgaria was under Turkish rule. They were voluntary associations formed by teachers and leading village "intellectuals". This was also the case in my time. Teachers, and more literate peasants joined as members of the *Chitalishte*, elected its boards and supported it financially. The municipal and the state governments had nothing to do with it. It was a private organization. Usually they would be tied to some cafene-owner, who would conduct his business and at the same time along one of the walls, would be some bookshelves and a table along them. Usually a few periodicals would be displayed there and read by visitors of the *cafene*. The place was usually packed. Our *Chitalishte* was tucked and attached to the business of Diado Tsaniu Hadjimitev, the building owned by him, given to the coffee shop owner, free of rent, but to be used as a public library, and a street leading down to our district. It was the most popular place, patronized by more educated and publicly concerned peasants. The word higher class would be misleading. People just stood there, talking to each other, discussing daily occupations, national events and international politics. Primitive, as these discussions were, they were pleasantly filling time. There was no radio and no television in those days. The first radio in the village arrived about 1938 and was placed in one corner of the *Chitalishte*. When the time for national news would come – 8.30 p.m., the room would be packed with peasants who silently listened to every word. For half an hour there was no word uttered by anybody. Only Diado Muiyno, the owner of the coffee shop, with difficulty would make his way through the crowd to serve his Turkish coffee.

There was a long table at one end of the room, perhaps ten feet, where current periodicals, reduced to shreds, were leafed through and read by eager beavers like me. One of these periodicals, the most interesting one, was the magazine *Priroda* (Nature), something like National Geographic, but in much poorer shape and form, with few illustrations, but fascinating information on popular science and natural events in the world. In the fall and winters of 1937-1939, this was my "cabinet" where I read and studied. Behind the table were two book shelves, glass-doors closing access to the books, containing perhaps four to five hundred books – literature, science, practical guides, foreign literature, agricultural information, and all sorts of literary works. Twice a week, a young man, Deniu Hadjimitev, would be distributing books for home reading, and every time I was first on line. A copy-book was the catalog listing the titles which I turned over back and forth a hundred times. And I read everything – from Jack London and Jules Vern, to how to grow ALFALFA, how to care for bees and other topics.

After Middle School I spent two years in the village tending neighborhood sheep or cattle. The Chitalishte became my school. We, the kids, were resented by the older people, because we occupied precious room for customers, on the table, but we were never chased out. Suffocating in the smoke of the "home-made" cigarettes, rolled in newspaper, we thrived, and we were growing, in the smell of garlic, and onions, and body odor (there were no showers, or bath-tubs in the winter and people bathed from summer to summer), but we were used to it, because everybody smelled the same way and it was not offensive. Offensive was the free release of gases, and of that there was plenty, because the average dinner was beans, and the beans were quite virulent in poisoning the environment, especially in a small room with tens of people packed like sardines. But where one was to go? The bars were frequented by drinkers, and both the smell of liquor and the loud arguments, often ending in fist fights, repulsed me to go there. So, I was the regular *chitalishte* bum.

When I returned to Zelenikovo, forty years later, I could not miss visiting my old beloved hangout – the *Chitalishte*. I did find the building a forlorn remnant of the past. The plaster on the outside was falling and never repaired. It had been turned into a store for distribution of bottled soft drinks. The old counter once used by Diado Muiyno to make coffee was still there, but was loaded with empty bottles and cartons. The floor was worn out. The walls were dirty and filthy, plaster was falling from all over, the ceiling was … still there. And where once the book shelves and the long reading table were, a pile of cases of empty bottles were stacked up to the ceiling. I took a look from the door, my heart could not tolerate this desecration of my old temple for learning, and moved away. Some young man, behind the counter, looked bewildered, with wide open eyes. He said nothing, and I cared to say nothing.

They have built a new *Chitalishte* now, on the same square, redesigned, expanded, after taking down some of the big houses, of the Kurtov family. Some sort of a small park had replaced the square. The new building seemed to follow a Bulgarian style of the 19th century, with large facade and long steps in front of it. I saw the several rooms with library bookshelves, and an impressive number of books. There is also a theater which is used for public meetings. But I did not see many people going into the Chitalishte. They had a librarian which happened to be the daughter of one of my first cousins.

During my visit to Bulgaria, from the middle of May to tenth of June 1996, for the Conference of Bulgarian scholars abroad, I was called by my sister-in-law Vessa in Sofia. She told me that the village was celebrating this year, on May 24, the day od Sts. Cyril and Methodius, the Apostles to the Slavs, the 125th anniversary since the foundation of the Chitalishte "G. S. Rakovski", the Public Library I described above. They had addressed an invitation to me to be present there. I could not. That same evening the President of the Republic, Dr. Zheliu Zhelev, was giving a reception for us, and I would not miss this opportunity. It was indeed an event, with the entire Bulgarian elite being there – from the Patriarch to anybody who was somebody in the academic world, the artistic and the political elite of the country. But after receiving the telephone call, I decided that I had to do something about this invitation. All of a sudden I remembered how much I owed to the Chitalishte "G. S. Rakovski". I sat on my hotel bed and drafted a telegram. I walked to the Central Post Office and sent it out. It read:

The President of Chitalishte "G. S. Rakovski"
Zelenikovo
Plovdivsko

I am deeply touched by your invitation to attend the 125th anniversary of Chitalishte "Rakovski". The Program which brought me to Bulgaria makes it impossibe for me to come to Zelenikovo. On the scheduled day we are going to attend a reception in Residence "Boyana". I am highly appreciative of the Chitalishte in my native village, where I was imbued with love for knowledge. During the 1937-39 it was my home, my temple and my university.
Yours: Spas T. Raikin

Next week, when I visited the village, I was told that the telegram was well received and applauded.

The third square, further up, was even larger than the one fronting the Church. This was the place where all holiday horos were danced. The entire village would come out there, men, women, boys and girls, and a myriad of children. Year after year, winter after winter, I used to freeze around this horo, until I was old enough and joined in it. I danced, and danced, and enjoyed it. Sometimes over two hundred people would be dancing this horo in a chain of people, holding hands together. The music was supplied by Pencho Radicala as a violonist, Vlaseto, the village tin-smith, with his clarinet, Andreshko with his big tupan (drum) and someone with a sort of a tuba. They would be paid for each dance by some young bachelor who would like to lead the horo, next to him would dance his bride-to-be girlfriend, then his father, his mother, his brothers and then... everyone would join in – usually boys and their girlfriends, then men and women, and at the end young kids like us who were just learning and made a mess on the tail. But it was fun, and it was a glorious celebration.

Every Sunday, between *horos*, and in summer days, the main street going to Rozovets, served as the *sturgalo*, as a place for promenade of the young boys and girls and anyone else who felt like walking back and forth a distance of one kilometer. Here, the young couples going steady would hold hands and walk, while all women of the village, mothers, would watch and then gossip who was going to marry whom. Those unattached, boys and girls, would walk two-three or more in a group, giggle, joke, wolf-whistle, scream, sometimes sing and make a raucous to demonstrate their presence. Gradually they would all find their partners for life, engage them in conversation, and show interest to each other, until they start walking alone, as couples. In summer time the whole sturgalo would move further up, outside the village, among the growing crops, and enjoy the cool evenings, and more. It would continue well after sun set, in the dark.

Fronting the square, with face toward the Main street, was a monument erected in memory of those who had fallen in World War One – a stone structure, some fifteen feet high, probably six-seven feet wide at the base and rising as a pyramid. The lower part was used as a double cheshma – two water-spigots, and above it a marble plate with the names of the victims carved in black, with a wreath on top, and two lions at its base. Whenever there was a civil ceremony, official celebration – Memorial Day, St. George's Day – Day of the armed forces, a Te Deum Service would be celebrated there, speeches would be made by officials and all school children as a crowd. The rest of the villagers, very few really, would stand on the side, but everyone stood at attention when the national anthem was played or sung. Inevitably there would be the widows of those who had fallen in the war, and one quite often listened to their wailing. These were patriotic occasions, which impressed me very much at the time. In their speeches the orators would decry the sad story of the Bulgarians under slavery in the neighboring countries and I, sort of felt moral indignation at the injustices done to Bulgaria. I suppose my nationalism, which has never left me since, was born there, since I, unlike the mass of school children there, listened attentively to all speeches. In my first visit in 1991, I felt obliged to stand at this monument, pay my respect to those unsung heroes and take a picture there. I was moved now, as I was in old times. It often crossed my mind that my father could have been on this marble plate. He had fought the war at the Doiran lake against the Anglo-French forces.

On the right corner of the square, at the main street, was a tavern where the dancers went to wet their whistles. On the opposite side of the square, across the main street, was the House of Mikhail Matevsky, once a candidate for member of the Parliament, an Agrarian, a mild character. When the partisans captured my village, July 21, 1944, a women partisan had led a group of her confederates inside the house to ask for money. As the story was told, he reached in his back pocket to get the keys for his safe, but this armed lady shot him in the chest, thinking, so was it claimed, that he was going to pull a gun. Ironically, he had been on their side. On the left corner of the square, at the main street, was the Municipal building. Relatively new, built in the early 1930s, it was the office of the mayor, of the tax collector, the village postman, and also served as local jail in the basement. That is where I spent some of my less pleasant days in December 1944. I will tell this story later.

The back side of the square was rising into a low hill where the house of my aunt, Raika, stood. When I would be freezing down at the horo, I would run there to warm up. Opposite it was the house of Parlapansky. I will have much to say about Georgi Parlapansky later on. He was the top Communist in the village, a shomaker by profession. His sister, Nedelya, had been married to my uncle, my father's brother who had died in 1928. I still remember how my father brought him from Plovdiv where he had been taken for treatment of pneumonia. In those days there were no antibiotics, and he had gone too young, perhaps 25 years old. His wife and my cousin Penka lived next to us, in my grandfather's house, for many years until they moved back to Diado Marin Parlapansky. I often visited them. My aunt's sister Drazha, was an old maid. A few years ago she was still alive – never gone to High School, but deeply involved in reading. She used to say, that if she was offered a room full of gold and a room full of books, she would have taken the books. We struck a close friendship with her and she started giving me books to read. It was from her that I got some book on history of philosophy prior to going to the Seminary, which I read with great interest. Somehow I never got close with her brother. She is the local historian who tracked down my ancestry. On account of these houses, I somehow felt at home when I crossed this square. After 1944, however, it became a place of bitter memories and I approached it with utter resentment. I am afraid, these feelings never left me, even to this days.

Numerous streets branch out from the main street to the right, leading to the river, and to the left, ending at the *poliana*, a stretch of uncultivated land, beginning from the northern end of the village and surrounding it all the way, from the west, to south, where it reached the river. At some places it was as wide as a third of a kilometer, and at other places – a mere 100 feet. It was the village green, covered from early spring to late fall with green meadows where flocks of sheep grazed all the time in one section or another. It was criss-crossed by many pathways and little roads leading in all directions where people and animals trudged to wherever they were going. It was also a place where we, kids, use to play all the time, summer, spring, fall and winter. During the summer families who did not have enough room in their backyards, had had their threshing fields all along the poliana. It was a place where people walked to do their chores in the fields, met other people, took animals for grazing and engaged in talks about their cares. It was a lively place. Near our neighborhood it had the distinction of displaying two Thracian mounds where we, kids, all seasons of the year, found all sorts of games to play.

There, midway from north to south, was the village cemetery. In olden times it was an open field of probably two acres land, where people were taken to rest in peace. Since it was not fenced, and people did not particularly like to walk through it, or graze their animals, it had much overgrown brush, all sorts of unpleasant weeds, like what was called – Donkey's Thorn and Sambukol – Magareshki Truny and Buzurniak. It attracted all sorts of animals left to roam the fields – donkeys and dogs. Old women were to be found all the time there, taking care of the resting places of loved once, but no special care was being taken of the place as a community concern. I visited there in 1991 to pay respect to my father and mother whose funerals I had not attended. In 1949, when my mother died, I was hiding from the local communists, and could not come anyway, because at the time I received the telegram, the funeral was already in progress. When my father was killed, I was in Greece, and later was told that he was buried during the night. I am not sure if they called a priest, or not.

Ever since 1991, when I was visiting there, I went to the cemetery. In Communist times, they had fenced it, so no animals could trample over the graves. And there were no dogs to roam around. I was told that sometime during these years, on orders from the authorities, all dogs had been taken to the river behind the church and shot dead. I never guessed why, but it seems to me it was a way of saving food. As I was walking across the cemetery I saw so many gravestones, and read so many names I was familiar with. Practically everyone whom I knew was there – people from my neighborhood, people from the village, relatives and strangers, prominent or unknown to me. Most of the gravestones were decorated at the top with a Communist star, indicating membership in the Communist Party, or simply as a sign of the times. Rarely I saw a cross. I spent a lot of time reading the names and remembering the people with whom once I was very close as friends, young or old. It seems to me that all those with whom I was in close relations were there. I saw the grave of my older brother, Petar, of his wife, Tota, and my beloved nephew, Slavcho, killed by a tractor while at work. He loved me so much that any time he saw me was calling me Chichko – my little uncle. His stone was covered with brush. There I saw the grave of my beloved teacher Boniu, where I stopped and layed some flowers on it.

The graves of my father and mother, side by side, were fenced with iron bars by my brother and cleaned of grass for my impending visit. There was no grave stone there. I ordered a stone and the next year I found it installed there. During my last visit I spent sometime looking for the grave of my close friend of our teenage years, Yovcho. I will have much to say about him later. He was killed on October 13, 1944 at Stratsin, in the war against the Germans. I could not find any traces of his grave. I was at the funeral and I long wondered why there was no mark of it. Later on it crossed my mind that he may have been exhumed and buried under the memorial in the center of the village. Shortly before September 9th he had joined the partizans.

Diagonally from the cemetery was our school. Or once our school. It was no longer serving that purpose. Now it was an orphanage of destitute children, and I would say, mostly gypsy kids from the entire county. It was built in 1928-29. Before that the old school was next to the Church, a two story decrepit structure, already falling apart. I was scheduled to start school in 1929, but after spending one period there, I ran away, to my home. I was frightened by the whole set up. I had heard horrible stories about the teachers, especially of the teacher who was going to be my class instructor, Diado Georgi Bradata. I was told how he would grab you by the ear, would twist it and call you "You cursed lice." I was terrorized by this thought. I developed instant stomach disorder out of fear and ran home. Nothing after that was going to take me there. My parents did not try to force me. Physically I was weak and fragile, and often sick, so I was not prepared to go to school. Only the teachers could not understand that and two of them, Dimitrina Vuleva and Mara Angelieva, often visited my home to scold my mother and father for not sending me there. But in 1930 I was ready to go to the new school.

It was a beautiful structure, one floor, large rooms, large windows, freshly painted and wooden floors soaked of some oil. The Director, Peter Videv Zidev, was watching over it like it was his eyes. We were not allowed to enter it in our street shoes, or tsarvuli, not to dirty it. We had to take them off outside of the class rooms where each one was assigned a small box in a wooden shelf. There we would keep a pair of home made slippers, *tirlitsi*, and walk in the class rooms. There were two buildings, the first was the elementary school, four grades, the next was the Middle School, the Progymnazia, three grades. The school yard was probably two acres, and was fenced with barbed wire. I need not say what a bedlam we would create during the short breaks between classes, running, screaming, laughing and often involved in innocent fights. I came to love this school. Compared to my house, it was a place to be enjoyed. After the first year under Dyado Georgi, who twisted my ear only once, to scare the hell out me, I never misbehaved again. The next year I was taken by a young man, Boniu Bonev, of whom I will have more to say. But my best memories of my childhood are associated with this school. Now, in 1991, I found it turned into an orphanage.

There were no more kids of school age in my village. All young men had moved away. There were left some five or six hundred old ladies, widows, and a few men, and no more than ten to fifteen school age kids. They commuted to Brezovo with the morning bus and returning home the same way. I visited the Orphanage. The rooms were adapted for this purpose. The classrooms were turned into bedrooms with tiny little beds, but clean and made in perfect order. I looked around, I visited my class rooms, I stood at the portal entrance of the building, which once was a crowded place. I remembered how many times, as a kid, I had recited little poems like:

Pencho bre cheti!	Pencho, do read!
Pencho ne chete.	Pencho does not read.
Pencho rabori!	Pencho, go to work!
Pencho pak ne shte.	Pencho again refuses.
Pencho potrasna	Pencho grew up,
Iska da yade	He wants to eat,
Nyama ot kude.	But nobody would give it to him.

The most glorious times in those days were the celebrations of St. Cyril and Methodius, the patrons of Bulgarian culture, having composed the Slavic alphabet in the 9th century. All of us would go to the fields, cut wild flowers, the *sinchets* (blue flower) being the most popular, bring them to school, weave long wreaths and decorate the facade with all colors and greens. Then we would stand there on May 24, recite poems and listen to speeches. This was a glorious day. I stood at this portal, remembering old times, while my wife was following me in silence. I had a brief chat with the caretakers of the kids, who, like all kids of that age, four to eight, screamed and yelled at the top of their voices. But there was something very sad in their eyes, in their voices. They seemed to cry out loudly, even in their naughty romping around. Some of them were clutching at my knees, and at my wife, as if begging to be adopted. It was sad. Came next Christmas, I sent to father Kolentzev a box of candies to take to them, which, I was later told, he did.

In old times the streets of the village were not paved and on rainy days, fall, winter and spring, were pools of water and mud. There was no street lighting and one had to take his chances in the dark, relying more on his memory where to walk, rather than seeing where he was stepping. It seems to me the distances were longer than they appeared to me in my recent visits. But one way or the other, all year round, they were an arena of action, full of people going to their chores into the village and out of the village, a scene of bubbling life activities. From early morning to late evening people were going to or returning from work, carts were pulled by cows, or oxen, donkeys loaded with sacks of roses, or mulberry branches for raising silk worms, sacks of potatoes or other vegetables from field gardens or bundles of grass for animals at home. Every morning the peasants would let into the streets, as they heard the voice of some respective tender, their animals – cows, or sheep, or pigs, or donkeys, or goats. One would be pushing a small flock of sheep, other goats, etc., etc. The most amusing animals were the donkeys in the spring when they would simply be loitering around until the donkey-tender would arrive, and in the meantime they would engage in their love escapades along the streets, for the hilarious pleasure and laughs of men, women and children milling around. And so, until all animals are sent out to grazing. In the evenings there would begin the return of the animals. They all knew their houses. Everyone was working hard to earn his living, to maintain his family, to care for his house and his little estate they had. So passed the days, the weeks, the seasons and the years. Every day the same in this endless cycle of life. Children were being born, old men and women were falling asleep in the Lord, young men and women were getting married and throwing big wedding parties where half of the village was involved, holidays were celebrated, baptisms, bethrothals, memorial services… And everybody would return to his home by evening, tired, sometimes satisfied, sometimes not happy of their day…

It was such a whirlpool of life that it is difficult to describe. Now all that has gone. In the 1950s they had collectivized the agriculture, all the animals taken out and cared for by the hundreds in their respective barns, most of the people having moved to the cities and the old houses left to deteriorate. In fact, the prevalent number of old houses were gone. Almost all of them have been replaced by new ones, all the barns and haylofts torn down, and what is left of the backyards or frontyards, turned into vegetable gardens. Inevitably, in front of every house now was sprawling on a wooden scaffold a grapevine, covering half of the space, growing special grapes, red, pink, blue and white, in long grape-clusters, sometime over a foot long, hanging there and ripening, perhaps five pounds, lucious and tempting. Then would come the tomato patch and the pepper patch, bordered by bunches of flowers. The most miserable remnants of the old times houses are those which in those days were considered to be most beautiful, belonging to the wealthiest families. Now they are ruins, sticking above the new one-level houses, plaster falling from all sides, uninhabited or visited by owners in the summer. It was a pity to look at these old time glories to be in such a pitiful state now.

11. A Rhapsody of Poverty. I. "White, Beautiful Little House…"

There was a little cute poem in the days of my childhood which was recited by everybody. I do not remember if it was written by Vazov or somebody else. It went like this: "White little house; Two linden trees in front of it; Here a mother's caress; I felt for the first time." This "Little beautiful house" was my refuge, my fortress, and my hide-out in my childhood. Only that my house was neither white, nor beautiful, but a small old ruin. I could paraphrase this song to read: "An old, decrepit ruin, falling apart under the weight of time, one scraggy pear tree and an old mulberry tree in front of it, here in abject poverty, I suffered, in my childhood days." It indeed was a hollow where misery reigned supreme. It was a relic from the 19th century. One wall was built of stone with the chimney in the middle. The rest of it was of tree branches, knitted around thin beams ("two-by-four"), plastered with dirt mixed with straw.

It seems to me our house was the oldest in the village and appeared to be cracking under a heavy roof of ceramic tiles, broken in so many places, never replaced. So that when it would stop raining outside, it would continue inside. Way back someone of the family, my father or my grand father, had purchased it from the Pekhlivanov's family, the grandfather of my school-mate Georgi Pekhlivanov. However, in recent years it had been discovered that the transaction had never been properly recorded and it was still deeded to the old Pekhlivanovs. My brother Stoiu had to go to Court, and Dyado Nacho Pekhlivanov to testify that a sale had taken place.

It was probably fifteen feet high to the overhang. There was a basement, under half of the house, some three or four feet deep in the ground. It was accessible through a door and a set of stone steps, if they were steps and not just a sloping ramp. It often happened that when it would rain, and the downpour was big, the water in front of the house could not run through the small pipe under the wall fencing the yard, some ten feet away, and found its way to the basement. So many times we had to dish it out with buckets. There were no pumps in those days. During the winter this basement was used as a barn where cows, oxen, buffalos, pigs and dogs would be accommodated, with all the excretions and urines smelling to high heaven. The room above this basement-barn, was our living-room-bedroom-kitchen-sitting and dining room. A wooden floor separated the basement from that room, with numerous cracks between the planks, sometimes as wide as half an inch, so that when the animals were doing their natural things, we could easily hear their biological activities. Well, it did not bother us. We were used to it. Neither the smell bothered us. It was part of our daily life. This lasted until my father was able to build a separate barn and the basement was free for storing winter foods. So, matters of hygiene, of cleanliness, or whatever one may suppose under these terms, were nobody's concern. My mother suffered, but there was nothing that could be done to change things.

The second floor was reached by a set of steps, five or six of them. More often than not they were loose and it would take long time to fix them. Such was also the case with railings which did not last long. We, the kids, often hung on them and their life was terminated. Fortunately, no accidents occurred on account of all that. The steps led to a small platform, more often than not boarded with railings, where a hallway turned to the left leading to the three upstairs rooms. There was no door separating the platform and the hallway, whose floor was of rickety, worn out wooden planks. At the end of the hallway, probably ten feet long, were the doors to the three rooms – on the right – our living etc. room, on the left my older brother's room, and in the back a utility room. During the summer, we ate supper in this hallway, as the three thresholds served as siting places, for three of us. The fourth one would sit on a small tripod stool, one foot or less high. The *sofra* – the "dining room table," usually called sofra in Bulgarian – but in our case it was called "*paralia*". It was a round table, one foot high, made of rough wooden planks. My mother would serve whatever she had cooked as soup which was our usual meal, in a wooden bowl, *kopania*. Then we all would grab a piece of bread and our wooden spoons and attack the soup until it was finished. It was the bread which was most of the nutrition we got. The soup just made it go easy, because, as a rule, it was dry and stale. If the meal was not soup, but something fried in a pan or baked in a baking dish, it was served in the pan (*tigan*) or the dish (tepsia, tava). Since we did not have forks, we used our fingers with pieces of bread. I did not see serving meal in individual plates until I entered the Seminary at the age of seventeen.

Our living room, as I already said, was our kitchen, our dining room, our bedroom and our sitting room. May be it was some ten feet long and ten feet wide. I cannot be sure of the size now. In the corner, right from the door, in winter time was set up a *kumbe*, our stove, a metal box of varying sizes. The front was a small door to feed the fire with wood. The top had one or two circular openings where three metal pieces, fitting in grooves on each one, so that one could use any size, depending of the utensils to be used – big or small pots for heating water or cooking. A set of pipes would rise on the back and taken out through the wall. When the wood was dry and inflammable, the flames often reached outside. On one occasion the house almost caught fire. My father extinguished it. In the course of the years the *kumbes* changed their size, forms and functions. The last one was so big that we could bake bread in it, banitsas (sweet breads of all kinds, and baked dishes). Before we acquired this "modern" stove, my mother used to make the bread home and then would take it to the bakery, eight loaves, lined up on a plank carried on the shoulder. My father would pay the baker, a cousin of his, by delivering him loads of wood for the furnace. When my mother fell sick, it all had to be done in our new stove. This mechanized bread-making at home continued for years, after I left home. In T. Zhivkov's times, they had built a modern bakery where the bread for several villages was being baked, in the city of Rakovsky, brought to Zelenikovo every day at 5.00 p.m. by a truck where a long line of people waiting for it. Occasionally, no truck would arrive and the peasants would go without bread. Subsequently they opened a bakery in Zelenikovo, where my niece and her husband worked. Now, under the new privatization processes it was closed again and the Rakovsky arrangement was resumed.

So, this room was our kitchen. During the summer the cooking was done outside, but during the rest of the year it was cooked on the stove in the "living" room, which also turned into "dining room", with the *paralia* moved inside. We actually sat on the floor, dispensing with the thresholds and the tripod stools. After supper, the *paralia* would be moved to the utility room. In the evening the living room was turned into a bedroom. In one corner slept my mother. In the other corners were we – my father and my younger brother. Sometimes there appeared a bed for my mother, with a mattress filled with straw or dry leaves on the floor. Very early it became my duty to bring from the utility room, from the *lavitsa*, the stack of our *chergi* (home weaved blankets from strips of old clothes). A very heavy *kozak*, made of goat's wool, spread on the floor, served as a "matress". We had no pillows, but some old clothes served the purpose. Over the *kozak* I would spread the *chergi*, lighter blankets, worn out rags, and we would cuddle under them in our street clothes. There were no sheets and there were no sleeping pajamas. For the first time I saw a sheet was when I entered the seminary. The *kumbe* was heating the room up to a point, and then we were on our own. In the morning I would drag all our sleeping accommodations back to the utility room and mount them on the *lavitsa*. The kozak was so heavy, I could hardly lift it. When my older brother moved out, must have been in 1938, I inherited his room.

In the wall facing South there was a small window, some two by two feet glass, two and a half feet above the floor. Here my mother sat at the daylight when spinning or knitting, or mending our clothes. As a little kid I very often sat by her, she would tell me stories and at some point started giving me lessons of mathematics. I still remember how she drew on the wall the number 2. It stood there for long time and I stared at it as if it was some mysterious sign. Well, the walls were not sacred and untouchable. Sometimes they were used for making "notes" about things – notches and scraches. Probably twice a year they would be "painted". The "painting" was done with *huma*. *Huma* was some sort of white clay mixed with sand, and made light blue with a chemical, called *sinilo*, blue die, which we bought from the store. The *huma* was "imported" to our village from somewhere in the mountains, where it was dug from the ground and made into small forms, like round bread loaves. It was sold for wheat, equal weight for equal weight. When the room was painted, it somehow improved its smell and one felt it like a new house. Over the small window was hanging a gas lamp, which was throwing dim light in the room during the evening hours. Too often the gas in the village was sold and there were times, for weeks, when the room was dark. Well, it soon was discovered that a vigil light would do the job. In a small cup my mother would pour some water, then some sun-flower oil, and attach a wick at a metal support, so the little flame of the vigil light would replace the gas lamp. During the war, this became the rule, rather than exception. Since I had moved to my brother's room, I had a vigil light hanging on a nail over my head. I adjusted it so close over my head that if I held a book while lying and exposing the page to the vigil light, I could read. There were occasions when I read all night. In one night I remember reading one 300-page book, biography of Alexander Stamboliysky, soon after September 9th. I suppose this book converted me to Agrarianism.

The ceiling of our living room, made of a finer grade planks, was pitch black, probably dating back to the times when the house was built. On the North side was a door leading to the *hambary* – some six subdivisions of space, good for a room, each one four by four feet separated from the others by lumber and serving respectively for storing wheat, barely, rye, corn, sunflower and whatever. From there it was easy to climb onto the attic. As a youngster I spent many hours there, looking for items of old times. I discovered a bag of Turkish *tapias*, documents, perhaps deeds of our properties, written in Arabic, with signatures so twisted that I was amazed staring at them. This only showed me that we, or I, were very close to the Turkish rule in Bulgaria, which had ended in 1878. In addition to that there was a small door at the right, leading to the utility room, which I used to bring in and take out our sleeping accommodations.

But the most important wall was that facing East. It was a stone wall – not all of it. Sometimes in the past there has been a fire place there (*ognishte* or *odjak*), hearth. It all was the predecessor of the *kumbe*. Over it would begin the chimney, with a periphery rounding the upper part of the *odjak* (*ognishte*), called *badja* or *kamina*. There were still remnants of its existence in older times on the roof, in the form of a pile of ceramic tiles. In front of it, the floor was still plastered dirt, made of clay, probably as a precaution against fire. It was some three feet by six or seven feet. Sometimes in the past my Father had eliminated the hearth, the *badja* and the chimney, and instead, had replaced it with a reasonable size window – four feet wide and five feet high. The window was tucked at the side facing the street, and the space between it and the end of the wall from inside was some two feet, the width of the stone wall, covered with a thick lumber, which served me as a desk. This was my favorite spot in the room. That is where I kept my school books, there I did my home work, and studied my lessons. It was facing east. Not far from us was the main street where traffic of all kinds of animals, men, and vehicles was moving up and down. What is more important for me is, that beyond the main street and the other section of the village, towards the river, was the *ruta*, the hillock and the ravine leading to the Bear's Hollow and Sakar Tepe. It was a wide opening to the skies. How many times I would lift my eyes from the books, look at Sakar Tepe, and the Bear's Hollow, also the Kutchukia, and the magnificent blue skies. There I would follow the sun climbing up the northern slopes of Sakar Tepe, and then returning back with the approaching of fall and winter. So many times, starting with all my dreams, I composed my ideas and my visions for happier days. There I was out of the way of everybody and could indulge in these dreams and contemplations. This special place was called *penjera*. I do not know what it meant. Probably it was some Turkish word. It is through this window that I saw and was enamored with Sakar Tepe and missed it so much when I moved out of the village.

At least twice a year I was displaced from my favorite place because my mother had to set up her loom there, a contraption where she weaved cloth for all purposes – shirts, pants, blankets etc. It was a rather large frame which took almost half of the living room. I risk to bore the reader with my detailed description of this monstrosity, and he may skip it if he is not interested, but having been for a few years a professor of Cultural Anthropology, and subjects of ethnography, I am personally attracted to such subjects where the daily life and people's needs and their satisfaction are concerned. I cannot skip it over. It was part of our life in those primitive times, and its inclusion in a study of ethnic cultures is a must. After all, preparing clothing for the family was as important as preparing food and securing a shelter for us. This was exclusively the domain of the woman in the family.

The process would begin with picking the cotton from the plants which every family raised for this purpose or cutting the wool of the sheep and the goats. The cotton was taken to a carder in some of the neighboring villages, where the seeds were separated from the "flower" and a fluffy ribbon-like stretch of wooly stuff was rolled into a big bundle. In the house the bundle was layed and split in smaller bundles. Each bundle, *kadelia*, then would be attached to a *hurka*, a wooden stick, often artistically carved, some three feet long. The woman would tuck one end of it at a string around her waist and hold erect under her armpit. She would then draw a thin string rolled with two fingers, attach it to the thin end of a spindle, *vreteno*, spin it and start the thread, which is progressively getting longer and longer and rolled over the spindle until it forms a ball in the middle. Then a second spindle is started, and so on, until all cotton is turned into thread.

The wool was handled the same way, except that after the shearing, it would be washed in the river, the clusters would be pulled apart and taken to the carder. Following this procedures all the thread from the spindles would be transfered to *massouri*, half or three quarters foot long cuttings from branches of sambucol, hollow from end to end, so that it could be placed on a thin stick, rotated and prepared for the next stage to be integrated into a complex arrangement of the threads for the loom. When all that was done, some of the thread would be dyed for coloring the cloth in a variety of geometric forms. The woolen thread was dyed by boiling it with branches of a bushy plant, called *mekish* growing in the fields of wild brush, or *smradlika*, growing in the woods. It would dye the wool brown color, so all the outer wears of the men in those days were brown. For the cotton special dyes were to be purchased. The women would select in advance the pattern of the cloth-colors and respectively split the thread for dyeing.

The next procedure was preparing the "foundation" – that is the length of the organized by color threads – some fifty to one hundred feet. For this purpose the *massouri* would be placed in a wooden tub, with a rock for weight, on a distance some thirty feet from the operator, who would be sitting behind a *krosno*, yarn-beam, and as she would turn the latter by using a stick inserted in the respective holes – there were four of them at a cross – the yarn would all eventually be transferred from the *massouri*, to the *crossno*. The operator would, more or less, distribute the colors to correspond approximately to the pattern.

The next step would be the *vdiavane*. This in itself was a complex procedure. Two pairs of *nishtelki*, heddles, would be suspended in a frame. One person would sit in the front and push her fingers through the heddle's loops in such a pattern that when the heddle is pulled up, alternatively one thread would be raised up and other pulled down, thus making an opening, a small tunel, through which a shuttle is thrown with another thread, the *vutuk* (U like in much). The heddle is about three feet long. The threads are distributed along the length according to the pattern. Once this operation is concluded, all threads are pushed through a reed *burdo*, which in turn is enclosed in the *vatali*, two thicker wooden planks, whith grooves where the reed is cradled and firmly held. Then the entire "foundation", passed through the heddles and the reed, is attached to another *krosno*, yearn-beam, on the opposite side, and the entire set up with the two yearn-beams, the heddles, the reed and the "foundation" are installed in the loom.

The loom itself was a frame of two side-sets, on the left and right sides, held together by four laths – two in the front – at the top and bottom, and two in the back – top and bottom. In the middle, on top of the frame, on a stick resting on the side-sets are suspended, over two pulleys through which a string is passed, holding two different heddles, tied to pedals on the floor, which are pushed alternately down, to pull the "foundation" threads, up and down, and open space for the shuttle to carry the cross threads, back and forth. As the shuttle passed through the tunnel, another pedal is pushed down, the heddles change positions, up and down, lock the thread behind and by pulling the *vatali* with the reed, the thread is firmly pushed to the cloth already made. The respective movements – pushing the pedals, throwing the shuttle through the tunnel and pulling the vatali are so synchronized and fast, that the work progresses accordingly. Still, thread by thread, it takes long time to finish the entire procedure. As the distance between the reed yarn and the operator gets shorter and shorter, she releases the beam and pulls the weaved cloth towards her. At the same time she turns the cloth-beam to roll, the finished product, over it. The release, and the arrest of the yarn-beam with the "foundation" is effected with a long stick positioned at the right hand of the operator. She just pulls the "foundation" enough, then push the stick in a hole at the end of the yarn-beam. At the same time, and with another small stick, she releases and stops the cloth-beam with the ready cloth.

All this continued for many weeks, because my mother had to do other things in the house and weaved only in her "free" time. So, I was displaced from my favorite place, the *penjera*, because the loom was set there – it was the only place where it could fit, and the light of the big window was essential for my mother. Now, one may wonder how I could be so familiar with all these details. Well, these procedures were performed year after year, twice a year, it was in our living room and since there was no other woman in the house, I was often mobilized to help with all sorts of things with all these procedures, except for the weaving itself. What I am amazed, however, at this point, is that I could remember all this detail some sixty or more years after such experiences.

I have to observe here that things did not go very well in our family daily like in times of weaving. My mother had to take care of too many things, she did not have a daughter, or other woman in the family to help, and her spinning, or weaving was falling far behind our needs. When she started going from one illness to another, which took her away from he work for weeks or months, and eventually being bed-ridden for life, we were caught in a state of misery. Our clothes were getting worn out, we were quite raggedy and in poor, intolerable condition. And as in those days one could not buy these things in a local store, where in the cities the cost of clothes was prohibitive for peasants, we had to hang on to whatever was left. My father was in such miserable clothes that sometimes, already in the Seminary, I felt sometimes crying. Things could not be helped. Gone were the years when my mother more or less was able to take care of some of our needs, but as times progressed things became worse and worse. I must admit that I see these things now, but in those days it never crossed my mind that our miserable appearance, where clothes were concerned, was the lack of a woman's hand to take care of things. My father, very capable in doing his field work, it appears to me, was not sophisticated enough to take care of his and our clothes. Probably if he had money some way might have been found, but money in our family was a scarce commodity.

And there was the utilities room in the back, straight from the hallway. It had a rather clumsy, old door, with an old fashion lock (lock?), called *mandalo*, a latch, three-quarters of a foot long metal iron handle, freely attached to a round bolt, affixed to the door, with a hole at the end, big enough to pass a walnut through it. I am describing this latch, because, besides its function to open and close the door, it also served medicinal purposes. If someone was frightened by some horrible event, an old lady would take two wooden spoons, would pass water through the hole, catching it with one of them, repeating the procedure three times, saying some magic words, and giving it to the frightened person to drink. I remember going through this magic once, after being frightened by a vicious dog. So, life in those days still had its primitive ways of handling crisis. Since I am on this subject let me mention another magic performed on me. Sometimes, as a kid, I had an infection below my elbow on the inside of my arm. It was swelling dangerously. I was taken to an old lady in the village, she drew a circle around the swollen part with ink, murmured some magic words, recommended some drink of something and let me go. Of course the wound took some time to heal by itself, but it was firmly believed that the old lady's magic words did the job.

The room was rather large and smelly of mildew. The entire floor was of dirt, and when it would become practically a ruin, would be plastered by a mixture of a reddish clay and cow's excretions, with some straw added to it. In the middle of the west side was the usual hearth, with the respective *badja* and *kamina*. It was an ugly, wide opening to the skies through the chimney. At the bottom there was a three quarter of a foot deep hole, two feet long and one and a half foot wide. Half of it was always full of ashes left there by the fire. The chimney, wide and black as it was, was reaching above the roof, was falling apart affected by its, who knows, may be one hundred years old life. It did not have a cap on the top and when it rained considerable amount of water was filling the hole below. In the winter we used to find quite a bit of snow in it. A beam was stretched in the middle of it, sort of a strong two by four, and an iron chain was suspended over the hearth. This chain was called *viruga*. It ended with a hook, where copper buckets would be suspended over the fire to heat water, or cooking pots. There was also an iron tripodion placed when needed over the fire, to hold a cooking pot, the *pirustia*. Next to the hearth, leaning at the wall were three or four heavy baked round clay *podnitsi* – sort of large trays, foot and a half wide, three inches deep. They were used to bake the bread. After the flouer is mixed with warm water in a tub, called *noshtvi*, left to ferment (a chunk of previous dough left in the *noshtvi* to serve as yeast to be mixed in the new dough being prepared) and in an hour or so made into big round loaves, called *kvasnitsi*, to fit in the podnitsi trays. The latter are placed near the wall, supported by three half a foot high rocks, and plenty of burning ambers pushed under it. It is covered by an iron *vrushnik*, a large enough lid, over which more ambers, mixed with ashes are put on, and in half an hour, the loaf is baked. Sometimes two *podnitsi* are set up for simultaneous baking. Usually, one baking session would produce four such big loaves, to last us for five or six days.

On the opposite side of the room were the *brashneniks* – two enclosures of lumber, two feet deep, three feet long and about five feet high, where the flouer was stored, after it would have been ground at the mill in Brezovo. One of the enclosures contained the flour for bread, the other – the bran, *tritsi*, used to mix it with straw for the animals in winter times. An opening at the bottom, closed by a movable up and down small lid, would secure access to the contents of the enclosures. On the top of them would be resting the *noshtvi* when not in use. Sometimes, when it would happen that we would be out of bread, or just to have a delicacy, a pita was made - mixing flour with water and spreading the dough into a pizza-type loaf, but not so thin, just about an inch. It would contain no yeast, would be placed in the hearth, covered with ashes, and a fire over it. It would be baked, pulled out of the ashes, dusted, and eaten while it was still warm. It was delicious. Well, it was! Twice a month the animals were treated with specially prepared meal by mixing the bran with straw from the threshing of the wheat, and salt melted in hot water. The animals would go wild when this *kurmilo* was served to them. I did not mention it above. Salt was used when making the bread. I also forgot to mention that when the big hunk of dough was split into loaves to be baked, they were layed down over special cloth pieces, long enough to accommodate four to five loaves. These were the *missally* which were important part of the household.

This room was also the place where our water was kept in two cooper buckets, appropriately tinned from inside and outside, to protect us from poisoning. But one of these buckets was used to boil water, and was all black outside. The other was white both ways. They both were hanging on the *vodnik* – a beam affixed on the wall, with four-five pegs extending forward, with groves at the end, where the handle of the bucket, the *vurbilo*, a strong semi-circle iron piece, would be hung. Always hanging on the *vodnik* was the *buckul*, the wooden big flask I have mentioned before, used for carrying water to the fields for drinking. If there were more buckets, they also would be hanging there. One of the big buckets carried probably two to three gallons of water. My mother would bring it from the *cheshma*, the neighborhood water fountain. She would have them hung on a bent wooden two-by four, carved in a special way, called *kobilitsa*. Sometimes I was assigned to bring the water and carried the big full buckets with difficulty, a few steps at a time. Unlike many villages, where the people still supplied themselves with water from the river, from specially dug springs, *kladenche*, scooping the water with small *kratunki* (Kratunka was a small vessel, a ladle, made of the fruit of a plant with the same name, growing on a vine similar to small squash, with long handle). When it would ripen, its rind would become very hard (as a thin coconut). Usually the tail of the fruit is trained while green, into a curve, and when it is ready one side of it is cut in a circle, and then it is used as a ladle. It was not much of a desirable water supply, because too often sand would find its way to the house. But our village had progressed into the 20th century in 1926 where water supply was concerned.

An alert mayor had conceived the idea that the water running from several brooks in the Yunchal could be captured with pipes, the pipes would be layed out in a deep trench to above the village and empty the water in a cement-built reservoir. From the reservoir the water was distributed to seven or eight water-fountains in different sections of the village. The water fountain was a cement structure, quadrangular, one and a half to one and a half foot, hollow inside, covered with a cement lid standing on a pedestal six by six feet. The pipes from the reservoir were still burried in deep trenches and branched out into the fountain, from inside. The fountain was some four feet high. Somewhere, at about the third foot, the pipe was led out and ended with a spigot. Under the spigot the fountain extended into a stand, where buckets, or any other water vessels, were placed over a grade to be filled, and then taken home. Indeed it was a modern water system at the time. Only one or two houses in the village had private fountains in their frontyards. There was plenty of water during fall, winter and spring, but in the summer, there was not enough of it. So the reservoir was closed during the day and about 6-7 o'clock would be opened, and people could then get their water. But sometimes the water would not last for everybody. So, every morning a family would rush to the fountain, to leave a vessel on a line to be first or so in the evening, not to find herself without water. During the day one could see these vessels lined up, so in the evening all women would know their turn. During the winter the pipes in the fountain would freeze, so a swat of straw would be dropped under the lid, would be lighted with a match, and it would defreeze it. Since there was quite a bit of water splashing out of the vessels, it would run through the drain outside and form a long streamlet, for the delight of the children. We slopped in the mud there, got dirty like pigs, scolded by our mothers, but loved to play with the water and the mud. Well, it was not so much of a problem, because the women were happy when the kids amused themselves.

Probably here is the place to say a word about one of our vessels holding water for drinking – the *stomna*. The stomnas were earthen ceramic pitchers. They came in all sizes, were produced somewhere in distant villages, and from time to time their craftsmen brought them to our village, selling them for wheat. I need not describe it, because the words – pitcher or jug – convey the idea what form they were in. There is only a little detail to be added. The handle of the stomna, one end attached to its round body, the other to the top, had a small canal from the body to the curve, where it ended in a nipple. The water would travel up the canal to the nipple when sucked by the thirsty individual. The use of glass for individual drinking in those days was not known – at least to the overwhelming majority of the peasants. We all drank water from this nipple, one after another, without fear of catching some disease. Sometimes the *kratunka* would serve that purpose, but it too, would be used by all, without any reservations for health reason. There was no water for showers and baths. Well, showers? Bath tubs? These words did not even exist in the vocabulary of the peasants. During the summer one could steal a secret place in the river and wash himself, but during the winter? Just wait for the next summer. At least most of the peasants did that.

I was on the subject of our utilities room. There is not much to say further, except to mention that along the wall to the north, with a small window, were lined up kegs and crocks, for sauerkraut, for cheese, for pork – *slanina* (bacon) etc. In the corner during the winter there would be a bundle of fifty to one hundred stalks of *prassa* (leek). This was probably the only raw vegetable that we ate during the winter. Also in a corner was a *dulap*, a cabinet, without doors, where-bottles of sunflower oil, jars of some other things, a jug of rakia – plum brandy – and all sorts of necessities for the kitchen…

12. A Rhapsody of Povetry. II. Our "Luxurious" Dining-Table Delicacies

While I am on this subject, our poverty stricken hovel we called a home, let me say a few words about our "luxurious" dining-room table delicacies. Our breakfast more often than not consisted of left overs from the previous day, if there were any. Most of the time it was *popara* – dry bread soaked in boiling water, sweetened by one or two spoons of *petmez*. The *petmez* replaced the sugar. It was sort of a sweet syrup. The sugar beets were placed under a primitive wooden press and the juice was boiled in large, four feet wide round metal pots until it would turn into a black molasses, then saved in earthen jugs for the winter. It was a rare occasion when the dry bread and the boiling water was sprinkled with fried butter and small pieces of feta cheese. It was delicious, better than the watered – down *petmez*. Sometimes the *popara* was replaced by *trahana* – little crumbs of yeast from the *noshtvi*, mixed with little flour, boiled in water and served as a sour soup, eaten without bread. Its sour taste made it edible. Another delicacy was the *ushmer*. It was prepared from flour, fried in a pan, and then some water would be added and boiled to desired thickness. The bread bites were dipped in and the ushmer was scooped. The sunflower oil made it tasty. A similar procedure was followed to make *volovarnik* – dried bread crumbs were fried in sunflower oil. They would turn soft, and tasty. The salt makes everything tasty. Quite often our breakfast was a pan of *kasha*. The kasha was made out of flour. It was sprinkled with water, then rubbed a few minutes, into small, rice-like crumbs, boiled and sprinkled with sunflower oil with fried small pieces of onions. Rarely it would be butter and cheese. A similar dish was the *kachamak*. It was made of corn flour, boiled in water to thickness much like mashed potatoes, then distributed into chunks in a flat pan, drops of petmez would sweeten it. Sometimes fried butter and feta cheese, would be used. Probably one of my favorite breakfasts was the *mikitsi* – chunks of dough, flattened to half an inch and in round pitas, eight-by-eight inches, fried in sunflower oil, then served hot, sweetened with petmez. In rare occasions *cutmi* would feed the family for breakfast. For this purpose flour is mixed with water almost to a liquid state, similar to pancake mix and poured over a *sach* – a black ceramic pan, thick at the bottom, shallow, placed over the *pirustia* in the hearth. The sach is oiled with lard, the mixture is poured over it, thin to one eight of an inch, and baked, being turned over on both sides. Some ten such *kutmi* would be layed over one another, as buttered cheese is sprinkled among them. When ready it would be cut to pieces like baklava, but in bigger chunks, and eaten with gusto. I would not forget to mention the *yuvki*. This was our home made noodles. During the summer the dough is thinned with a *tuchilka*, a smooth stick, two and half feet long, inch thick, in round, up to two and a half feet sheets, thin one sixteenth of an inch, baked dry in the sun and then crushed to pieces, similar to potato chips. The chips are then boiled a minute and then served with cheese or some oil. Well, they were not bad.

Now that I have mentioned all these delicacies, which were very often made also for dinner, one may wonder if I have made them. Yes, I have made them all. When my mother was sick for extended periods of time, and when she became bedridden, I assumed the business of cooking. She would watch me from her bed, give me instructions what to do and when, I would follow her instructions, and put a meal on the table. I do not have much recollection about lunch receipes. From early spring to late fall, the lunch was eaten in the fields, at work or grazing the animals. There was a *torba*, a bag, different sizes, made of cloth, with a thick string in a loop, so that it could be, usually it was, hung over the head of the side shoulder. Inside the *torba* there always was a wooden small dish, *zahlupka*, with lid, tightly screwed in a groove, filled with ground salt and another for *merodia*, a mixture of salt and some crushed spicy herbs, most of the time *chubritsa*, I believe savory. Other delicacies carried to the field with the bread in *torba* were often feta cheese, in a larger wooden *zahlupka*, or a raw white thick piece of bacon. To all that most of the time there was a head of onion – sometimes, in the absence of all others, there was only the salt and the onion. When I was tending sheep or cattle I always had in my *torba* newspapers and books for reading. It was my passion. Sometimes the onion was replaced by garlic. When I was in Bulgaria in 1991 nobody missed to remind me that I may have had a bag empty of food stuffs, but always filled with books.

Our supper was not much diversified. Almost daily, we had bean soup for dinner. The beans were boiled for some two or three hours, close to the end a head of chopped onion would be added for taste, then salt would be sprinkled in, and during the summer – some vegetables – peppers and tomatoes. It was mostly water and if you caught some three or four beans in your wooden spoon, it was an event. But there was a way to improve the taste of it. It was the special *kavardiska* which made the difference.

The *kavardiska* was prepared by frying a chopped onion in two-to-three table spoons sunflower oil, adding two spoons of flouer when the onions turn reddish, stir the flouer until it turns brown, lace it all with some red pepper for color, pour some of the soup in the pan, stir the whole mixture and pour it in the rest of the soup. The soup is then boiled for two-three minutes, and it is ready for serving. It is used practically for every meal that is cooked for dinner. It makes every dish very tasty. A few years ago a friend of mine, in Stroudsburg, Bai Dimitar, a retired bachelor, who has cooked for himself all his life, complained that he cooks bean soup, but it never taste as his mother used to make it. I explained to him how to make *kavardiska*. Ever since, he could not thank me enough.

During the winter one of the frequent dishes made was the *prassa* (leek), cut into pieces to one inch. After it is split, then boiled in already prepared *kavardiska*, is sometimes improved with *djumerki*, fried pieces of bacon. I already mentioned the lentil soup. We did not have *progresso* in Bulgaria. The lentil beans were grown in the fields in sufficient amount to meet our needs. Another soup was called *bulgur* – crushed wheat grains, boiled in water, with added salt and *kavardiska*. It was not very popular, but in difficult times there was not much of a choice. All these soups were greatly improved with homemade vinegar. In the spring we had our first taste of vegetables. As the nettle showed its little green heads, we would go down along the river, pick them carefully, fill a bag, boil it and sprayed it with some fried oil, in the form of puree. It was delicious. Such was also the case with the *kisselek*, sorrel, a sour plant which retained its taste even after being boiled to a soup. A similar plant, cut into puree, was the *lapat*, I could not find it in my dictionaries, but it was a broad-leaf weed, almost a foot and four inches wide, cooked either as a soup, or as a puree, and improved with *kavardiska*. A similar soup was made out of *lobuda*, goose-foot. It is a garden weed, silvery color, grows high up to two feet, but is cut for soups while young. Let me finish with the *tarator*. It is made of water, some eight to ten cloves of garlic, crushed to liquification, vinegar, salt and, if available, cucumbers cut to small pieces. A better tarator is made of yogurt diluted in water, or turned into Airian (*Ai* pronounced as *I*, and *ia* pronounced as *ya*). A more substantial dish, re-served for holidays, was the guvech. An earthen pot, known by the same name, is filled with chopped vegetables – potatoes, tomatoes, pepper, onions, greenbeans in the summer, and, when available, some chunks of meat. During the winter the most favorite meal was *kokal sus zele*, a pork bone with remnants of pork meat, baked in the same pot or in a flat metal dishes, with sauerkrout. I should have mentioned the number one meal of the Bulgarian peasants – soup of beans – sometimes cooked as pilaf, but most of the time as soup. All soups in Bulgaria are called *chorba*, the Turkish word for soup.

Should mention that we ate very little meat. Every village household would raise a pig during the summer and slay it before Christmas. The slaying of the pig was a festive occasion. Neighborhood men would join in, four or five of them. The poor animal would be taken to the backyard, would be wrestlled to the ground and … Nevermind! A fire would be built, then the carcas would be dragged over it to burn the hair, exposed to the flames. There would follow the scrubbing and cutting the body, head first. The ears would be immediately grilled in the fire, salted and cut to small pieces for *meze* – appetizer – and everybody would share in the cartilage. Then the lower part of the body, the breast and stomach sections, cut, cleaned and given to the woman of the house who would prepare out of it the *usmianka*: meat, onions, fat, potatoes etc. and by the time the entire procedure is completed, all participants and the family would sit down and feast on the tasty dish, of course, preceeded with plum *brandi*, *rakia*, and wine, drinking from a common bowl, passing from man to man. After the piece for *usmianka* is cut, the insides of the pig are opened, the kidneys are removed and grilled in the fire for another *meze*. The intestines are cleaned, turned inside out and the large intestine given to the woman of the house. She would prepare special mixture of rice or bulgur and liver and lungs cut to pieces, cooked in advance, with spices, and stuff the cavities into a *kurvavitsa*. The *kurvavitsa* is cooked sometimes later for the family in sauerkraut. The lean red meats are set aside for stuffing the small intestines, with spices, into *nadenitsi* (nadenitsi – a kind of a long hot dog). The white fat is cut to *snupi*, pieces of one foot width and one and a half length, and layed in special kegs, as a layer of salt is spread between them. This is the bacon which is used for preparing several types of meals, and some is saved for the summer, when raw pieces are sent with those who would work on the fields, for lunch. Much of the fat is melted into lard which is used for cooking purposes and baking of goods. Special care is taken of the head. It is kept for "New Years Eve", St. Basil's day, January 14th. It is boiled and turned into *pacha*, a white jelly-like mass, with pieces of meat inside. One thing that interested us, the kids, was the bladder of the pig. It would be taken out and given to us, the kids, after being cleaned. We would blow it up into what was called *guida*, and played with as a ball.

Another festive occasion was the slaying of a lamb on St. George's day, May 6th. It was sort of a sacrifice for blessings the agriculture during the forthcoming summer and abundance of the crops. Every year, on this day, I would lose usually my favorite lamb. It was slayed in the morning and something was done with its blood. I do not remember what. It was skinned, cleaned up, filled in with all sorts of food stuff, including its chopped intestines and livers, placed in a big pan, decorated with a green branch from the trees and taken to the bakery. Early in the morning all doors were decorated with green branches of the trees. After the lambs were baked, they would be taken on the upper end of the village, by all families, together with some special big *kvasnik* of bread, lined up in a circle – one should imagine what a big circle that was, accommodating every family of the village. When all families are there, the priest would celebrate a Te-Deum service for the health and the prosperity of the people and then moved from family to family to sprinkle holy water over the lambs and stuff. Then the woman would tear one of the shoulders of the lamb and give it to the priest's assistants, dropped in a big sack. Likewise, the priest would cut a piece of the bread and it is also dropped in other sacks. It was all done for blessing of the household. After that there would follow a common dinner on the polyana. Must add that the lamb was so tasty that we eagerly waited for this day. St. George was the patron of the family farming. The State

76

too had adopted it as a special day for the armed forces. It was designated as the Day of Bravery in war times and for commemoration of the soldiers who had died for their country in the wars. A Te Deum service would be celebrated by the priest on the central square of the village where all school children would be brought to attend the service and listen to the patriotic speeches. In the cities there were military parades, following the Te Deum Services.

Other occasions when we would have meat was when by some accident a sheep would break a leg and had to be slayed, or a chicken, for some reason, found itself injured. Quite often that was when some cow kicked it behind her. For our family, on some occasions my father would bring home a cow's head, given to him free of charge by the village butcher. Usually such heads would be disposed by throwing them to the dogs. People would not look for them. It was a humiliation. My father did not think of it this way. The head was boiled in a big bucket, then all meats and brains in the cavities taken out, and with the brine left, cooked again, adding salt and garlic. Well, it was tasty, even though all people looked at it with apprehension. My father did not care. Well, these were rare occasions.

I have missed some details in this story. The *pacha*, the pig's head, was cooked with reasonable amount of garlic. Also, for the same occasion, "New Years day", celebrated on January 14th, as per the old calendar, which was also the St. Basil's day, the family would traditionally have for dinner *tikvenik*, a special kind of pastry made of thinly grated pumpkin meat, rolled in home-made filo, then layed spirally, beginning from the center, in a roumd pan, to the rim of it, two feet in diameter. The bottom of the pan is laced with thin lard and the whole thing is baked in the oven. Before putting it to bake, a coin is stuck somewhere in it, and dogwood brnaches, with different number of buds, are layed on the top of it, pointing from the center to the outward rim. Each branch is named after the family possesion – sheep, cows, harvest, plowing etc. After it is baked, the entire family sits around the *paralia*, the father would spin the pan. Everybody is watching which branch would stop pointing to him or her. It would mean that during the year each one would be responsible for the subject assigned to his branch. The luckiest one would be the person who discover the dime in the piece of pastry after it is distributed. The pastry itself was very tasty. The Bulgarian pumpkins, unlike those for the American Halloween, are very sweet.

Of course, there was another kind of *banitsa – tukmanik*. It was baked whenever there was some special occasion, or no occasion at all. The dough was flattened, cheese and butter was sprinkled on it, then it was folded, flattened again, cheese and butter sprinkled again, for several layers. Then it was placed in the round pan, flattened, baked and served. It was very tasty. During the war, when our family prospered, where cheese and butter were concerned, I made it very often and, taking it to my father in the fields, he would call everybody around to taste the good banitsa, which *Pankata* (me) had made. Many people reminded me these things during my visit to Zelenikovo in 1991.

The reader may think that I am writing a cook book. I am far from having such intentions, but having touched on the subject of this part of my past, I could not resist mentioning all that was running through my mind. After all, writing about it, I am reliving my past experiences and could, and would not interrupt the flow of my memories. I learned this craft of cooking out of necessity. They say that life is the best teacher. I have experienced this "education" on so many stages of my life.

13. Rhapsody of Poverty. III. Our Village Backyard

In front of our house we had a small flower garden – to call it a nice name at least. During the summer, some sort of a fence would divide it from the backyard and my mother loved to plant flowers and care for them. It was my duty to water them. I loved them too – geraniums, nasturtiums, snow-drops, crocusses, chrysanthemums, *popadiyki*, (asters), snapdragons, and what else. But quite often someone would leave the rickety gate open, some animal would invade it and make a good graze of it. My mother would be crushed and would cry over the leftovers. It was one of those disasters that took away the little joy and pleasure which this garden would provide. Some twenty feet in front of the house was a brick wall, some six feet tall, to guard the property from invaders, and animals, with entrance door made of carelessly nailed planks. Outside, ten feet into the street, was a mulberry tree. Sometimes in June or July it produced sweet white mulberries, which we anticipated every summer to ripe and supplement our poor diet.

Some forty feet west of the house was the *plevnia*, the stock house, where all the straw from the summer threshing, all the hey, and all that could be used for forage was stored for the winter. At one end, near the street, it extended with a shed, open in the front, an extension of the *plevnia*, where the sheep in winter time were enclosed, and fed. East of it, sometimes in late 1930s, a barn was built for the cattle, so that the basement of the house was relieved of its use, which, even though we had grown used to it, did not appreciate very much. The space on the east side of the backyard was reserved for storage of wood for the animals, long oak tree branches, with their green leaves piled in a *listnik*. The space in the middle, perhaps forty by forty, was the backyard. In the summer it was the night enclosure of the animals, prevented from escaping by a fence and a gate made of long wood branches. During the night the animals would perform their natural duties there, and in the morning it was for me to take a shovel and clean it all. Most of the time slapping it on the house walls, or the brick front side street wall, sort of plastering of the walls with inexpensive materials to protect them from the rain and ruin. In late July the entire backyard would be cleared and after we acquired the *viyachka* – the contraption to separate the grain from the straw – we moved the threshing field from the *poliyana* home, so it facilitated much of our work.

In the winter it all turned into a pool of mud and we had to place stones to jump from one to another to go and do our myriad chores during the day. On one side of the backyard, close to the house, we had a pear tree, *petrovka*, after the day of Sts. Peter and Paul, July 12th, when its pears would ripen and give us delicious fruites. The tree was very old, but its pears very tasty, and we eagerly waited every spring for its blessings. I came to love pears and still cannot have enough of them. Further up, behind the new barn, up to the street in the west, was an empty lot which for many years was a rose garden, then the roses were uprooted and we started planting tobacco. It was convenient. Later, after my older brother got married, in 1936, in 1938 we all joined and built him a sun-dried-brick house. I remember making the bricks – dig a pile of dirt, mix it with straw, take it with buckets to special forms made of planks, then pull the planks after a couple of days, and let the sun dry them.

When I returned in 1991, I learned a sad story. Sometimes in the 1970's my younger brother, who had inherited the middle part of the lot, apparently considered to be my share, returning from work in the fields, found that my older brother had moved the fence in the middle and split the property in two. Stoiu felt that, well, the lot could be divided by the two brothers, and let things go. But after my older brother had died, may be in 1980 or so, of heart attack in his sleep, one day my younger brother found the fence moved further down by my nephew and his wife, so that the part reserved for me was taken over by them. Well, I never intended to claim that property, but I felt that whatever was done was not right. And so did my brother and went to Court. The Court decided in his favor, but my nephew and his wife appealed the judgement and took to the Court as witness one of the leading village communists, Georgi Kraevsky, a dentist, one of the brothers of Mitiu Kraevsky, whom I am going to write about later. He testified in Court that Stoiu had a brother in the United States, a fascist. When the judge heard it, he ruled in favor of my nephew. He gave my brother the right to appeal, but Vessa, his wife, stopped him. "Let them have it", she supposedly said. "We do not need it. If this is going to fill their eyes!" Vesa, my younger brother Stoiu's wife, confronted Zlata, my nephew's wife, across the fence. Zlata reportedly replied: "He (that is me) does not exist any more for Bulgaria." She was referring to my exile as adversary to the regime. So, Stoiu dropped the case. All this made me fume at my nephew and his wife, and I gave them very little attention when I visited the country.

But let me go back to the backyard. I had my pets all the time. As far back as I remember we had a dog, some sort of a mutt, all white, medium size, Stoicho. Stoicho was so vicious that no kids dared come close to our door. He was succeeded by Komatan, a large scrawny dog, always hungry – so his name, meaning a glutton for a morsel of bread. He was succeeded by Mecho, another mutt, yellow and quite friendly. But my favorite pets became two kittens: Fedya and ... I do not remember his name. We found them on *Tenyov Giol*. They apparently were born to a stray cat, in the fields. In July we were harvesting the beans when I saw them running up the elm trees. I climbed the trees and managed to catch them, I tied them in a bag and tossed it on the top of the cart, over the pile of plants. It was a hot day. When we arrived home I picked up the bag, opened it and found them half dead. They were all sweat, as if they have been dipped in water. I took them into the house. I dried them up, and then... they became house pets. They loved me and I loved them. I would be reading in bed and they would be resting on my chest and my shoulders, purring endlessly in pleasure. I would be sitting at the window, reading or writing, and they would be resting on my two shoulders and again purring and purring. In this world of adversity, in which I lived, until I left the village for good, in 1946, they seemed to be my only and most loyal friends and I loved them. I do not know, but I had always had weakness for animals.

Among my cows, the first calf that was born was white like a goose, I named it Patka (Goose). She was so attached to me that wherever I was, when grazing them, she would be always next to me. If I walked in the backyard, as soon as I appeared, she would get up, if she was lying down and come to me. I would be standing and she would be licking my arms, and reach for my face, which I did not much appreciate. But, if I left my bag somewhere, she would help herself with my bread and let me starve for the day. Well, I often rewarded her with whatever I could come across in the fields, including morsels of bread. The same was with the sheep. I always had some of them following me wherever I went. This is how I acquired my love for animals, and I never lost it. All my animals were my friends. I was emotionally attached to them. I found it hard to separate from them. My most tragic experience was when I had to take my little dog Nikki, a few years ago, to the vet to put her to sleep. She was old, was becoming incontinent and could not take care of herself. I returned home, I burst into unconsolable cry. I have never done so.

At one point Ruby tried to console me: "I will call, if she is still alive, we will get her back." No! It was too late.

As I am writing about Nikki, I am concerned how I would face the day when the same fate comes to separate me from my present dog, Sandy. Sandy is about 15 years old. About 12 years or so ago we were eating breakfast with Ruby in a McDonalds and reading the local paper. There was a large picture of her, placed there by the ASPCA. Ruby was just recovering from a serious surgery and asked if we could adopt that dog. I could not refuse. We went to the place, saw her standing at the wire enclosure, a small terrier, about 25 pounds, with a face like a fox, and sparkling eyes as begging to be saved. In no time she was in our car. Her previous owners had moved away and had left her for adoption or destruction. Her name was Cyndy, so we settled for Sandy. She soon "adopted" us and became our most precious pet. She became my shadow. For so many years she follows me anywhere I am in the house, even joins me when I shave and shower. I may climb the steps one hundred times up and down in the house, and she runs with me one hundred times. I go outside, and she runs out the door with me. She barks her head off when visitors come, sort of greeting them and stops barking when they sit down, only to resume barking when they get up to leave, "asking" them to stay. Between me and her there seems to be a language and she all but would talk to me with her barking tones. She is my most loyal and most affectionate friend. When I sit to eat, we eat together, as I give her some morsel of bread or meat after every bite I take. She has her paws on my laps and reminds me every minute not to forget her – whether we have guests and eat at the dining table or the kitchen. Ruby tells me that I had spoiled her. Well, it is true and I do not regret it. In all my life I have never met a more friendly face. Well, this is how I have been with my animals, in Bulgaria and in America. Pets are the most precious thing a man may have.

There is one more detail about outside facilities in the backyard, but I will not mention it. I would not even appeal to my readers to use their imagination, because even the most fertile mind would not be able to visualize its deplorable condition. So, let sleeping dog lie.

14. My Parents, the Martyrs

So, this is where we lived – my father, my mother and my two brothers. My father probably was born about 1890, or about, because he served in the Balkan war. He had a brother Stefan, who died in 1928, very young, of pneumonia, I believe. I vaguely remember his death.

Other than that, my father had three sisters – Raika, Nedelia and Plina. I was closest to Plina, but often visited Raika and Nedelia, and their families. My grandmother, Ivana, was from the village of Chekhlare. She lived alone until about 1935, and I loved to huddle around her. For some reason she did not like my mother very much, and rarely they communicated with each other. I used to take food to my grandmother every day. I sensed that on account of her my father and my mother often had violent arguments. Aunt Plina also did not have much relations with my mother. It may be that my grandmother may have had some gold pieces of money, which she had given to Aunt Plina, because when she died, sometimes in the 1970s or 1980s, she had left them to someone who took care of her. This is what the gossip was. Be that as it may.

My father had a rather large family. They all seemed to be well off, while his family, we, were on the poor side. He struggled to make ends meet, he was rarely dressed in decent, old fashioned clothes, and more often than not was raggedy and looking like the poorest man in the village. He was medium height, probably 5'7", round face, hazel squinting eyes, perhaps his vision was failing – as it turned out my older brother and I had cataract – greying hair, shaving probably twice a month by the village barber. There were no razors at that time, except the barber's, a sharp knife attached to a bone handle. Very few people had this instrument and could have the luxury to shave themselves. He had a strong voice, baritone, and loved to sing. Mostly he yelled at the animals, and to us when he wanted us to do something. He was a workaholic. He never stopped from doing something. When he would finish with the season's field work and the care of the animals at home, he would go to the fields and cut thorns – I have not seen their equivalent in this country, very dangerous to catch your clothes, (you cannot free yourself easily) – then load them on the ox-cart and take them wherever a fence was to be built. In the spring he would be planting poplar and willow trees, or digging wild apples or pears in the woods, plant them on our property at *Tenyov Giol* and when they start growing, in a year or so, he would graft them to domestic apples and pears.

He never rested on Sundays, never went to church, but would grab the axe, or the fork, or whatever tool, and head for the fields – winter, spring, summer and fall. He was illiterate. He had gone one year to school, or two, knew how to write his name, but nothing more. I never saw him to take any paper and read. His marriage to my mother was his second one. His first wife had died in childbirth, she was from the village of Kolachevo, or later known as Radomir. My mother had also been married before, in some distant village in the district, her first husband probably died, without children, and she had returned to Rozovets where my father took her from. It may have been an arranged marriage. He had served in the Balkan wars, and then in World War One as an artillery man, responsible for transporting the guns and taking care of the horses, near Salonica, at the Doiran Lake. When the war was over he simply grabbed his horse and returned to my village, some one to two hundred miles. The government claimed the horse and he surrendered it. All I remember from my early childhood is, that he had two oxen – Sariu and Pencho. Perhaps this was his bad luck. Oxen do not breed, and so no milk or additional animals are added to the family, for extra income. It was only when he exchanged Sariu for a cow that things started moving up. Pencho was a nervous Nelly and dangerous to go near him.

There was one side of my father which I, as child, did not appreciate. He was ever ready to scream and yell, and sometimes become violent – be it with the animals, be it with the family. The most frightening moments I witnessed was when he would jump on my mother, for whatever reason. I do not remember at all what they argued about, but these were fearful scenes. She would retreat to the corner of the room, hide her head in her hands over her knees and cry. He would rant and rave, while I would be frightened to death from his threats, cry and beg him to stop. Sometimes all this would end by my mother escaping from the house and going to her brother in Rozovets. We would be left in the house, and for two or three days there was a deadly silence all over. Sometimes he would turn to me and ask me to go to Rozovets and ask her to come back. Sometimes I did it on my own. I would go there and both of us would explode crying. Then I would beg her to return and she would refuse. Eventually she would agree and I would reassure her that he would not do it anymore...until the next time.

But these unhappy events have paled in my mind. There was so much good in my father that it has overcome whatever weaknesses he had. When my mother fell sick in subsequent years, he never failed caring for her, always making sure that someone of us would be there when she needed us. When I went to the Seminary, he never let me down, he suffered so much humiliation, and lived in such poverty, passed through so much pain begging relatives and strangers for the last penny needed for my education, and never, but never, uttering one word of regrets. Sometimes I felt like crying for the humiliations that he was suffering on my account.

These were bitter memories, the most trying moments in my childhood, and I have not forgotten them. All this was in the order of things in those days. They happened in the neighborhood all the time. There was a Diado Marin, who did the same to his wife and she would spend days with us. I suppose that was part of the village culture. I never saw my father crying, but furious – so many times. When I was arrested in December 1944 by the new Communist authorities, he had dreamed that I was beheaded. My mother, who also had never seen him crying, told me that after this dream he had exploded in inconsolable cry. That afternoon I was freed from the jail. The poor man! He had a terrible death. On May 7, 1952, the anniversary of my escape from the labor troops, my regular military service, he was killed – the same way he had dreamed me – by a village common criminal as I was told. I learned of that on October 22, the same year, when I received a letter from my aunt in Rozovets. I never believed the story which was put out. He, supposedly, had yelled at a kid which was interfering with his duties as shepherd of the collective farm sheep, the kid went home, complained to his uncle, and he went out in the field and ... A year ago, when I received my Dossier from State Security, I learned that the assassination had coincided with the receipt of a letter sent by me through the International Red Cross, to notify my father that I was alive. They had maintained until that time that I had been killed when crossing the border. I do not know whether it was a coincidence or my letter had led to this bloody reprisal. If it is the latter, I have a heavy burden to bear for the rest of my life. Nobody would ever understand what these satanic forces controlling the communist minds have sought to get with the murder of this poor, worn old man. My recent perusal of my Dossier left me with the impression that the murder had taken place about the time the letter was received. I will go over the xeroxed materials and will publish them.

Ever since, the memory of this event makes me shudder. When my brother came to America in 1990, the first thing I asked him that we do not discuss, was the death of my father. It was, and still is so painful for me, that I rather would not talk about it. When I was in Bulgaria and we were returning from the cemetery, my brother pointed to the site, some one mile away, where the murder had been committed, I turned my head away, I could not look at it. It is only this spring that I asked my cousin, Maria Darzhalieva, to see if an attorney would dig into this and make me a copy of the court case. The murderer had been sentenced to two years and then freed after a few months. She discussed my request with an attorney, Eldarov, but he told her that cases so far back are destroyed and nothing further could be found.

The better side of my father, besides his hard work, was that he would easily flare up but the next day he would forget it all, as if it had never happened. I met so many people in the village who remembered him as a generous man, that he had distributed freely to all the produce of his gardens, his watermelons, his melons, his tomatoes, his apples and whatever he had in his hands. He was remembered as the good Diado Todor. As years passed, I came to appreciate him more and more, forgiving him the sharp edges that sometimes hurt me, and more than anything else, remembering his sacrificing all that he could to see me continue my education, after I entered the Seminary. More than anything else, I remember with deep gratitude the days when I and he worked in the fields, alone, and took our lunch hours in the hot summer days.

My mother was another story. She was a stranger in the village. She was from Rozovets and somehow never integrated, or was not allowed to join the Zelenikovo women. This made for a lonely life for her, even though she was at every *sedianka* in the neighborhood, or every *horata* (evening and day time group working – spinning, or knitting, or wool-spreading), and all that brought the women out there. Very shy, she would most of the time listen and speak little, but participating in the social life of her friends. She must have been very pretty when she has been young. I remember a large portrait of her, in elegant dress and hairdo which distinguished her from the ordinary peasant women in the village. For one thing, she never dressed like the rest of the women in Zelenikovo, with their bell-shaped woolen skirts (*sukman*). She was wearing old, but modern skirts. The sophistication of Rozovets was all there, but my family was not the right place for her. She seems to have been always scared to communicate with the world. She was surrounding me with gentle care and watched over me as if I was a helpless little creature.

She did her house chores – cooking, spinning, mending always socks, pants and shirts, weaving, cleaning in and out of the house and keeping things in order. Her problem seemed to have been the non-acceptance by the rest of the family of my father, but she had to live with it. Much of the arguments between her and my father started with some remarks made by somebody against her and she had to defend herself as well as she could, always loosing the arguments before the forceful and physical domination from him. Much was expected from her, but she never had the wherewhital to meet the expectations of life. In the summer she would work in the fields as all women did, and, weak as she was, by the end of the day she could hardly keep up the pace. She was rather tall, 5'9", slim, sometimes with half a smile on her face, which had lost its beauty of younger times, as she had appeared in the portrait I remembered, and wrinkled all over. She was very often sick, which I could not diagnose and attributed to her physical weakness. Sometimes in the late 1930s she fell really sick, could not get out of bed and appeared to be dying. There was no doctor in the village to tend to her. A Russian exile, supposedly a doctor, was a joke, distributing pills for everything. There was a *feldsher*, some sort of a pharmacist, who served the medical needs of the village when there was no doctor. There was nothing they could do for my mother and she was waning slowly, reaching the point where she could not get out of bed. It happened that one Sunday the village was visited by a team of doctors from Plovdiv, some sort of a voluntary group to help the sick peasants, but the latter had to come to the school for consultation, where the doctors settled. My mother could not go, she could not walk. I went to one of them, described the situation and he agreed to come with me to the house. He was a Jew, Buko Aladgem. For me he appeared to be an angel of mercy. After examining my mother, he suggested that she should be taken to the hospital in Plovdiv, otherwise she would die. If I recall, he diagnosed her illness as peritonitis.

Somehow we managed to take her to Plovdiv. She spent three weeks there, and then they told us to go and pick her up. I did this mission. I was told to go there at 2.00 o'clock, and though arriving in the city about ten, I had to hang around for the appointed hour. When I arrived at the hospital, they had already signed her out, and had placed her in a hallway, where she had sat for hours. In any event, I took her to the bus, with a taxi and we arrived in the village. But she was so weak, that nothing seemed to help. She needed good food, but there was nothing that we could offer. Fortunately, she did not have objections to suck raw eggs. As I discovered that, I started looking for our chicken's nests, discovering eggs here and there, brought them to her and she consumed several of them a day. In no time, she regained her strength. This encouraged me, and I continued my nursing her. It helped. In a month or so, may be longer, she was on her feet and little by little started working in the house.

A much more serious situation developed probably a year or so later. I saw how, while she was in the backyard one evening, a ram, behind her, she being unaware of it, with all force slammed her in the back and she fell to the ground. At first, it did not seem to result in anything, but subsequently she developed pain in the pelvic area and especially in the joint of her hip. The situation gradually deteriorated until she was unable to get out of bed. She was sort of paralysed, her leg bent upward, she was unable to straighten it out, experiencing unbearable pain in the joint of the pelvic area and in the knee. She was frozen in bed. There was nothing we could do, but care for her, serving her natural needs as inexperienced nurses, using a wooden pan, taking it out in utter disgust, but there was no other choice and we got used to that.

It all fell on me, and then on my father and my younger brother when I was in school. I believe that it has been degenerative arthritis, but we could never get her to a doctor, and we were told that nothing could be done to help the situation. Since I was mostly in school, it was my brother and my father, in addition to all that they had to do to keep life going. In the absence of a woman in the house, the miseries which we had suffered before, because of poverty and lack of enough helping hands, especially a female hand, our daily life, our appearance in clothing and our food and cleaning situation fell to the bottom of the pit. But we pushed on and on, silently caring for my mother. After 1946, when I had to hide all over Bulgaria and could not return to the village, to save my own life from the communists, who were after me, the burden fell on my brother and father. On February 17, 1949, I received a telegram in Sofia that she had passed away and the funeral had already taken place.

I was shocked, I was devasteted. I could not suppress my grief, hid in my bed, pulled the blanket over my face and gave in to tears and inconsolable cry. Students noticed me, came around me, tried to find out what had happened, but I brushed them aside. I did not want to talk to anybody, and continued my agony. Sometimes later, an older monk, Father Sofrony, later Metropolitan of Rousse, came to my bed and tried to soothe me. Finally I was able to tell him of my personal tragedy. He talked to me for a long time and did not leave my bed, keeping others from crowding around me. I will have to say about him more later. Just to mention here, long after I left Bulgaria, he had become Metropolitan of Russe, and in his old age had lost the use of his legs. He died last year. I often asked myself, and wondered what had caused my mother's death, but I never had a definitive answer, except to think that her heart might have given up after so many years in bed or may be some pneumonia had settled in and had brought her life to an end.

Many years later, in the United States, when I had to explain to my wife my family history, I avoided telling her all that suffering. I just told her that my mother died of pneumonia. It was too painful for me to go into detail. In fact I have never shared with anybody my personal tragedy of those years, my love for my mother, and the suffering she went through. It was in 1992 that Ruby learned of it in a conversation with Vessa, the wife of my younger brother. The latter, unknowingly of my secret, had started a conversation and had told Ruby the whole story. In moments of difficulty, Ruby felt that I had lied to her. I could not defend myself otherwise, except to say that all these experiences were so painful for me, that I could not relive them by discussing them with anybody, and preferred to keep them to myself. There are some inner sanctums in the heart of people where they would not admit anybody. This was, and this still is my inner sanctum, and I would not discuss it. I could write about it, as I am doing now, but I would not vocalize it. It is so painful. I went to see my mother's grave, and my father's, in 1991, when I first visited Bulgaria, after forty years of exile. I had the priest celebrate a Memorial service, I had a bottle of wine to pour over the two graves, candles to light and flowers to place upon them. There was no stone over their graves, but they were fenced with wrought iron. I arranged that a stone be placed there, as it was.

I have never erased my mother's image in my memory. I loved her from the bottom of my heart. She never hurt me in any way, she suffered with me and I suffered with her. One great moment has left a lasting imprint on my mind. I was preparing to go out to the *horo* where the whole village was, I was a teenager. I was putting my best clothes, tying a tie, fixing my hair to be as handsome as I could make myself. She could only lift her head, looked at me, proud of me, gave me a smile of infinite pleasure which she was experiencing in this agonizing position, and I felt good that I could please her so much. Because, after all, I was a handsome young man, slim as I was, but grown up enough to attract attention. My poor mother, I never stopped praying for her soul, all these years in exile. She was a church going women, she took me sometimes to church and we prayed together, never knowing what was ahead of us. It is far back behind me, but it still torments me, the fate both of my father and my mother. When I published the five-volume collection of my works abroad, in 1993, in Sofia, I dedicated it to them with the inscription: "To the martyr's passing away of my parents – saints – Todor and Minka."

Our family tree, on my father's side, goes long way. Drazha Parlapanska in 1991 gave me a chart which she had prepared, having collected the appropriate information. She was the village historian. An old teacher, retired before my time to go to school, I believe by the name of Diado Todor, had kept some chronicle of events for as far back as he could. He had left this chronicle to his nephew, who turned it to her. She had studied it and had turned it over to some museum in the Academy of Science in Sofia. It is from there that she had obtained much of her information. This chart is included here. (See p. 8@). The bigger family branch had been known, and is still known, or was known and continuing in my village, under the name of Raikovsky. Drazha had traced our family roots to 1777, founded by my great-great-grandfather Petar and his wife Vela. My great grandfather, Todor, had been born in 1810. He had died sometimes about 1860, murdered by the Turks with an iron fork, at the threshing field. So my grandfather Petar had become an orphan about that time. Since he played with the neighborhood kids, and did not have a father, they refereed to him by his mother's name, Raika, which in the Bulgarian language, on female line, is rendered as RAIKIN. This bit of information was not known to Drazha. I remember it from my grand mother, Ivana. With this story I helped Drazha to fill in the gaps in our name origins.

In 1992 I learned that my nephew, Todor, and his wife had changed their last name from Raikin to Petrov, after my grandfather, Petar. I suspected that they had done this, abandoning our family name, on my account, not to be associated with me, the political refugee in the United States, to erase any connection with me. I felt offended and when I heard that their son, Slav, on occasion of his wedding, which I attended, was referred to in the civil ceremony as Petrov, I left the celebration. Later I wrote to Zlata, that our name, Raikin, was sanctified by the blood of our great-grand father and my father, and that changing it is an offense to their memory. She tried to tell me that it was my older brother who had wanted it this way. My brother was already dead, and dead people do not tell stories. In addition to that, they, my nephew, specially his wife, had become active members of the Communist party, so any explanations were of no consequence for me. Some years later I heard from Zlata, that they will register their grandson, son of Slav, as Raikin, when he goes to school. She was seeking ways to appease me. I also heard that as things stood now, the name of Raikin was going to come to an end. My other nephew, son of Stoiu, Todor, has two girls, and my son Todor is not yet married. Zlata's daughter, Totka, has grasped the situation and told her mother that I was right to make an issue of the name. But I gave them all to understand that the name Raikin is sacred for me, and anyone who drops it, is no longer member of our family.

15. My Brothers

Of my older brother, Peter, I have little to say. The difference of age, between us, ten years, made it difficult to effect emotional ties. We were growing separately. But the time came when I sort of adored him – for his protective indulgence to me. I was proud of him, having an older brother. There was a tradition in Bulgaria to have a *sobor* in every village when the patron saint of the local church was celebrated. People of all neighboring villages would gather there to take part in the festivities, and especially to dance the *horo*. This was the *sobor*. In my village this was the day of St. Petka, Oktober 26th. Merchants of all kind would bring their goods to sell, especially non-alcoholic drinks: "red water" – water sweetened with sugar and coloring die; "boza" – some sort of a fermented millet flour, mixed with water; fruits – apples and pears and candies, especially halvah, a firm chunk of sessamy seeds flour, loaded with sugar and some oil. It was a delicacy, but was expensive. Those of us who could not afford it, would beg those who would buy it, for the paper in which the chunk was wrapped, to lick it. On such one occasion, looking at the other kids buying this or that, I spotted my brother in a company of boys and girls, went behind him and asked him for fifty cents. With a big smile, he reached in his pocket and gave me a coin of two levs. He was pleased to treat his kid brother, and I was happy to have such a generous old brother. I was thrilled and never forgot this gesture. He caressed my head and told me to go and buy myself whatever I wanted.

GENEALOGY OF THE RAIKINS FAMILY
Original clan:

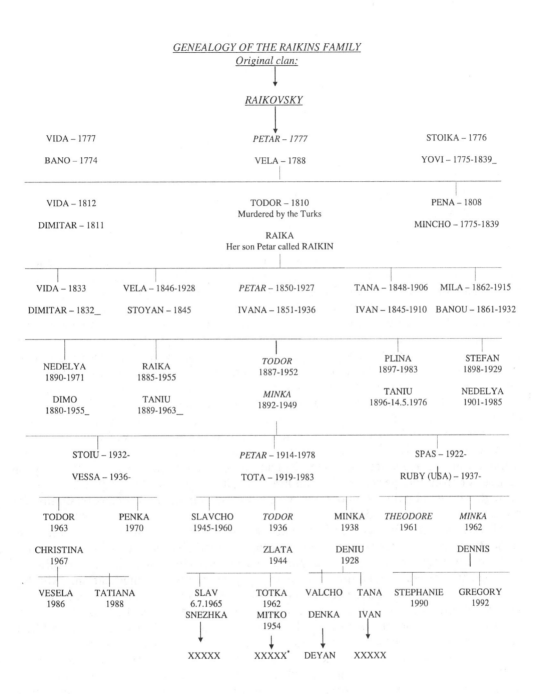

RAIKOVSKY

VIDA – 1777   PETAR – 1777   STOIKA – 1776

BANO – 1774   VELA – 1788   YOVI – 1775-1839_

VIDA – 1812   TODOR – 1810   PENA – 1808
      Murdered by the Turks
DIMITAR – 1811   MINCHO – 1775-1839
      RAIKA
      Her son Petar called RAIKIN

VIDA – 1833   VELA – 1846-1928   PETAR – 1850-1927   TANA – 1848-1906   MILA – 1862-1915

DIMITAR – 1832_   STOYAN – 1845   IVANA – 1851-1936   IVAN – 1845-1910   BANOU – 1861-1932

NEDELYA   RAIKA   TODOR   PLINA   STEFAN
1890-1971   1885-1955   1887-1952   1897-1983   1898-1929

DIMO   TANIU   MINKA   TANIU   NEDELYA
1880-1955_   1889-1963_   1892-1949   1896-14.5.1976   1901-1985

STOIU – 1932-   PETAR – 1914-1978   SPAS – 1922-

VESSA – 1936-   TOTA – 1919-1983   RUBY (USA) – 1937-

TODOR   PENKA   SLAVCHO   TODOR   MINKA   THEODORE   MINKA
1963   1970   1945-1960   1936   1938   1961   1962

CHRISTINA   ZLATA   DENIU   DENNIS
1967   1944   1928

VESELA   TATIANA   SLAV   TOTKA   VALCHO   TANA   STEPHANIE   GREGORY
1986   1988   6.7.1965   1962   1990   1992
      SNEZHKA   MITKO   DENKA   IVAN
      1954

XXXXX   XXXXX*   DEYAN   XXXXX

---

* Names of children not known.

87

One could never catch such moments, but when he does, they leave a deep impression and strengthen the emotional attachment. So, I never stopped loving my brother... But he was sort of impatient with the animals when he was plowing, and sometimes punished them with the *osten* when disobeying him, they would run in circles and he would continue hitting them. I did not like it, but then when I started plowing I found myself doing the same thing, until I realized that this was not the way to act and started to change. In fact this change affected all my behaviour, in all instances of life where patience appeared to be the better way to do things. My older brother married sometimes in 1936 and brought his bride home. In fact she was quite pregnant and the wedding took place not in the church, but in our home. His bride, Tota, turned out to be a very nice lady, peasant as she was. For a while she took many of my chores in the house. But it did not last long. They wanted to have their family, we managed to help them build a house at the end of the lot, on the street parallel to our main street, and my father gave them one third of the fields which we owned, to cultivate and make a living.

It was very difficult for them to start their separate family. Soon they had another child, Minka. The first, Todor, was born in our house a few months after the wedding. Then there was a third child, Slavcho. My sister-in-law was all but disowned by her parents for marrying my brother, but I guess, there was no other choice after what had happened. They did not receive help from the other side. We struggled as we could. Then I had to go to the seminary and spend only the vacations in the village. I was not too close to see what was happening. Much later Minka confided to me that there were many occasions when they would not have bread to eat, and her mother would send her to my father, who would immediately guess what was the matter, would call the crying little girl, would fill a bag with flour and send her home. As the years passed, I became very closely attached to Slavcho. He was constantly around me and called me "*Chichko*", diminutive of "*Chicho*" (uncle). I came to love the little kid and played with him for his infinite pleasure. After I left the country, I learned the sad news that at age of fourteen, while in training as a tractor operator, he had fallen on the ground and the other guy, on purpose or from inexperience, had run him over. It was a family tragedy which I lived through with great pain. I do not know whether he was another victim, as retaliation for my escape, or just an accident.

Minka, the sister of Todor and Slavcho, was growing very thin and tall and looked, at that time and now, very much like my mother. Sometimes she had developed a tumor in the brain, I saw the x-ray of it, apparently arrested, which had caused her epileptic seizures. It is my understanding that it is being controlled by medications. She has married a young man from the Kraevsky family – there were so many of them – Denue. They had a big house near the church. Sometimes in childhood he had lost the sight of one of his eyes, but had been a hard working decent man. They had two children. Minka was sending me pictures of Vulcho way back. He married two years ago and they have a child. Hope to see them when I go there next year. He is very gentle and well mannered young man. Their daughter, Tana, has married a very industrious and responsible young man in Brezovo. During our first visit to Bulgaria, he took us around with his car. At that time he was working for the local telephone company as a technician. Presently he is the Manager of the telephone station in the city of Rakovsky. For some reason, Minka is not on good terms with her brother Todor and Zlata, his wife. Minka told me much about the generosity of my father in times past, even in his very poor state.

As years have passed, my nephew Todor had joined the Communist party. In addition, I learned that in his old age my brother had not been treated with mercy, that on one occasion he had been tied by Zlata and my nephew to a pear tree for some reason. All this left a bitter taste in me. He was afflicted by cataract very early and some doctors had botched the surgery. I could never understand and know how much he had suffered. But this was another dark side of our family life, which still plagues me in my memories. When the news reached me sometimes in the early 1980s that he had died, I did vent my pain in uncontrollable cry at home and in the church at the memorial service.

My younger brother Stoiu, named after my grand father on my mother's side, is a jewel in my tormented soul for so many years. I remember his birth, I was ten years old when my father sent me to fetch my grandmother from the *horata*. I listened in utter frustration from outside. For the next years, until 1939, I was constantly taking care of him while around the house. After I left for school, he was more or less on his own, helping my father as he could at the age of seven. As my mother fell sick, it was him who took care of her. During my summer, spring and winter vacations I would relieve him of some of his duties. But he still was a kid. After 1946, when he was already fourteen, I practically never saw him. It was only when I was going to serve my military service that I was home for a few days, and in the beginning of March, before leaving for Balchik, I went to the fields to look for him to say good bye. I found him on Baidachka. All I remember is that I embraced him, and kissed him on the forehead above his eyes. This kiss for forty years was burning on my lips. As if I felt that this was going to be our long, long separation.

After the fall of communism in 1989, I brought him to America for a month. For the first time I learned of his sufferings in Bulgaria. He was the one called to witness the death of my father and bury him during the night. At the time of my escape, May 6, 1951, he had been under arrest. At a theatre presentation in the village someone had slipped a gun in his pocket and the militia immediately moved to search and arrest him. He had been tried, imprisoned and served in Belene concentration camp. He related to me the story how one day he was being taken for questioning and the guard led him between cages of vicious dogs which wildly showed him their teeth. The guard had told him: "Do you see these? If we open the doors of the cages, not a piece would be left of you."

Apparently they could not tag anything political on him and after serving his term, he was released. He did not have a penny to pay his ticket to go to the village. He went to a militia station where my cousin from Rozovets, Yono, was working, to ask for help. Yono directed him to go and find his wife at home, who fed him and gave him bus money to go home. In the village, it was his girlfriend who was waiting for him. They moved in the empty, unfinished house – without windows, without doors, but were determined to make it a home. They did. Not much later the local tax collector came and told them that they owed 300 leva taxes, otherwise the house would be taken from them. He went to some-one of the Pundev family and begged for a loan. The man had a heart and readily helped my brother. He paid the taxes and subsequently repaid the loan. He was ever so grateful for the help. The last time I asked him if any of my relatives, my brother or my nephews, ever visited him in jail to give him a piece of bread, a word of encouragement or a few leva. No! They had never sought him, and after he returned to the village, they never said hello to him. But he survived. During one of my visits to Zelenikovo I asked for Grozu Pundev, who helped him pay his taxes, to thank him, more than just with words, but he had passed away and his family scattered all over Bulgaria, with no trace of them.

When Stoiu and Vessa had returned to Zelenikovo, she has attempted to get a job in Rozovets. They had opened a factory for padlocks there. She had to fill an application. When she reached the question if she had relatives outside of Bulgaria, she could not lie, because all the world knew of me. She showed the application to the man administering the test and asked him what to write. He picked her paper and told her to leave. She then reluctantly went to work in the pig farm in Brezovo, ironically, built on one of our better parcels of land near the river. My nephew, Todor, son of Stoiu and Vessa, showed me his trudovak's work book or something. On the very first page, where identification data was entered I read in one of the columns: "Has uncle in Chicago, USA." Well, I was not in Chicago, but still my being in America prevented him from seeking better chances in life. He served his military duty as *trudovak*. But still, there he learned the craft of plumbing and after finishing his service continued to practice plumbing for some construction company. My niece Peppa, daughter of Stoiu, had been told in not uncertain terms, that the only job she could hope for was to go and work on the pig farm. Tender and elegant as she was, this was the last place for her. Vessa told them that she will take care of her in the house, but will not let her go and clean and feed the pigs. In recent years she married a young decent boy from the city of Rakovsky, who is an experienced baker and manages the bakery in the village. Soon Peppa joined him there, and now both of them bake the bread for the village. This boy, Stanimir Gagov, is a Catholic, and his last name suggest to me that he may be from the family of Prof. Gagov, a Catholic priest in the Vatican in the 1950s, publishing in those days a periodical, "Emigrantski vesti", *Emmigrants' News*. He was also a prominent figure in exile politics in those days, and professor in some Vatican college.

As I write these things now, and as I reminisce over the past, it is weighing heavy on my mind how much pain I have caused to my people with my escape from Bulgaria. Stoiu has reassured me on numerous occasions that I had done the right thing escaping from the country. He is convinced, as I was, that had I stayed there, I would not have been left alive. So much I have been hated there. I felt happy, that after forty years, I could see them again and pay them back for all the suffering I had caused them – not with money, but with the love which I had for them, and they had for me. We were able to meet again. I will leave this world with my boundless love for my little brother of old times and our rejoining together in the autumn of our present life. During all these forty years, especially the first ten years, I knew nothing about him, whether he was dead or alive, and what had happened to him after the death of my father. Much later I learned that he had joined in my older brothr's family, but it had not worked well, and very soon he had been on his own. My memory is not quite clear what happened to him after his arrests, his prisons and concentration camps, including his service as trudovak. I never stopped praying for him and lighting candles in the churches I visited. I have already told the story of the conflict between him and my nephew over the part of the backyard, which was supposed to be my share of the inheritance. It hurt me when I heard of it: not that I had any claim for this piece of land (God forbid! I never thought of that), but for the greed of my nephew, behind whom I saw the hand of Zlata, his wife, and for the injustice suffered by Stoiu. Zlata's words, that I did not exist for their State, which had entitled them to my share of "inheritance", revealed their, my closest relatives, political condemnation of me. But, be that as it may. All this left a bitter taste in my mouth, though I never lost affection for my older brother Petar and his wife, Tota, the mother of Todor.

Sometimes in the 1960s I received a letter from Stoiu, from somewhere in the Rodopy Mountains, in a mining area. Later he moved to the village of Sestrimo. He somehow had specialized in dynamite blowing of rocks in the building of tunnels, or blowing up mountains for aqueducts. He had become such expert, that the engineers and the chiefs had exclusive confidence in him in setting up the charges and connecting the wires, that sometimes let him supervise young engineers. I should say that he, as my older brother, never had schooling, hardly reaching third grade. But he had mastered the art of dynamiting and was paid well, until he had retired and got a better pension. Then he returned to the village, to fix the house – one of the best houses in Zelenikovo – and work on the farm taking care of baby calves. All these years I never knew what was happening to him, and if anybody, after the passing of my father was in my heart constantly, it was Stoiu. If I have been so far five times in Bulgaria, and if I go once there again, it is for him. Should he pass away, I am not going to step there, especially in my village Zelenikovo. I suffered there too much pain to have any feelings of loyalty for anyone, except for the beloved little corners at *Tenyov Giol*, *Baidachka* and *Chervinatsite* with their mighty, but now castrated oak tree... and *Sakar Tepe*.

16. The *Wunderkind*

But let me go back to my childhood in Zelenikovo. It was not a happy experience and my memories are not very exciting. My mother showered me with her tender cares and moral support when I messed up things. I do not remember any occasion when she punished me for anything, or else I would cry so much that she would rather not do it. With sweet talk, she advised me what to do and what not to do, what was right and what was wrong for me to do. Too often I was sick, never knowing what was wrong with me. I was slim like a stray dog and apparently my immunity was down to zero. One thing that comes to my mind was that I was chronically undernourished and whatever I was eating was a poor substitute for a growing child.

When I had to go to school, in September 1929, in the old school, I was seized by panic. Too much has been said by older kids that the teachers were some sort of monsters who punished left and right, beating the students, etc. I was scared to death to go there except once in the beggining. It frightened me out of my wits, and after the first hour, during the intermission, I ran home, huddled in the corner and told my mother that I shall not go again to school. She did not force me. She tried to persuade me to change my mind, but I was adamant, I was crying and felt that I was going to throw up. Soon I was in bed sick. Neither did my father show any interest of forcing me. He and my mother had not gone very far to school, and so was the case with my older brother, as also was going to be the case later with my younger brother. Education, School, did not mean much in my family. Two teachers of the school kept coming to our door to pressure my parents to send me back to their cares, but without success. I remember one occasion when they, Dimitrina Vuleva and Mara Angelieva, arrived in the neighborhood. They found me at the knees of my mother in the "horata", eating baked pumpkin and bespattered from ear to ear. They started at her with their lecture why I should be going to school. She was embarrassed and told them to leave me alone. She told them that they should see for themselves that I was not ready for it as yet, which, I suppose, was right. But they continued infrequently these visits and torment my poor parents. They were the wrong teachers to send. Their reputation as the severest educators, were well known to the entire village and only their sight at our door sent me to hide wherever I could. Dimitrina was a tall, little hunching, with severe eyes, and expression on her face, and a city accent with a pitched voice. Angelieva was of Macedonian background, middle height, slim, hysterical and not quite liked. They were not encouragement for me to return to school.

When the next September came I was no longer resisting. The school was moved to a new modern building. My teacher happened to be Diado Georgi Bradata (Grand Father Georgi the Beard). His reputation was not very pleasant either. I have already mentioned that. He was an old Socialist, and as the custom for the socialists in those days was, had a well trimmed beard. He walked with a cane, and in a year's time, retired. As to Dimitrina I heard last year that she was still alive, but Angelieva had committed suicide sometimes in the 1960s. During my last visit to Zelenikovo I saw the exhibit of the 125th anniversary of *Chitalishte* "G. S. Rakovsky". One of the exponats at this exhibit featured the sister of the great Bulgarian poet Dimcho Debelianov, photographed with one of the local teachers, Zlata Parlapanska, whom I met in Sofia at a dinner in her son's home, now 93 or more. I vaguely remember that she was a teacher in the village, but I never had her. It happened that only three days earlier I was in Koprivshtitsa, and visited the home of Debelianov, with the famous statue of the poet's mother, sitting on a stone waiting for the return of her son from the front where he was killed. One of his best known poems began with the strophe: "To return to your father's home..." It is a classic in Bulgarian literature.

The poetry and its powerful impact is lost in the translation, but in the recitation of an actor, as it was recited by an immigrant in the Unites States, during our visit there – do not recall his name – it is a heartbreaking performance. So, the sister of this great Bulgarian had been for many years teaching in my village shortly before my time. Debelianov, incidentally, was killed on the Salonica front.

It turned out, that Diado Georgi Bradata retired at the end of my first grade school year. Only once, I do not remember the occasion, he pulled and twisted my ear, without saying his so much repeated words: "You cursed lice." His place was taken by a young man from my village – Boniu Atanassov Bonev. We hit it with him right away. Energetic, all immersed in studies, deeply infected by the virus of Tolstoism and Christian Anarchism, he took over our class in the newly opened school building at the west end of the village. There were two classes of each grade in the elementary scool. The other class was taught by Dimitrina Vuleva. Was I glad that I was not in her charge! Very soon it became clear that the two classes were in some sort of a competition, or it was between Boniu and Dimitrina: who was a better teacher. Very soon I was going to become a bone of contention in this rivalry, and bring on myself the wrath of Dimitrina. I emerged as intellectual force in Boniu's class and she could not stand it. Or may be I just imagined that. But she never showed any sense of friendliness towards me.

Boniu was deeply involved with his ideas. He was not simply a teacher, very effective in his presentation of his subjects, which in the 2nd, 3rd and 4th grade, all were taught by him, and he did not change class until we graduated from the primary course, the 4th grade, but also very much a missionary for his ideology. Very soon he discovered my intellectual curiosity and started giving me extra literature to read, at this early stage just folkstales. But by the end of the second grade he was leading me to more serious materials. He gave me to read *The Gospel, adapted by Tolstoy for children*. It was a fascinating story for me. I loved the parables of Christ, and somehow Christ became my hero. I saw that he was always on the side of the weak, and of the poor. He was teaching moral principles in inter-personal relations which captivated me, and I translated these principles in a social system of values, which elevated my spirit to dream of an order of things where people of my class were dignified and humanized.

I came to see the world in a different perspective, a world which needed changing, that this change was to proceed along Christian lines, of love, of peace, of brotherhood, of mercy, of compassion, of sacrifice, and humanism. As I was progressing, he introduced me to other pamphlets written by Tolstoy and I devoured the Christian philosophy of the great writer. Then he introduced me to the writings of Gregory Petrov. Petrov was another Russian writer. He had been actually a priest who eventually was excommunicated by the Russian Church, as was Tolstoy, for his revolutionary, in fact his ideas of translating Christianity into practical life. I read all that Tolstoy had written along these lines in the second half of his career as writer, and all that Petrov had written. These works were translated into Bulgarian and were circulated among Christian anarchists. One little booklet of Gregory Petrov made the strongest impression on me. It was entitled "The Great Men". One of these great men was Socrates, but there were more than a dozen among them. I do not remember the others. What they had in common was that they all started their lives from humble backgrounds, and with perseverance and hard work, had achieved very high positions in service to society. Somehow, this idea breathed optimism in my mind that I, too, could rise up somewhere, to make a difference in the world. It was public service that strongly attracted me, where I could implement my ideas.

I remember a story, the name of the author escapes me, where a young American boy became aware that two trains were put on the same track running from opposite directions. He tried to alert all sorts of responsible officials, but to no avail. In the last seconds he talked to a station master where the trains were to crash into each other and gave him an order to stop the trains. The station master asked who was giving this order, and the young man said: "The Secretary of Transportation". The trains were stopped. The catastrophe was averted and the young man became a hero. So, for the good of society, even a falsehood is permitted. Well, as I went up and up, upon graduation from the seminary, a classmate, Ivan Apostolov, gave me as a parting gift the *New Testament*, a recent edition and signed it: "Go to the world and be a second Gregory Petrov." I had this *New Testament* with me in the military service, where I escaped from in 1951, but did not take it with me. However, I have never forgotten it and the Gospels which I read and reread.

It is said that the basic human personality is formed at an early age. I could with certainly say, that my basic ideas were formed during these young years, in the second, 3rd and 4th grade, guided by Boniu Bonev. The basic ideas which I was to hold and promote, in all the transformations that I had to go through in all my life, were the ideas of the Gospel which I absorbed in those years. Who knows, may be all that I have achieved in life was due to this young man, Boniu Atanassov. When I visited the cemetery in Zelenikovo and passed by a grave with inscription of his name, I stopped, placed some of the flowers which I was bringing for the graves of my parents and prayed for him. This made a deep impression on the priest. He knew Boniu Bonev. He had been going to Church and, it was already a well known story, that he had lost his mind. But what this mattered to me? Nothing! I was ever so grateful to him.

So, it was Boniu Bonev who turned my life in a direction never foreseen in my family. He is one of the key men, out of half a dozen other incidental persons on my path of destiny, who played a decisive role in my life. It was he who inadvertently charted the course of my life. If he did not recognize my potential at this early stage in my life, I could have easily slipped into the mold of all my relatives of illiteracy. It was he who planted ideas and principles in my life which stayed with me for all the decades that have passed, as light beams on the high roads of my moral, social and, yes, political directions. It is him who won me for the doctrines and the teachings of Christ, as presented by Leo Tolstoy and Gregory Petrov, as unknown as he is lost somewhere in the history books, or never mentioned at all. Boniu was my teacher of *Zakon Bozhiy* (The Law of God). Tolstoy was a further expansion of Boniu's lessons. Further than that, there was the inscription in my local church, up near the ceiling – "Do not do to others what you don't want them to do to you!" I am translating it word for word as it was written on the wall. These were my first lessons on Christian Ethics which have been imprinted on my young mind. First with my grand-mother, and then with my mother, I attended services in the church and stared at this inscription, left behind by some humble priest. We lighted candles in the church, celebrated the religious holidays, like *Voditsi*, (Theophany). The priest was visiting our home to bless it with Holy Water. On such occasions my mother covered the floor of our common room with her best *cherga*, large home-made blanket, which she carefully preserved for such occasions. She did not know much about and of religion, and did not explain to me anything on the subject. She was just following the traditions, and I was by association bound to them. So it was from Boniu that I learned my Christian ethics, and decades later were to strive to integrate them in my political views.

When I moved to the Middle School (pro-gymnasia), 5th grade, we were taken over by other teachers. My class teacher now was Angelieva. From the very beginning she made it clear that she did not like me. The reason was that I was the valedictorian of Boniu's class, and she and Dimitrina were dead set against him. So, by association, I was on her black list. As soon as I showed from the very beginning my superiority over Dimitrina's students, Angelieva turned against me with vengeance. In one class, she assigned us to compose a story. At one point she came over to me, looked at my paper, grabbed it, cut it to shreds, dropped them in the wastepaper basket and asked me to write it again. She did not like my handwriting. She said that it looked as if it had been written with a goose foot. She was right. My writing was atrocious, but when I finished rewriting it, it looked the same. She said nothing. To this day my penmanship is still the same. Sometimes I myself could not read it. When I started writing English, it was all over again the same. My students in America often asked me to spell it vocally, so they could copy it. I guess, it will never change. Soon Angelieva got it that I was the best student in class, and stayed so to the end, but she never gave me credit for it. The day I graduated from the Middle School – it was the terminal grade for mandatory schooling – on my way home, barefooted, with a patched up shirt, running to grab my bag and join my father in the fields, she was standing near her home on Main Street, stopped me and asked to see my grade card. I showed it to her, and it was all A-s and one or two B-s. She looked at it, folded it, gave it back to me and made a comment: "I always knew that you were a good student." But when she took over in my 5th grade, she did all to beat me down, favoring Dimitrina's students, especially my neighborhood pal, Georgi Natchev. She was a hysterical and nervous woman. Poor soul! When my brother came to America he told me that she had committed suicide.

Indeed, I was the top student in my classes, from the 2nd to the 7th grade. I listened to the teachers, I kept notes, I read all in the text books, and I read anything that fell in my hands.

In the beginning in the Middle School I was interested in the sciences – physics, chemistry, botanics, zoology, geography and especially history. Mathematics was one of my strong subjects. On one occasion, we were given a home work to resolve several problems. Some resolved part of them, other more, still others less, but there was one problem nobody could figure out. The teacher assigned it to some three of us, the best mathematicians, to work together. We tried, we turned it in every possible way, but the answer eluded us. After three days of trying we gave up. The morning before the report was to be given, it clicked in my mind and arriving in class I proceeded to offer my solution. It was correct.

On an other occasion, in physics, we were studying the steam engine. I was called to the blackboard to make a drawing of the basics and explain its functioning. I did a good job and point by point made everything clear. At end the teacher asked me how the wheel will continue revolving. I looked at the drawing, concentrated a little and it hit me: There must be a lever which would be attached to the turning wheel, so when it complete the cycle, to pull or push the wheel forwards. And I added to my drawing where the lever would be attached. I figured that all by myself. The teacher, a Plovdiv woman, look at me and then to the class and said: "This is a perfect presentation of a lesson." I was pleased of myself and proud of my performance. And so was I all the time that I had a chance to show my knowledge. In fact these were the brightest instances in my otherwise unhappy life, where I had to cope with hunger, poverty and miseries. Another occasion, where I was so pleased to express myself was at a teacher's conference. I had to make a presentation of the eruption of a volcano. A mound of sand over a table, with some chemical in the centre at its base, lighted by me, let the eruption of smoke and sand go up one foot.

My glory as a gifted child and as the best student in school was all over the village. My classmates looked to me for help. When I was walking the streets, people stared at me and openly greeted me with my successes reported to them by their kids. But at home this did not seem to be noticed by anybody. Neither my mother, nor my father, or my brother, ever mentioned it. For them I was doing what I was supposed to do in school. People congratulated them outside of our house for the miracle child that I was, but they shyly ignored it. What good was it to them and to us in the family when we just struggled to survive? There was no occasion that I remember when they would congratulate me, or encourage me, or express a pride with me. As if nothing extraordinary had happened. When the school year would end and the parents of all students gathered at the school to see their kids recite poems or be cited for their achievements, my parents never came to see me. And I did not expect them to be there. They did not have decent clothes to wear, and I, likewise, did not have a decent shirt, or pants, and never anything that was called shoes. I was always barefooted. They did not come to my graduation where I was distinguished as the student with the highest academic grades. When my name was called, I ran in there to the Director, Zidev, picked my diploma, he congratulated me, shaking hand with me and staring in my face with a smile. After that I was running home. That morning my Father had instructed me where to find him and what I was going to do, and I was on my way to do my family duties. I have to say, that I did not study, as I did, for good grades, but this was the only way I could stand before the world and assert myself, that I, too, was somebody. That was the only gratification which I derived from it all. I never did it for something that was going to be useful for me in the future. In fact I never thought of the future, because I did not see any future for myself, except the future in Zelenikovo – with the sheep, with the cattle, and with our house – a ruin, rather than a home, but a sweet home as it was.

17. My University Under the Village Overhangs

So, this is where my education ended – 7th grade. There never was a talk in the family for continuing in high school. I did not pressure for such resolution of my future. I had accepted my future as a settled matter, as something that was unavoidable. My father did not have decent clothes on his back. We rarely saw any cash in the house, so how could I dream of going to school? The High school was in Plovdiv. That was the only High school in the district, some 35 miles south-west of my village. There were no prospects for me to continue my education. It was expensive. One needed to rent a room, or live with some family, then pay for his food, for his books, some school fees, and all that life needs to go on. A school uniform itself would cost a fortune, and without an uniform – cap, "*kurtka*" (a tunic buttoned to under the chin) trousers and a winter overcoat in style for high school students. Only few peasant boys' families could afford it. The local priest took an interest in me to apply for admission to the Orthodox Seminary and for a scholarship, but a vital document for the scholarship was necessary – a certificate of poverty. I tried to obtain this certificate from the municipal authorities, but it was denied to me. In the land register it was recorded that my father had some 63 decars – some 15 acres land, and this disqualified us as poor. They could have made this information available, they could have added a brief statement describing the special hardships the family was going through, and it would have been the truth, the Seminary may have considered it, but they stopped at the mere statement of land ownership, and under the circumstances, I could not pursue the matter further.

An alternative to further studies was sometimes considered: to make me an apprentice to a tailor, so I could master the craft and became a tailor, but the local tailors refused to accept me. They had family members or relatives to consider first. Because of my physical unfitness I could not handle other profession. In the meantime I was now full time farm worker with my father – plowing, planting, cultivating, harvesting, threshing, tending sheep, cows and all domestic work. Some seasons during the next two years I was the neighborhood shepherd, or cow and calf caretaker, taking the animals to graze in the fields, things which I have already described. When my mother fell sick, of peritonitis, and almost died, I took care of the house work. She would instruct me from her sick bed how to cook, how to make "kavardiska", how to make bread – pass the flour through a cieve, then mixing it with warm water and yeast, leaving it to rise, then making it into loaves and baking it in the new stove which we had acquired. Cleaning the house, washing clothes (rarely) and taking care of my younger brother were extra duties. When I was needed in the fields, it all would be on my shoulders. I hardly had time to stop and rethink my life prospects.

All the time I used every minute I had to read, anything that fell in my hands. During the fall of 1937 I learned of some correspondence course in bookkeeping. I signed in and did all the work. At the end I was given a certificate of having completed the course, but there was no job of that kind for me in the village. So, I was destined to stay in the village and continue doing what I was doing. In the fall, I spent the long evenings with other young boys wandering around the central square where the taverns, the bars and the *chitalishte* were. We were not allowed anywhere except in the *chitalishte*. We would occupy the long table in front of the two bookshelves, and incur the displeasure of Diado Muino, who was serving Turkish coffee to his customers. We were occupying the seats, reading the one or two periodicals or left-over newspapers.

Most of the time we would huddle in small corners along the buildings, protected from the winds and the rain and talk. We discussed all that interested us, all the time eating sun-flower and pumpkin seeds. When the first phonograph (gramophone) arrived in the village in a tavern, we would glue our faces, taking our turns at the window to catch the words of the single record which was played, and played again and again.

During the first weeks of November, we had already formed a small group of friends, youngsters, not fitting with the adults, and not fitting with kids. We would huddle under long overhangs to escape the rains, or small corners to escape the winds, and spent long hours in idle or serious talks. Soon our group was joined by a high school dropout, Taniu Raichev Grozev, lame with one leg, walking with a cane, perhaps five or six years older than us, but wise in the ways of life, who would command our attention. He had a lot of stories to tell, all kinds of adventures, true or false, quite often of sexual contents, and very often touching on social and political issues. He was a leftist, though his father was one of the few well to do peasants and sometimes in the past leader of the Agrarian Union in the village. Taniu made our long, autumn evenings – rainy and windy, unpleasant and boring – amusing and interesting. At least I appreciated this sophistication. He appeared well educated, no non-sense man, adult among us, midway between childhood and adulthood. We all participated in the discussions, whatever the subject was. Thinking of these evenings much later, I was inclined to call them our free university under the overhangs along the main street. By all means, he was a left-wing oriented, but at that time this did not mean anything to me. He never mentioned the Communist party or communism, but stressed the ideology of cooperativism.

At that time the Communist party was outlawed, but as I was later to learn, it was conducting its activities with front organizations. Not allowed to have a legal youth organization, it was covering up under Youth Co-operative Groups – young men for co-operative movement. So we did form such a group under Taniu's leadership. Communism was never mentioned. I got quite enthusiastic about the co-operative movement. I started reading literature on the subject and soon I knew all about Robert Owen and Schultze Delitsch. I prepared a referat (a paper) on Cooperativism. I liked the idea. It was going all along my ideas for changing the lot of the peasants. It was an ideology worth spreading around. We asked the leaders of the *Chitalishte* to allow me to present that paper as a public lecture in this "Home of Culture". I was allowed to. One evening, the *Chitalishte* room, packed as a can of sardines, cigarette smoke thick as a dark cloud, and smells of peasant humanity mixed to the point of poisonous gas, I was standing there, just 15 years old, thin, with no beard or anything to suggest maturity, reading my manuscript. All present there were silent, listening attentively, so that if a fly buzzed, it would have been heard. There was the mayor sitting at the end of the table next to me, smoking his cigarette. He was a retired military officer, appointed from the central government. There were some of my teachers, and so many of the "intellectual elite" of the village. I was not perturbed, I was not shy and I boldly stated my views. When I was finished in 20-25 minutes, ready to sit down and wait for questions, the Mayor stood up and proclaimed: "There will be no discussions."

So, there was no discussion, but I was thrilled that I could state my views. Subsequently I was congratulated by so many, older people for whom I had respect. They were all Agrarians or communists. I believed everything that I had said. It was a triumph that I had cherished before, and now I had accomplished it. In those days I was fired with idealism and fantasies for social change, though I could not channel them in one or another current political or social philosophy or ideology. I just felt that something had to change, that the existing order of things was no good and that I was its victim along with others. My most painful realization in those days, though it never articulated in my mind as a cause for political action, was the poverty of my family, and above all, my bitter disappointment that I could not continue my education. I was victim of an order of things which had to be changed and I saw cooperativism as way out of poverty for the hard working peasants. I was reaching for something, and cooperativism appeared to be the answer. I did not realize that Taniu Raichev, and the Youth Cooperative Groups at that time were a tool of the Communist Party to carry on its subversive activities. I would have realized that much later, but for the time being, I found myself involved in it. Much later I was to learn that cooperativism was also a key item in the ideology of the Bulgarian Agrarian movement, where I eventually found my political home, and still believe in it.

Taniu Raichev was still alive in 1991 and when I was passing by his house and my brother pointed it to me, I tried to knock at the door to see him, but apparently nobody heard to open it. During my visit last year, 1995, he was already dead. It was by chance that I, looking for the priest in the cemetery, found him performing a memorial service for the 40th day after his death. I regret not having seen him. I was told that he had become some official in the municipality, Tax Collector or something, that after that he had been thrown in jail or concentration camps, but had survived. I understood his leftist leanings, but it did not bother me. I wanted to know as much about everything as I could, and if he was a leftist, at the time it did not mean much to me. He probably was a communist and was assigned the task to influencing us, the younger.

Sometimes during this difficult period in my life I befriended Taniu Dangov, a tailor, much older than I was, but deeply involved with the ideas of Anarchism, some different brand that the one Bonio taught me, but I could not tell the difference. He too was an adept of Gregory Petrov and Tolstoy. He often quoted the great Russian writer's stand for equality of men. In his words, there could be no equality because, at the end, there always will be a need of someone to clean the toilets, and those who do that job could not in all fairness be equal to those who did not. He gave me the book of Prince Kropotkin, *Bread and Freedom*. I read it and reread it, and was greatly impressed by this Russian aristocrat anarchist.

The difference between Taniu Raichev and Taniu Dangov was, that the first did not attack the state and the authorities as institutions. He was only implying that they should be changed and replaced by a better system of government and social organization. Taniu Dangov was anarchist and argued for the demolition of all authority. His ideal was a society without these institutions. They were simply practicioners of violence and oppression. Much that I shared with him, all criticism which he had for them, I could never fathom as to how the society would and could function without any authority to enfocre order and justice.

There was Drazha Parlapanska, another one of my intellectual friends who gave me a number of books to read on human behaviour. I was growing in my understanding of the world, but, somehow, I did not feel like adhering to one single philosophy of life. I was critically examining everything, trying to find my own way in the world. Somehow, I was charting an independent position between all of them. I never severed my relations with the priest and sometimes even attended church services. It is probably on this account that the Raichev's group did not trust me, as I was going to realize. At one point they had appointed some sort of a rally of Youth Cooperative Groups in the district somewhere in the nearby mountains, I was told of the event, but when it came to let me know where exactly it was going to be held, I was simply not invited. It probably was a conspiracy of the Communist party, where only tested loyalists were to meet. Still, I was moving in a leftist direction, but never found myself in a communist environment or associated with ideological communists.

Sure enough, much discussion occupied our evening sessions around Taniu Raichev concerning communism. But something did not sit very well with me. The collectivization of agriculture did not attract me. I was emotionally so attached to our small pieces of land, of *Baidachka*, or *Tenyov Giol*, and I was again emotionally so much attached to my sheep, and my cows and my animals, that I did not see how I could part from them in some collective farm. Besides, I had read enough of the miseries which communism had brought to Russia, the human sacrifices, murders, assassinations, oppression, etc. that it did not appeal to me.

I was aware of the existence of communism and communists in our midst. Sometimes in the 1930s they would hoist their red flag on May Day at some inaccessible place, like the highest poplar tree, or the Church belltower, or even at the municipal building. Then we, as kids, would watch how the authorities would struggle to remove it. The Communist party was outlawed and the hoisting of that flag was an act of defiance. There was this case where someone had carved, rather cut into the stem of a poplar tree the three mysterious letters CCCR (USSR) and as the tree was growing from year to year, the cuts were emerging as more and more visible and readable. As kids we stared at these letters while some better informed would explain to us their meaning. Somehow, I had been always negatively conditioned to view communism. I did not relate communism to my personal misfortune, not being able to go to school. I did not see in communism a solution for my personal problems. At this time I would have accepted help from any source.

I had never asked myself until recently why didn't I join the Communist party. If there ever was a case for anybody to turn to communism on account of poverty, and on account of an unjust social order, my case would be classified as a perfect fit. This question sprang in my mind about a year or less ago when I started drafting my notes for this memoire. I myself was surprized by the answers which came to my mind. Communism was supposed to appeal to the downtrodden masses, it was for the poor, for those who had nothing to loose but their chains. Society had always been divided into two classes: the class of the exploited and the class of the exploiters. These two classes were in permanent struggle. Now, I was experiencing on my back this division of society to privileged and underprivileged, and there was no question where I belonged. Why didn't I join the Party and fight for the overthrow of the order which had destined me to drag myself, perhaps for life, after sheep and cows, to live in perpetual poverty?

All of a sudden, seeking an answer to these questions, for my own satisfaction, and reviewing all circumstances already described, I came to the firm conclusion that I, in fact, did not belong to the social class of the communists, that my social class was the class of the pariah, a subclass, sort of an Indian cast where I was destined to live without prospects of ever getting out of it. My conclusions were based on the knowledge of the social and economic status of the Communists in my village. These were the only communists whom I knew, and socially and economically I did not belong to their class. The most prominent communists in the village were the family of Mitiu Kraevsky. Mitiu had graduated from high school and had studied at the University in Sofia. They were the wealthiest farmers in the village. Their house was a palace, if compared to our hovel. As it turned out later, during the war, Mitiu became Mayor of the village. When the Communists came to power in 1944, he continued to serve as Mayor, showing that all that time before, he was a communist. I will have to say more about him later.

And so were the cases of the other village communists. Georgi Parlapansky, the leader of the partisans, brother of Drazha Parlapanska and my aunt, was a village shoemaker, lived in the biggest house in the village, at the most prominent place, and was not exactly a poor man. Another one, Koliu Merazchiisky, a classmate of mine, had come from a well to do family. It turned out that much later evidence had been found that he had been a secret agent of the police and had actively participated in the searches for partisans, thus loosing all his power and privileges. He came to me, when I was arrested, to lecture me that they had given us a little freedom, so we could show our horns. And so was the case of others – the Pekhlivanov's, of our neighborhood. Our neighbor, Stahia, was known under the nickname "Komunata". He had been in America, had returned with some money and had built himself a big house. As it turned out his son, Koliu, was to become my bitterest enemy and has caused me much of my miseries. So were the Mincho Kurtitsata (the mole) whose father was a court clerk in Brezovo, and a number of others which names I now do not remember. But most of those who went to study in Plovdiv, leaving me behind, returned as communists. They all were from well to do families.

My father was walking around with raggedy clothes, I was most of the time without shoes, not to repeat other aspects of my life. Sometimes we sat around the table to eat without bread, sometimes rushing to harvest the barley, still green, to get our first bread. We depended on the chickens laying a few eggs, to take them to the local convenience store to buy the most essential things. I graduated from the Middle School as valedictorian of the class, but for two years I could not go to study when they all left me behind. I ran after sheep and cows of the neighborhood, sometimes hiding my miseries in shame and humiliation. They were walking in the village as Counts in their school uniforms. Did I belong to their class? No! They were all communists, except for a few of them with whom I retained close friendship.

This was the communism which I learned about in my own village. I did not belong to their class. It is interesting to study the social class of all those who joined the party and later the partisan movement. Most of them, in fact all of them, were from the upper middle class in the Bulgarian conditions. What did I have in common with them? Nothing! I substituted for my formal education free reading. I read day and night. When there was no gas lamp I would hang a vigil light over my bed and read from dusk to dawn. I read in the fields, behind the sheep and in the forests behind the cows. And all the time I dreamed of a world of social justice. How many times I secretly cried in helplessness of my personal destiny? So, it was all that consciousness that I belonged to a subclass in our society, which I could not break away from, that kept me out of the Communist party and kept the communists away from me. I was constantly dreaming of a new social order and to a some extent had come to think that it could be achieved by way of cooperativism, not by way of communism. Communism seemed to me as another version of the existing order, replace one set of exploiters with another one – not a social order where the working men would be in charge.

I did not need to study communism in books and papers. It was in front of me. Most of those who had gone to high school, returned to the village as communists. And they all had come from families of considerable means, in any event far better off than my family. They lived in houses which were the top of the line in the village. I was living in a hole. They were all going to schools and universities, while I was shivering around sheep, cattle, in cold, in rains and winds, in snow and mud, and all sorts of indignities. They ate better than I did with my popara, and petmez, and ushmer, with my rageddy pants, side by side with my father who was even worse dressed than I was. We were shlurping the lobuda soup. It was not a life. The communists did not know and did not experience this life. My young friends were from a privileged class. I was left with my vague ideas and dreams of a better world, but I did not see it being prepared by the communists. This is why, when everything is said, I did not become a communist.

There was more that kept me at a distance from Communist commitments. I was a nationalist in my own way. Not that I had been indoctrinated in the school on nationalism as a political philosophy opposed to communism. I had grown up, from early childhood, with much concern on national issues. For me this nationalism was an integral part of my personal philosophy for justice. I had been impressed by speeches at the celebrations in front of the war memorial in my village. I had been deeply convinced that Bulgaria had been administered a terrible blow of injustice, where fellow Bulgarians had been left under foreign oppression, and as human being deprived of elementary human rights. I had not heard of any nationalist organizations, least of all of nazis and fascists, as exponents of nationalism. It was my sense of human justice to which all men were entitled, that dominated my thinking. I firmly believed that all nations, all people were entitled to such justice. But in our street-corner lectures nationalist themes were never discussed, and I did not bring them up. There was much discussion about nationalism in general, but nationalism was attributed to the capitalist bourgeoisie and was regularly scoffed at. All bourgeoise governments were equally reprehensible and no excuses were admitted for Bulgarian nationalism, because it was said to be the same as Serbian or Greek nationalism. This did not sit well enough with me, but I let it go. I did not engage is disputing such views. But I was very much with the group, when questions of colonialism and imperialism were brought up and England and France were outrightly condemned.

Looking to the broader world, I was not quite impressed with English and French democratic systems when so many millions were held in colonial servitude. In fact they were the English and the French who had dealt the blows of injustice to Bulgaria. Of course, this was the time when Germany was emerging as a challenge to England and France, and especially to Soviet communism. Nothing much was known to me, or could I have known, about Hitler and Germany at that time. Our newspapers, in our controled press, presented the issues favorably to Germany. In 1938 Germany swallowed Austria, but Austria was a state of Germans. The question of Sudetenland soon occupied prominent place in the news, but it was presented as an issue of justified German claim where 3.5 million Germans were held by the Czechs. The issues of naziism and fascism were debated, the communists were ferociously against fascism, but on the scales of universal justice, the Germans seemed to be justified, where I was concerned. And when the Nazi-Soviet pact was announced on August 23, 1939, I thought that that was the wisest international event. Still I did not particularly care for naziism, or fascism, or communism. For me the matter was one of justice, and the justice was on the side of the Germans. A Russian-German understanding, in my view at the time, was the best solution of all problems, or would have led to such solution. As to Bulgaria, in the rapidly growing international crisis, I was seeing the best policy to follow would be a policy of neutrality.

In the meantime, I was fully involved in my father's farm work – spring, summer, fall and winter. Life was going on as usual. But not without some bitter frustrations for me. The most humiliating days for me were the times when my classmates would return for vacations or semester breaks to the village and I had to meet them. Sometimes it happened that they would be coming home from the bus station when I would be leading my sheep flock back home from the fields. It was painful for me to look at my friends of before, now in their smart uniforms, while I, ashamed of my raggedy appearance, was trying to cover the big holes on my pants at the knees. I would try to use my bread bag, or some coat or something, to cover myself, which not always was successful. My embarrassment would lead me to depression, and not once I would go home, retire in some corner and cry out of desperation. When I would put something more decent and joined them for a walk through the village, I still looked and felt out of place, sometimes barefooted, while they were walking in their shining black shoes. They respected me for what I have been, but I felt out of place. They were all urging me to go to school, but I knew that it was out of my reach. I felt humiliated, I felt lost in the world, and I did not know what to do. Whatever pride I had from the past, was evaporating with every day that passed.

I could not help thinking what I once was, and what I had become. Worse: what was going to become of me? I must be fair. My friends of the past, who had gone to school did not turn their back on me. They did not snub me. They were as friendly as they had been before. For them I was the same young man with extraordinary gifts, and they often urged me to do what I could to re-enter school. Some of them, coming from wealthy families, had sought me to tutor them before they went to study. They made it through 7th grade by the skin of their teeth. One of them was Luka Georgiev, whose grand father had been a village *chorbagjia*, village elder at the service of the Turkish oppressors. I tutored him. I did not ask and they did not offer to pay me. Neither did I expect it. It was out of friendship. But he did not last long in High School. I believe he repeated one or two classes. What happened after that I have forgotten. It is only in 1991 that I learned of his tragic fate. He had become an alcoholic and had been hit by a truck and killed in Rozovets. Poor guy! He was such a good soul! I did not take people on account of their wealth or social position, least of all of their intellectual abilities. They were good friends or strangers. With Luka, we were just good friends, he was not the only one who failed in school. Others did also fail to graduate, but the world was big and they found their place in other alleys.

Two years passed after I graduated from the middle school. Socially, intellectually and politically I was wandering aimlessly, but always focused on reading and dreaming for a better world. But in the spring and summer of 1939 things started moving.

18. "Lazy Bum."

By the spring of 1939 I had come to the conclusion that I had to seek some way out of my predicament, that I had to take things in my own hands and cast away this resignation and submission to destiny. I could not rely on my father or my family to find this way out for me. Not because they would not desire it, but the poor people did not know how to help me. I was given plenty of advice from well-wishers what to do. It all came down to suggestions to leave the village and seek a job in the city. I was ready to take this road. But where was I going to start? I have never been in Plovdiv. Where was I going to go? Fortunately, someone came out to help me. I do not remember the name of the man. He was in Zelenikovo and was visiting there from time to time. He was some twenty or so years older than I was. He offered to house me and find me a job if I went to Plovdiv. He was working for some sort of an agricultural institute or a veterinary service, with his place near Maritsa and our bus station. I accepted his invitation and held to it as a drowning man to a straw. A week before St. George's day I arrived in Plovdiv and settled with him. A few days passed but nothing happened. I was loosing hope. One day he was visited by a young man from my village. I overheard their conversation. The young man was screaming at my friend: "Why do you keep this freeloader here? Throw him out!" My friend was trying to persuade his guest that, after all, we have to help each other, and that if nothing turned out in the next few days, he would let me go back to the village. I was shocked by the protestations of this young man. He was related to the village Tax Collector, the same man who had refused to give me a certificate of poverty in 1937. I do not know if there was some bad blood between the two families, mine and his, but my attempt to secure a job in Plovdiv, was hanging on a thread.

It turned out that the next day a job was found for me. The owner of a shop selling coffee beans, ground if requested by his customers, needed an assistant and had agreed to take me. I immediately informed my family to send me a *cherga* to move to the home of this coffee merchant where I was going to live. The next morning the bundle arrived and I was taken to my new quarters. The employer pointed to me a hallway, narrow, some five six feet, and long some 20 feet. On one side was a toilet, the rest was used as a storage room, but I was not going to be too demanding. I sighed with relief that at last I was going to have a job. After bringing my sleeping arrangements to the house, I was to report to the store and begin work. He put me in a corner to observe the operations. Customers would come and place their orders for fresh ground coffee. I noticed that he always had a bag on the scale ready to go. The customers asked for pure coffee, and he assured them that he had just ground it, pure and fresh. It was not pure. I noticed that he was mixing it with peanuts or some other grains. The name of the store was "Mocha", my favorite coffee today.

It just happened that this afternoon was the eve of St. George's Day, a big holiday in Bulgaria. About six p.m. some military units led by music started marching outside, I suppose, they were rehearsing for the parade the next day. All present in the store, including the owner, went out to watch. I ventured to go near the door to see too, but in a second he turned to me and shouted: "Back to your place!" Well, the day was over. Back in his house he gave me something to eat, which I do not remember what was, but managed. Then I spread my old *cherga* on the cement floor, and, exhausted, soon fell asleep. In the morning I was awakened by his movements around me, washed my face and he called me to assist him. He had already baked one big pan – six-by-six-by-two pan of coffee, still steaming, gave me a long stirrer and showed me how to stir the production back and forth so it does not burn. I started, I continued perhaps for half an hour, not knowing when to stop.

The coffee seemed to be sufficiently cold, and I stopped. He flew from somewhere, furious at me, and shouted with the top of his lungs: "Why did you stop, you lazy bum.

You will destroy my coffee." I was shocked! I resented being called a lazy bum. His words were *"Murzelanska zodia."* I did not expect that. I did not appreciate it. I began doubting the sense of the entire affair with me seeking a job in Plovdiv. My mind was set to go to school. If my job in the city was to be so humiliating, and if I was to be treated like garbage, "Thanks, but no thanks."

When the whole procedure was over and finished he told me that I could go and watch the parade. I did! It was such a display of the Bulgarian military might, there were long speeches, and my patriotic sense was somehow uplifted. All the time, however, I was rethinking my position, my job, the insults of this coffee merchant, and started asking myself if it was worthed to continue. I immediately decided that I could not take all that. I rather be in the village than subjected to such indignities. So, I returned to the house, quietly rolled my *cherga*, made a bundle of it, and headed for the bus station. There were some friends there, high school students. I told them what had happened. They did not judge me, but they most forcefully urged me to make a final effort to arrange for studies, to go to high school. It was with such thoughts and such determination that I left Plovdiv. I saw that a servant to some idiot businessman I could not be. At the same time I was firmly determined that I could no longer run after sheep and cows. Something else had to come. The entire episode in Plovdiv opened my eyes. I felt that in a few days I had grown up sky high, that I had to take my destiny in my own hands, that Zelenikovo was a past story for me. I could not expect anything from my father and my family – not that they would not allow me to quit the village, but the poor people did not know how to help me. When I unexpectedly appeared at the door in my home, with the *cherga* in a bundle under my arm, my parents understood what had happened, without asking any questions. They did not reprimand me. The next morning everything was going as if nothing had happened. My life was returning to the tracks which I had tried to escape. But suddenly it all changed.

III

THE CHALLANGES OF THEOLOGY, WAR AND POLITICS

"There is nothing, nothing
certain but the nothingness
of all that is comprehensible
to us, and the grandeur of
something incomprehensible,
but more important."

Leo Tolstoy. Prince Andrey
Bolkonsky at Austerlitz,
War and Peace, Modern
Library, XIX, pp. 267-268.

1. Reluctant Seminarian

After returning from Plovdiv, by the end of May, I was appointed assistant to the bookkeeper of the Consumer's Cooperative. This was the time of harvesting the roses and they needed extra help. My service was over after a month, a month and a half, but through this brief tenure in the office of the cooperative, I came in touch with many people and apparently impressed them with my work and my potentials.

Sometimes in August our village was visited by Georgi Govedarov, Chairman of the Foreign Relations Committee of the National Assembly. I believe it was a period preceeding national elections. He was a candidate for membership in the National Assembly. He had asked those present at the meeting what he could do for the village. One of the leading political figures in the village, Koicho Manevsky (the "professor"), prominent in the Social Democratic Party in the Plovdiv District, has asked him to help a gifted young man in the village, who was being wasted taking care of sheep and cows. Govedarov had reacted at once with his readiness to help. Since at that time it was known that scholarship for study could be provided only by the church, for studies in the seminary, and he was told to help there, he had pulled one of his business cards, wrote there a short note to the Rector of the seminary, asking him to take care of me and had given it to Koicho. It did not take long until this news reached me, with exaggerated prospects, but it was enough for me, to respond immediately with my wish to follow up.

In a few days, I and Koicho were on our way to Plovdiv and visited the Rector, Bishop Flavian. He was given the card of Govedarov, flipped it between his fingers, and advised that I should apply for the already scheduled competition tests. For forty seats there were over 150 candidates. I composed in a hurry an application and submitted it to the office. I was told that in order to qualify for scholarship, I had to submit a certificate for poverty. Again the same story. I knew that I would not be given a scholarship on the basis of such certificate, but submitted it. The time for the tests came. I had to sing a song, so they could verify my musical abilities. I did well. I sang a song well known in Bulgaria: "Tell me, tell me, little white cloud" etc. There was a test on mathematical skills, which I had no difficulties, and the test in Bulgarian language. For the last test they read us a short story "King Trajan had goat's ears". We had to write the same story in our own way.

A week later, the first days of September, we had to go for the results. As we were walking down the Djumaya street, Koicho turned to me with some skepticism. "Well, you may be accepted in the school, but the war just started and perhaps in a few weeks the schools will be closed, and then who knows what will happen." I was not discouraged. I told him that this was my last chance to go to school and I will not miss it, if I am accepted with a scholarship. Then, let things take their course. As it turned out, I was accepted, without scholarship, the war went on and on and I graduated on July 29, 1945. But let me go back to the day and time when they announced the winners of the two full and four half scholarships, amounting to one or half-annual fees. My name was not mentioned. After that they immediately started reading the names of those who were being admitted. I was the first to be mentioned. What good was it going to do to me? I could not take advantage of it anyway. The money, some 8000 leva a year, was too much for me. My father could not afford it. I felt that my last chance to go to school was fast disappearing in front of my eyes. I could not restrain myself and broke down crying. Tears were running down my cheeks.

The scholarships went to two students – one from my village, Tsaniu Arabadjiysky, and the other to the son of the chief priest in one of the Yambol churches. Tsanko was the son of a small businessman, who collected eggs from the peasants and then resold them to dealers in the city. They were not exactly poor, in fact they were ten times better than we were, but how they got the certificate, I would not know. As it turned out his father was a secret communist, and as it came out later, during the war he was serving the partisans of the Communists resistance as a courier. After the Communists came to power, they became big in the party. Tsanko never stepped in the church thereafter. He has never shown to the annual meetings of my classmates. One of them called him "Chudak" (an odd fellow). The others with half scholarships also eventually turned communists. The Yambol boy, Yanko, was not exactly a poor kid, but his father, a prominent priest, had a strong pull somewhere in the Church circles. But I was disqualified and had to fetch for myself.

As they were still announcing those who were being accepted, a young handsome man, with a chunk of hair falling over his eyes, was negotiating his way through the crowd and was asking: "Who is Spas Raikin?" I stepped towards him and identified myself. He grabbed my hand to congratulate me for the brilliant success on my tests. He could not help noticing the tears in my eyes. He asked me what was the matter. I told him that even though I was admitted and had performed well on my tests, I could not come to school because without scholarship I could not afford it. He started encouraging me. "You try to pay part of the fees now, and latter we will see what could be done." That was Dimitar pop Tomov. Last year he was still alive, but the Alzheimer's had taken its toll on him and I was told it was pointless to try to see him. But I was encouraged. Perhaps not all was lost, may be after I signed in, some way could be found to subsidize my studies. It very soon was to prove to be a vain hope, but for the time being I decided to follow his advice.

I returned home and told my father what had happened. His face dropped. I was crying. He was looking down and thinking. I told him of the encouragement I got from PopTomov. We still had one thousand levs. While I was working for the Cooperative Office, I somehow had persuaded him to borrow 1000 levs. For some reason he left it with me. He looked at me and told me that I could use that money and he would try to borrow from relatives another one thousand, and then… "We shall see." I did not insist on going to school. I had made up my mind that it was beyond the means of my father. But he did not let me down. And there was the advice and the promise of Pop Tomov. On September 14, 1939, I entered the Seminary.

One thing is sure: I did not go to the Seminary because I wanted to become a priest. It was generally assumed that when one enters a Seminary he is preparing to serve as a clergyman. I entered the Seminary, because I had no other choice to get an education. Because of my age – going on 17 before the fall semester was over – I was disqualified for a secular high school. The Seminary was not too squeamish about it. For me it was either the Seminary, or nothing. It was important for me to start, and then come what may. Besides, having engaged in what may be called self-education, the problem of religion clashed openly with my understanding of the world.

I had delved in science so much, and I had read a lot about Greek philosophy, that the religious, that is the biblical explanations, seemed to me to be an out and out mythology. Very soon, after starting our studies, I found myself arguing with some students on the subject. These young kids believed that since they were in the Seminary, they had to take the fundamentalist view of the Creation, and I openly challenged them. It looked like I was pushing some sort of atheism, and nobody realized that better than myself. I was, however, aware that in view of my being in the Seminary as my last resort, I was searching for some sort of a modus vivendi, some understanding of the Bible which would square off with my personal predicaments. A liberal interpretation of the texts was not far from my search for truth. In fact, I have never stopped searching for a truth which would insure me against accusations in atheism.

When I entered the Seminary, questioning the Bible was a dangerous expression of opinions. Fundamentalism was assumed to be the way to go. Liberalism in its interpretation was admitted but as a last choice. That is where the few elements of faith which were still surviving in me found a refuge. My faith in God was not touched, but my faith in the Bible was shaken. In fact all my life, in the years and decades to come, I have searched for a solid fundament of my religion. One year, after discussing the Bible as a literary and religious document of the ancient world in my class of History of Civilization, a student came to me, after my lecture, and asked me if I believed in God. I had to assure her that I was attending Liturgy every Sunday, which was true, and sure, I believed in God. I am not sure if I satisfied her curiosity. My lecture had led her to believe that I was a non-believer. As I was struggling with these difficult issues in the course of decades, plowing through history, science, religion and philosophy, I fortified my beliefs in God, and religion became a fortress of my convictions. When I retired, I conceived the idea of writing a cathechism for the intellectual where I would have stated my arguments for my religious beliefs. I did begin writing and I composed an introduction, but I never had time to enter a systematic exposition of my Orthodox faith, to work out the doctrines and the liturgical practices in a manageable and understandable way for an inquiring intellectual mind. Still, in my *Introduction*, I argued my answers to the questions which had plagued me as an young seminarian with considerations derived from philosophy. However, in those days I saw these answers in general outlines, which helped me, indeed, to free myself from the doubts which troubled me. Still, I did not decide to move for ordination, in one form or another, but did not rule it out for some future date. But I became a firm believer and defender of religion whenever such questions were raised.

So, I escaped the pitfalls of atheism, overcame that crisis in my intellectual growing up – not for opportunistic reasons and as a hypocritical cover up, but as a unshakable inner conviction in the truth of the religious beliefs. Having become an outspoken advocate of religion in those days, 1944-1945, I automatically qualified for the status of reactionary and anti-communist. Slowly, and methodically, I reached the limits where science led me and could not go further. From there I followed the paths of philosophy, and when I exhausted its ways, I passed to religion. Very early I came to the conclusion that science, philosophy and religion followed different methods in establishing their truth – science and philosophy relied on experiment and reason, while religion was based on faith. Where experiment and reason, logic, reach the limits of their possibilities, there religion takes over. From here I deduced the conclusion that when science and philosophy reach their limits and still attack religion, they end up in self-destruction, enter the realm of nothingness – to use the terminology of Jean Paul Sartre. And where religion assumes the role of science and philosophy, that is the methods of experiment and logic, it also exposes itself to be clobbered to death by science and philosophy. When scientists enter the field of religion to destroy it by their methods of experiment and logic, they are beating in empty space. When theologians enter the field of science, to argue their doctrines with science and philosophy, they expose themselves to certain death. So, my conclusion was that religion could not be proven by science and philosophy, nor could it be refuted by science and philosophy. Science and philosophy are the kingdom of experiment and logic – religion is the kingdom of faith. So, the Bible is not a textbook to teach science and philosophy, but a book of faith, where the ultimate judge is the individual with his personal mental inner vision of the world. Nobody can read any single mind as to what kind of faith is stored there, and how it interacts with reason and logic. Religion uses metaphors to translate its concepts in terms which could be communicated to others, but not grasped in their roots in the individual minds, where they are stored in the domain of faith, inaccessible and unexplainable in material form. Having reached such conclusions, I had no difficulty to refuting the arguments of my opponents on the atheistic or rationalistic side.

Recently I intervened in a public discussion on this subject by writing a letter to our local newspaper, Pocono Record. In the midst of a national debate over teaching of evolutionism and creationism in America, I argued:

August 21, 1999.

Sir:

The continuing debate between creationists and evolutionists, in the nation and on the pages of your paper, is based on a fundamental error on both sides: they fail to distinguish between the sources of truth of science and religion. Science derives its truth by experiment and logical reasoning, while religion is based on faith. In this respect science has the better part of the argument. The claims of theologians that science also relies on faith are a misapplication of half truths. Science may not have all the facts, as Francis Bacon in the 17th century argued, but it does not need all the facts to make valid conclusions. Isaac Newton formulated the law of gravity, which in his time was not verifiable if it was operative on the moon, but after Allan Shephard landed there some twenty years ago, the validity of this law could not be disputed. There is no way for religion to prove the Immaculate Conception or the Resurrection, except to accept it by faith alone.

When theologians and ministers of God invade the field of Science as they question the concept of evolution, they expose themselves to be clobbered to death. Likewise, when scientists invade the field of religion to disprove truths based on faith, they are chasing the winds. Bertrand Russell, quoted by Robert Boyd (Poconc Record, 8/21) that "we started somewhere, we do not know where; we are here, we don't know why; we are going to some great oblivion, we know not whither," quite appropriately answer their quests. Religion and science are not juxtaposed in parallel, to be reconciled (this was the mistake of Thomas Aquinas); they are a continuum where science is inexorably advancing in the field of religion, and religion is endlessly retreating to infinity. As science will never reach infinity, there always will be room for religion – now and in the future. Science will never reach infinity on account of its physical limitations imposed by the structure of the brain, and religion will never disprove science because material facts are outside of the territory of faith.

History, Archeology if you like, confirm the continuum relationship between science and religion. There was the time when religion was man's science, man's knowledge of the world. After the span of six thousand years of civilization, with all the advances of science into the field of religion, the problem of understanding the world in scientific terms is not far from where it started, and religion still holds its own. But there is no question that in our times it is loosing its ground, because fundamentalists substitute religious truths derived by faith for scientific truths based on experiment and logical reasoning. This is where the theology of creationism leads, and where science is winning the millenial war in the human mind.

David Hume, a Scottish philosopher of the 18th century, argued that our knowledge is nothing but a bundle of sensations which we do not know what they are, we could not prove or disprove the reality of their existence, and consequently end up in what is known as the doctrine of agnosticism. This philosophy leaves to science to study the sensations, called by Emanuel Kant, a German philosopher of Irish ancestry of the same century, PHENOMENA, and left the door wide open for religion to fill the other end of the continuum by faith alone, which Kant called NOUMENA. So, in the world of phenomena, science, from there evolutionism, is a king. Then, let science teach evolutionism based on experiment and logical reasoning, and let religion teach creationism – each in its own temple – science in the school, and religion in the church. They are irrelevant and mutually exclusive if taught in the same place.

Their sources of truth being different, Science is at home with its terminology, but religion, in order to express its ideas, has to fall back to material metaphors (exampli gracia: "Our Father who are in Heaven"). The very term "creationism" refers to physical processes, where God, a spiritual reality, engages in activities incompatible with His nature, which will be unexplainable, unless it is understood in metaphorical sense. There is a Latin formula – ex nihilo nihil – which, if applied here will mean that God, as nothingness in material sense, will create nothing in that sense. Kant would have called the opposite of that paralogism – beyond logic. We recite the Nicaean in Creed every Sunday: "I believe in one God, Maker of Heaven and Earth," but what we believe in is a mystery which is beyond experiment and logical reasoning. Religion is a mystery, and if we reduce it to science, it will be a false science and will cease to be religion. Tertullian, one of the early Fathers of the Church (c. 230 A.D.) has said it right: Credo, quia absurdum est. (I believe, because it is absurd). For science creationism is an absurdity; for religion – it is a faith.

Ultimately the conflict is resolved in the mind of every individual. After all Charles Darwin proposed and argued for the theory of evolutionism, but this did not lessen his belief in God and His powers.

This analysis would be incomplete if I do not add a few words about my understanding of the concept of FAITH as a source of valid truths. Faith is a firm inner conviction of the existence of God and His creative powers. But God exists, as a truth rooted in faith in the sense of an immaterial reality, and on account of that cannot be object of scientific study. In this sense the so called ontological proof of the existence of God, as stated by St. Anselm of Canterbury of the 12th century, is inapplicable. Anselm formulated this proof on the assumption that nothing exists in the mind which is not in existence outside of the world. And since the concept of God as power superior to all powers of the world, or as a being more perfect than anything in the world, etc., and such realities are not perceived by the mind from the world of senses, but it is in the mind any way, then His existence should be presupposed as perceived through the mind by faith. It is an open question how this immaterial reality comes into contact with the material instrumentality of the mind. Indeed, Anselm uses logic in order to prove the existence of God, existence of God. But it is a false logic. It is an arbitrary attribution of sense-manufactured concepts - Allmighty God, All-perfect God, et al. - to a concept, the concept of God, which is rooted in faith and is beyond any logic. Thus, rejecting the experimental attainment of knowledge in this case, he settles for the instrumentarium of logic, a philosophical method of rationalization. But this is not a knowledge acquired by faith. Likewise, inapplicable in this case is the doctrine of St. Augustine of the so called "illumination". According to St. Augustine, God from the very beginning of the creation, has illumined the mind of man, as the sun illumines the material objects in the universe. So, some people are illumined by God and they accept this illumination as a relation which thus becomes their faith. They need not look for Him by way of science and philosophy. He is in direct contact with the mind of man where it generates the faith in Him. Those who do not believe are simply excluded from this process of illumination. If this is accepted, then God is a capricious supreme being which illuminates some people and leaves others outside of His benevolence. This doctrine then justifies Calvin's doctrine of predestination – some people are destined for salvation, others for perdition at the time they are born. This leaves no room for freedom of will, where a man by his own will may walk in God's ways, or may negate God, and all that is left is inconsequential for the salvation or perdition of man. Such a view is demeaning for God. One could rather accept Henry Bergson's view of intuition, where man alone penetrates into the mysteries of God, and not charge God with a choice of the "elected" and the "rejected."

This is how I rationalized my religion way back, in very general terms, and in the course of the decades built upon it my believes. Having entered the Seminary, I had to dig in theology and absorb as much as I could from the subjects I was going to study. I did and in no time I was at the top of the class. But in later years, reviewing my experiences as a seminarian, the way the theology and other subjects were taught, I inevitably came to the conclusion that they, the teachers and the administrators, could have offered and implemented a much better program than that which we were offered at that time.

2. Illiterate Theologians

The most important subject of studies during the six years course, naturally, was the Holy Scriptures. It was taught by a Russian emigre, Ivanitsky, a short stocky old man, speaking Bulgarian with a heavy accent, which we soon became accustomed to. We did not have a text book, or if we had, I do not remember it, or as much as I vaguely remember, was well above our level of comprehension. We were not given an overall knowledge about the Bible and biblical history. He kind of followed his own syllabus, read and interpreted selected texts and offered some conclusions which were utterly incomprehensible for a beginner. He was teaching the subject in a way which would be appropriate for graduate students, not for young kids. We were never asked to read as a whole one or another book of the Bible, let alone the entire Bible. It all ended as an incoherent study of dribs and drabs with heavy concentration to prophesies of the coming of Jesus Christ. When the time came to study the New Testament, it was in the spring and summer of 1945, after we returned from a long vacation beginning in October, 1943, on account of the war. So, our study of the New Testament was reduced to virtually nothing.

Our teacher in Psychology, Logic and Philosophy was also a Russian émigré – a priest, Father Mikhail Shishkin. He had been an officer in the Russian White Army who opted for the priesthood. But he was highly educated, competent, well organized and effective instructor on all subjects which he taught. He had an accent, but had mastered the Bulgarian language perfectly. Still, it took us time to get used to it. At one point he pronounced some word incorrectly and I, God forgive me!, exploded laughing. He gave me a stern look and I froze in my place, not that I feared him, but embarrassed. Many years later, when I was going to teach in the American universities, I had to experience such moments when students sometimes reacted to my accent, insignificant as it was. In this respect there are situations which cannot be avoided, no matter how much one tries. Someone had said that a refugee is a man who has lost everything, but his accent. Father Shishkin's mastery of the subject of psychology did not seem to have advanced beyond Aristotle. He, it seems to me now, wanted to use the great classical philosopher as a underpinning of our Theology. His "Psychology" had nothing to do with what the subject of Psychology in our days is as an academic discipline. His teaching of Logic, was the Logic of Aristotle. I was much affected and benefited from his lectures for all my discussions in years to come. His was the "Formal" Logic, but for me it was a system of rules of correct thinking which made an enormous sense. Otherwise, he was a man of impeccable manners. His priestly cassock, sort of cape, with wide, very wide sleeves of the best cloth was fitted to him in a perfect way. When he entered the school, at two minutes before eight o'clock in the morning, that was every day – with his punctuality he apparently imitated Emanuel Kant – he walked like an aristocrat and proceeded straight to the class room. When he was delivering his lecture he stood at the podium like a Roman senator, holding the two corners of the "Cathedra" and alternating his focused eyes at every corner of the class room. His diction in delivering his lectures was artistic. Many years later, there were moments in my lecturing to my students, when I caught myself posturing like him, and delivering my lectures in his style, which pleased me. There was a dignity in his performance and I must confess that he was my favorite teacher. Philosophy, under his penetrating presentations, became my favorite subject in the Seminary. The poor man! I met him on the Djumaya in Plovdiv after September 9 – I could not say when. He was pale, depressed and frightened. He could hardly speak. He had been in the hands of the Soviet military for questioning. He was an old White Guard officer. I never saw him or heard of him again.

Not every teacher was well prepared or well oriented in his subject. We had to study Church-Slavonic language, but our gramar and syntax had much to be desired. The emphasis was on reading the liturgical texts, especially the Psalms. Not much attention was given to translation. Probably it was assumed that since it had much to do with the Bulgarian language, it was understood. But it was not. The archaic grammar of the language was difficult to apply when reading the texts. In fact, almost all the texts which we read in class and in the church during services remained incomprehensible. How one could understand the Troparion for the First Tone *Kameni zapechatanu ot iudei...* translated from the grammatical form of Dativus Passivus, which did not exist in the Bulgarian language? Some of the key hymns of the Divine Liturgy, like *Izhe Heruvimi Taino obrazuiushte* – (We who mystically represent the Cherubims etc...) were beyond our comprehension. I must admit that for the first time I was able to understand what this hymn was all about was when I read it and sang it in English in the United States in the late 1970s. Such also was the case with the entire Liturgy and all the readings. When practicing in Church, we just mechanically sang or read, not understanding the meaning of what we were reading. We graduated from the Seminary as mechanical phonographs. No different was the case with our study of Byzantine Music which was prevalent in our churches in those days. Our teacher, Mircho Bogoev, taught us simply to memorize the tunes. It seems to me he used the class as an audience where he would show off his own skills in twisting the tunes, similar to Arabic singing. He also was impeccable in his suits, but in his teaching it was simply a matter of memorizing, singing and nothing more. There were all sorts of terms there – Irmos, Kondakion, Katavassii, Troparion, etc. They were never explained and I never understood what they were and how things fitted in the order of the worship until I joined our multiethnic Orthodox Church in my town in America, with our English language and our choir music following the tunes of the "Western" composers, the Russians. In the Seminary we were going through the motions, but the deep mystical experience was somehow escaping us. No wonder that religion in Bulgaria, specifically Orthodoxy, has lost its attraction with the masses of the people, and it continues to this day, even more than any other time.

When the times came to master the liturgical services, not practicing them in the church, but comprehending them in class as to their structures, their evolvement and their history, that was in 1945, it all was so condensed that we hardly got to anything of substance. We were going to graduate as illiterate theologians where it mattered most – the liturgical services. All the times we spent attending Vespers, Matins and Divine Liturgies, without a piece of paper in hand to follow, and no participation whatsoever, we just drifted in thought about anything but not with the contents and the spiritual height of the prayers. Sure, we took our turns to practice in singing the responses, one by one, or two by two, but it would come so rare. We were much more concerned not to make an error in our reading Church Slavonic, or singing the Byzantine tunes after the models learned in the classes of Mircho Bogoev, than to grasp the deeper spiritual meaning of it all. It was our mechanical performance which was evaluated, not our personal emotional experience of the mysteries and the mysticism of the rituals. All that was left to be developed in later times, when someone would be ordained and concentrated his efforts in the daily worship. To make things worse, it was a tradition that in the beginning of a course they would select two altar boys, who would serve in the altar throughout the six years course, and familiarize themselves with the service – to be inherited by others after they graduated. From our class they had selected two little angels, both of them known as Ivancho, both short and angelic kids, one from Tanturi, Turnovo District, the other from Tsarevo, or Malko Turnovo, near the corner of the border with Turkey and Black Sea. They became experts in the religious service, but the rest of us – illiterate observers from outside.

No different was the case with our teacher in "Western Music," Mr. Kochetov, another Russian émigré. He was a "Music man," highly competent and experienced in his field. He taught us the theory of the western music, with all the notes in the scores used in the West, unlike the Byzantine music which notes were very similar to Arabic writing, with all little twists and turns, dots, commas, hooks and small notches. He was the director of the Seminary choir for which he selected students from all classes, the choices going to those with better voices and musical ear. He too would select some student conductor to lead the choir for three years, until he would graduate, and then pick a successor. This way the rest of the members of the choir gained no experience in that part of our education. In those years the trend was, in the cities, to replace the Byzantine music with "Western Music," and choir directors were badly needed. But the Seminary was producing very few with some experience. In addition to the Seminary choir every class had a choir, where most of the students participated. But here again a conductor would be selected early in the course and he would direct the class choir until the class would graduate. The rest of the students were deprived of the opportunity to try their hand. As it turned out, when they graduated and found themselves serving in one or another church, they had to master this art from no previous experience. In communist times, those who became priests had to train their own choirs and teach their members without having prior experience of conducting. Of course I had no ambition in that line of liturgical performance. Good as I was with my singing and a member as I was to every choir, the class and the Seminary, it did not attract me. My interests were more intellectual than practical involvement in the worship, be it as a altar boy, be it as a choir conductor. And may be, because of my non-involvement, I lost interest in these activities. Had this been a requirement and had all of us been included in these activities, I might have developed more interest and experience, and, going further, might have aspired for the priesthood with greater readiness than as it happened to be the case. Very few seminarians opted for such service, and my explanation is, that the overwhelming majority were not included in the practical aspects of the regular daily and weekly worship. We got more practice as psalts during our vacations when we were obliged to participate in the service with our peasant old singers and readers, than we were in the Seminary. I came to such conclusions much later, even just now that there was a gap in our specialized preparations for church service. If I am making these critical observations now it is not just to be critical, but in the hope that they may prove helpful for the Church authorities who still subsidize these schools, to re-evaluate their practices, if they have not changed them during the past five decades from my times.

3. The Salgunjiev Affair

My favorite subjects taught in the Seminary were History, Geography, Philosophy and foreign languages. In the beginning we had to choose between French or German. I chose French, "The language of culture," as I thought, only to realize in future times that I needed German more for my special studies in History than the French. I had no idea who was going to teach French. As the first class came, a young very handsome, elegant and humble man, impeccably dressed, with pleasant smile and very shy, entered the room. He was like a pretty dove, with an angelic disposition. Very soon he was tonsured as a monk, grew his beard and conducted the class with poise and attention. In future years he rose to the rank of Arkhimandrite, entered the church administration and I lost sight of him. He was an inspired religious poet. His monastic name was Seraphim. In resent times I became aware that he had written some essays on theological matters, in the realm of spirituality, which are presently very popular in Orthodox circles in the West, translated into English, published and widely distributed. When I was in Sofia, in 1991, one evening while we were packing to leave Sofia the next day for some visit in the provinces, I received a telephone call from a nun, who said that she was taking care of Arkhimandrit

Seraphim and Arkhimandrit Sergei (a class mate at the Theological School). They were sick and infirm. My presence in Sofia had already become known on account of newspaper publications where my intervention in the crisis in the Bulgarian Church had become widely known. The nun pleaded with me to come and see them. I found it impossible, at this late hour and out of fear to use taxis at that time. In a few years Seraphim died. I regretted that I did not take a chance that night to go and see him. Later on I learned that he and Sergei had withdrawn from active service in the church administration, having objected to the introduction of the new calendar in church service in 1968. They just continued to stay in the church, without attempting to defect from it because of its departure from the traditional calendar, which, indeed, was an honorable way to express their objection to official church policies. But I got a solid foundation for further study of French language from Father Seraphim.

It was the second year that my study and interest in French got me in trouble. Father Seraphim's place was taken by an old man, thin like a straw, but with younger looking face, stooping somehow, Mr. Salgandjiev. With this man, I found myself climbing the steep slopes of my Seminary Golgotha, but also learned a lesson which was going to last me for life. I never forgot it, I never talked about it to anybody, and it is for the first time that I am discussing it publicly, after it was mentioned to me, to my unpleasant recollection, by one of my classmates of those days, Father Christo Christov. Salgandjiev apparently was a protégé of some higher up in the church administration, or some political friends. My guess was, and it is, that he may have been involved in the IMRO, the once considered Macedonian terroristic organization. He was probably imposed on the seminary to earn a living in his old age. But he was no teacher in his place. Instead of concentrating on the subject, he was indulging himself in long meaningless stories during class time and from time to time suggest what to read in the textbook. I was exasperated and looked with envy to our classmates who had chosen to study German and were advancing very fast. One of them was Father Christo whom I mentioned. It often happened that students sitting on the back benches while he was going on with his stories were playing cards and after class bragged about it. I was frustrated. I often thought of how difficult it was for my father to find money to send me to school and how this old man was wasting my time. I thought much about the situation and came to the conclusion that something had to be done about it, to terminate this joke of a teacher. But what was to be done? It seemed to me that the honorable way to help the situation was if he resigned. But who could suggest that to him? It was up to him and he would not move. With all my naiveté, I decided to do that by sending him an anonymous letter.

At the end of the first term, the last day before going home, I sat there and wrote that letter. I do not remember how I phrazed it. I do not remember how I found his address. After the Christmas vacation, the very first day, he read that letter to one of his classes. He proceeded to read it in all classes. When he read it in my class I felt as if I was sinking to the bottom of this world. In no time they all figured it out that I was its author. The entire student body was made aware of it. But nobody approached me to talk about it. All of a sudden I was left alone, without friends. They all kept away from me as if I was sick of some plague or something. The week was passing and I had not exchanged even a word with anybody. Only my closest friend, Neophit, from the village of Vruv, District of Vidin, talked to me, but the question of the letter was never touched. During breaks between classes everybody was going out for a walk. I stood there, at the window, looked down to the frontyard, depressed to death and wandering what was I going to do. The teachers all knew about it. Nobody came to talk to me. At least if someone had approached me, to reprimand me, to counsel me, to ask for explanations... Not one whisper! The administration kept numb on it. I had to live through this humiliation alone, to burn in my own fire. Never, never in my life, before this incident and after, have a felt so helpless, so depressed, so low in some bottomless pit as I did during that cursed week. I had brought all that upon myself. All my justifications for writing this letter now evaporated. I still felt justified to protest the abuse of my time and sacrifices to be in school, but I started asking myself why I had to take such responsibility to correct the injustice which affected, after all, everybody? Pain, intolerable pain was gripping my soul. Three days after all that, I decided that something was to be done to come out of this predicament. I decided to seek Mr. S. at his home and apologize for my foolishness. I hardly could wait for Saturday or Sunday, when we would be free in the city and flew out of the door. I was almost running to his place, near the Business School. I found the house and knocked at the door. He answered and let me in.

I do not remember how I started my apologies and how I ended. The gist of it was that I was apologizing and begging for forgiveness. All the time he stood there and looked at me with icy hostile stare in his eyes. He made no move to utter a word to put me at ease, to show that he was extending to me his forgiveness for my impudence in writing that letter. I had to drink the bitter cup to the very bottom. When I finished my excuses he looked at me and asked only if any teacher had a hand in it. I assured him that I did that alone, and that only two people know of it – me and him, that it was my own initiative. He led me to the door and said: "Now you go, and let us forget about all that." I left his home feeling that a heavy burden had fallen from my shoulders. Monday morning he started his classes by telling the students that the author of the letter he had read to them the previous week had gone to him and had apologized and that he had forgiven him. This episode ended, but I had to feel its scars for long time after that. Little by little my friends came back to me. Many of those whom I had been helping before in their studies returned to my circle of socializing. Well, the teaching of French did not improve, but some things that had scandalized me before were not repeated. At the end of the year Salgandjiev retired – on his own or otherwise, I never heard about it. I never saw him again. With this also ended the study of French language under the program of the Seminary. What puzzled me at that time, and puzzles me even to this day, is the strange attitude of the administration and the faculty of the school. Why did they let this scandal go without investigating, without calling me on the carpet, why they just stood there and did nothing? My impudence most certainly was subject to penalty and even the worst penalty – expelling me from the school. Did I step on someone's toes with my action? If they expelled me from the Seminary did they fear a public interest and reaction unfavorable for the school? Was any action in this case representing some danger? My regrets for what I did are mixed with this puzzle about the school's attitude. But I was glad that it was over and it had no consequences for me.

I did not feel that I had performed some heroic deed. Neither I wasted any time feeling guilty of some crime. I only felt that my zealousness for knowledge had led me to commit a stupid act and expose myself to dangers which I would be the last one to take such a chance. What was left behind it as an ineffaceable track, was my decision to never again repeat this way of pointing to the faults or the weakness of anyone. If I had to say something to that effect, I had to stand up and speak directly to those concerned, good or bad, not hide behind anonymity, for as long as I was convinced in the validity of my cause. I never forgot this lesson, and when I entered public life, especially with my publications, I never hesitated to place my name over my criticisms of high and mighty or low and vicious opponents. I never mentioned this epizode to anyone. I would not have discussed it here, if it was not mentioned to me by my very good friend Christo and probably I would never mentioned it in the future. But now that it was out, I decided to bring it out. Recently I addressed a letter to the only surviving Seminary teacher of those days, Dr. Konstantin Tsitselkov, at that time teaching German. I asked him if he could tell me why they did not react at the time to my senseless behavior. He never answered me directly, but had told a classmate, Ivan Tinchev, that he does not remember anything. To conclude this regrettable incident in my Seminary days, I want to add that I continued to study French on my own and by the time I graduated I was able to read French books without a dictionary, which I never forgot and still enjoy reading it. The first book I read from end to end was Alfonce Dode's *Lettres de Mon Moulin*, and quite often, when the occasion presents itself, to quote from it.

4. Behind the High Thick Walls

Meanwhile I never let my studies go down. During the second year, when the students considered as leading in grades were falling one by one, after their initial successes and personal connections between parents and teachers were worn out, the ragged peasant boy from Zelenikovo climbed to the top of the class. Following that, until graduation in 1945, I continued on the top semester after semester. At the end of every term the Rector announced the top students of every class, and my name was never missed. So the entire student body and all the faculty had come to recognize me for my academic successes, sharing this distinction with five other students, one from every class. I was anticipating this event with much satisfaction. It was a public recognition of my efforts to do my best. Indeed, there was nothing else by which I could demonstrate my existance and this recognition was a sweet reward for my labors. After all the miseries which I had suffered, and I was still suffering, following me at every turn of my life, I had something to be proud of. This was the only moment in my life which allowed me to look at the rest of the world with a sense of personal dignity and honor. I was absorbing quickly and easily the lessons and digesting fast, and storing in my mind everything that was assigned for reading. Still, I was working hard to overcome all obstacles and occasional hitches. When I would finish my preparations for the next day's classes, I indulged in writing of essays on a variety of subjects – social and historical problems, as well as philosophical issues, which ended nowhere, but I was constantly trying to clarify for myself matters which intrigued me. There were times when I asked myself why was I putting so much effort in everything, especially when I saw some of my labor inapplicable anywhere in my life, as study of classical languages, for instance – Latin and Greek. But even if I did not see justification for that, I pushed and pushed to go ahead. I did not realize at that time what a widening of mind these studies represented. In a class of Latin language I had an incident which proved of invaluable benefits for me. One day the teacher, Mr. Agopov, was quizzing my friend Ivan Borodjiev – we sat on the same desk throughout the course of our studies. Having him stand in front of the class, as was the case in all subjects in all classes, he was asking him to identify the grammatical form, I believe of the pronomenon Huius. He did not know it. I whispered to him, I believe, Pronomen Demonstratif – if this was the case – and he repeated it. Mr. Agopov could not miss my intervention, called me out to stand before the class and assigned a punishment for my impertinence: "For this," he said, "You are going to memorize in a month's time, the entire first paragraph (one page) of the speech of Cicero against Katilina and will recite it in class." Well, I did. I translated it to know what was it about. The day came when I recited it, with pathos, from the "Cathedra." The piece fascinated me. It was a masterpiece of a speech. Much later in life, I came to the realization that in my writings on political subjects, in my polemics with opponents, I was following the style of Cicero – if this is not saying too much, or imagining things. But I did think this way. So, it was not a waste of time.

Some of our teachers were preferred, while others were somehow resented. I was not much impressed by our teacher in homiletics – art of sermon writing and delivering. He was a Deacon, brother of the future patriarch Schismatic after 1992, Pimen, Yoan Nedelchev. He was not any kind of an intellect or impressive speaker, if he was a speaker at all. I never heard him delivering a sermon. Apparently he was appointed there under someone's protection. At that time Pimen was abbot of the Bachkovo Monastery. As a lecturer he was boring and intrusive in a most unexpected time. When lecturing he walked the whole hour between the rows of student desks and we had to turn our heads to follow him all along, until we got tired and then just stared ahead of us. It was in such moments that he would call someone who was not looking at him and ask a question, or to contribute the following word which he was going to say. This of course would catch us most of the time unaware of the question or of his intended word. There would follow a moment of silence and embarrassment. Such instances were unexpected and humiliating for anyone who was singled out. My relations with him were on a very low level.

Besides Homiletics, Nedelchev taught us in the second year of the course the subject of history. But there he was utterly incompetent and was killing the most important subject for me. At the same time, another teacher, Bogomir Bossev, an ordained priest – I will have to say more about him later – was teaching Geography and was fascinating, with the broad knowledge which he showed and the delivery of his lectures. Once I, inappropriately, suggested to him that he should take over the teaching of History. They were friends with Nedelchev, and he had told him of my suggestion. Nedelchev told me bluntly, privately: "So, you do not like my teaching history, do you?" This set him against me throughout the course of my Seminary days. When Homiletics came, he was already set against me. As it happened, there was a competition in class for the top student's place. There was a student from Sliven, I do not remember his name, who was his favorite, and when determining grades, Nedelchev gave him the A's, and to me the B's. It did not help his favorite, but it hurt me – not in my cumulative average, but psychologically. I came to resent him. But time came when my capabilities were to be judged on the basis of my performance as a preacher. My first sermon was to be on the subject of "Anger." I had a good example to describe. One of our teachers-supervisors, Father Gerasim, was physical example of a man in anger, when he was angry. In my sermon I described the angry man so well, that it impressed all students and most of all the rector, Bishop Nikodim, later Metropolitan of Sliven. He went on checking my grades in Homiletics and when seeing my B's, he called Nedelchev on the carpet, and ever after, my grade did not go below A. As to Father Gerasim, I have to add that we were close friends, in a way. He had a following of favorites and I was one of the group but otherwise he was somehow irritable when things did not go right or someone misbehaved. When angered he would turn red like a lobster.

I have to interject here some material of recent origins. After I wrote my recollections of the Seminary life up to this point, I sent my unedited notes to some of my former classmates for their information. I sent them also to Dr. Tsitselkov. I did not hear from any of them, and if my hunch is correct, my observations were not liked. Only one of them responded, but in language which suggested to me that I have offended them.

I answered the criticism of my classmate in a long letter dated 9 September, 1999. I presumed that he would circulate it among the rest of them. I explained that the Seminary was very close to my heart and that if I indulged in criticizing its programs and practices of some fifty years back, I did it out of concern that it should review these programs and practices and introduce the necessary changes in order to better fulfill its mission as a church educational institution, training its students for pastoral duty. I pointed out that out of twenty nine graduates of our class, only three or four had entered the priesthood and that if the Church had to invest its resources to train future physicians, or lawyers, or historians, it was of no help to her, admirable as the success of our classmates outside of the Church was. Someone had to address this issue. I have been involved in Church affairs abroad, I have observed how prepared for church service students of other denominations are, and that if our Church was to meet the challenges of our times, it needed to take a look at its seminaries if they were meeting its expectations. I stressed that our nostalgic memories of old times should not stop us from pointing out the weaknesses of our Seminary education, if there were any. I also indicated that since 1992 I have been very much involved in the problems caused by the schismatic movement on the side of the Church. Even though in exile, I had most vehemently attacked the church leadership for its accommodationist policies with the communist government. The current criticism of that leadership by the schismatics, and by the political authorities of the country, for the stated reasons, is unfair, because they themselves are guilty of the same failure to oppose communism in those days, and they had risen to professional and academic positions by actively supporting the communists, not by opposing them. I further explained that if my articles and interviews are published in the organ of the Bulgarian Socialist Party, formerly the Communist Party, it is not that I have joined them, but because, on account of my position on the church question, the papers of the governing party, the Union of Democratic Forces, would not publish them. I wanted to make my opinions known, and I would accept any avenue open to me. I concluded my letter with the observation that the nasty remarks of the organ of the UDF about the Bulgarian Orthodox Church and its leaders make me suspect that some anti-Orthodox and anti-Bulgarian elements may have risen to power there and from there are directing their campaign against our Church. How my explanations have been received, I do not know. Likewise, I would not retract my criticism and my observations about the Seminary because people who for decades, indeed, had stayed outside the church, are now taking a different view.

But be that as it may! I will continue my observations on the subject the best way I remember them.

Some four weeks or so after I entered the Seminary, having paid only half of the first semester fees, I was called to the office of the Rector, to be told by him to go home, and unless I came up with the payment of the balance, I could not return. So, my Seminary education was to be terminated only a few weeks after I had passed the hurdle of being let in. I was stunned and I was shaken to the bottom of my mind and soul. I had no choice. I walked out of the school and soon saw myself crossing the bridge of Maritsa River and heading for my village, following the Brezovsko shosse road. I had no money for a bus ticket, so I was going to make it walking – 45 kilometers. The news which I brought to my father that evening was not any joy. The money had to be found. He did not step back and advise me to forget about it. That evening he had gone to the village butcher and had sold him a two-year old calf, to be killed for its meat. He was keeping it to replace its old mother. The third day I was in the bus to Plovdiv, utterly depressed and discouraged, but determined to continue in the Seminary. I was readmitted. This was only the beginning of my travails where money was concerned. This story was to be repeated time and again several times. On one occasion after I was sent home for the same reason, my father sold a big wine butt, made for him by a neighbor craftsman in exchange for oak planks good for two butts – one for himself and one for us. When there was nothing to sell, my father begged the relatives for loans. Incidentally, these loans were never repaid. I remember one occasion, when, standing near the *chitalishte*, staring around me to dissipate my pain, I saw him making his way through the crowd, with his physical misery, with head down, going to visit my aunt to beg again for yet another loan. I knew where was he going and how he felt about it. He was sinking in debts and in self-esteem. In his crumpled old fashion hat, and his poor clothes, unshaved – this was the rule in the village in those days for most of the people. They would visit the barber infrequently. He was walking as if he was forcing himself to make every step forward, as if he was approaching his Golgotha. But the money was found and I returned to the Seminary.

So, this is how my studies in the Seminary began and progressed during the first three years. In 1995, when I attended the 50th class reunion in Plovdiv, I was told by my classmate Stefan Minchev, village of Resen, Turnovo District, where all these years he had been a teacher of History, that during my first eviction from the Seminary, they had discussed collecting money to help me, but nothing had come out of it. What would have come when they all were no older than fifteen? But the thought that they were concerned and ready to help, pleased me. On another occasion I went to the diocesan Metropolitan, Kiril of Plovdiv, later Patriarch of Bulgaria, and begged him for assistance. I was given 1,500 levs. In 1943 things changed for the better. The war had entered a critical phase. Food supplies in the country, especially cooking oil, had been almost exhausted, and at that time our family possibilities had improved. Our cows and our buffalo were producing plenty of milk, I was almost daily churning it and producing plenty of butter – the churning was done by hand in a *butalka*, a narrow keg-type of a vessel, a foot and quarter wide at the bottom and three-quarters at the top, four or so feet high, with a stick from bottom to the top five or six feet long. At the bottom it had a round wooden plate, one to two inches narrower than the *butalka* itself, held above the upper opening by hands and lifted and pushed down in rapid succession in the milk. After half an hour or less of continuous operation, the butter would separate from the milk, would be taken and stored for cooking or for sale. It was a tiring job, sometimes I had to climb on a little stool to be able to operate it, but the income from the butter was rewarding. Some former immigrant to America, who had made it somehow, had moved to the village on account of the bombings. He was native of our village, but he had people in Plovdiv who were prepared to pay for our butter good money. If he was making money out of that, I would not hold it against him. The thing is that I was making good money to support me in the Seminary.

The regime in the Seminary was very strict as to discipline and daily routine programs. We would be awaken at 5.30 in the morning, wash, dress, make our beds and go to the class rooms for study period by 6.00 a.m. After one hour we would go to church for matins, then to breakfast, and by 8.00 a.m. we would start classes, until 1.00 p.m. Then we would have lunch in the vast dining room, where the table had already been set, ten men to a table, five men to a *baka*, a pot filled appropriately with food. One of the five would grab the ladle and distribute the food – everybody watching if he is being fair to everybody. As it would turn out some of the men would favor themselves or their friends, with bigger portions, but since everybody would have his turn to serve, it was very rare that any scandal would ensue. Two of the teacher-supervisors were at hand and maintained order. The meal would start with prayer said by the Seminary's chaplain, preceded by the respective hymn sung by the entire student body, and would end with prayer. The most popular dish was the bean soup, especially in lenten seasons. After that we had two hours free time, to walk, to play or read on our own. At 3.30 we would have a study period in the class rooms, where all our books were kept all the time in the drawers of our desks. At 5.00 p.m. we would have half an hour break, then back to study until 6.30, then to church for vespers, supper and following that – free time until 9.00 p.m. At 9.00 we were to be in bed, in a dormitory which sometimes had as many as 80 beds. At a certain time the lights would be turned off, but conversations, would continue for a while. Sometimes jokers would entertain the whole dormitory with their eccentrics, until a teacher-supervisor would show up, ordered all to keep quiet and gradually we would fall asleep. There were three teacher-supervisors, usually low clergy of the monastic orders who took their turns during the week. The food was nothing to talk about. Sometimes it was good and nourishing. This was the daily routine. It was good. Our life was well organized and our time well utilized for concentration on our studies.

The Seminary was enclosed with thick walls, over seven foot high, with no windows or place to be able to see the outside world, except the front door, where a wrought iron gate allowed us to have a glimpse at people passing by, traffic going on and life still going outside, behind the thick walls. The doorman, Bye Methody, was always there as a cerber, never letting anyone out, unless the teacher supervisor had issued a pass. But opposite the school, visible from all rooms of the building and from the school yard, was the Northern side of Nebet Tepe, a steep on our side hill, rocky and brushy, with numerous pathways leading in all directions. During the spring, summer and fall, that side of the hill was the hiding place for young lovers and provided the seminarians with quite an entertainment, often causing them to wolf-whistle and throw some cat calls. Every Saturday, however, we would be let free to go in the city for two hours, and on Sundays for four hours. But there was always the temptation to climb over the walls at evening time, go to town and return late during the night. Sometimes such escapades were secretly arranged with the cooperation of some teacher – supervisors who, one way or the other, would receive some rewards. Sometimes students were caught in the act of escape and punished with taking away the privilege to go out on Saturdays or Sundays. I never ventured to break the law in this respect. Where was I going to go? I never had money for the most essential things, and what use was it going to be this risk for me? And, let me admit, I feared being caught. Well, when the time came for me to make the big jump, I did it in a big way, escaping from "military" service and crossing the border into Greece. More on that later.

Very soon after entering the Seminary we split into small groups of friends – friendships which were going to last throughout the course of our studies, and well beyond

122

that. Warm memories of these friendships still hold our feelings, years, and decades after our Seminary days. My closest friends of those days were Neophit of the village of Vruv, District of Vidin. Later he became a priest, but passed away during my exile years. My other close friend, with whom we sat together, sharing the same desk in the class room for as long as the Seminary was in Plovdiv, was Ivan Borodjiev. He did not return to the Seminary after September 9th, 1944, and for some time I lost track of him. Later I heard that he too had become a priest, had passed away early. Later I heard that he was still alive, but I could never get the full story and what had happened to him. I believe I was still in Bulgaria when I learned, or met him, but to find out that he had joined with the communists. Probably my best friend was Yanko Boev of Assenovgrad. He was from a wealthy family. His father was operating a Wine Factory or something. My impression was that he was half Greek, his mother being a Greek. At home they spoke Greek. Yanko was a man with a golden heart, we sought each other's company. He was a man of class, of manners, of deep religious feelings, and sophistication which attracted me to him. He often received packages from home and when I happened to be in the room where we had our little private cabinets, he would call me to share with him his goodies. I will have to say more about him. My other very close good friend was Nako, Atanas PopIvanov, from Malomirovo, County of Elhovo. Nako was very religious young man, his father was a priest. Our beds in the dormitory were always side by side. In the evening, before going to bed, he would kneel on the floor, rest his elbows on the bed, open the prayer book Chasoslov, read psalms and pray for twenty minutes. It would look as if he was falling into a transe. But some things always change. He did not return to the Seminary when communism came, had gone to secular High School, and then to the School of History of Sofia University. There I met him again, but this was not the same old Nako, engrossed in mysticism and prayer. Nako had become an ardent atheist and communist. I remember that we argued a while when we met, and then never met again.

Soon, after I entered the Seminary, I became aware of the existence of a student philosophical society. Once in a while, they held public discussions on philosophical questions, open to all seminarians. I was fascinated by these exercizes in intellectual confrontations and regularly attended the meetings. More than any other group the young philosophers packed the classrooms where they were held. The young man who led the group was an Atanas Gashtev, from the village of Shiroka Luka in the Rodopi Mountains. From the same village was the Metropolitan of Sofia, Stefan, later Exarch of the Bulgarian Orthodox Church and the most prominent personality in pre-war, war time, and early Communist Bulgaria. Gashtev was a phenomenon in the Seminary. He had already published several essays on philosophical subjects while still a junior seminarian. He was a fanatic follower of a Proffesor in the Sofia University, Dimitar Mikhalchev, Editor of *Philosophical Review*, exponent of the philosophical views of some minor German Philosopher, Remke. I do not remember now anything of the *Ramkeanstvo*, except their interpretation of international events, where Mikhalcheff argued for the inevitability of World War II, which was then in progress. His basis for such views, as I remember, was that Germany, as an industrial nation, deprived of her colonies in 1919, and with world markets closed to her by the colonial powers, had been economically suffocating, and had no way to survive, but to break up the iron ring around her, which was possible only by way of war.

It happened that in the spring of 1943 a group of six leading seminarians from the upper classes were expelled from the school. One of them, their leader, was Gashtev. It was said that the Seminary officials had accused them of subversive leftist politics. Their expulsion was a shock for all of us. I never learned what exactly was their political affiliation but the charge was serious enough. During the summer the Metropolitan of Sofia, Stefan, had intervened and in the fall they were readmitted. After September 9th, Gashtev had entered the clergy as a monk and had become assistant to Stefan of Sofia, under the name of Antony. For a while he had led a group of clergymen of pro-government, pro-communist orientation, and had pushed for reforms in the Church. Nothing came out of it. On some occasion he had been sent abroad on some mission of commercial nature, with considerable amount of money. He never returned to Bulgaria. Many years later I was able to discover that he had attempted to emigrate to the United States, but Metropolitan Andrey had refused to sponsor him and he had gone to Chile in South America. I was to track down some of his philosophical publications in Spanish, which I could not read. I do not know more about him. But this entire affair was going to have important reflections on my life. I will discuss it later. As an afterthought I felt that by denying him admission to the United States, Andrey – he revealed to me Gashtev's attempt to come to America – had deprived the Bulgarian community abroad of a promising public figure. Andrey and Stefan were no friends, to say the least, and Gashtev had paid for this bishop's quarrel.

Besides the philosophical society, there was also an Abstinence Society in the Seminary. While I was under the influence of Boniu and Tanu Dangov in Zelenikovo, I had become a convinced advocate of it, opposing smoking and drinking, though in my times after I graduated from primary school, as a young village man, in times of big holidays, did not exactly follow my principles. In moments of depression, with friends, I tested the blessings of the country life. But in the Seminary I reverted to my earlier convictions, though I never smoked in those times. This was to take hold on me much later as it was with difficulty that I dropped it. The Abstinence movement was controlled by leftist seminarians, and I found no special reason why I should not join in. But in the fall of September, when I joined the Legionair organization in the school – I will discuss that next – I was picked by the Legionairs to stand for the presidency of the society, so the Legion takes control of it, as against the leftists. I was elected, and, as president, I took possession of the typewriter of the society. I do not know if this was my blessing or my curse. Soon I was an expert typist, and up to this moment of my life, I have become an addict to it. Sometimes my wife has fits of exasperation for my continuous clicking and threatens that when I pass away, she would throw the typewriter with me in the grave. She made all my friends aware of this idea, and they all laughed at the joke. Sure, during the past few years I have some difficulties with my hearing. I read somewhere that continuous noise may cause that. But even if it is true, I would not blame my typewriter. It has been my companion for life and my life of the mind. I do not see how I could have afforded this tool in my hands for the following ten years.

So, my "abstinence convictions" were not something that made me a fanatic opponent to drinking. During the two years between schools, 1937-1939, I often found myself in a company with teenage friends, with whom we would visit family celebrations of name days. Birthdays were not celebrated in those days. Name days, the names of the patron saints, were the big thing. Biggest of them all was the day of St. John, January 20th, *Ivanovden*, because there were too many Johns (Ivans) in the village. So, we had to honor many of them and wherever we went, we were welcomed, and had to drink wine from a common bowl. Here little, there little, until our heads began to spin. On one such occasion we had been preparing to entertain the village by performing a play, being the actors. I was assigned the key role. But by 5.00 p.m. I was so drunk, that I reached home with difficulty. My father met me, looked at me and said: "What have you done to yourself?" I could only tell him that that evening I was going to the stage and had to hurry.

He saved the situation. He brought me a big bowl of sauerkraut juice and asked me to drink it. I did take a few big gulps and a storm exploded in my stomach. In another ten minutes I was cleaned of all that was in my digestive tracks, my mind cleared and in one hour I was ready to go. I may still had retained some of the inspirations of the day and the alcohol influence. I performed on the stage with such an ease and with such an artistic spirit, that long after that I was going to be congratulated for my performance. During these two years a small bar near the *chitalishte*, catering to young people, quite often made us enjoy the taste of the Bulgarian *slivovitsa*, with boiled dry beans, with some spices for *meze*, hordeuvers. But when I entered the Seminary, I reverted back to my abstinence. Well, there was not a chance to indulge in such activities, and so, when I was invited to become President of the Abstinence Society, I had no scruples to hold me back.

5. SBNL (*Saiuz na Bulgarskite Natsionalni Legioni* – Union of the Bulgarian National Legions – to be referred hereinafter as SBNL)

Far more important and serious for my age organization in the Seminary was the SBNL. It was a political organization. There was only one such organization in the Seminary, and it was the SBNL. There was no room for communists there, and I could swear that there was no secret communist group in our school. But the Seminary had won its unenviable reputation of being a fortress of the Legion.

The Legion had appeared as an youth organization in the early 1930s, probably under the influence of the then rising national socialist movement in Germany. A number of some conservative nationalist organizations had popped up and had failed in gaining popular support. It was for University students to take up the cause of Bulgarian nationalism and spread their message and organization to the High Schools. The leadership of the Legion was soon taken over by a Law student, Ivan Dochev (still alive as far as I know, in his nineties). Having fled from Bulgaria days before September 9, 1944, he has spent time in Germany, and then in Canada, until, sponsored by me, he settled in the United States and continued as leader of the Bulgarian National Front, the reincarnation of the Legion in exile, founded in Munich, Germany, and then expanded in North America. He is still one of the most controversial figures in the 20th century Bulgarian history. I have been associated with him in late 1950s and early 1960s. I will have to say much more about him in the following volumes of this Memoire.

He is the son of some low level military officer and his nationalism may be traced to his family origins. He and the Legionairs soon emerged as the most fanatic anti-communists and nationalists. In time they organized mass demonstrations every November 27th, the day the infamous Peace Treaty of Neuilly was signed in 1919, to protest the despoiling of Bulgaria by the Allies. It did not take long for them to clash with the communists and it will not be an exaggeration to say, that for the following years, 1931-1944, the Bulgarian High Schools and universities were turned into a battle ground for control of the Bulgarian youth – by the Legionairs or by the communists. It has been reported that over 150,000 Bulgarian youth have passed through the rank and file of the Legionairs. The battle between the two groups was not only along ideological lines, but also in physical confrontations on streets, class rooms and school yards.

My brief membership in the Seminary Legion was to bring me later, in exile, to prominence – Secretary of the Central Committee of the Bulgarian National Front (BNF) – a new name of the Legion abroad, following the communist takeover of the country, and editor of its organ "Borba." I will have much more to say about that later.

After one has said everything about the SBNL, good and bad, all facts considered, it was a poor imitation of the fascist movement in the West. The Legionairs had adopted the ideology, rather semblance of ideology, of the National Socialist Party in Germany, though after the war, attempts were made to deny this connection. At least I, while I was a prominent figure in the Bulgarian National Front in exile, made such efforts, with the secret desire that the large following of the organization abroad might be integrated in the Bulgarian Agrarian movement, to which I had belonged in Bulgaria, and to which I adhere to this day. It was an impossible metamorphosis which I was after, and it was doomed to failure. I found myself out of the movement early in 1960s. Its old leaders were dead-set against the Agrarian movement and my efforts were destined to fail. But nobody could deny to anybody the right to redemption, even if it is conceived in a hopeless cause. In the final analysis the old Legionair traditions prevailed and I had to resign from the BNF.

In the late 1930s and early 1940s the Legionairs tried to attrack prominent national figures, as General Nikola Zhekov, Commander-in-Chief of the Bulgarian armies in World War I, as Hitler had Ludendorf on his side. Zhekov was invited to meet the Fuhrer in Germany where he received considerable amount of money to carry out political activities in Bulgaria, under the banner of the Legion. Ivan Dochev was also received and paid by Hitler, but everything fizzled into nothing. Much hope was put on General Christo Lukov, in late 1930s Defence Minister, dismissed by King Boris out of fear that he might carry out a coup d'etat to impose a dictatorship under his leadership. (Bulgaria was already a dictatorship under King Boris). This too did not work. The civilian followers of Dochev rejected Lukov (He was assassinated on February 13, 1943 in Sofia by an operative of the communist underground) and split: one wing led by Ivan Dochev, the other by Ilia Stanev. But one way or the other, the Legion remained an organization primarily of High School and University students. An attempt of Dochev to stand for election to the parliament had ended in a fiasco.

A number of leading lawyers, after being elected as members of parliament, posed as Legionairs, but never formed a parliamentary group as a legislative tool of the Legion. The organization called for broad economic and social reforms, but all they accomplished was to popularize some well known slogans, as "The lands should belong to those who cultivate it." This slogan, as I was to discover later, was coined by Alexander Stamboliysky. With such demagoguery they successfully lured young men of the low classes, out of reach for the communists, with promises for drastic reforms and changes. There was a vast reservoir of youth who one way or the other were tied to the existing social and political order, but had been made aware that something had to change, in order to preserve their own privileged position by spreading the blessings of life to broader masses of people, especially those who were on the sidelines and could easily be attracted to communism. The Legion seemed to offer them such alternative. There were many young people for whom nationalism was a moral philosophy. Where the communists vented their derogatory abuses on sensitive nationalist issues, the Legionairs seemed to offer an escape into a land of justice. Where young men for one or another reason had fallen under religious influence, and communism had nothing to offer but mockery and blasphemy, the Legionairs seemed to represent the only alternative.

So, the young idealists joined in the Legion in the high schools, strengthened their beliefs there, and, after leaving the schools and the universities, plunging in the waves of life where concerns of another nature overwhelmed them, drifted away from the youthful attractions of the Legionair's movement. Most of the 150,000 or so who passed through the school of the Legion, soon forgot about it. There was no political organization of adult Legionairs to accept them. The Government, in fact, had outlawed all political parties, and had on a number of occasions prohibited the Legion, but this prohibition was not enforced. It was recognized that it was doing a good work in fighting communism in the schools, but this was all that it was allowed to do. In the schools they battled the young communists, who eventually would join the Party in its underground activities, and later recruited as partisans. But the Legionairs had nowhere to go. However, having been indoctrinated in anti-communism and nationalism, even without an organization after reaching adulthood, many of them would keep that faith for life.

After the communists came to power in 1944, they, the Legionairs, were singled out as the most dangerous enemy. As insignificant a part of Bulgarian society the Legionairs were before September 9, 1944, after that date, the communists, by singling them out for most severe persecutions, built up their reputation. The more they were persecuted and attacked as anti communists, so much more the Legionairs emerged as the avantguard in the struggle against communism. Communist writers published books on SBNL, constructed some systematic philosophy and ideology of the Legionair movement which few people were aware of, and even fewer followed it, let alone to continue expounding it after the communist takeover. When I read these books much later, in exile, I pondered the gullibility of these writers, and was not sure whether they believed their assumptions or not, whether they were making a mountain out of a mole hill, or they believed that the moon was made of green cheese. They in effect were turning a garden scarecrow into a giant monster. Who knew that the Legionair movement in Bulgaria had been such an orderly system of ideological concepts, some sort of a major philosophy? Actually, the Legionairs prior to the communist takeover for some ten or fifteen years, were street packs of brawlers. When the communists took over, having not forgotten their street fights with the Legionaisrs, and their ideological anti-Communism, made them the most respectable fighters for freedom, against the oppressive communist regime. Since nobody picked up the banner of anti-communism, the Legionairs, with their established credentials in the past, filled this void on the Bulgarian political stage in exile and monopolized this cause. Where once a Legionair was seen as a fanatic nationalist and anticommunist, otherwise an empty head, after 1944 he came to be considered as a front line soldier against the red slavery.

However, it should be pointed out, and it is the truth, that as much as the Legion was not a serious organization before communist times, it turned out to be a training ground for dedicated nationalist leaders and anti-communist activists in exile. In fact, if it was not for them, nobody could have represented the anti-Communist cause abroad among the exiles as forcefully as they did. The rest of the Bulgarian political groups and individuals were all tainted with some form of past collaboration with the Communist party or government. Some of the Legionairs were recruited to work for the Western intelligence agencies in their war politics, trusting in their Legion's credentials. Those who kept their allegiance to the old Legion were never accepted for such services, except when they were needed as canon fodder to do the dirty work. Some of the leading personalities of the Legion's movement, for personal benefits, did cooperative with these agencies, but were never admitted to titles and positions of honor and public recognition. This was the case with Ivan Dochev, Kalin Koichev and Dr. Georgi Paprikov.

The problem with the Legionairs after the war was that they had been involved and ideologically compromised as anti-semites. Anti-Semitism in Bulgaria had never grown deep roots. Most of the Bulgarians had never really met a Jew – except in the big cities, where there existed small Jewish communities. If there were any Jews in the administration, in the educational institutions and in the professions, they were so few and of high reputation, that no one ever pointed to them as a threat to the nation or the national economy or culture. Anti-Semitic slogans raised by the Legionairs at their meetings, or on some other occasions, sounded somehow as empty phrases and left no lasting traces or attract public attention. It was all taken as parroting of foreign anti-Semitism. The Bulgarian Jewish community, some 50,000 strong, was well integrated in national life that if it was not for the German Nazis, no one would think of singling them out for abuse. If there were any Legionairs engaging in anti-Semitic demonstrations, they were rather isolated cases and of no consequence. Things changed during the war, but, it was the government, not the Legionairs, who devised and implemented its visibly anti-semitic policies.

6. Who Saved the Bulgarian Jews?

As it turned out, when the Jews from all over Europe were being transported to the death camps, their Bulgarian brothers and sisters were saved from the fires of the Holocaust. Government officials, undoubtedly proded by the Germans, had gathered them and loaded them on railroad cars, ready to move out. They waited for order from above. Such an order never came. It would have come from the King himself. It has been said that Metropolitan Kiril of Plovdiv, later Patriarch of the Bulgarian Orthodox Church, told the King that if he ordered the trains to move, he would lie on the tracks and let them go over his body – true or not true. If all this is an indication how much influence the anti-semites in Bulgaria had – and this includes the Legionairs – it is a testimony that it was next to zero.

All through the years, from 1944 to this day, the question as to who saved the Bulgarian Jews, has not ceased to be in the center of our political debates, all parties claiming the credit for this noble act for themselves, while denying it to their opponents. The issue has become a political football. Those of the right give all the credit to the late King Boris, and use it for their own rehabilitation, to cover up their war-time collaboration with the German nazis. Those on the left attribute this credit to the nation as a whole for its resistance to the nazis, arguing that the nation was behind them, except the few "fascists". By claiming the credit for King Boris, the rightist belatedly try to curry favors from the war-time Allies, and from the Jewish community at large, and conveniently interpret this pro-nazi orientation of Bulgaria as imposed on him by Germany, going as far as to insinuate that he was poisoned by the nazis after his last meeting with Hitler, when, all things considered, he apparently died a natural death. The leftists, on the other hand, to deny to their opponents this credit, have never stopped pointing out to his collaboration with the nazis. The historical truth is, that both, the "fascists" and the communists, distort the facts by preaching half truths. In 1943 King Boris was, after all, the man who could have ordered the trains loaded with Bulgarian Jews to leave the country for the extermination camps, and at the same time he was in a position to stop them by not giving the order. Ultimately it was he who had to decide whether the Bulgarian Jews were going to live or die. He did not issue the order that would have taken them to Germany, and the 50,000 were saved. A shrewd and calculating statesman as he was, and under public pressure to save them, where the Bulgarian Orthodox Church and a sizable number of members of the National Assembly took the lead, he saved the Jews, and saved the honor of the nation. But his overall policies, deliberate or forced on him, were pro-nazi, however, they are explained. In this respect both sides, the rightists and the leftists, tell only one side of the story. Who is right and who is wrong in this debate is determined by the values of those who would judge the past. If saving 50,000 Bulgarian Jews is more important than winning the war by Hitler, then King Boris was a hero. If winning the war against the nazis was more important that saving 50,000 lives, then King Boris was a quizling, nazi collaborator.

Until the fall of communism in 1989, this debate was a dormant issue. The rightists argued their case – all for their own political rehabilitation – while the leftists denied them this comfort. But after the fall of communism it all changed. The newly proclaimed Union of Democratic Forces (UDF), emerging as a governing coalition in 1992 and in 1997, to this day, having assumed the political heritage of King Boris and seeking to ingratiate themselves with the Western war-time Allies, tirelessly try to claim his saving the Jews as a credit for themselves and appear very sensitive when this credit is not recognized, or not sufficiently appreciated by Jews and other factors in the international community. In fact, for the good of Bulgaria, they were joined by many leftists in claiming this credit for their country, not necessarily advocating royalist sentiments in the framework of internal politics. When the Holocaust Museum in Washington D.C. was officially dedicated, and heads of states from all over the world were in attendance, the President of Bulgaria at the time, Dr. Zheliu Zhelev, noticing that no credit was given to his country for saving the 50,000 Jews, walked out to protest this injustice. This issue re-emerged on the international scene just recently, in the summer of this year, with a vehemence that shook the political stage in Bulgaria.

Years ago some grateful Bulgarian Jews, and other public figures, had erected a memorial to King Boris, Queen Ioana, the late Metropolitan Stefan of Sofia, and the Vice-President of the National Assembly Dimitar Peshev, for having led the campaign to save our fellow countrymen from the death ovens of the nazis. The memorial corner was named "Bulgarian forest". But somehow a protest campaign against the memorial was instigated by Jews from Macedonia and Greece. They charged that King Boris had allowed trainloads of 11,000 Jews arrested by the Germans in the territories formerly parts of Greek and Yugoslav Macedonia, some of them occupied by Bulgaria after 1941, to be transported through Bulgaria to the concentration camps. The fact that Bulgaria, and the king, had no jurisdiction over these territories, was ignored, but public opinion in Israel was incensed and some unknown individuals went to the "Bulgarian forest" and poured tar over the bust of King Boris. Following this incident, some Israeli agency, caring for such monuments, ordered that the entire memorial be dismantled. Symbolically this meant denial of recognition of the Bulgarian contribution for saving Jewish lives. The furor in Bulgaria reached a crescendo when, in addition to that, it was revealed that leading members of the Bulgarian left, had written letters to the President of Israel also protesting the existence of the memorial as a tribute to King Boris. One of them was the Vice President of the National Assembly, a prominent scholar, formerly President of the Bulgarian Academy of Science, and President of the previous National Assembly, Blagovest Sendov. The governing majority of the National Assembly, the UDF, lost no time to move and strip Sendov from his office as Vice-President. His support of what was termed as anti-Bulgarian campaign, was reprehensible for the UDF parliamentarians and they pulled him down, as the National Assembly in 1943 had stripped Dimitar Peshev from the same office for standing for the Bulgarian Jews. It is not clear who was behind this campaign for removal of the Bulgarian Memorial in Israel, formally attributed to Jews from Bulgarian occupied Greek and Yugoslav Macedonia. But it inferentially placed the Bulgarians in the unenviable position of anti-Semites. The ruling right-wing UDF coalition in Sofia, fervent admirers of King Boris and his regimes, lost in their efforts for rehabilitation – for the dismay of its sponsors – the United States and Western Europe. With this black mark on their face they are no longer good for their own sake for the Allies, except as servants, maids for the powers that are today in the West, begging for mercy, not as proud Bulgarians. This course of events has not benefited the Bulgarian leftists either. They still have to cope with their slavish subservience to the Soviet Union and their miserable policies in pre-war and post war Bulgaria which they still foolishly admire and glorify.

7. My Personal Contacts with Jews

I personally had a few contacts with Jews. My first meeting with a Jew was my "employer" for a few hours, the owner of the "Mocha" coffee business, the man who called me "Lazy bum" and made me drop him as a hot potato. But at that time I did not have the vaguest idea about anti-Semitism or something like that. I just took him for a rude city businessman, which could be anybody, and ran out on him. I interpreted the whole affair as a personal, not as a racial incident. Up to this moment, and for some time to come, I had never heard about anti-semitism, and if I had, apparently, it did not impress me. Sure, I had already read the "Protokols of the Elders of Zion," – I do not remember who had given it to me during my difficult years, 1937-1939 – but the little pamphlet appeared to me to be some sort of a stupid story, apocryphal literature for some international Jewish conspiracy to dominate the world. It never crossed my mind that anybody was taking seriously this trash literature, of this yellow, falling apart brochure. Much later I was to realize that it is being used as a Gospel by the anti-Semites.

My future contacts with Jews proved to be happy occasions and good friends who helped me in my difficult days and for building my future – not for reasons of winning me for some cause espoused by them, but on purely human, personal, friendly or official terms. One such occasion was my meeting with a Jewish doctor, Buko Aladjem, whom I have mentioned before. I have already told this story. The Government was pressuring all city doctors to set aside some days, to go to the villages and see how they could help the peasants. Such doctors would come to the village and the peasants would go to them for consultation. It happened that when such a group arrived in my village, my mother was very sick, I believe, of peritonitis. She was bedridden and could not be moved to the school for examination. I wanted to take a doctor to my home, which they were not, as a practice, going to do. I made my way to a tall man, serious looking, bespectacled and pleaded with him to come to our house. I described the condition of my mother. He half listened to me, but I was surprised when he took my hand and asked me to take him home. I did. He established her condition and advised that we should take her to the State Hospital in Plovdiv where he was practicing. He was going to make the arrangements for admission. When I took her to Plovdiv, he did take care of her and after three weeks, I went to take her back home. She recovered. I was grateful to this man. He was a Jew.

After the war, soon after the Russians had occupied Bulgaria, I had another chance to meet a Jew. For some reason I was traveling by train from Sofia to Plovdiv – may be this was during my early years of study at the University. There were eight people in the compartment. They all engaged in lively discussions on all sorts of subjects and the Jews were soon brought into the conversation. Some of the company expressed anti-Semitic views. One of the group reacted excitedly to the offensive remarks. The exchange went on and on. At a point I, having kept quiet all the time, intervened in support of the Jew. I said: "After all, Jesus Christ was a Jew." The Jew, sitting in the opposite corner, diagonally across me, jumped and extended his hand to shake mine. I did not think this was such a big thing. After all, I had just stated a historical truth, but for him this admission seemed to be a great favor. Much later in my life, in 1950s, I found myself among a number of Jewish men and women in the Ecumenical Institute in Geneva. I was showered with favors. In the miserable condition that I was in – my poor English language, my isolation in this international crowd, my statelessness and lack of prospects of any future, or knowledge where I was going to be the next day, I was taken under the wings of one woman, translator for the Institute during international conferences. For two years I had no better friend than Ilse. She was a German Jew who had grown in England during the war, having immigrated there long before 1939. She was some ten years older than I was, but somehow adopted me as her brother. I was treated the same way by one of the professors, also a Jew, Suzanne de Dietrich. I will have much more to say about them in the next volume of my reminiscences.

Another chance in my life where I had benefited from a Jew was in New York. I had fainted at a party in Far Rockaway. It is a long story why this thing happened, I will discuss it later. My friend, Mr. Galabov, where the party was held, called a neighbor, a doctor, to see me. He had recommended that I should be taken immediately to a hospital. When they offered to pay him, he refused to take money, because it was the Sabbath. And finally, it was a Jew, Dr. Kurt Winer, Chairman of the History Department in Stroudsburg University, who gave me a permanent appointment as a professor, and I made Strousdsurg my home and my career for three decades. So, where I was concerned, I never had any reasons to carry on anti-Semitic biases and feelings, though I have always been critical of events in the Middle East, where the historical developments after the war were not quite creditable to the Jewish people.

So, whatever anti-Semitism was inspiring the Legionairs, it was far, far from representing any effective force to have consequences in Bulgaria. It was nothing more than parroting the Germans, and never impressed the thousands of Legionairs. The leaders exerted themselves, but nobody was listening to them. Those who later were to blow up all that, and seek anti-Semitism in the Lagionair movement, simply forget the general mood of the Bulgarians: lack of xenophobia. No Bulgarian shares, or has shared at any time in the past xenophobic feelings, and this includes anti-Semitism.

8. How, and Why I Became a Legionair

So, it is understandable why the Seminary would become a fortress of the SBNL. The fanatic anti-communism, article number one of the movement, was the firmament over which the security system of those electing to study there were destined to build their mental, social, professional and economic future stability. Communism was their mortal enemy, the anti-communism was their salvation. Where in every other ideological and political system one could find cracks of insecurity, the anti-communism was a pure, unadulterated defense mechanism to rely on. When I first broke the news that I was going to enter the Seminary, my close friend and confidant, student in the Business Gymnasia in Plovdiv, somehow distraught – he was deeply involved with the Communist youth movement – Yovtcho, with a sour expression on his face, commented: "All they (the seminarians) are Legionairs." It is for the first time that I heard this word and asked him: "Who are these Legionairs?" He answered: "Wealthy people's bastards." This was the first lesson I learned in my political education which shocked me. Sure, I did not belong to that social class. I felt that I was being thrown in a den of wolves. These words left a deep impression on me. But very soon I was to find out that the overwhelming majority of the seminarians were not coming from wealthy families, in fact most of them were middle class and poor young kids, so I did not feel so threatened or misplaced in some well defined class of wealthy kids. The qualification Yovtcho gave to the Legionairs was more applicable to the dedication of the members of the Legion to their middle class values than to the class interests of the wealthy. But be that as it may. When I arrived at the Seminary, my mind was firmly set against the Legionairs.

Sure enough, a few weeks after we settled, someone from the upper classes came to us and told us that a speaker was going to address us on some very important matter. We all were invited to come to the meeting after supper. We did. In a few minutes there came a bunch of upper class students, with serious expression on their faces, sort of marching to the front, two of them standing at the right, two at the left of the "cathedra," where their leader occupied the podium. We were told that there was this organization of the SBNL in the Seminary, that most of the seminarians were members. There was a bitter attack on the communists and their atheism, there was much talk about nationalism and injuries Bulgaria had suffered at the hands of the Allies and much praise for Germany and Hitler. Political democracy and English plutocracy were subjected to much scorn and abuse. I was skeptical about everything that was said. Yovcho's words were still resonating in my mind, though I was impressed by frequent references to badly needed social reforms in

Bulgaria. At the end we were invited to join. Some class "leaders" had already been picked and we could make our registration for membership with them. Other meetings were appointed for organizing the class Legion. I thought it over and decided not to join. The "Bastards" word was deeply implanted in my mind. May be seven to ten, may be less, of my classmates did not join. And when the Legionairs of the class held their meetings in the class room, we huddled in corners all over the Seminary and found some more interesting things to do, rather than listen to some pudgy pompous students from the upper classes, inexperienced and boring with their clichés. The Seminary Legion held its meeting once a month in the big hall used for official celebrations. All class legions would be lined up in military formation, half of them at one side lengthwise, others on the opposite. They would exercise until the Seminary Leader with his lieutenants would arrive, saluting by holding his right hand in nazi fashion, marching in front of the line on the left, then on the right, then posing in the middle and screaming: "For Bulgaria!" The Legionairs in stiff military pose and extended Nazi-style hands would respond" "For her we live!" ("Za Bulgaria!", "Za neya niy Zhiveem"). This ritual would be followed by a speech, then marching military style in the school yard. It was all phony, but the seriousness with which it was carried, made one to prefer staying out of sight.

Every Legionair had a badge – a round burgundy glazed circle, surrounded by a golden ring, crossed with a golden arrow broken in two angles. It was the size of a dime. By this badge the young Legionairs recognized their fellow Legionairs from other schools in the city, and when they met on the street, each one would raise his right hand nazi style and salute the other with "For Bulgaria," and "For her we live." The ordinary Legionairs did not have uniforms, but the leaders of the national organization sported brown shirts, ties and britches. The speeches of the leaders were not God knows what. Patriotic exclamations, one-sided reviews of international events, praises for Hitler and Germany, and bitter denunciations of the plutocrats of the colonial powers. There was always something to be said about the "new order" but one could never understand what this new order would be. Communism was the desert for every speech and it was virulently attacked and denounced. Those who listened to the speeches, the ordinary Legionairs, thought that it was all great and applauded with "Hurray" or made it appear that they were impressed. For us who happened to be watching from the side, it was all a charade, aping foreign models, empty rhetorics, nothing of substance, and we often called it "wooden philosophy", while for those enlightened with membership it appeared to be a sacred gospel. We did not consider ourselves less nationalists, or less interested in international events, to add also – less concerned of the problems, social and political, that plagued our society. As far as I was concerned, I stayed out of all this until the fall of 1943.

Things changed in the course of 1943. I have already mentioned the group of Gashtev, some five or six of the top students in the Seminary, who were expelled from the school in the spring. Nobody explained what had happened, what was their infringement of the rule under which the school operated. This happened only one or two weeks before the summer recess. Rumor had it that the Legionairs had discovered them to be a part of some political, leftist conspiracy. I never learned the true reason. But I believed what was said, that the Legionairs had a hand in it. Needless to say, this planted the fear of God in all of us, who had stayed out of the Legion. The expulsion of the Gashtev's group froze us to our seats. As soon as we returned for classes in September, the Legionairs started exerting every kind of pressure to have a one hundred percent membership in the Seminary. Somehow I managed to place myself as their target.

There was a Seminary Newspaper, "Svetogled" (Outlook), which the Legionairs had taken control of. They had planned to publish the first issue dedicated exclusively to political articles along their line of thinking, to make it an organ of the SBNL. I could not have known that. I had decided to submit an article on the subject of Dante. I wrote my piece and gave it to our teacher of Literature, Mr. Lilov. Somehow it was known that he had leftist leanings, at least looking unfavorably to the Legionairs. As a teacher of Bulgarian Language and Literature, he was advisor to the editorial board. He had objected to making "Svetogled" a Legionair's paper. He told the Editorial Board that my article had to be published, to have at least a little something that was not political. He had his way and my article was published. This was the first article I published in my life. The leaders of the Legion in the Seminary had been furious at my going behind their back to spoil their plans.

As this was boiling behind the scene, my friend of Assenovgrad Yanko Boev, President of our class Legion, with whom I maintained a close friendship, came to me with some advice: "Why don't you join in," said he, "don't you see that they all are wooden heads? Why do you need such troubles?" All this was coming from Yanko, the class Legion's leader. I had known all the time that he was not taking all this seriously, that he had taken this leadership position reluctantly and because it might have gone to someone else. He did not care about it. I did not doubt his sincerity and his concern for me. He had the best intentions to save me unpleasant consequences. To that he might have heard something that he did not want to tell me. He further told me that they were planning to take over the Abstinence Society, and if I wanted it after joining the Legion, I could have the presidency of it. This offer was virtually meaningless for me, and I would not have sacrificed my independence. I was much more concerned that along the paths of the invisible moves of the leadership of the Legion on a school level, someone might stage some intrigue against me and others and we end like Gashtev's group. The Gashtev's group was reinstated in the Seminary. They were all returned, probably as the result of intervention on their behalf of the Metropolitan of Sofia. I may have already spoken of that. He, Stefan, was of the same village where Gashtev came from, Shiroka Luka. Who knows, may be they were relatives. But if it came to me, I knew nobody in such a position who would intervene on my behalf. Knowing at what cost my father was supporting my education, what humiliations and deprivations he had suffered on account of me, I could not do that. Besides, there were many occasions, as the rumors were going, where Legionairs had physically attacked non-Legionairs. It may be that all my fears had been exaggerated, but I could not overlook all these things. In those days everybody was looking to save his skin and was looking for a cover. How many communists were joining non-communist societies just to hide themselves?

So, I joined the class Legion sometimes in September, 1943. In a couple of weeks I was "elected" President of the Abstinence Society. I already mentioned it that I took possession of the typewriter of the society which became my constant companion for the rest of my life. It was of enormous help for me. Probably the most important thing that happened now was that I was given a few pamphlets outlining the programs of reforms which the Legionairs were committed to. I must confess, some of their ideas appealed to me, especially where questions of reforms were concerned. I was virtually illiterate on party platforms for changing things in the country. It soon appeared to me that together with the silliness that was going on in the organization there were some ideas that had to be looked into. It was the silliness that still bothered me – boring meetings, speeches, marches, parroting the Nazis, which nobody was taking seriously any way, unless one, and there were some, who had taken seriously all of it. My biggest challenge came when I was asked to lecture to one of the lower classes. I had not expected that such assignments would be given to me, but they sure came. My first lecture – well it was also the last – I managed to deliver without going to the ideological or political slogans of the Legionairs. I simply decided to talk on the roots of Bulgarian nationalism. This threw me right in the middle of history, and the subject was as common and as comfortable, as simply one calling himself Bulgarian.

At that time Leader of the Seminary Legion was Atanas Dinchev, from some village near Plovdiv. He had never impressed me with some political ideas, or any profound thought on any subject. Simply, he was the usual pompous sloganeer and there was nothing in him that could inspire me in my new incarnation. In 1995 I met him in Plovdiv, at our class reunion celebrating the 50th anniversary after graduation. He had come to see us. He was an old man, obviously a ruin of the communist times. I would have expected him to have been killed by the communists, but apparently the communists in his village were not as savage as those in Zelenikovo. He had survived. After my escape from Bulgaria my favorite aunt, Plina, had heard an explanation that I have been subjected to persecution because I had been a millionaire. She could never understand how I, poverty stricken, could have been a millionaire. So much for the popular knowledge of the Legionairs. But, for good or bad, my joining the Legion branded me for life as a follower of that movement and my life experiences from this point on were to take an entirely different course.

IV.

WAR. UNDER THE BARREL OF
THE SUB-MACHINE GUN

"Blessed are you when men shall
revile you and persecute you, and shall
say all manner of evil against you
falsely for my sake. Rejoice and be
exceedingly glad, for great is your reward
in Heaven: for so persecuted they the
prophets which were before you."
St. Math. 5:11-12

1. The War in Our Backyard

I entered the Seminary at a time when the war started – a few days later. As the war raged in Western Europe, moved to the Balkans and bogged down in the boundless expanses of Russia, we continued with our studies without anybody disturbing us. Sure there were shortages of all kinds which affected us. In the spring semester of 1943 we had to do without bread for two days of the week, replacing it with boiled potatoes, but we never missed a class. It looked like this was how it was going to continue, until sometimes about the end of October, American and British planes bombed Plovdiv. It was not much of a bombing, some of their bombs exploded in the subburbia, while we, frightened by the screeching alarms, huddled in the basement of the building and took advantage of some of the provisions stored there for the Seminary kitchen. The whole thing lasted for three hours, but it was enough to throw the school authorities into a panic. Nobody was prepared to take a chance with our lives, some three hundred young-sters in one single building. One bomb would have sent us to Heaven. Such was also the case with all schools in the country. The next day we were told to pack all that we had and head for home. This was to be a *sine die* vacation.

The events which led to this abrupt end of our education had passed us without much notice, as if nothing had been happening. On March 1, 1941 our government had signed to join the Tripartite Pact of Germany, Italy and Japan. As they were affixing their signa-tures to the papers in Vienna, the German tanks were already roaring over the gravel roads of Bulgaria. The next day they were in Plovdiv and settled in the country as if they were at home. Much later I had to learn that Hitler had ordered this operation of his troops on Bulgarian territory, "with or without the consent of Bulgaria" and that his army poised on the Bulgarian border was upward from 700,000. His Order "Marita" was is-sued sometimes in December 1940. At the time his armies were being deployed to attack Russia, but the failure of the Italians in Albania against the Greeks, threatened his entire right flank. Greece could have become a staging area for the British armed forces while he would be engaged in Russia. That necessitated his crossing through Bulgaria. It would have been a heroic stand if the Bulgarians had refused to comply, but it would have been in vain, and would have resulted in the destruction of the country. Mighty France had gone down in two weeks, so, little Bulgaria was going to stop the German machine? The British-American attempts to persuade the Bulgarians to resist, in the light of the strate-gic and military facts on the ground, often presented as a noble alternative for the Bulgar-ians, were meaningless in those days. Besides, what moral grounds were there for the Bulgarians to sacrifice their country and their people to defend Greek and British inter-ests, and French interests to that, when from the capitulation of the Bulgarian army in September 1918, to these days, Bulgaria was treated as a criminal and robbed of her lands and wealth? There was not one single argument that could have any validity to convince the Bulgarians to stand up to the Germans, except that if the Allies won the war they would have dealt with Bulgaria again as a criminal. But the end of the war was far from being visible on the horizon, and at this time everything pointed to a German vic-tory. In retrospect, even if the Bulgarians had acted on such assumptions, it would have not affected their fate at all. They still would have fallen under Soviet domination, as did Poland, Czechoslovakia, Romania and Yugoslavia – their protégés in Eastern Europe in the inter-war period. As it turned out, the policies of the Bulgarian government in those times proved to be the best of the bad choices presented to them, though I was skeptical of the whole thing in those days. My preferences was for Bulgaria to have declared neutrality and stay out of the war, keeping in mind her previous interventions on one or another side.

The Germans settled in Bulgaria as our guests, not as our occupiers. They were admitted in the country with our consent and it is fair to say that to the very end they behaved exemplary as guests. There was nothing that could have turned the people against them, except for political considerations held by the supporters of the Allies. They moved around the country until the summer of 1944, not bothering anybody, minding their own business. We became used to them. The German soldiers who were stationed near the entrance to our building in the Seminary guarding the entrance to the auditorium, which they had transformed into depot for military ammunition, soon became a permanent fixture. We were passing day and night by them going to our chores and they, with stony faces, behaved as if we did not exist. They never communicated with anybody and nobody ever approached them.

I had developed an attitude in those days, or rather I viewed the German cause in the war, as justifiable. As our press presented the facts, all actions of the Germans preceding the war seemed morally right. Much was made by the crushing defeat which the French had suffered in the plebiscite for Saar in 1935. The occupation of Austria in 1938 was treated as a unification of one part of Germany to its main body. As I was to learn later, in 1918 the Austrians had voted to join Germany, after the dissolution of the Austro-Hungarian empire, but their vote was vetoed by France and the Allies. The Sudeten question was treated the same way: there was no moral justification to hold three and half million Germans under the rule of small Czechoslovakia. In both cases, in Austria and Sudetenland, as was the case with Rhineland, the Germans seemed to have the better part of the argument. In 1919 they appeared to have been administered a blow of injustice, especially in the light of the Fourteen Wilsonian points. I held firmly these convictions at the times, and it was not because I had joined the Legion. I held these views before the war started. For me it was a matter of moral justice, not a matter of great power interests. This is where I was seeing the parallels between German and Bulgarian interests.

Turning point in these international developments was the conclusion of the Nazi-Soviet Pact of August 23, 1939. It was a shocker because all the time, in the summer of 1939, it had appeared that a war between the two countries was imminent, that the ideological conflict between communism and anti-communism was moving to the front stage in international relations. For me this treaty was a welcome blessing, because it removed the inconvenience to choose between Germany and Russia, between Hitler and Stalin. This treaty was a convenient way for Bulgaria to stay neutral among the powers which mattered most to her. Soon after the signature of this treaty both powers prevailed on Romania, and Bulgaria received Southern Dobrudja, which, incidentally, today remains part of Bulgaria. When I found myself involved in debates with anybody on foreign policy, communists or nationalists, I advocated the position that Bulgaria was too small to take sides between the great powers, and that it was to her advantage to remain neutral; that this was the best policy which she could choose to follow. I held these views to the end of the war. The Bulgarian gains in Thrace, Macedonia and Dobrudja were made during the period of German-Russian cooperation between 1939 and 1941. Where Bulgaria was concerned, this was restoration of her injured justice. Later on I was to come to the conclusion that always when Germany and Russia were allies, Bulgaria was gaining her justice, and anytime they were opposed to each other, no matter which side we were on, we were the loosers. This is what happened in 1878, this is what happened in 1915, this is what happened in 1940-41*. As to the future, as seen in 1941, before Germany was to attack the Soviet Union, there was no crystal ball to predict it. As it turned out, in 1944 Bulgaria switched sides, retained her gains in Dobrudja, on account of the USSR but lost Thrace and Macedonia, more on account of England and the United States, than on account of Russia.

It is with such understanding of the international situation that I returned to Zelenikovo after the bombing of Plovdiv. All high school students were sent home too. We all, of the same generation, were all of a sudden dumped together and there was no way that we could escape the events of the year. Well, things had already changed dramatically after June 22, 1941, when Hitler attacked the Soviet Union. My position that the Nazi-Soviet Pact was a marriage contract concluded in paradise, was destroyed. Until that day the communists were too quiet to attack the Germans, and the anti-communists dropped their accusations of the cruelties perpetrated by Stalin. June 22, 1941, Sunday I believe it was, as I was returning from church service, a neighbor, a communist, met me on the street and inquisitively asked me: "Did you hear the news?" "What news?" I asked him. "Hitler attacked the Soviet Union this morning," he said. So, that's what had happened. Sure, the Pact between Hitler and Stalin was taken half seriously in Bulgaria and all indications were that a showdown was coming, sooner or later, and I was convinced that it was going to end this way, but, somehow, I wished that it did not happen. I felt comfortable with the situation as it was. But now it all changed. I was not surprized, but I was disappointed, or rather, I should say, I felt that the ground under me was splitting. My entire mental set up on the international situation instantly collapsed. Until this turn of events I did not have to choose whom I was going to back, I advocated a policy of neutrality for Bulgaria, but now it was all over. One had to choose: Germany or the Soviet Union. I was not prepared to choose either. Somehow I felt that Germany was acting irresponsibly. At the same time the Soviet Union was a threat to the social, political and ideological order I had grown in, and as uncomfortable as I was with it, there was something ominous coming with communism which I could not verbalize, but I feared. Above all, for some reason, I felt that Germany was going to lose the war with Russia.

Such was the confusion in my mind when I heard the news. I sensed that the man who told me that news was concealing his gloating over what had happened. He knew that I would not be so happy to hear it. Indeed, I found myself in a predicament, which I would not be able to come to terms with for the rest of the war, until Bulgaria was occupied by the Soviet army in September 1944, and then, from there, things were to turn for the worse. I had never felt sympathy for Hitler, though the media at the times were spewing cavalcades of hymns of him as the apostle of anti-communism. He was not my hero, never mind all the anti-communism which he was preaching from Germany. I could not wish him well in this war. Indeed, thinking of that day, June 22, 1941, sixty years later, I reluctantly admit to myself, that the news threw me off balance. I found myself instantly in a state of confusion and a sense of betrayal by history overwhelmed me. I could not see any perspective in the future which could help me to organize current and future events in a manageable optimistic order. Not much time past when news bulletins started arriving for spectacular German victories, hundreds of thousands of Russian armies encircled in giant pincers and captured by the Nazis. The Germans were inexhorably marching to Moscow and visions of its imminent capture did not seem so unrealistic. A friend of

---

* This seems to be the case now, in the year 2000, but the game is not yet over and if things continue as they are, the future of Bulgaria is very dim, indeed.

mine, a classmate, Oleg, son of a Russian émigré, confided in me that the day when Moscow would be captured, the Legioanairs were going to stage mammoth demonstrations in Sofia and Plovdiv. I could say nothing to question his expectations, because the German war machine was rolling to the Kremlin, the seat of Stalin. Yet, I did not feel that it was going to end in such a way. The vast Russian land, in my mind, had seen many invaders, from the East and the West, and had defeated them. As hopeless, as the situation looked like, in September and October, 1941, it was not over as yet. I simply could not say one way or the other what was going to be the final outcome of this giant confrontation between Germany and the Soviet Union, until the surrender of Marshall Paulus in February, 1943, at Stalingrad, when the German armies were turned back. It was in such a state of mind, that I never expressed confidence in a possible German victory in my debates with the young communists in the village.

Things became more complicated when on December 13, 1941, following the Japanese attack on Pearl Harbor, Bulgaria declared war on the United States and England. It was a cold Saturday afternoon. We had free day to go to the city. The streets were virtually deserted from the cold. Suddenly, a small newspaper boy appeared on the streets screaming the big news: "Bulgaria declares war on the United States and England". He was running from one man to another wherever he saw one. I bought my paper from him, "Yug" (South), and read the big news. That morning the Parliament in Sofia had an extraordinary session and had decided to declare war on the two great powers. As it turned out, that war was declared under German pressure, justified by the Tripartite Pact to which Bulgaria had subscribed.

It took me some time to digest this news. I did not see why Bulgaria had to declare this war, and of all things on America and England. Where the Bulgarian armies were going to meet the Americans and the English to fight them? It all was a nonsense. How we, the Bulgarians, were going to match our forces against the Americans and the British? Didn't we learn our lesson from world War One? Why repeat this foolish adventure? This was a war between the great powers. How was Bulgaria going to tip the balance between the Great Powers? How did we allow ourselves the luxury to solve the problems of the Great Powers in the wider world? As to America, I knew nothing much except the admirable Wilsonian fourteen points, which were embedded in our brains as a new Gospel for a better and just world. Well, this Gospel was ignored after the first war, but its principles had received great attention in Bulgaria and if anyone ever spoke of America, it was in the sense of these principles.

So, we were being thrown into a war against the country, which, in a way, had been on our side, on the side of justice. England was another story. She was a colonial empire and at that time colonialism was seen by the Bulgarians as a slave system. This is how I looked at it. There was no room in my personal concept of justice for colonialism, wherever it was, whoever practiced it. To that, England and France had taken over the German colonies, and the consequence of that was they had deprived Germany of access to row materials and markets for her industry, and whatever the sins of the Nazis were – in those days we had never heard of a holocaust and crimes against humanity perpetrated by the Germans – the British and the French, in a way, shared in the responsibility for the war. I believe I have mentioned the debate in Bulgaria, where Prof. Mihalchev – in the Seminary Atanas Gashtev – propounded the idea that the war was inevitable, on account of the fact that, deprived of her colonies, Germany was going to suffocate economically. Ergo? The war was a battle for survival.

So, viewed from Bulgaria, the war was justifiable. Still, this was a war between Great Powers, and Bulgaria had no place there. Why she had to expose herself to become target of any of the warring parties? That I had good reason to resent the political click which was governing Bulgaria, I never had any scrupples about it. For me the regime in Bulgaria was a system of glaring social injustice. I was suffocating under the social conditions in the country, going through so much humiliations and struggles to keep my head above water, at the same time when things seemed to be coming so easy for everybody around me. On that score I was prepared to support the Legionairs – not because I had some sympathy for them, and not because of fear of reprisals alone, but because they attacked the social injustices in Bulgaria. All things considered, this last condition made me feel better for my compromise. It was much easier to attack social injustices in Bulgaria from the anti-communist positions of the Legion, rather than if one stayed out of it. Since the Legionairs were fanatic anti-communists and boldly attacked the social injustices in the country, and since I had joined them, I had a free hand to attack the regime without being taken for a communist. Given these circumstances, while Bulgaria was out of the war, it was easy for me to defend a policy of neutrality, even though the Legionairs, being fanatically pro-nazi, pushed for intervention in the war with military participation on the Eastern front. But now, after the bombing of Plovdiv and my return to Zelenikovo, I was no longer in contact with any Legionairs. For all practical purposes I was out of the Legion and was free not to support, indirectly, the German war. I could, and still vehemently supported the thesis for Bulgarian neutrality. It did not bother me, because I did not have to account to anybody. At the same time, under the cover of the Legionair label, I freely propounded my criticism of the social injustices in the country. My only tie to the Legionair's philosophy was my nationalism, which I was ready to defend in any circumstances.

It is on this subject, on the intervention of Bulgaria in the war, the declaration of war on the United States and England, where I parted ways with the Legionairs. But as I said, I did not have to account to anybody for my views, I felt free to follow my own understanding of things. I did not see why Bulgaria was to be exposed by the Government to dangers which in the past had led her to catastrophe. By defending the position of neutrality for Bulgaria, I was disasociating myself indirectly from Germany. I had not reacted to Germany's crossing through the country in March, 1941, for the simple reason that we could not stop the Germans and that if Bulgaria refused to admit them on her territory, she would have sacrificed herself for the interests of Greeks, Serbs and the British, the same countries which for twenty years since WWI had been crunching on the bones of my country. I could not, in good conscience, condemn the actions of the Bulgarian Government and advocate a policy of national sacrifice for the sake of Greeks, Serbs and the British.

The way it had worked out, was all right with me. No Bulgarian armies were called to fight the Greeks and the Serbs, no Bulgarian army fought the Germans. Their presence in the country was not in Bulgaria's power to prevent. After the Germans attacked the Russians, our territory was used, in a way, against the Soviet Union, but again, we could do nothing to prevent that. Above all, Bulgaria did not declare war on the Soviet Union. The Germans were all over the country. In Sofia they had to pass by the monument of the Russian Tsar – Liberator, they frequented the imposing Cathedral of St. Alexander Nevsky – a memorial for the Russian soldiers fallen for the liberation of Bulgaria – but they never asked the Bulgarian government to destroy these symbols of Russia, or to force the

Bulgarian government to break diplomatic relations with Moscow and send her army to the Eastern Front, as Hungarians and Romanians did. The position of Bulgaria, as far as I was concerned, was uncomfortable, but all things considered, was a choice between the bad and the worst.

But the Bulgarian declaration of war to the United States and England, was another matter. Whether it was demanded by Germany or not, whether there was any chance to avoid it, I do not know. It was presented as a free choice of the Bulgarian government, or an obligation under the Tripartite Pact. The first was possible, the second was not true, though it was presented this way. Under the Tripartite Pact, Bulgaria had assumed obligation to enter the war if any of the three axes powers were attacked by a third power, thus excluding those who were already in the war. As it happened it was Japan who attacked the United States, not vice versa, and it was Germany which declared war on the United States, in solidarity with Japan. It was all contrary to the Tripartite Pact, and where Bulgaria was concerned, she was not under obligation to go to war against England and the United States. But in the whirlwind of the events these little details were ignored and little Bulgaria declared war on the United States and England. At first I took this to be a bad joke, an un-necessary joke on the field of international politics where the big stallions were blasting each other. The Bulgarian proverb for such contingencies goes: "When the Stallions are kicking each other, the asses suffer." This is where Bulgaria placed herself, between the stallions on both sides. For a while it all looked like a joke. Much later our leaders in Bulgaria called it a "symbolical war." But following Stalingrad, February 1943, when the Germans started "shortening"the front, it became increasingly clear that eventually this shortening may reach the frontiers of Bulgaria, and then we had to pay the accounts of our Government again, as it had happened in 1918. There was no doubt in anybody's mind that the dark cloud which was rising in the East, was moving in our direction. When the bombs started falling over Plovdiv, the delusions for "symbolic war" and for a joke evaporated overnight. I packed my things and headed back to Zelenikovo, leaving behind me my brief Legionair career and emerging on the stage of Zelenikovo in an awkward position where politics was concerned.

2. The Political Landscape on the Eve of Communism

The political life in Bulgaria after 1935 was for all practical purposes dead. The personal regimes of King Boris, whatever present day royalists and right wingers may say, were nothing less than dictatorship. The King was the dictator, assisted by a personally selected click of advisors and administrators. All political parties were outlawed and all leading political figures excluded from participation in the making of State policies. The regime came to be known as "Regime without parties" (Bezpartien regime). The King sounded the individuals he wanted to appoint as Prime Ministers, and they were all either from among loyal diplomats and military officers, or experts in the respective fields. If someone attempted to carve a piece in public opinion, he would be dismissed. Two elections were held for National Assembly in the late 1930's, but it was made sure that only convenient political figures would be chosen, even if they were going to be in opposition to the regime. Some of the former political leaders managed to get elected, but the majority of the members were supporters of the King's men. When the Assembly appeared to be getting out of control, the King dissolved it and had the police elect him a National Assembly obedient to his ministers. In a very general way the system produced two kinds of opposition: the right-wingers were led by Prof. Alexander Tsankov, who overthrew Alexander Stamboliysky in 1923, and having suppressed with brutality an attempt of the Agrarians to resist and subsequently an insurrection led by the communists, had won himself the unenviable title "The Bloodsucker." On the other end of the spectrum, a small group of democratic representatives – Agrarians, Social Democrats, Democrats, Zvenars et al. – stood as formal opposition, but were utterly ineffective. When the Government declared war on the United States and England on 13 December, 1941, they were shouted down and no one was allowed to say even a word. To their credit these men stood firmly in defense of the Constitution and of democracy until the fall of the country under communism.

The rational of the politics of King Boris was that he was trying to "protect" the country from adventurers on the right and the communists on the left. Tsankov, on the right, even before the coup of May 19, 1934, executed by the same people who made June 9, 1923, Kimon Georgiev and Damian Velchev, had emerged as leader of a "Social movement," parroting the Italian Fascism. He himself had began his career in politics as Socialist, as Musolini had. This movement, if allowed to grow, and take over the country, would have established a quasi fascist dictatorship. The King behaved as if he was guarding the country from such a regime. If there was going to be a dictatorship, he was going to be the dictator. In later times the regime was to be characterized as monarcho-fascism, but, in fairness to historical truth, it was not fascism in the exact sense of the word. Maybe paternalistic dictatorship would be closer to the truth. In trying to prevent a fascist takeover, Italian or German style, he was successful. It is to his credit that Bulgaria was spared the political extremism of Tsankov or the Legionairs had taken over. In view of the developments that took place in 1941, when the Germans entered the country as "guests," Bulgaria would have fared much worse during and after the war. On the other hand the King's justification for his dictatorship was to save the country from communism. In conditions of political democracy, for sure the Communist party and its allies, some of them bankrupt politicians on the left, considering the social conflicts in the country, would have won parliamentary elections hands down. King Boris did not trust the political regime of the center and on the left to be able to stop communism. In this respect his regimes were a defense wall for the class interests of the Bulgarian bourgeoisie, small land holders and capitalists. Detestable as his dictatorship was, as a dictatorship, for the historical moment it appeared to be playing for time and maintaining the social and political statusquo.

For sure, the King and his ministers were aware that by their system they had created a political vacuum, and they themselves presided over this vacuum. They realized that they needed organized public support for their regime. They tried to fill this vacuum by several initiatives, but they all fizzled. They attempted to organize a "Public Force" – the English translation kills the positive inferences of the terms used in the Bulgarian language, "Obshtestvena Sila," but there is no better word to render the meaning of the Bulgarian terms. This "public force" was to be a government political association, not a political party, to stand behind the government which basic philosophy was reflected in its nationalism and royalism. It did not catch fire. An attempt was made to bribe Ivan Dochev to join this force, and he, an opportunist as he was, accepted. He was appointed President of a new organization, under the name "Agrarian Associations." The intent was to pre-empt the social base of the Bulgarian National Agrarian Union. This attempt was silly, and Dochev only showed his silliness. At the end all efforts to fill the political vacuum in an otherwise sophisticated Bulgarian society, failed.

But having failed with the adults, the proponents of the "Regime without Parties," turned to the Bulgarian youth where Legionairs and communists had already occupied the political landscape and were batting each other. Neither the Legionairs, quasi fascists, nor the communists were of their liking. In this case the communists need not be taken into consideration at all. They were enemies of the Government. But there was much to be gained to try to pull the rug from under the Legion. The government needed nationalists and royalists, and on this score the Legionairs were both – nationalists and royalists. They were otherwise known as street brawlers, immature for serious political work and public service, and were thus excluded from consideration. However they aspired for political status, sort of a fascist type political party. Neither they were prepared to, nor the government expected them to support it. This is how the ruling click under King Boris conceived the idea of creating a youth organization after their model – to be nationalist, to be royalist, to be non-political, and to be a loyal supporter of the regime. Such an organization was created, under the name "Brannik," (Defender), equally opposed to the communist REMS and to the Legion. The staff of the organization was paid by the government. "Brannik" chapters were set up in every High School and in every class. Very few, if any, "Brannik" organizations were established in the villages to attract the young peasant boys. The "Branniks" were supplied with dark-green uniform – smart ironed pants, boy-scout type of jackets and a beret, leather belt with a stripe over the shoulder. Like the Legionairs they saluted each other raising a hand only from the elbow. One could say this was the Bulgarian version of Hitler Yugend. But unlike the Legionairs, they had no philosophy and no ideology whatsoever, except for their nationalism and their royalism. They were exact replica of the political profile of the personal regime. This is why they were organized: to be the social force behind the government. The organization was open to all young people, regardless of their social background. It was understood that membership in the organization was open to every young man if he so desired. The young people, boys and girls, were free to join and free to stay out, but membership automatically entitled the recruits to privileges and preferences for advancement in careers and education beyond high school. It is understandable why a great number of the youth flocked to the organization. Parents of the upper classes, state and public officials, and the young men and women themselves, those who had only careers on their mind, but no political convictions, saw in the "Brannik" an opportunity for advancement and privileges, in addition to the status going with the uniforms, which gave them an air of importance.

The Legionairs and the communists were a political opposition, from the right and from the left, while the Branniks were supporters of the regime. But since it was a government organization, opened to all wishing to join, very soon it became a shelter for subversive elements to secure themselves a cover for their clandestine activities. I was quite amused when I saw many of my communist opponents in Brannik uniforms. The leaders of the organization were happy that they could show a large number of members in their chapters. It justified their salaries and privileges. In my village, my favorite teacher

Peter Dimitrov, was appointed as organizer of Brannik. I never suspected him of any political orientation. My bigger surprise was when on September 9th he emerged as leader of "Zveno" group in the local Fatherland Front. All this held no attraction for me. It represented the existing social and political order. In this order I was a pariah. Excluding the communists, I saw myself closer and closer to the Legionairs, though in my heart I was a politically independent man. The political set up in Zelenikovo, in the absence of an organized Legion, gave me a free hand to battle both, Branniks and communists, under the cover of a presumed membership in the Legion. This protected me from being seen as a communist, or as being a blind supporter of the regime. Later I realized what a dangerous position I had put myself into, when the communists came to power. Fortunately, I managed to make some moves which were going to save me. For this, later.

Side by side with the Legionairs and the Branniks, there appeared also another kind of an organization – both for the adults and for the youth: the "Ratnik", RNB, with their slogan: "Fight for the advancement of the Bulgarian Cause" ("Ratnitsi za napreduka na bulgarshtinata" (RNB)). All over the country one could see graffity with their initials on the walls. The Ratniks were extremists along all lines that were followed by the Legionairs. They claimed to be Hitler's favorite group in Bulgaria and their emphasis was much more on anti-Semitism, with a sprinkling of slogans for reforms along the lines of fascism and naziism. President of the organization was a Veterinary doctor, Prof. Assen Kantardjiev, and the Minister of Interrior, Gabrovsky, was supposed to be their man. Since Gabrovsky was the top police man – all internal security agencies were under his control – one could surmise that the Ratniks were a police sponsored organization. Their presence in the country and among the young, however, was more on the walls with their graffity scrawls in the Bulgarian political zoological garden. I heard rumors that the village priest and Koicho Manevsky were Ratniks, but I was never sure of that. They never talked to me about such a movement. I could surmise that they both, being my friends, would have attempted to recruit me.

A third person in my village, who, as rumor had it, was a Ratnik – "Gocheto", as we called him – Georgi Nachev Pekhlivanov. We grew up together as kids and classmates in the Middle School. Unlike me, he had gone to study in the Business High School, and was two years ahead of me. He was in Dimitrina's class in the elementary school, and was somehow my rival in the Middle School when the two classes were merged into one under Angelieva. I soon overpowered him. But we remained close friends and spent many evenings together, discussing all sorts of issues. Somehow I noticed his left wing leanings, though he skillfully hid behind generalities. He went to the Army in 1942 and there they sent him to the School for Reserve Officers. Coming to the village for his furlows, he cut a smart figure with his lieutenant's uniform and long sword hanging on his side. It did not intimidate me and I made nothing of it. It just was not my style. I did not care a bit about the military. It would have been the last thing I would have desired for myself, so, to be jealous of him? No! We continued to be friends, but I always felt that he was somehow duplicitas. At some point I heard the rumor that he was a Ratnik. In my estimation he was more of a crypto-communist than a Ratnik. I took the liberty to talk about my doubts to Koicho Manevsky, who also was rumored to be a Ratnik. I told Koicho of my suspicion that he was a communist. Koicho was somehow intrigued but said nothing. After September 9th, it came out that Georgi had been working for long time with the communists, but had taken cover under the Ratniks. Only a few days before September 9th, he was stationed in Rozovets with a military batalion in the school. The

Partisans had attacked it, he and the man under his command resisted them, but apparently he was in collusion with them. Though the incident ended with no results, nobody was killed, there was something suspicious in it all – until Georgi Pekhlivanov emerged on September 9th with a communist label. There was to be a brief confrontation between me and him, when I had fallen on my face in December 1944.

Other than those things, I had never read or discussed anything with anybody on the Ratnik's affairs. It was only recently that I read somewhere that Assen Kantardjiev, the presumed leader of the Ratniks, had died somewhere in the United States. He never, never appeared anywhere in the exile world as a political activist on behalf of the movement with which he was associated. This makes me think that the whole Ratnik affair was a put up job, some sort of a trap set up by the police. In any event, the Ratniks were an insignificant distraction in the Bulgarian political world, especially where the youth was concerned.

Much more serious was the challenge of communism among the Bulgarian youth, organized in the REMS – RMS (*Rabotnicheski Mladezhki Saius* – Workers Youth Union). It was the youth organization of the Communist Party. Like all other parties, it was outlawed, but unlike all of them, it never ceased to maintain its structure – its Central Committee, its Regional, District and local committees and its youth organization. It was all done clandestinely. Its leaders were in Moscow – Georgi Dimitrov, Vassil Kolarov, Vulko Chervenkov et al. Its leaders in Bulgaria during the war ended in prison – Traicho Kostov, Anton Yugov, Dobri Terpeshev, Tsola Dragoicheva et al. Virulent anti-fascists and anti-nazis, their objective was the bolshevization of Bulgaria – a copy of the Soviet Union with its system of total dictatorship, total socialist economy and total imposition of the Marxist philosophy on all aspects of cultural life. They adapted their tactics according to the changing conditions in society, working hard to entangle non-Communist political leaders and ordinary citizens, with demagogic espousal of democratic ideas, and then turning them into instruments for the attainment of their ultimate political objectives.

In 1942 Georgi Dimitrov from Moscow launched his "Fatherland Front Declaration," which, if interpreted literally, was nothing more or less than a political commitment for establishing a democratic government in Bulgaria. But it could not deceive anybody. Anybody who understood what the objectives of the Communist Party were, knew that the Declaration was like a spider's web, where the flies would land for a rest and then would never be able to free themselves until they are devoured. One of those who had fallen in the spider's web, duped by the communists in 1944 and participated in the toppling the government of the Agrarian and Democratic Parties, was Nikola Petkov. In 1945 Dimitar Gichev, leader of the moderate, majority Agrarian party, at a conference where his faction and that of Petkov were reunited in opposition to the communists, told the delegates of the Petkov group: "In 1944 we saw the Fatherland Front to be what you saw it in 1945." In 1947 Petkov was hanged by the communists. No reputable politicians had responded to Dimitrov's appeals to join the Fatherland Front, except Petkov, represented in exile with the Allies by Dr. G. M. Dimitrov, and Kimon Georgiev and his group – Velchev, Kazassov, Venelin Ganev, Petko Stoyanov, Tsviatko Bobchevsky – all members of the gang which had toppled Stamboliysky in 1923. They all, one by one, stood in opposition to the communists at one time or another, were thrown out of the Fatherland Front, and saw their hopes for democracy in Bulgaria under communist tutelage wilt and die.

In this context of the political landscape of Bulgaria, the RMS – the Youth branch of the Communist party – played a key role in shaping the political future of the country. It supplied the party with a large reservoire of idealists as sacrificial lambs for its struggles to undermine the regimes of King Boris. It was an illegal organization of committed young men ready to jump in the fire when called to duty. They were the one who filled the ranks of the partisan movement. It was not difficult to recruit them from among high school students and village youths. The way things were in Bulgaria, the prospects for making careers after school, were very limited; the social conditions in the country clamored for change and reform, and the policies of the ruling government were leading nowhere. The big promise, the Soviet model, was too attractive, to be ignored. The yearning for a new world of social justice found a satisfaction in the illusions perpetrated by little known apostles of communism. Unsuspecting village young men were easily tripped in the web of these speculations by socialization in personal friendships. The choice was simplified: it was rich against poor, capitalists against workers and peasants. Fascism and Nazism were for the "wealthy bastards." Once caught in this web, the young men acquired status by being "somebody" in their small youth groups. Nothing could take them away from the influence of communism. The old society had nothing to offer them. It is a fallacy to claim, that the Party and the RMS were insignificant force in the country in those days. They were the only force organized against the regimes of King Boris.

The RMS members, guided by the Party, used every means to hide behind some fronts, always seeking some innocent and neutral associations as a cover for their subversive activities. I already mentioned the Youth Cooperative Groups, where I was an active member once. In the village of Rozovets they had found a cover in a Christian Youth Society. It was organized by my former teacher in the Seminary Bogomir Bossev, who had been fired from the Seminary for his connections to the Gashtev's group and was made a parish priest in Rozovets. All Remsists in Rozovets were members of this Christian Society. After September 9 Bossev shed his cassocks and became Director of the newly opened High School in Brezovo. After a few years he was thrown out of there after some girl students had been impregnated by him. In later years I have seen information that he had authored pamphlets or books propounding the doctrines of atheism. The Christian Youth Society, after him, came out as a remsist organization. Other organizations where pure idealism attracted young people were all sorts of Abstinence Society, Vegetarian Society, Esperanto Clubs, et al. Through these groups the Party and the REMS recruited their rank and file members. Where Zelenikovo was concerned, though among the young returnees from high school there were some following one or another line, there were no organized groups on the basis of their special interests, which meant that we all were dumped together – Legionairs, Branniks, Remsists, crypto-Ratniks and independents.

3. Yovcho[*]

So, when we, the returnees from the city, found ourselves in the village, we socialized, we debated all kinds of issues, some of us defending leftist positions, others rightist positions, nobody called his opponents communists or fascists publicly, but privately

---

[*] In my D.S. dossiers one of the agent reporting on me has used this name as a pseudonym. I have a pretty good idea who he is.

suspected who was what. All of us, taken together, were the future intelligentsia of the village. We were all fresh from school classes, still too young to split in hostile camps and managed to hang together, in spite of our differences along ideological lines. We talked about philosophy, history, politics, religion, social issues, science and all that interests a young inquiring mind. Publicly we were all independent; privately we were either pro-communists or anti-communists. Naturally, I found myself defending positions which were opposed by the leftists. How could I defend religion, being a seminarian, without being opposed by the marxists? How could I defend nationalism on historical grounds when the leftists saw it as fascism? How could I talk philosophy and raise questions which remained unanswered by the Marxists, when I was trying to lay foundations for my own views? How could I take independent stand on political issues where the leftists had taken positions opposite to my views? I had only one backer, and it was my neighbor Todor, son of an invalid from the war, who was receiving a State pension, and, apparently his son thought that it was mandatory for him to be on the rightist side. I did not have such reasons, but I had ideological presuppositions, which put me there. Todor was a nationalist and had joined the Legion long before me. Yet, they looked at him more benevolently, because, after all, he was not a seminarian, and the seminarians and the Church were viewed as the arch-enemies of communism. I had joined the Legion only weeks before returning to Zelenikovo, but how could I explain myself? It was my views that dug the trench between me and them, not that I was in the Legion, though the Legion thing made things worse. I was attending the church services. I had to, because my student grade book had to be signed by the priest, testifying that I had regularly attended Church service. It did not matter for the leftists. They saw in it a political statement. Sure, I did not have any more contacts with the Legionairs, but my views on so many issues coincided with their views, and all that made me enemy number one for the fanatics in communist circles. But, in spite of all this, I had managed to win the friendship of one of those among them, who, otherwise, was better educated and with broader views – Yovcho Denev Filchev.

Yovcho was a free spirit young man. He was a Remsist, in the Business school in Plovdiv, but was open minded. His communism did not prevent him from friendly relations with anybody. As a matter of fact, he had lost the confidence of the other young communists in the village, because he was not stranger to drinking once in a while and when he was drunk he would spill quite a few beans which they did not want to reveal. So they kept him at arms length. We became close friends, we freely discussed politics, the war, society and all that bothered us. We often had opposing views, but we still remained good friends. At the same time, he was still on good terms with the other side. I remember one of our political discussions on the peasant question. By now I had read some pamphlets published by the Legionnaires, and had been impressed of their criticism of the existing order. It was a masterful combination of clichés and slogans, but somehow they appealed to me. I had them digested quite well. Somehow I was distrustful of the real intention of the Legionnaires, I was still convinced that they were defenders of the old order of things, and that what they were writing did not sound convincing for me of their intentions. Still I had absorbed their criticism of the existing order of things, the need for reforms, the defense of the peasants, etc. When I was discussing with Yovcho my understanding of the peasant situation, he listened attentively and at the end interrupted me: "You are an Agrarian (Zemedelets)" he said. This was the first time when I was called Agrarian. Sure, I had heard this word before, but I was absolutely ignorant of the Agrarian movement in Bulgaria. Much later I was to remember that in the room which I had inherited from my older brother, above the bed, there were two photographs plastered on the wall, without frame, probably 10" by 14", which I was to recognize as those of Al. Stamboliysky and Petko D. Petkov. So my older brother had been an Agrarian. But he never discussed politics with me. I was ignorant of the Agrarian movement because of the strict censorship. It was the forbidden fruit in the political garden of Bulgaria in those days. So, this same Yovcho, who had warned me in 1939 that the Legionairs were "Rich man's bastards," now recognized in me the Agrarian. So, it was by way of the Legionairs pamphlets that I had been exposed to an ideology which I could accept on the peasant question. Perhaps it was this initial introduction to the Agrarian ideas, that had helped my easy transition from a Legionair to an Agrarian after September 9, 1944. So, I learned Agrarianism by way of Legionair's literature.

Never fully integrated in the Legionair's world, but infected by its Agrarian platform, somehow I began looking at Agrarianism as the answer to all my questions. I returned to my pre-seminary days when I was all excited about cooperativism, and started thinking along these lines. In my mind the phrase "The land should belong to those who cultivate it" had taken deep roots. I detested communism, which was to take the land and all that made up the peasant life, and dump it in the Kolkhose. I was emotionally involved with every fiber of my being in the landscape of the peasant life: the Tenyov Gyol, the Baidachka, the Giul Dere, the animals we had – cows, calfs, sheep, even chickens – each and every one having its name, responding to me when I called them, following me, or I following them in the fields, making sure I grazed them the best way, assisting them to have births, suffering with them when they were somehow in bad shape. All that was imbeded in my soul, even though I had been away, in the Seminary, for several years. How could I accept an order which would take away all that from me, to look at them all as milk, or wool, or meat producing objects? For those who have not gone through this kind of life, my thinking is a silly romanticism, but for me it was a real life.

So, I could not digest communism. My school friends on the leftist side had not really spent time on their farms, they had gone from Middle school to High School. With me, it was different. I had spent two years on the farm, and it had become my second nature. The other system, capitalism, which allowed the land owner to come to the threshing field, wait for the grain to be separated from the shaft, and carry away half of the labor of my father, did not sit well with me either. It was a glaring social injustice, and I was continuously searching for a way out of it. Communism promised to take it all out of our hands and transform us into general laborers supervised at every step by party officials, to deprive us of the deep psychological delights which we experienced as free men – in misery, but free men. This freedom and independence was paid for with much hard labor and pain, but the feelings of being on your own over all that you could steal from this life of misery in poverty, was worth preserving. Later on when communism did come, and turned the entire village into one feudal estate run by party officials, this truth was manifested in its ugliest face.

Going back to my cooperativism, pointed out to me by Yovcho as Agrarianism, even if I learned about it in the Legion, where I first came across the principle "The land to those who cultivate it," put me in opposition both to communism and ruthless capitalism. So, there was another way out of this predicament – not the communist and not the capitalist. The way to come to this new model was through legislative reforms. Thus, under the influence of Yovcho I was slowly moving away from the mind set of a Legionair.

We came closer and closer to each other, which was not taken too well by the communists.

In line with this thinking, I should mention that in those years, in 1939 or 1940, on the political stage in Bulgaria appeared Ivan Bagrianov, Minister of Agriculture in the Government of Bogdan Filov, architect of the personal regime of King Boris. Bagrianov delivered a series of sharply critical speeches in the National Assembly, subjecting the existing social order as a cruel betrayal of the peasants. I was tremendously impressed by these speeches. They were a shocking revelation of the misery and the injustices suffered by the peasants, exactly as I had been seeing the state of affairs in Bulgaria. Older Agrarians in the village were likewise full of praise of his courage. Well, he was kicked out of the Government and replaced by some other non-entity. Subsequently, he was sometimes suspected of having been a Legionair, as well as an crypto-Agrarian. He was the answer to my prayers, but he did not last long. He was called back as Prime Minister in June 1944, to prepare Bulgaria for getting out of the war, but failed – not because of his failure to act. He did act, but the political and international set up had simply overtaken him. It was too much, or too little and too late to change the course of Bulgarian and international events. For his attempt to prevent the rise of communism to power, he was executed in the beginning of February 1945 by the pro-soviet regime established in the country under the bayonettes of the Soviet army.

So, there I was: anti-communist by conviction, Agrarian in my heart, and Legionair in public eyes.

4. Between Scylla and Haribdis

Thus, I found myself between two poles – the legionairs and the communists – with the Branniks on the side. Communism was advocating revolution from the left, the Legionairs were pushing for reforms from the right. The Branniks were for the status quo. All this was taking place while the old political parties were fading away. So, I stood between the two alternatives – the communists and the Legionairs. The Legionairs were patriots, nationalists, reformers and defenders of Church and religion. The communists were dead set against nationalism, for radical social and economic changes not to my liking and were outright atheists. The Legionairs and the communists were looking to the future, each one in their own way, the Branniks were looking for the present and the past. The Legionairs and the Branniks were looking for German victory, the communists – for a Soviet victory. I was not convinced that Germany was going to win the war, but I was not thrilled by a Soviet victory either. By 1943 I was convinced that Germany was loosing the war, but somehow this conviction did not affect my general attitudes on the matters under consideration. Whatever was going to happen with the war, it was beyond my control. Standing between the two alternatives, I could not change my thinking. To all that, the Germans were in Bulgaria and behaved correctly as "guests", as if they were people of another planet, and even if they were not to our liking, they might not exist for us. What the future was to be, in case of a German victory, we had nothing to fear from it, in fact we had much good to expect where Bulgaria was concerned. On account of such presumption, never articulated in any specific scenario any way, I simply did not waste my time speculating about it. The worst which might have happened was that the existing internal political and social order would be preserved, with possibility to change it by way of reforms under pressure from society. Here the Legion was in its place, though I did not give it so much credit as to rise as a political power of such dimensions where it would lead in implementing such reforms. But these reforms were not perceived as radical and painful departures from the old system. Communism, on the other hand, in case of Soviet victory and occupation of Bulgaria, was a much greater threat. With the prospects of a Soviet victory, already looming on the horizon, I ought to have taken a much more serious view than I was actually prepared to.

In the discussions which we, the returnees from Plovdiv, had, I unintentionally – by the positions which I was taking – exposed myself as a Legionair, though I was careful never to argue my points from their pedestal, never, but never even insinuating that I was speaking as a Legionair. My growing up into the Legionair's philosophy was based upon the principles of nationalism, anti-communism and program for social reforms. I could defend these points without any reference to the Legionair organization, as I did, indeed. They, the Legionairs, did not have monopoly neither on nationalism, nor on anti-communism, or on a reformist program. The communists played a decisive role for my firm standing on these principles. I was easily recognized as standing on the Legionair's side. They saw my views as Legionair's ideology and would not let me forget it. The more they attacked me, the more I defended myself and my ideas. They never tried to understand me, they had me as their target for attack, in my presence and behind my back, and I never gave in. Yovcho alone had seen in me an independent way of looking at things. The war and its possible resolution moved to the backburner of my thinking. I was virtually convinced that Hitler was loosing the war, but my nationalism, my anti-communism and my religious orientation as seminarian and as theologian, for all practical purposes, held me in the ideological parameter of the Legionairs. I could never abandon my hope and my insistence that Bulgaria had to be satisfied where her injured justice was concerned – the recovery of her lost territories, where millions of Bulgarians were subjected to denationalization and oppression, with their human rights – to use a contemporary, present day principle – trampled over.

Obviously, this, under the circumstances, was possible only in case of German victory. We, the generation after World War I, were raised in the spirit of the Wilsonian 14 Points. Even though they had remained a dead letter, I never gave up my faith in them. But I did not have confidence in England and America to implement them. It would mean taking territories from Greece and Serbia, which had sacrificed themselves for the Allied cause. The broadcasts of BBC to Bulgaria were viciously anti-Bulgarian and could not inspire any positive response among the Bulgarians except those who favored the Soviet Union and the Anglo-Saxons. One of their commentators, an old Bulgarian lady from a prominent family of the past, Vlada Karastoyanova, was shouting from London that the streets of Sofia would be plowed and we would be planting potatoes there – a fate no better than that of the Carthagenians after the Romans captured their city. All things considered, I could not expect any consideration for our just claims over Thrace and Macedonia from the Western Allies. Where the Soviet Union was concerned, I saw the Bolshevization of Bulgaria as an inevitable calamity. At the same time the country was being thrown around like a boat in a stormy sea. After the bombings of January and March 1944, I came to understand Dobri Bozhilov's words, that we were being punished for following the iron logic of our foreign policy. I only hoped, against hope, that the storm would pass over us and we would be left to order our country in our own ways. I personally hoped that the powers would recognize the predicament of Bulgaria when she submitted to Nazis or let herself be destroyed in 1941.

So, this is how things were moving along during my unscheduled vacation, October 1943 – September 1944, when all the fire in the world fell over our heads – over my head, I should say. Uncertain in my Legionair's standing, certain in my anti-communism, with no taste for the Brannik's accommodationism, I was fluctuating to one or another direction, as if I had fallen into a labyrinth without an exit. Outside of my small world of Zelenikovo where my ship, doomed to destruction, was maneuvering between Scyllas and Haribdis, there was the stormy ocean of the world where in a gigantic confrontation the worlds of fascism and communism were battling for life or death. The capitalist world of democracy and plutocracy, hand in hand with the communist slave empire of Stalin, like an alliance between the angels and the satanic powers of the devil, were circling over the nazis for a final grab of their remnants. We, the Bulgarians, and I among them, found ourselves faced with a choice to sink in the ocean with the wreckage of Germany, which was an inevitability, or to be devoured either by the beasts seeking retribution for our crime of self-preservation, or be burned in the fires of the red conflagration, whichever reached us first. The clouds of the great war were rapidly rising on the horizon in the East and the horizon in the West, which left us no place for escape, except to hang in there until the Germans pulled out of the country and let us go down the drain.

Germany was loosing the war. The Soviet armies were crushing the Germans over the frozen expanses of Russia and the Ukraine and soon were to land on the Danube. The Western powers were dilly-dallying in the Mediterranean and, in all appearances, we were to be gobbled up by the Soviets. Only the blind could not see this course of history. It is doubtful if they did not see it, or had no other choice but to pretend that it was not there. My feeling is, contrary to all charges otherwise, that those who were governing Bulgaria at this time, saw the imminent inevitability of Soviet occupation, but had no idea how to extricate us from the holocaust coming to us. When, on Christmas day, bombs started falling over Sofia, and bombed out of existence the illusions for a "symbolic war", it became clear that we were not going to be spared either by the "Perfidious Albion" or by the red demon of Kremlin. As Hitler had crossed over the Danube into our land, so would the red legions of Stalin visit us, "with or without" our consent. The iron laws of war are applicable all the time, by all the warlocks, anywhere in the world. Such was the case of Belgium in 1914 and 1940, and such was the case with Bulgaria in 1941. Why these laws were to be different for us in 1944? Bulgaria was not going to be allowed to pass between Scyla and Haribdis without being crushed, and me with it all.

Before King Boris died on August 28, 1943, there still was a smoldering hope that he still could play some card and save us from the worst fate, though betting on this card had not much of a chance. The savage attacks on the monarchy by the communists left very little to hope for. After the king died, things became hopeless. I just followed the events and waited for the finale, whatever it was going to be. Only the communists were confident, and with every justification, that History was turning their way. No one in the rest of the world – politicians and ordinary people – had any inclination to bet on England or the Americans to come and save us, as BBC broadcasts sometimes ventured to spread false hopes. On the top of it, the entire country was on fire, and there was very little time left for contemplating the future. This fire was, no more and no less, a civil war.

5. In the Middle of the Fire. The Partisan Movement

While we, the young "intelligentsia," were debating and fighting over theoretical propositions, the fire of the communist led partisan movement was raging around us. Shortly before the beginning of the school year, September 1943, an incident occurred near my village which changed everything. A group of five policemen were passing through Zelenikovo, on their way to Rozovets. They have been met by the top communists, among them Georgi Parlapansky, and invited for a few drinks. The few drinks had turned into a night, "on the village." The Communists hosts treated, the policemen, mobilized peasants enjoyed themselves and stayed over, to continue their mission the next morning. The next morning, on their way to Rozovets, they were ambushed somewhere above the village, their guns and their clothes taken away and their dead bodies left on the road. The same people who were treating them the evening before, had attacked them and had gone underground, to the mountains, to wage a guerrilla war, which was already under way in the country. That same day, having learned of what had happened, the police authorities in Brezovo had dispatched other men, to arrest the families of those who had committed the hideous crime and interned them far away in the country. They included my aunt and my cousin Penka. They were taken together with the father of Georgi Parlapansky and Drazha, my aunt's sister and my friend. The whole procedure had lasted one hour or two. I learned of all that had happened that evening when I returned to the village from the fields. Many times after that I remembered that day and had thought that if I was there I would have gone to the chief of the police in charge, would pose as Legionair and would ask to spare my aunt and my cousin. But I was also always aware that they would not separate from the old Diado Marin and Drazha. These were difficult times. This is how the partisan movement involving my village began.

After the war we learned that all the time, Zelenikovo had been hiding the partisans in private houses. Secret communists had housed them in the cold days and nights, and nobody knew about them. One could not speak on these matters because nobody knew who was what. In the course of time it more or less became known who was supporting the partisans, but nobody dared to speak about or report anything to the authorities. It was like playing with fire. So it all appeared that nothing was happening. As things were going to be revealed later, it happened that the Mayor of the village, Mitiu Kraevsky, a graduate of a Business University Institute, had himself been a communist and had worked with them all the time. His family was one of the wealthiest in the village. He was the only one who had gone to study. His brothers stayed on the farm. A few years ago, when I visited the village, I saw one of them still alive. Day in and day out he was taking some 15 or 20 dirty looking sheep to pasture. I met him but I never spoke to him. He was a pathetic figure, a remnant of the old glorious family of the Kraevski. Mitiu had already passed away. Another brother had gone to study after the communist takeover and had become a dentist. It was him who was called to testify in court against my brother Stoiu when my nephew had occupied "my share" of our property and told the judge that I was a fascist.

It is not easy to write about these things, but the times we lived in were not easy either. The Mayor, Mitiu, a tall and handsome man, some two months before September 9, had shot to death the brother of the priest, on the pretext that the man had questioned his order. The village was shocked of this bloody affair. Mitiu was in command of the night patrols of the village and the man killed was serving his call of duty. No one dared to comment on this murder. On Sunday, after the burrial of the dead man, Mitiu was in church, attending the Divine Liturgy. In those days attendance of church by public officials was a sign of support for the government. This was a supreme hypocrisy. I was singing the responses to the priest's invocations, the mayor murderer was standing two feet away from me, and the priest, in the altar, was celebrating the Divine Liturgy. How grotesque all this thing was, and what a soul of steel this priest must have had to compose himself in the face of his brother's assassin?! He, the priest, himself was killed a month after September 9th. How this murderer had the courage, and the temerity to stand in the church, to make his cross, with blood on his hands? This was the climax of inhumanity that I have seen in my life. Well, there was more to come. For it later.

Sometimes late in the fall of 1943 the village was occupied by the army, with assignment to seek and destroy the partisans. In our region they were organized as a "Christo Botev" Brigade. The Municipal building was used by the military as their headquarters, while the officers occupied the first floor of the House of Mikhail Matevsky, opposite the municipal building. Military officers moved in and out of these buildings day and night. From there they coordinated the actions of the military units. So, our "vacation" *sine die*, was running in this atmosphere. It soon looked as if all that was a part of our daily routine. I was perfectly aware that the communists saw in me a fascist, but did not think that it was such a serious offense. What crime was it if one held opinions different from others, including political views? I did not see any danger for me, or I should have, but in time one gets used to such attitudes. For me it was some sort of a sport to formulate effective arguments on any subject which would have come up in our discussions. For as long as I stayed out of touch with any officials, which I did, and did not represent a threat for anybody, somehow I blindly ignored the clouds darkening over my head. However, in the midst of all that a few of the young companions in our circles had come to hate me with all the vehemence that sometimes overwhelms a person. One among those was my neighbor, Koliu Pekhivanov.

6. Koliu Pekhlivanov Wanted Me Dead

Koliu was two or three years younger than me. I believe he had gone to High School and outrightly flunked, but was involved in our group. Their house, opposite ours, was one of the better homes in the village. His father Stakhia had been in America and had made some money. Back in Bulgaria he had managed to build this house. It is still standing there. His nickname, Stakhia and his son, was "komunata" – the communist. He probably was expelled from America in the postwar years in the times of the "red scare." Koliu's older brother, Guncho, older than me, had gone to serve his military duty in the "Bulgarian Air Force." Returning to the village, he was quite a hit. We had grown in the same gang of neighborhood kids, and did not have any special feelings one against the other. In 1991 Guncho was still alive, in the same house, a decrepit old man, loaded with the burdens of diabetes. Everyday, at exactly three o'clock, he was mounting his bicycle and going to the center of the village, meeting old men like him, chatting there for a couple of hours and returning home. When I arrived, he crossed the street from his home to greet me, but could hardly talk. A few years ago he had become very sick and was taken by some of his kids away from the village. Maybe by now he is dead. There were no particularly good, or particularly bad relations between our fathers, but my mother and their mother were close friends. Their mother, Baba Stakhiytsa, and my mother often sat together, worked together and went to church together. I was always welcomed by her in their home.

But Koliu, he was another story. He also played with us in the street as kids. We were almost every day together and nothing happened between us at that time to pit us one against the other. I remember no personal conflicts with him. In 1943 he had turned into a different man. Towards me he had an attitude as if I had robbed his inheritance. He looked at me with blistering hate. I never asked myself why, I simply went along with that and never spoke with him at our gatherings. It was not a secret that he was a fanatic communist, and had contact with the partisans in the underground. Later I was to learn that on one occasion he was going to meet them in the forest, while his mother was running after him with hysterical cries pleading with him to come back, but he had ignored her. Much later I learned that after September 9, when the Remsists had been discussing plans to extend a hand to those outside of their circle, and my name was mentioned, he had categorically vetoed it, threatening that if they accepted me, he would quit the organization. Moreover, when they have been discussing who of the "fascists" in the village were to be killed, he had proposed my name as one of the projected victims. When I went to study in the university, he had commented: "He ought to have gone with those in the Yunchala, not to go to study in the University." Recalling all these events of the times I am wondering what may have prompted him to feel so bitterly against me. My only explanation is that he had taken his communism so seriously, that I must have emerged as the worst challenge for him. Otherwise, nothing ever happened between him and me to cause such a hysterical hate. When I returned to Bulgaria in 1991, he was nowhere to be seen, so I could notice him. It was during my visit in 1995 or 1996 that he was sitting on a bench in front of a tavern in the central square, resting his chin on his cane, that my brother pointed him out to me. He was staring at me, but never made a move to come and greet me. I often wonder, would I have extended a hand to him if he approached me? I probably would. In those days of euphoria, when I had forgotten and forgiven to everybody, with feelings of oblivion and forgiveness for all that had happened, I probably would have *.

In any event I learned that he hated his older brother Guncho. They had no relations with each other for long time – for reasons that their father's house had been left to Guncho. He would go to the Thracian Mounds outside our neighborhood and stare at his old home for hours. It is said that he had somehow lost his capacity of sound thinking. This spring my cousin Maria sent me a e-mail letter telling me that he had died. My family had asked her to do it. So, that is how his death has been considered important for me, that they wanted me to know. I could only say, that much of the misfortunes which followed me, leading to my escape from Bulgaria, may be traced to that man. In my memories for most of these years, he had never figured anywhere, for good or for bad, but during these past few years I was more concerned for the pain which he had caused to my relatives and for me. Before ending this epizode I must mention that during the summer of 1944, one evening, while I and Todor, my neighbor, were going home late in the

---

* When on May 15, 2000, I was examining my State Security dossier I saw a hand-written report by him plotting to help them find me in Bulgaria. There was a rumor that I have been seen there.

night, from a side street, along the house of Georgi Lukov, we were attacked by a fusillade of rocks, being thrown at us. It became known to me that it was him who had organized this ambush.

My second nemesis in the village was another Koliu – Koliu T. Merazchiysky. We were classmates in the Middle School, and, I must say, rivals for the first place where grades were concerned. We were never close, but during the school years, and after I stayed behind from them in Zelenikovo, 1937-1939, nothing really happened between us, except the resentment which he appeared to have for me. As far as I was concerned, he did not matter to me one way or the other. I thought that it was his father, a suspected communist, who was jealous of me for beating his son in school grades. When I went to the Seminary and got involved in their high school group, we inevitably clashed in our debates, where he openly advocated communist views, and behaved somehow haughtily towards me. It did not bother me. Just keeping away from him was enough for me. Somehow he was the first to have a phonograph among the young people and when he would call all others for dancing parties, he never invited me. This was a painful experience for me, to be singled out for isolation, but there was nothing I could do about it. He was actively working for my isolation from the group. Yovcho was also deliberately ignored.

Koliu Merazchiysky rapidly emerged as the leading communist among us. He was an arrogant man, much stronger physically than I was – his nick name was *bureto*, "the little butt" – because he was on the fat side. They lived at the other end of the village, and appeared to be one of the better to do families. He was emerging as the intellectual leader of the young communists among us. He hated me and did not conceal it. In our public group discussions, we often clashed. He too, no less than the other Koliu, was doing everything to aggravate my situation after September 9th. I had no doubt that his attitude towards me was the result not only of our diametrically opposed views, but a follow up on our rivalry for better grades in the Middle School. But one way or the other, he had become something of a leader of the Communist youth group. He, together with the other Koliu, had categorically opposed my acceptance in the RMS, which I had never tried to get into, any way. It was later that I learned of their opposition. Our first, and last face to face confrontation, took place in the basement of the municipal building in the summer of 1946, where I was held prisoner of the authorities. I will say more about that later. He walked down the steps, looked at me, did not say much, but I remember most distinctly his last words: "We gave you freedom to show your horns, and now we will smash you like bedbugs."

Many years later, when my brother came to America, I learned a very different story. In the seventies, someone, leafing through the police records of war times, had discovered that this Koliu had been arrested by the police, threatened, and offered to be released if he would cooperate with them. He had accepted and had signed a statement to that effect. Subsequently, he had been called to join special anti-partisan armed units, "hunting battalions", and had joined in actions against the partisans in the mountains. With this discovery, he had been immediately suspended from all his high party positions, of his job, thrown out of the Party, and returned to Zelenikovo. In Zelenikovo he has been living as a recluse, hiding from everybody, a prisoner in his home. He never appeared anywhere where I could see him when I returned to Bulgaria. This story, which I learned from my brother, was later confirmed by others in the village. How he concealed his services in the police from the other communists, I do not know. Apparently he had not participated in the actions in our region. But it may be that his hostility against me was a sort of cover for his clandestine cooperation with the police.

These were my worst enemies in the village, actively working against me, reporting me to the higher ups in the Party in Zelenikovo. Apparently, they believed them, since the two were their most fanatical and dedicated supporters.

7. My Dilemmas

But let me go back to the unscheduled Winter "Vacation." From month to month it was growing more and more unpleasant with the terror which was raging all around us. The troops stationed in the village quite often went out in the mountains and forests to seek and destroy partisans units. At the same time the war from the East was coming closer and closer to our frontiers and the fate of Bulgaria was becoming more and more inevitable. The communists were already testing the delight of soon coming to power. Somehow, I was perceiving this imminent change in academic terms: the storm will come and will pass over us, and we will continue our day to day life. Today the power was in the hands of the fascists, tomorrow it will be the communists. So what? But it was not so easy to ignore the coming wave and to feel helpless to prevent it. This wave was bringing so many unknowns. This was one way of looking at it, but it was impossible to foresee that it was threatening to blow us away physically and politically. If not consciously, most certainly by intuition, I was seeking some way to come to terms with the inevitable course of events, to prepare for myself some cover. But there was no exit from this dilemma. This dilemma, and whatever scenario I would settle for, was always slipping out of my hands.

I recognized two insurmountable obstacles between me and communism which I could not overcome in order to adapt to the future order of things. My inner religious set up stood between me and them. This was the core of my soul, of my spiritual world, of my system of values intertwined with my religious beliefs. It was difficult for me to close this door of my inner life and to plunge into the ocean of communism where there was no room for religion, there was no foundation for my system of values. Atheism was a dogma for the communists and their coming to power implied the demand, and imposed closing of this door. How could I accept closing this door and became an atheist in order to survive and live in the new world which they have swore to create? How could I compromise my spiritual integrity and hypocritically pretend that I have renounced my religious convictions? Where the communists were concerned, it was not enough to accept their social, political and economic system. This acceptance made sense for them only if it was rooted in the intellectual foundations of their total philosophy. I was faced with the choice between a hypocritical parody of communism, while preserving in secret my inner convictions, and an open declaration of my religious beliefs. In the first instance I had to settle for a double life, to live in two worlds – one open and another concealed. But how long one could maintain his mental balance in such circumstances? Perhaps for some it does not present any problems, but for me it was impossible. With the second choice I would condemn myself to a social, economic and political ostracism, which would doom me to the periphery of life. My entire nature was rebelling against either of these choices. I could not live a double life, and I could not be isolated from the rest of the world.

The other obstacle was my nationalism, my Bulgarian nationalism. The communists had deafened the world with their proclamations of internationalism and their castigation of Bulgarian nationalism as a bourgeois chauvinism, a Great-Bulgarian chauvinism, an ideology, which, in their vocabulary, was looked to as a spiritual disease from which we had to be cured. But for me Bulgarian nationalism was a question of espousing an elementary moral justice – denied, and being denied to the Bulgarian people. This Bulgarian justice was badly wounded and I felt morally obligated to defend it. For the communists Bulgaria was a criminal, and every Bulgarian nationalist was, ipso facto, a criminal, following in the paths of the great-Bulgarian chauvinism. My nationalist convictions were deeply rooted in my soul. I had perceived nationalism as an universal philosophy founded on principles of international justice – not as an ideology of hostility among nations, for imposing the rule of any nation over any other nation or part of it. The communists saw nationalism as a poison which was killing the class conscience of the masses, diverting them from the class struggle.

For me nationalism was a question of elementary human rights with universal application. I personally experienced their view on nationalism when an older than me young man, Sava Mangov, brother of my friend Stefan, at the bus station in Plovdiv, ripped off a tricolor, a miniature Bulgarian flag, from my lapel. He threw it in the mud. Soon after this incident he was arrested as a member of some Communist conspiracy, sent to jail, coming out of it two days before the Communist takeover, becoming a colonel in the State Security service, indulging in alcoholism and dying in disgrace. With his brother Stefan we hanged together around the anarchist Taniu Dangov. Years later, actually after 1991, I began wondering whether Stefan was hanging around me as a friend, or, on account of his brother, was seeking protection from persecution. Who knows? He alone could answer this question. It is this little incident with Sava, which demonstrated to me what nationalism meant to the communists. This incident occurred on March 3rd, the day of the Liberation of Bulgaria, 1878. I could never understand how these people, the communists, could ignore in such an arrogant manner the Bulgarian injured justice. To that, it was Russia which had liberated us, and the Soviet Union was a reincarnation of old Russia. As it turned out, decades later, the communists were to turn around and join with the nationalists to celebrate March 3rd as national holiday. Where they eventually came, I was there way back, when the Germans were all over Bulgaria. But in 1943-44 we stood, on this issue, on opposite poles. Probably I would have accepted all other of their policies as an inevitability, but their atheism and their anti-nationalism turned me against them. There, they were reaching with their fingers to the core of my soul.

While I was going through this internal struggle, consciously or unconsciously, the communists were already seeing the end of the war and the triumph of their cause. I do not know why, perhaps because of my nationalist views, I did not look to the Western Democracies as my life-saving device. I did not see them coming us as saviours. I clearly saw that Bulgaria was going to fall under Soviet occupation. Within myself, I was terrified by this inevitability. I felt in exasperation my helplessness. I was against an armed rebellion to liberate Bulgaria from the Germans. I never looked at the Germans as occupiers of Bulgaria. Simply, I looked at it as a presence of the forces of a great power on Bulgarian territory which we could not have prevented, nor could terminate it without risking our own destruction. Least of all Bulgaria had any moral obligation to sacrifice herself by resisting the German war machine in 1941 to save Greece and Serbia, neighbors whom I have seen as criminals against Bulgarian justice. I saw the Communist partisan movement as a military guerrilla action in support of the Soviet Union, whose ultimate intention was to sovietize and communize Bulgaria. I never looked to communism as a possible solution to my personal problems with poverty. Somehow, all talks about collectivization, socialism and communism, where the wealthy were to be dispossessed, their lands and capitals expropriated, never impressed me at all. I was more perturbed by the expectation that the power was going to pass from the hands of the people of the royal regimes and fall in the hands of the communists from whom I expected nothing good, in fact, the worst.

Under the circumstances I felt that it was better for Bulgaria to wait that the events take their course to their conclusion, and then, without exposing herself and the people to painful excesses, to join in the new order of things. This is what I understood in those days as neutrality, but amidst the raging passions of political extremism, there was no room for such neutrality. In one of my sermons in the church I outlined the reasonableness of such a policy. Without mentioning the partisan movement in that sermon in the church I stressed that under the existing conditions it would be better if we patiently and in peace waited for the end of the war, so that nobody of us would suffer in those turbulent and insecure times. Later on I heard that the mayor at that time – he was a retired military officer, I believed from "Zveno" political group – had liked my sermon.

The communists had not liked my sermon. Somehow, with this speech I had taken the position of the government, which was far from my intentions. I was following the logic of my own thinking, not anything in support of the government or the Germans in the country. Much later, when I was studying this period and became aware of the views of the Bulgarian political leaders of the opposition to the Governments of King Boris, from Gitchev to Mushanov and Petko Stainov, I discovered that they had been taking exactly the same position which I advocated. Instinctively, following the dictates of sound reasoning, I had taken the correct position, and to this moment I maintain that this was the best policy for the times to be followed by the country, regardless of how anyone would characterize it. This was least of all, a policy advocated by the leaders of the Legion. The Legionairs were pushing Bulgaria to intervene in the war on the side of Germany against the Soviet Union.

8. The Bloody Severed Head

But my "sound reasoning" was suffering a crushing defeat amidst the horrors which were taking place around us. In the vicinity of my village, in the Balkan, the Partisan Brigade "Christo Botev" was reigning supreme. Many peasants of Zelenikovo supported them secretly with whatever they could. In those days, I do not remember exactly when, one of these secret helpers, Taniu Gidikov, was arrested by the police and was either killed while under arrest or thrown alive, or half-alive into a deep well. Later it was reported that his bones had been found broken. When this tragic news became known in the village, his relatives arranged for a funeral service. It happened that that day, when I was walking up the main street, about to pass by the Church square, I saw the funeral procession led by the priest, singing "Holy God...," followed by a small group of women and men, some twenty or so relatives, going to the cemetery. The brother of Taniu, Slav Gidikov – later to emerge as the second most important communist partisan from the village – was carrying a suit of his dead brother on his hands, as if this was the body of the man killed by the police.

I was stunned. I stopped, let the procession go, made my cross, and then continued on my way. This same Slav Gidikov, in 1946, when I was targeted for persecution, on my way to Sofia, had cornered my father in Plovdiv and had told him at the bus station: "Sooner or later we are going to catch him (me). He will not escape from our hands." He had advised my father to prevail on me to stop studying theology. My father, an innocent old man, scared of what may happen had answered him: "How could I tell him anything? He is an educated man, he knows more than I do. I am not supporting him. He supports himself." But the murder of Taniu Gidikov remained hanging as a fiery sword over Zelenikovo.

During the month of March, in a shooting confrontation between the partisans and the army, a young, 18 year old boy, had been killed. The military had decapitated him and brought his head to the village. Human senselessness and cruelty know no bounds of decency. The bloody head was exposed at the sidewalk before the headquarters of the army, at Matevsky house, in the street dust. The idea was that all those who wanted to join the partisans, or whatever, had to know what was expecting them. In those days, in January 1944, we, the "vacationing" students, were ordered to meet once weekly in the local school for lectures and discussions. This particular day I joined a group of 10 or 12 young high school students. Yovcho was among them. Someone suggested that we go and see the gory scene. Most of those in the group were leftists, and as it was to turn out later – Remsists, members of the Young Communist League. It dawned on me much later, that in effect that suggestion was meant to pay last respect to the young hero of their cause. I had no objection to go. So we moved up to the square to look at the head of the young partisan. It was splattered with blood, curdled all over, with a chunk of long hair stuck over his face. A swarm of flies was hovering over it. Taking a look at what was there I felt that my stomach was turning up. Yovcho, who was walking besides me, asked me: "What do you think of that?" I whispered with difficulty: "This is a revolting bestiality." The boy had fallen as a soldier on the battle field. Why did they have to subject him to this indignity? Even more, why did they have to expose themselves as being capable of such inhumanity? I was devastated by this spectacle of brutality. I had turned white, and returning from the scene of this atrocity, just kept silent.

We went to the school for our lecture. After a while, the leader of the "Branniks," my former teacher Petar Dimitrov, was to deliver a lecture. He did deliver his lecture. I half listened, still shocked of what I had seen. Midway of the speech, I glanced out of the window and saw on the street another tragedy staged by the same people. A gypsy, flanked by two soldiers with fixed bayonets, was being escorted to the cemetery. He was dragging behind him, tied to a wire, the same head, which an hour earlier we had seen at the military headquarters. I could not keep my eyes on this sacrilege before me. The teacher continued and finished his speech. I did not listen, I was thinking of the head of the dead young man being taken to the pit where the peasants threw dead animals. But I heard the end of the speech: "This is the way we have to follow up." Without thinking, still under the spell of the horrible scene on the streets, I murmured, loud enough to be heard by the whole room of people: "I could not follow up this way."

Surprised, the teacher shot back at me: "Why can't you follow up this way?" "Because I do not have shoes," I replied. The class exploded in applause. Those who applauded were communists. I do not remember how I came up with this answer, and what I wanted to put in it. It was true that I did not have shoes. I was wearing what in those days were sold to poor people as shoes: wooden soles, with skin tops, in the form of a shoe, one grade above of what we called in those day *nalumi*. May be this was the softest way to react to what was happening on the street. I was, in effect, raising a social question, the question of poverty, to which I was a victim, and to which all were familiar with. No one could dispute the veracity of my statement and keep me responsible for it, in other words, to arrest me for questioning the social foundations of the existing order. I do not remember how this meeting ended but apparently it made a deep impression on the communist youth. Much later, in 1947, this incident was recalled to me by Marin Pekhlivanov, in a note in the margin of a letter which I had sent him. I will publish that letter later in this book. But this mindless remark, which was actually escaping the challenge to speak on the subject of what I had seen on the street, caused the communists to soften their attitude towards me. But I did not move to exploit this accident. What was I going to exploit? I was up to my knees sinking in insoluble dilemmas.

9. Temptations by the Devil

This is how things moved in the spring of 1944. In the beginning of June the village and the fields were devastated by a fierce hail-storm. The entire harvest appeared to be destroyed and the peasants were faced with starvation for the next year. Somehow the government had come out with a promise to pay for their losses. At that time I had gone to the municipal office and had asked someone, I do not remember whom, for a possible job there. It happend that they needed a clerk to receive the reports of the peasants of their losses, subject to compensation. The job was given to me. I started working diligently, I was recording all information which was brought to my attention. When I did not have somebody with me, I was asked to do other office chores. The payment was meager, but I was happy even for that. I was performing on the top of the line, and when my assignment was finished, I was retained as a clerk. As I have mentioned it before, the municipal office was filled with military officers and soldiers, coming and going for their work. The only telephone in the village was there, and they were all the time at it. The Mayor, that time, was Mutiu Kraevsky. I have already mentioned him. Among other things, he was the one to arrange for the night patrols to look for partisans. Once, I remember opening a draw in the office, and discovering there all the plans as to where night patrols were stationed. I close it and never looked there again. Anybody could open and see these secret security arrangements.

One day Mitiu called me to his office and confided that he had a delicate mission for me. He asked me to go to Todor Zlatarev, a heyward, ask him to give me his binoculars, then go across the river in the growing young forest of pine trees. I was going to watch for partisans and come and report to him. I was stunned. This was outside the line of my duties. Besides, I immediately recognized this as a political assignment, as a direct involvement in the intelligence units of the army looking for partisans. I did not have the courage to turn down his order on the spot. An order is an order. I went out looking for Todor Zlatarev. His wife answered the door and I asked for him. She seemed scared and I am sure, she lied to me that he was not home. I did not think at the time that this could be a trap, to send me there to watch for partisans, to be captured by them on the spot and dispatched in any way they considered advisable. I was thinking more in these terms in that I was being used for political purposes which I was not willing to accept. This was a deadly serious affair and I did not feel like involving myself in the struggles with the partisans. I was crushed under the burden of this assignment.

Not finding the heyward, Todor, I headed for the Chervinatsite, where my father was working that day. I found him there and told him the entire story and my misgivings. I told him that I was going back to the village and I would resign my position. He looked scared of this whole thing and made no objections to my intention to quit. I returned to the office, knocked at the door of the Mayor, was admitted in and told him of my failure to find the heyward and the binoculars. He most pleasantly told me not to worry about it and continue with my work. But some two hours later, he called me again. He said that he did not have anybody available and wanted to send me to tell my neighbor, Marin Pekhlivanov, to come and see him. Marin's and our houses were adjacent to each other. I knew that Stahia, his father, was at the bottom of some game with the communists. He was an Agrarian, and from old times was nicknamed "Gitchev," the name of the leader of the Bulgarian National Agrarian Union. But Stahia had moved to the extreme left and was cooperating with the communists. Later I was to learn that he was even more involved with them that I could have supposed. But now, as to this assignment, I did not see anything unusual in it. I knew that Marin was one of the leading Remsists, but I did not put much thought in that. So, I went to look for him. I knocked at the door. His mother answered and I told her that the Mayor wanted to see Marin. She exploded in hysterical cries. At the house entrance appeared Baba Marinitsa, grand-mother of Marin. She also cried out hysterically. I left. Apparently there was something terrible that this summons meant to them. I returned to the office, puzzled of all-these happenings. In one hour or so Marin came to see the Mayor. The minute he stepped in his office, from the hallway, not passing through our secretaries office, he broke down crying. I could not help overhearing. The Mayor was trying to calm him down. He calmed down. I did not hear anything that transpired between them. But I read in this whole thing something that did not augur well for me. That evening I thought over all these events of the day. The next morning, going to work, I walked straight in the office of the Mayor, and told him that I was resigning from my job. I did not try to explain why. For myself I felt like if I had stepped on a snake. He did not try to disuade me. Simply he said O.K. and that was all that there was. I clearly saw that if I had continued working in that office and was given such assignments, I was exposing myself to mortal danger. It was a question of fighting the partisans, and I would have nothing to do with that. Marin may have been involved with them. The Mayor had probably called him to talk to him about something connected with the whole affair. His family may have known about all that. That's why so much crying and hysteria. All these things were going through my mind when I made the decision to quit my job. This was my last connection with the authorities. From the Mayor's office I went to Baidachka, where my father was working and told him that I had quit. He was relieved. Somehow I felt that a burden had fallen from my shoulders. He gave me my assignment in the fields and thus I re-entered my old life.

10. Under the Barrel of the Submachine Gun

On July 21, 1944, I had taken two lambs for grazing out in the fields near the shosse (the road) to Brezovo, very near the village. While they enjoyed themselves I was lying down, absorbed in reading Henry Senkiewicz masterpiece *Quo Vadis Domine*. The sun was close to setting behind the horizon when I heard a loud voice coming from the Thracian Mound, calling all people to go immediately in the village. It was one of the haywards. At the same time the village cryer was beating his drum in the village and the church bell started ringing continuously in a manner signifying not a religious service, but some sort of an alarm. I picked up my book and led my lambs towards the shosse. When I reached the embankment, three young men jumped from the trench (I had come from behind them), looked surprised, as if I was a threat to them. They had sub-machine guns in their hands, were somehow raggedy looking, but seeing me with a book and lambs, just pointed to the road and told me "Hurry up to the village." I realized that these were communist partisans and that the village was occupied. I felt like a little speck at this time, I had come face to face with reality, which until now I was aware of, but it seemed too distant from me. I proceeded to the village. In our neighborhood I saw women and men gathering along the road, and my friend Todor, somehow bewildered, told me that we had to go to the meeting, that the partisans had occupied the village and he urged me to hurry. I took my lambs home, washed in a hurry my face, put another shirt and rushed out to join Todor. We proceeded to the center of the village. Many people were walking along. All in some sort of a confusion and fear of what was happening. I was scared in my heart, but tried to hide it and like all people around me walked briskly to the village square. At some point another young man joined us, and three of us, barefooted, in wonderment, followed the crowds. Midway on the road to the public meeting, I saw three partisans coming opposite us with submachine guns pointing ahead of them, marching fast, with fierce expressions on their faces. As they came at about twenty feet facing us, they raised their submachine guns, pointing them at me. The leader, Parlapansky, blurted an order in my face: "Stop! Hands up! You are under arrest!"

It sounded as if he had held his breath a whole day for this minute. They proceeded to reach us. We all had stopped with our hands up. Parlapansky waved the other two of my company to continue on their way, while one of the partisans approached me, asked me to put my hands behind my back and he tied them up with some hard rope. They ordered me to turn around and walk back. One of them placed the barrel of his submachine gun at my back. I glanced around and saw a large group of women and men from my neighborhood, their faces dropped, watching in silence the drama taking place in front of them. A swarm of kids had gathered and watching me being arrested and tied. Many years later, a man, one of my godfather's family, also named Spas, by now an old man with white hair, told me that he had watched the arrest from the side as one of these kids. "I was deadly scared," said this man. I was walking ahead of them. At one point I was pushed with the barrel of the submachine gun and ordered to walk faster. When we reached my neighborhood, from distance, I saw my father standing on our street. I never asked him, and he never mentioned it, if he had seen me.

At a curve of the street, the partisans moved to enter the house of the Kraevski – Peniu and Petar – one of the well to do people in the village. I was pushed inside too. They were met by the old Baba Petkovitsa, their mother. She stood there greatly perturbed. They asked her for bread and cheese. The two partisans walked with her into the house. I used this moment, alone with Parlapansky, collecting all my strength, if there was any left, though shaking, to tell him: "Georgi, I am not what you think I am. A month ago the Mayor asked me to find a hayward, to take his binoculars and spy on you in the pine-forest. I refused and on that account, I resigned my position as clerk in the municipality." He had not heard of that. I continued: "If you want to know the truth about me, ask Yovcho. I kept company with him and he knows everything." Parlapansky was listening attentively to me, with an eye on me. When a partisan reappeared outside, he told him to untie me and told me to hurry up to the meeting, adding that he will verify my story.

I fled out of the door and ran to the meeting, but I felt how my legs were giving out. Soon I reached the square at the municipal building. I could not see my friend Todor anywhere. So I stopped in the crowd. Someone was making a fiery speech about the "Fatherland Front". At one point I was pulled from the back. Yovcho was standing be-hind me, excited, breathing hard. He nervously whispered to me: "Come on! Let us go! I am joining the partisans!" I did not know what to say. He did not know what had tran-spired with me some twenty or so minutes before. In confusion, I told him: "You go! I do not feel ready." He did not try to persuade me. He rushed home to pick up his things. He did join them. This was our last brief meeting. I saw him again on September 9th, again for a few seconds. Our last meeting was when he was brought back to the village for his funeral in October. I will discuss that later. I believe I have mentioned it already.

The meeting continued for one or two hours. In the center of the square, where the leaders were making speeches, a bonfire out of Municipal archives was burning and casting a light at those surrounding it. There were five or six men with their hands tied behind their backs, "enemies of the people." One of them was an old man, Diado Spas Kolev. He was working in the Municipal office as a man in charge of requisitions for the armed forces. From his position, with hands tied behind his back, he had volunteered an advice to them: "Do not burn these books. Tomorrow, when you come to power you will need them." They had laughed at him. Later on, when they came to power, they called him back to assume the same position and do the same work, but for their government. He was the most competent for that work.

When the meeting was over, I returned home. My father was already sleeping. He had not gone to the meeting. I picked up a blanket and walked out to the backyard. I layed there on a pile of hay, tossed and turned for a long time not able to fall asleep, staring at the millions of sparkling stars and felt my defeat in every fiber of my soul and body. Something snapped in my soul. I was feeling how a cloud of darkness was descending over me. If I ever had any illusions that I could handle the transition to the new order which was inevitably coming with the imminent storm, when the communists would come to power, it all evaporated now. I still felt the rope around my wrists. I still felt the touch of the submachine gun on my back, and stared at the barrel of that gun pointing at me. A feeling was setting deep in me that I would never forget what I lived through that evening, and that under any conditions could I join the satanic forces which were now rampaging in the country. I could not see how I would handle these forces which apostles held me one step from execution. Whatever would happen to Bulgaria, I would be first on the line in the hands of the people who had pointed their guns at me. I was no longer a free man. From this moment on I was going to dance the way they wanted me to. But I would be playing this game not because I would have enjoyed it, but out of fear of them, not out of respect for them and their system. I saw ahead of me wide open the gates of death, or, in the best of circumstances, the gates of hypocrisy, with disgust in my soul.

All this was igniting a fire of indignation and a feeling of helplessness in me. And fear! A horrible fear hovered over me. This was only the beginning. I was hopelessly outplayed by the forces of history, I was besieged by enormous odds and helplessly stared in the eye of the devil in front of me, trying to discover in it my future fate.

Sometimes, late in the night, I must have fallen asleep, for a short time. When I awoke in the morning twilight, the memory of the evening events hit me like a ton of bricks. I still felt this blow. Everything was painfully turning in my mind. I was trying to see it all as a misunderstanding, caused by slanderous information given to Parlapansky. I was turning over in my mind my attempts to convince Parlapansky that he was making a mistake in arresting me, trying to structure some defense for myself, bringing out the instance when the mayor was trying to send me to look for them, how I had refused and had resigned my office as clerk, suggesting that he asks Yovcho about me... I was trying to convince myself that it was not all lost, that I had something to say in my defense and dispel the charges of my enemies. I could explain my ties with the priest with my obligations as a seminarian, that I had no political commitments to him. I was turning over in the haystack and reluctantly concluded that the time had come for me to play a hypocritical game. This is what the communists were doing. This is how they had all joined the "Brannik" and then some of them ran out with the partisans, in their "Brannik" uniforms. I was going to explain my membership in the Legion as a ploy to save myself from persecution, as, indeed, this was the truth.

The third evening after these events I was awakened at midnight by a knocking at the window with a branch. I looked out and saw under the window three armed men. One of them was Parlapansky. They all had their sub-machine guns. Georgi made a sign to me to come out. In a minute we were all standing in front of the house. While we were shaking hands Parlapansky was already apologizing for what happened the other day. He had thought of me as a fanatic Legionair and had me arrested for that reason. He had spoken with Yovcho, and Yovcho had confirmed my assurances at the time of the arrest. He has told him, that I am actually an independent who could work with them. Then he delivered a long monologue on the "Fatherland Front" as a political coalition of democratic parties. Among the leaders of the non-Communist parties he mentioned Nikola Petkov of the Bulgarian Agrarian National Union and Kimon Georgiev and Damian Velchev of "Zveno." It was for the first time that I was hearing the name of Petkov. I knew that Georgiev and Velchev had executed the coup d'etates in 1923 and 1934. I was not much impressed by this combination of parties, but who was I to protest and comment? I was simply agreeing.

Parlapansky went to say that there was a place for everybody in the "Fatherland Front." When they would take power there would be freedom for every man, the church included. If anyone believed in God he could freely confess his faith and attend church services. I explained my relations with the priest as mandatory. As a seminarian, I was obliged to attend church regularly and he would certify that by signing my student book at the end of every vacation. But I had no political commitments to him, which was true, except for the fact that we shared the same views on communism. I was closely associated with him and the church – not along political lines but in the context of our church work. We were friends and I was trying to help him organize a Youth Christian group. This was not a secret. I was representing this too as part of my duties as a seminarian. The priest was a dedicated pastor and the church was my spiritual home. It was very difficult for me to turn my back on him. But I would not reveal these things to my interlocutor with submachine gun in front of me. This was not a time for expressing personal feelings for anyone which might jeopardize me. All I was thinking was how to save my skin and keep my religious feelings to myself. It was for the first time that I put my religion on the back burner, under cover, and preserve my faith in God. So, what was my choice in this moment? Simply, I was to stand there and if I had to lie, to do it without feeling of guilt.

I found myself in an impossible situation. This was only the beginning. Very soon I was to realize that this way of covering the truth and following the path of lies during the next few years, was to be my only way to save my life. The only redeeming thought was that they, the communists, also practiced such deceptions. Was this a sin? If it was a question of violating one of the Ten Commandments of the Bible, taking it all literally, it was a sin. In the future I was to be confronted with the conflict between two moral systems of values – one of a common ethical sense of right and wrong and one which was legally defined. I would very often point to the example of Jean Valjean, Victor Hugo's hero in *Les Miserables*, hiding in a convent where the Catholic nuns were known to never lie. When Jxavier came to look for him with his police, they lied to him, they would not betray the good man that they knew Valjean to be. Likewise, I have often used the case of the Catholic nuns who assisted a doctor to perform an abortion in an incest case, while being told that this was a surgery of appendicitis. The Doctor trusted them to keep secret this prohibited procedure. He trusted in their integrity to view things in a moral perspective and so they would not reveal him to the authorities. This is the story in *Peyton Place* by Grace Metalius. Much later, thinking over these questions, I concluded that the Ten Commandments were not a true criterion for moral values, that much higher to these Commandments were the Beatitudes in the New Testament. So, in this difficult moment in my life, I had to save myself from Parlapansky and to rethink my actions later. So, I did not think I was betraying anyone and I was not committing any sin. Rather, I was maneuvering to evade the blows which otherwise might annihilate me. I explained my friendship with Koicho Manevsky the same way. He had helped me to enter the Seminary. This was all.

Parlapansky started discussing the war. Germany was loosing and would soon capitulate. Already there was a rebellion against Hitler. On July 21 an attempt was made to assassinate him. I had already heard that … from them. When I was arrested and escorted under the barrel of their guns, I heard them talking about it. I had not heard of it before. It had happened that very day. I had nothing to say against this. It was obvious for any one who followed the events. I tried to introduce some humor in the conversation by referring to the German communiqués of every day shortening the front. By the time they shorten it enough, they would be in Berlin. They accepted that. He then turned to Bulgarian affairs, against the regime of King Boris and Bogdan Filov. I had no objections either. I was unhappy enough of these regimes in order to defend them. I would not dare, any way, but I shared their views. In my mind all this was nothing but a charade, but I tried to act as serious as I had to be. What could I say to these men armed to their teeth? After all, history had turned to their side. Inside of me I was shaking, realizing that my life was in the hands of these people. Inside of me it was crossing my mind if all that was a genuine expression of trust, or it was a trap for me. I was expecting that they would ask me to join them on the spot and wondered how I could decline. But no such proposition was made.

I was shaking in the night's cool air and prayed that the conference would end as soon as it could. The tension which I was feeling crushing me was unbearable. I felt that any minute I was going to collapse, but forced myself to be calm and follow the tyrades of Parlapansky. The other two partisans never said anything. I did not know them. They were from other villages. At the end I was left with the impression that I was accepted by them, that I was not considered an enemy. Still, this did not give me solarce, because I felt otherwise, but had to agree and node with my head to signify agreement with them. Otherwise, my soul was burning in a hell. Frightened, confused, helpless, without any hope and terrified, I was playing for time – until times cleared up. In this tense moment I felt like running to the end of the world to save myself of this torturous conversation. At the same time Parlapansky was babbling and was pulling me more and more into his world. I was thinking what Koliu Pekhlivanov and Koliu Merazchiysky would have to say.

So, while listening to Parlapansky I was turning in my mind all other dangers waiting for me at the hands of the Remsists. Now I was saved by Yovcho, but they did not have much respect for him. The only difference at this time was, that Yovcho was a partisan and they were not. If it came to a choice, I could not say who would prevail – Yovcho or these young men. As things were to develope later, Yovcho was killed on the front at Stratsin and I could not longer have his support. I was going to be at their mercy, and, as it turned out, I did not find mercy in them. But, at this time, for a brief period of time, between July 21 and September 9, I had some respite, in the most critical days of my life. I still felt the barrel of the gun touching my back, or pointing in my face. So, I had to suffer the long night conference with Parlapansky. Finally, the conference was over and we parted on a good note. The partisans vanished in the darkness of the night and I returned to the house. I stretched on the blanket to catch some sleep, but my eyes would not close. I felt like I was hanging on a rope which was tightening around my neck. The victory of the Soviet Union over the Germans on the southern front was all but here and the coming to power of the communists was only a matter of weeks. Much was said about Americans and English, and those who had declared war on the United States and Great Britain had put all their hopes on them, but somehow I did not have any realistic expectation that they would save us from the avalanche which was rolling towards us from the Northeast. It was so strong, so overpowering, so impressive, and the communists were so embittered, that I could not see escape from our imminent fate. The storm was heading in our direction and I saw no place where we could seek shelter. So, there was nothing left but to adapt to the inevitable and hope for the best.

All these events of July made something snap in me. I never for a moment forgot the barrel of the gun pointing in my face, or the rope cutting into the flesh of my wrists. A paralyzing fear was seizing my mind and I was seeing nothing for me in the future. It all came down to a sense of my freedom to think and to decide my questions of life or death. Even though I was left behind by the partisans a free man, I felt like I was already a prisoner in their hands. A year earlier I had joined the Legion to save myself from the pressures of the good-for-nothing people whom I had never taken seriously in the political field. Now I was faced with the necessity to make another adjustment. There was some difference though. As a Legionair I was free to attack the existing order of things and advocate a new order, even though I never fully understood what this order would be, I felt my freedom of action unimpeded. This is what I was doing, as indicated above. But things now were shaping in a different form. I had to march under the barrel of a gun. Soviet occupation was looming over the country and this meant that the barrel of this gun will never stop being pointed at me. Where I was concerned, the communist regime began on this fatal July 21 day. The occupation of Zelenikovo was a signal in what direction the country was going. What I clearly saw ahead of me, and I could not digest, was loosing my freedom to think and evaluate things with my criteria where truth, justice and freedom were concerned. In my youth I was a captive of these principles, I had given myself to seeking these basic principles and I had never compromised my conscience along these lines.

Now it all was lost. I had no choice but to give up and to seek a way and means to preserve my freedom. This freedom was the air which was supporting my inner life, my sense of honor and dignity, my spiritual security and stability. Otherwise I saw myself as a dead corpse. Would I have a chance to survive in this world? The days and nights were passing in a way that made no sense to me, in the absence of hope for salvation. After the events of July 21 I was living in a fog. In Sofia they had called back to session the National Assembly and the newspapers were full of reports on the debates going on there. It was for the first time since I had begun following national politics where prominent national leaders were hurling back and forth accusations and justifications as to why the country had been thrown in this predicament. A Bulgarian delegation was dispatched to Cairo to seek an armistice from America and Britain, but somehow this entire thing seemed hollow for me. The country was going through a fool's paradise. When these negotiations with the allies collapsed, the Government of Bagrianov resigned and a government of the leading political parties of the B.A.N.U. – Kosta Muraviev, Dimitar Gichev et al. who all the time during the war represented the opposition to the regimes of Tsar Boris – was formed. It was all in vain. The Bulgarian delegation in Cairo was not offered armistice terms. Apparently this had come too late, and was too little to change anything. Moscow was delivering blistering attacks that the new government was still serving the Germans and was giving its overwhelming support to the partisans who pushed to take over.

11. Tightening the Rope Around the Neck of Bulgaria

Be all that as it may, after the midnight conference with Parlapansky I felt a little relieved. I saw a small crack in the wall ahead of me which I had to try going through and to calm the storm raging over my head. It was not going to be easy. At this time Bulgaria was agonizing on the threshold of its deathbed. The rope around her neck was tightening and it was crystal clear that her fate was inevitable. During the months of January and March American and British planes had burned down and destroyed Sofia. From Radio London, BBC, Vlada Karastoyanova, from a prominent Bulgarian family, was spilling venomous curses and promising that the people of the capital soon were going to plant potatoes in their torn apart streets, appeared to have been fulfilled. On 28 August, 1943, King Boris lay dead and the last hope that he would somehow maneuver Bulgaria out of danger had gone with him. The right wing politicians in Bulgaria today still try to blame the Germans for the King's death, trying to present him as a martyr, and thus exhonorate themselves from responsibility for having pushed the country to the brink of disaster. In the numerous studies and theories of this death there is only one that makes sense – an article in *Historical Review* by Prof. Ilcho Dimitrov, arguing for a natural death, which sounds convincing. The election of the regents after that death – Prince Kiril, brother of Boris, Bogdan Filov, Prime Minister and General Mikhov – using procedures not sanctioned by the Constitution, appeared to be only the last move of the governing click to retain power in the country. Prince Kiril was not a party in making the policies which led to this debacle. He was known as a playboy, rather than a politician. His inclusion in the regency was only an attempt to add prestige and legitimacy to the institution. This reaffirmed the policies followed thereto, but exposed him to mortal danger. But he assumed this position, endorsed the policies followed by the government, and in less than two years had to pay for it with his life, executed by the communists in the beginning of February, 1945.

Chief architect of these policies was Prime Minister Bogdan Filov, who appeared to have believed in the victory of Hitler. He was of the very few who shared this belief. He did not see the catastrophe hanging over the head of Bulgaria and never tried, except in the last moment when it was too late, to change course. General Mikhov was a military man who operated under the command of the civil government and never moved to correct things. In Bulgaria people looked to the military as wooden heads who have no understanding of things political. So, he gave a shoulder to the blind men who were leading the country in these critical days. King Simeon at that time was a five-year old child and could not have understood that his Throne was on shaky grounds in the stormy winds raging over and around Bulgaria. The Prime Minister who succeeded Filov was a jolly good old man, Dobri Bozhilov, an expert on financial matters, but apparently, a lightweight where national policies were concerned. He was a pawn in the hands of Filov. When the Americans bombed Sofia on Christmas day, he went on the radio to address the Bulgarian people, and, almost crying, he uttered these words which I already quoted and have never forgotten: "Bulgaria was punished because it followed the iron logic of its foreign policies," or something to that effect. The words "Iron logic" stuck in my mind. I have quoted this phrase on numerous occasions, having accepted its unquestionable truth. Still, this admonition was not going to shake the pillars of international politics in those days.

Bulgaria was doomed. By the spring of 1944 the Russian armies were poised to invade the Balkans, waiting for the spring to start moving southward, along the Black Sea, and through Romania. The allies had smashed the Germans in North Africa and Italy, now led by General Badoglio, had joined them. The Americans were battling the German rearguard at the walls of Monte Cassino. Against the background of these events, King Boris had sounded the leader of the Social Democratic Party, Krustiu Pastukhov, about forming a new left of center government, or whatever, but nothing came out of this demarche. When it became abundantly clear that things were moving towards catastrophe, the Bozhilov government was dismissed and Filov called his old nemesis Ivan Bagrianov, out of retirement to form a new, transitional government. In 1938-39 Bagrianov was speaking the language of Stamboliysky and for this reason was thrown out. Rumors had it that he was an Agrarian. Others thought him to be a Legionair. He was neither the one nor the other. He was a maverick in Bulgarian politics, a common sense man of the palace guard who was kicked out by Filov. Bagrianov attempted to attract left wing pro-Soviet individuals to his government, including Prof. Doncho Kostov, a prominent biologist who for many years had taught in the Soviet Union and was the darling of the Bulgarian communists. But Kostov did not bite, and after a week or so, keeping a position in the cabinet open for him, someone else was picked and the whole Bagrianov strategy proved fruitless. From then on, it was only a matter of time when the whole political system in Sofia would be overthrown.

The most painful problem for Bulgaria was the partisan movement. Bulgaria was in a state of civil war. How many partisans were in the mountains and what kind of force they were is an academic question. They had paralyzed the country, spreading fear and terror in the countryside and in the hearts of those who sided with Germany. They had tied the Bulgarian army at home, though it has not been established how much this confrontation mattered for the one or the other side. King Boris effectively used it as an argument to decline Hitler's demand to send the Bulgarian army against the Soviet Union. But when the Soviet armies broke the German lines at Kishinev and the red legions of Stalin started their march to the South, there was very little hope left in the country that the Soviet occupation could be prevented. In these circumstances Bagrianov allowed that the National Assembly, which mandate had already expired, be convened and a furious debate on Bulgarian foreign policy was widely opened. It turned out into a hysterical polyphony of accusations, explanations, apologies and protestations, but nobody seemed to have anything constructive to offer for meeting the threat hanging over the country. I followed in detail all discussions and I learned a lot about what had happened in Bulgaria since 1923. All attacks seemed to have been directed against Alexander Tsankov, the leader of the coup of 1923, who was probably the last person at this time, who would desert the German camp. He could at any time be called by the Germans, appointed Prime Minister, and take charge of Bulgaria, to send the Bulgarian army against the Russians. That the Germans did not make this move indicates that they no longer cared about Bulgaria, that it was irretrievably lost for them. The only move which the Bulgarians could make now, was to seek an armistice with the Allies. Bulgaria was not at war with the Soviet Union, so they had no cause to go to Moscow, unless they realized that this was the only course open to them.

All this time Radio "Christo Botev," the Bulgarian voice of Moscow, run by Bulgarian communists, was castigating the rulers in Sofia, the Filovs and Bagrianovs, demanding their overthrow. In 1942 Georgi Dimitrov, the top Bulgarian communist in Russia, had issued a "Declaration of the Fatherland Front" and urged that all power in Sofia go to that political formation. But the communist front in the country had failed to persuade any significant political leader to join the Fatherland Front with them. The Dimitrov declaration was calling for a broad democratic coalition to oppose the government. But it met with total rejection from the political world in Sofia. Leading political parties, favoring the Allies at this time, were the Bulgarian Agrarian Union led by Dimitar Gichev, the Bulgarian Democratic Party, led by Nikola Mushanov, the Social Democratic Party led by Krustiu Pastukhov, and a sundry of smaller political groups. To be sure, since 1934 all political parties had been outlawed and in fact did not exist as national organizations, but their leading personalities were well known and widely respected. By midsummer, 1944, a few of the well known political leaders, motivated more by pragmatic considerations in these confusing times, faced with the inevitability of the events rapidly approaching the country, than representing any political organizations, had joined the Fatherland Front. In the first place these were Kimon Georgiev, from the political circle known as "Zveno", "Link", and his alter ego, Damian Veltchev. The Coup d'etate of 1923 was the work of these two men. So was also the coup d'etate of 1934. They were professional conspirators overthrowing the governments of Bulgaria. As it was the case, in 1923 and 1934, the coups were pro-Allied actions, favoring France, England and Yugoslavia. Another one in their group was Dimo Kazassov, one time a Social Democrat, but in fact an opportunist without firm political convictions. A left-wing Social Democrat, Gregory Chesshmedjiev, had also joined the Fatherland Front.

But by far the most important politician who joined in, was Nikola D. Petkov, a prominent Agrarian who in the 1930s had practically moved from one Agrarian faction to another. Now he had joined the left-wing leader Dr. G.M. Dimirtov, who had managed to escape from Bulgaria and during the war had formed the Bulgarian National Committee "Free and Independent Bulgaria." He had the support of the Allies and was broadcasting to Bulgaria from the Middle East. Petkov was his spokesman in Bulgaria. At the same time, the most respectable political leaders in the country – Dimitar Gichev, Nikola Mushanov, Krustiu Pastukhov, Atanas Burov et al. steadfastly refused to join the communists under any circumstances. No temptations of any kind attracted them. They had recognized the true nature of the Fatherland Front – it was a front of the Communist Party, with a clear objective: the bolshevization of Bulgaria after a Soviet occupation. For them the partisan movement and the Fatherland Front were the avanguard for the sovietization of Bulgaria and they would have nothing to do with that. Anybody who added his name to their initiative was becoming an accomplice to the Communist Party for Sovietization of Bulgaria. Their hopes focused on America and England and expected them to beat the Russians to Sofia and save Bulgaria from communism. They were bitterly deceived in their expectations.

So, Bagrianov had sent a delegation to the Western Allies to Cairo, led by Stoicho Moshanov, to seek an armistice. This mission failed. Moshanov was led by Dr. G.M. Dimitrov to meet the Allies' representatives. One would have expected that Dimitrov, as the only and most prominent Bulgarian political exile with the Allies, might have prepared a way for extricating Bulgaria from the predicament she was in. It was a deception. The broadcasts from the Middle East left that impression, but when everything is said and considered, he was nowhere near the center where big politics were made. The Allies had governments-in-exile for Poles, Czechs and Yugoslavs, but for the Bulgarians they had allowed only a "National Committee." Apparently Dimitrov was never consulted on their future plans about Bulgaria. In Cairo he simply turned a tourist guide for the Bulgarian delegation – until the Bulgarians were told to talk to Moscow. So, there it was, the Bulgarian case was going to be settled in Moscow.

There was more in this Bulgarian diplomatic fiasco. It was orchestrated by the Allies long before the Bulgarians had decided to dispatch their delegation to negotiate. In the second half of May and the beginning of June, Winston Churchill had taken the initiative to persuade Roosevelt to agree that they both give to Stalin free hand in the Balkans. After some hesitation, and behind the back of Foreign Secretary Cordell Hull, Roosevelt had given his consent. So, in the month of June, while the Russians and the Germans were battling at Kishenev, Stalin was told that the Balkans were his for the taking, and that meant Bulgaria. No one in Bulgaria knew of that arrangement. In fact I am not sure when was it revealed that it existed. Maybe it was Cordell Hull who revealed it when he published his Memoires in 1948. So, our delegation was wasting its time in Cairo, while both London and Washington knew that the Balkans were already promised to Stalin. It was going to be a "temporary" arrangement. In the future Stalin was to muse that the army which occupies a certain territory will decide on the political and social system in the conquered lands. Much later a great deal of ink was to be spilled to blame Yalta for the fate of Bulgaria and other Balkan countries. Yalta simply confirmed what the wartime strategic decisions had led to. At Yalta Churchill and Roosevelt had Stalin sign high sounding declarations for freedom and democracy, when they knew that Stalin understood one thing, they understood another thing about the meaning of these terms. After Yalta they argued that he had not kept his word, that he had cheated them. In fact, they simply washed their hands of what they already had done. It was long before Yalta that we have been surrendered to the Soviets. It was initiated in June, 1944, and then formalized in Moscow by Churchill with his famous percentages agreement, on October 9, 1944. Yalta took place in February, 1945. So, Bulgaria, at the time of Yalta, had for long been in the paws of Stalin. While Churchill, Roosevelt and Stalin were conferring at the east coast of the Black Sea, the Bulgarian communists were executing princes, regents, ministers and members of the Bulgarian parliament. So, when the Bulgarians were trying to obtain an armistice with the Americans and the British in August, 1944, they had already been sold out. They should have understood the grammar of international politics at the times, and instead of seeking understanding in London and Washington, to have gone strait to Moscow, declare unilaterally the end of their war with the Americans and the British and throw themselves at the mercy of the Russians. They did not do that and had to pay dearly for their misreading of world politics. It was this misreading of world politics that sent the Bulgarian delegation to Cairo, instead of to Moscow, dilly-dallied there without any success, until they were told that they could not do anything for them and told to go to Moscow.

At the same time the Russians had occupied Romania, which prudently surrendered on August 23, and the Russians armies were poised on the Danube, ready to cross into Bulgaria – just like the Germans in 1941. Having failed to obtain the armistice in Cairo, Bagrianov resigned as Prime Minister and the Regents had to appoint a new ministry. Their last resort was to seek the traditional democratic pro-western parties – the Agrarian Union, the Democratic Party et al. But here again, even on the brink of the precipice, they chose to play small times politics. Instead of calling on Dimitar Gichev, the recognized leader of the BANU, or Nikola Mushanov – the veteran leader of the Democratic Party, they entrusted the premiership to Konstantin Muraviev, former minister in the cabinet of Stamboliysky with close ties to the palace. The new government, formed on September 2, was definitely democratic,definitely pro-western. Great efforts were made to win the Fatherland Front to join in the government, but the Fatherland Front, proded by its Moscow god-fathers – Georgi Dimitrov and Vasil Kolarov – categorically refused and blasted Muraviev and Co. as fascists. They wanted an all-out Fatherland Front government.

Bulgaria was agonizing. The Soviet invasion was imminent, the new government freed the political prisoners, it made it clear that it was breaking with the past, but nothing would satisfy the Fatherland Front. On top of that, on September 5, in the evening, the Soviet Union declared war on Bulgaria. One hour or so later – it is disputed who did what when – Bulgaria declared war on Germany. By that time all German military units were out of Bulgaria. For a long time arguments were to fly all over the Bulgarian political landscape that the Soviet Union did wrong declaring war on Bulgaria, while all the time, during the war, Bulgaria never declared war on the Soviet Union. The logic of this argument is reasonable to accept, but in the framework of inter-allied relations at the time, it is pregnant with negative inferences. Had not Russia declared war on Bulgaria, she could not have moved to occupy it, and could not sit on the green table to negotiate peace terms with her on the same level as the Americans and the British. Having declared war, she was free to move into the country, to occupy it, and then sit at the negotiations for a peace settlement. As things were to turn out, this presence of Russia on the diplomatic settlement with Bulgaria proved beneficial for the Bulgarians. It was Russia which thwarted the demands of the Greeks for territorial expansion at the expense of Bulgaria. Likewise Bulgaria retained southern Dobrudja, taken from Romania in 1940.

This is how September 9 came about. With the Russian armies advancing from Romania into Bulgaria, the Bulgarian government, having ordered its armed forces not to resist the Russians, had in effect opened the door for the Fatherland Front operatives to seize power in Sofia. A few military units, won to their side by Kimon Georgiev and Damian Velchev, proceeded to the Defense Ministry where the headquarters of the Government were, killed an officer at the door, and the revolution was over in a few minutes. One or two hours later Kimon Georgiev announced to the nation that the old regime was toppled and that the Fatherland Front had taken over. At the same time partisan units entered the capital and marched to the center with their submachine guns, with their raggedy revolutionaries, with their public meetings and vitriolic speeches and with shouts heard everywhere: "Death to Fascism." In Zelenikovo the "revolution" was carried by Parlapansky and Co. At 8.00 a.m. he stormed the citadel of fascism, the municipal building, with a revolver, ordered all clerks there to stand up, told them that they were to receive orders from him, and to continue their work until further orders. In no time the old Mayor, Mitiu Kraevsky, arrived and they all celebrated the victory of the Fatherland Front. The same take-over had taken place everywhere in the country. Day one of the Communist era.

V.

SEPTEMBER 9, 1944.
AT THE GATES OF HELL

"Through me one enters the sorrowful city;
Through me one enters among the lost
race. Justice moved my Maker;
Divine power made me, supreme wisdom
and primeval love. Before me were no
things created except eternal ones;
and I endure eternally. All hopes
abandon, you who enter here."
Dante, Inferno, iii, 1-9.

1. September 9, 1944: a Day to Remember

Where was I on September 9th? I was at home. It was Saturday. That morning my mother mobilized me to help her set the threads of the "foundation"for the loom. This is work which women do, but she had nobody to call on, so I had to fill in. She would be sitting on one side, would pass her fingers through the heddles, would tell me what threads I had to give her, depending on the pattern of the cloth she was going to weave, and sit there for hours. So, there was I sitting, listening to her instructions, separating the thin threads and killing my boredom with thoughts on anything that came to my mind. One thing that was always on my mind in those days was the war, the politics of the times, and the imminent future that was looming on the horizon.

Events were moving so fast that we could hardly take our minds off them. One thing was sure: communism was coming in our direction and there was nothing to stop it. After the incident of July 21, I was resigned to the idea that it soon would be here and there was nothing I could do to avoid the consequences of it. I was isolated from everybody and everything. I was continuing my going "uptown", just to show myself to the world, that I was still there, that I was not running away, that like everybody I was prepared to face it, whatever it was to come, but it all was a façade. Nobody dared to approach me to talk to me. Neither did I venture to approach anybody. It was as if I was a leper. It was known to all that I have been "arrested," that I was branded as "enemy" to those who were coming to take over, that once they were in, who knows what was going to happen to me? Nobody knew of my night conference with "them", nor I could tell anybody. I myself was not quite sure of the modus vivendi I had come to establish with Parlapansky and how long it was going to last. People were, in a way, afraid to socialize with me. Only Stefan Mangov continued to hang around me. His brother, Sava Mangov, was in jail as a functionary of the Communist party.

Stefan was very much interested in intellectual conversations and in politics. We had good time with him. He did not abandon me. We never discussed his brother. Much later it dawned on me that he might have sought my company to protect himself from reprisals. There was also Taniu Dangov, the Anarchist-Tolstoist and vegetarian, proponent of Esperanto. He was well known for what he was and nobody suspected him in collaboration with the "fascists". He was open to meet and talk with anybody. There was much that I had in common with him in our discussions on social and political matters. It was all the legacy of my pre-Seminary days, and I was not pretending to have altered my views. In these difficult days they were my only companions, if and when they were out there, with whom I met and talked freely. I was not, and I repeat, I was not trying to convert them to my nationalist and religious views. But considering the fast moving events, they were not often to be seen, and after purchasing my newspapers, I would just return home. There was also a young boy, Ivan Aksiysky, son of Tsaniu Aksiysky, otherwise known as "Daraka", the carder. He was impressed by our discussions, was not much into our level of knowledge, but was a good listener. Besides, I was very close friend with his sister, something that may have figured in our associations, and I suppose, his parents were not displeased of it, to say the least. After all, my intellectual superiority in the village was common knowledge and Tsaniu Daraka was one of the more sophisticated among the rest of the peasants. Sometimes in the past, must have been before I went to the Seminary, he had opened a small bar, very popular, and quite often we, my young company, would stop there, enjoy a *rakia*, plum brandy, and his delicious hordeuvers, boiled dry beans with chopped onions and parsley laced with vinegar. It was quite a popular hangout. As it turned out, following September 9, he all of a sudden emerged as Secretary of the local Communist Party. I was to presume that he was already deeply involved with the communists prior to September 9th, without anybody suspecting it. Much later it was to turn out that he had come out in opposition to the communist regime, had been in concentration camps and alone in the village had refused to enter the Kolkhoz.

I was to learnt much later that when on October 6th, the party leadership had considered the people to be murdered that evening – I will talk later about it – and my name had come up, it was him who had stood to prevent it. He had argued that I was too young, a child, to be punished so harshly. The community would resent it, and, after all, I had not done anything that justified even the slightest punishment layed at me. If I had some ideas different from theirs, I might change and be of use to them. So, my name had been dropped, and probably I owe my life to him. I wonder, if my friendship with Ivan and Vidka had something to do with it. In 1991 I met Ivan at Central Square. He was sitting with several old men … old – they were my age, but looked twenty years older. The conversation was going with all this way: "Do you remember me?", "No, I do not remember you." They told me their names and then I remember who they were. At the end of the bench was an old, ruin of a man, hardly any teeth in his mouth. He refused to give me his name. He insisted that I recognize him. I was listening to his speech and all of a sudden I recognized him. I recognized the staccato in his way of speaking. This was Ivan, all right. But what a change? The once vivacious, handsome young man that he was, was now a pitiful remnant of the past.

But during the weeks preceding September 9th, I rarely met these friends of mine, and when I went uptown, and did not see them anywhere, I just went home, read the papers and enjoyed the safety of my lonely life. All that was passing through my mind while I was helping my mother. I was splitting the threads and delving in all alternatives ahead of me in the new conditions, that were to develop in the upcoming few days. Such thoughts never left me, indeed. No matter what I was doing, I was constantly concerned that dark days were ahead of me. I was killing time with reading, day and night, under the vigil light over my head on the wall, or taking care of the animals in the fields. I felt cornered at some invisible wall, and sought escape in any work that I was assigned to do. Helping my mother at this moment, as tedious and boring as it was, at least it was a distraction from my loneliness.

But then something happened. It must have been about 9.00 o'clock a.m. when I heard some commotion in the street. I lifted my eyes to look what was happening … and who do I see? Parlapansky! He was heading towards our house, with a submachine gun on his shoulder, flowers sticking out of the barrel. I stood at the window and looked at him. There were a lot of people already on the street and some kids running around him. He noticed me at the window, waved with his hand and yelled: "Come on, what are you waiting, we have taken over. Hurry up to the Central square for a meeting." He was accompanied by a young partisan, also armed with a submachine gun, decorated with flowers, all over, gun and body. I did not wait for a second "invitation." It was not an invitation! It was an order. This ended my work for the day. I left my mother to continue alone, and fled to the street. By that time the partisans had already moved further up the street. I was on my way to the square where the municipality building was. People from all over were rushing there. By the time I arrived, half of the village was around. Lots of people were entering the municipal building and leaving. Here and there occasional gun shots were heard – more of excitement than anything else. People all around me were commenting that a new government had taken over in Sofia. Soon, I found my friends – Taniu Dangov, Stefan Mangov and Ivan Aksiysky. For the rest of the day we were to hang on together. By now it was quite clear who had been helpers of the partisans and the honor they were given to by all who congratulated them. Before arriving at the square, I had stopped at the little store of a young enterprising man, some distant relative, to buy something. It was packed with people. Someone, I do not remember who, came to me and tagged a red ribbon on my lapel, telling me: from now on this is the color to wear. Indeed it was. Much later, in America, as a professor of history, when I would be discussing the French Revolution and mentioning how Louis XVI was decorated by Lafayette at the Paris City Hall with the revolutionary cockade, two days after the fall of the Bastille, and had told him: "Take this, it is going to go around the world," I could not help not remembering this incident and share my feelings with my students. I felt uncomfortable wearing this red ribbon the whole day, knowing full well that it was a joke on me, but I could not take it off. Everybody was wearing it, and if I did not, it would look like a defiance. This was the last thing that I needed this day.

At about eleven or so, someone climbed the steps at the entrance of the municipal building and announced that a procession was to form on the main street and march to the place, or the grave, where a young man had been killed, a partisan, near the Yunchal. We joined in the procession. Hundreds of people were in the crowd. At the end of the village, the column left the road and followed a narrow field path, amidst brush, thorns, potholes, rough ground and dry grass. In one hour or so the place was reached, a large clearing, here and there with stunted bushes, and the leaders, partisans, lined up at one end, probably the grave or the death spot where he was killed. I did not understand, and did not inquire if this was the young man whose head was exhibited in the summer at the street or not. I felt that the less I asked and talked about anything, the better for me it was. One could easily be misunderstood or misinterpreted and with all this shooting going around, who knew what might happen? Listening to the speeches, and following the ceremonies, I was shaking inside of me, but tried to keep up my composure. All the time I feared that someone might meet me, point his gun at me and finish me there. Does anyone know what a devil may be resting on his shoulder? All that time I had to smile to one or another person, to exchange greetings for the happy event with one or another individual I had never spoken to or being in close relations with. They all spoke of the freedom which had shined over Bulgaria. I had the red ribbon on my lapel and many people took it as a sign that I was with the new regime. It was, after all, my insurance against suspicions. I felt that this was a hypocrisy, but what other choice was there? I could not deceive those who knew me better, but so did everybody in this crowd. Only weeks before that I was spattering the dust under my bare feet with hands tied behind my back and machine guns pointed in my face, and now I was displaying a red ribbon on my lapel. Well, I did not wear a coat, but there was a place on my shirt for that. My best protection however was that I was in a company which was beyond reproach where communism was concerned: Taniu Dangov and Stefan Mangov. Sometimes during the day Taniu confided in me: "These people, parading their submachine guns and shooting for what is and what is not proper, will not take us to a good end." I did not protest, but just noded in agreement.

As the speeches ended, with all the curses peppered with "death" for this or that and the shootings in honor of the fallen man, the "Memorial Service" was over and we had to march back. Instructions were given to march to the lower end of the village, where trucks were to pick the partisans to take them to Plovdiv, so we could give them farewell greetings. In one hour or so the entire procession reached the specified place, and lined up on both sides of the street. We were all exhausted, hungry and thirsty, but nobody complained. There were some ten or so trucks parked there. At one point the partisans appeared from the village, marching down in formation of three men to a line, to give the impression of a large number. I did not care to estimate their number but there must have been about one hundred men. All of them were armed with submachine guns. It was a raggedy assortment of characters, unshaven, their hair not cut God knows for how long, in a diversity of uniforms or civilian clothes – military, police, or mixed up, even high school caps and jackets. They all looked exhausted, undernourished, coarse, weather beaten, anxious, frowning and depressed, but all the time trying to give an air of joy. It was a pitiful sight of unhappy men, with all the glory which was bestowed upon them for their heroic fight. They lined up, feet away from us watching them from the side, awaiting for order to climb into the trucks.

My luck, right in front of me, I saw a boy whom I knew, and he knew me very well, Koycho Koev, from the village of Babek – three miles North of Zelenikovo. We had met so many times at the bus station in Plovdiv, hanging around there to meet people from our villages. He was student in the Business High School. We had argued with him to our heart's content on all matters of politics, religion and whatever. He was a short fellow, a little plumpy, with sharp eyes. He knew perfectly well where I stood and considered me enemy number one to the communists. Our eyes met, he perked up, raised the barrel of his submachine gun at me, without taking it off his shoulder and through his teeth threatened me: "Either you walk the right way, or this is going to come to you." He was gripping his submachine gun, I had my hands in my pockets. There was nothing I could say to him, except to nod with my head that I was getting his message. Fortunately for me, at this instant orders were given and they all moved to get on the trucks. If this had continued longer, I do not know how it would have ended.

Many years had to pass. In 1991, when my brother came to America, I heard of the unhappy ending of Koicho. He had fallen out of grace with the Communist party. He had been sent to jails and concentration camps. A few years later, meeting with the present Vice President of the Republic, Todor Kawaldjiev, at the Agrarian Center on "Vrabcha 1", I received from him as a gift the books of the prominent Agrarian writer Petko Ogoisky – two volumes of "Notes on the Bulgarian Sufferings," where he describes his experiences in the concentration camp of Belene. Ogoisky had shared the same camp with Koicho, known to them as "the Green General." I read in Ogoisky's book about Koicho: "Partisan, commander of Partisan battalion somewhere in the Sredna Gora was a Koicho Koev. Here, between the cold walls, he was meeting the sunsets and sun rises, psychologically distraught and confused in his incredible state. The "Green General," as he was known to all after someone had pictured him on a wall "newspaper" hanging in the premises, had been pushed to the brink (of mental breakdown). Every day he was to be administered a cold shower on the ground floor, together with a Stefan Ludev, a Legionair fuhrer from Gabrovo." (Tom I, p. 198) Further down Ogoisky quotes some prisoner, Raiko Mustaka, who comments that in the West, Palmiro Togliati (Leader of the Italian Communist Party) and Moris Torez (Leader of the French Communist Party) have been publishing newspapers of thirty or so pages, and still complaining that there was no freedom in their countries, and he, Mustaka, had continued: "What an insolence! Koicho Koev is right that we have to rampage at least a month in the country against the communists with two submachine guns and a sword in our mouth when they fall from power..." So this was the man, that is what happened to the man, who on this day, September 9, 1944, was pointing the barrel of his gun at me.

I vaguely remember that somewhere during the day, most probably at the trucks, I met Yovcho, but we could not have spend more than a few seconds with him. He looked as if he had fallen from another world, loitering as if he was dizzy, haggard, concerned, somehow lost his good looks, with ugly expression on his face. It had lost its freshness which once used to radiate around him. We exchanged a few words, small talk, and separated, never to see each other again alive. On October 13th he was killed at Stratsin in the war against the Germans. Most of the partisans stayed home, did not go to the front, scattered around the country to solidify their power. Why had they sent Yovcho to the front? Was he expendable? Had he fallen out of grace with the upper ups? This was to affect me. In him I had lost my best defender. In my most critical hour he had saved me. I must have already mentioned the fact that when his coffin was brought back to Zelenikovo for burial, his mother had sent for me to attend the civil memorial service. When I appeared, she leaned over his coffin and cried out: "Yovcho, Yovcho, if you had stayed with Spas, you would not be in this coffin now". At this time I have to mention that when I was arrested in December, a few weeks after the funeral, while I was being questioned, I mentioned his name as my friend who knew me best. Then one of the young communists, he was a little older than me, Tsaniu – I do not remember his last name, a vicious little thing – interrupted me, jumped at me and blurted out: "Do not blaspheme with the name of Yovcho. If he had been a friend of yours, I will go and spit on his grave." I went silent.

Observing the crowds during the day, I thought of the traditional "royal" holidays when the priest would celebrate a Te Deum service at the War Memorial near the municipal building, where, attended by public officials, one or two classes of the school for public, while very few people would stand, at a distance, and follow the ceremony. Now the whole village had descended there to honor the partisans, but I had the feeling that no one was feeling comfortable, everybody was scanning the place with concealed looks, talked very little, and if they ventured something, it was of the nature of flattery to the new rulers. I did not see expressions of genuine enthusiasm, except in the people who were known from before as communists, or sympathizers. There were many for whom this all did not mean anything and only waited to see what would come out of it. They had not compromised themselves with the old regime, nor had they given an ear to the communists. They were indifferent, they had nothing to fear, and acted as independents.

At last this day was over. We returned home. I was physically exhausted and emotionally drained. I had not eaten a thing during the day, but my stomach was tightened into a knot. I felt that any moment I would faint. After so many hypocritical smiles and forced accommodation to the new order of things, all the time to be on the good side, I felt my soul crippled and empty of all feelings of honor and dignity. I was overwhelmed by fear of the unknown, of the next day. The red ribbon on my lapel reminded me that I had become a slave to new, merciless masters. In front of me was the new world which would either smite me to a precipice of nothingness, or I had to reduce myself to nothingness and accommodate to it. How could I accept this new faith of communism which was being rammed down our throats without consulting us if we desired it or not? How was I going to digest this slime of cliches, threats, brutality soaked with blood which knew no line of separation between oppression and freedom, which called the oppression freedom and the freedom slavery to capitalist exploiters? One could have suffered all that without pain if he was born to be a chameleon and changed not only the color of his hair but also the color of his skin under it. But how a man with a sense of honor and dignity could deal with this barbarian power without experiencing spiritual death? In order to accommodate to the new world which was being borned this very day, September 9, 1944, a man had to be reincarnated, if there is reincarnation, to be born a communist without remembering his past life, and so to completely extinguish the pain in his soul. Unless there was such a reincarnation, there was no place for me in this new political order. In my soul I saw myself defeated by the events, rejected by the society of the new people, unable, even with disgust and fear, to accommodate, for which I had no inner tuning to that effect and for such attitude. I was standing at the beginning of a long road of torments and pain in my soul, and search for a way by which I could extricate myself from this precipice.

2. "St. Bartholemy's Night Massacre"

On September 9, 1944, Bulgaria was seized by an unimaginable political hysteria. One could hear all the time and everywhere the curse: "Death to Fascism," "Death to the Fascists." Rumors for murders and assassinations of people were circulating all over, as if the whole nation had lost its human soul. It was whispered about mass murders. Military officers were reported being killed at sight on the streets. Public officials and prominent intellectuals were said to have vanished, murdered or in hiding. Businessmen and ordinary peasants were having the same fate. The editor of the most prominent conservative paper, "Zora," Danail Krapchev, was found hung in a public bathroom in the city of Dupnitsa. Jordan Badev, prominent literary critic, and the editor of the comic newspaper "Shturets," Raiko Alexiev, had either committed suicide or had been murdered. The country was turned into a butcher's block. In the streets, in the cities and in the villages bloodthirsty gangs were roaming and calling for murdering more and more people. In this outpouring of hate, who could be concerned for guilt or innocence? The nation had lost its human sense and was behaving as if it was attacked by a pack of wolves. I still remember the headline, on the first page, on the first column of a new newspaper – "Svoboda" (Freedom): "Daylight has shined!" At the same time, as I saw it, this was the sun set of darkness. I kept hearing the curse: "Death!", "Death!", "Death!" This call caused those who were hearing it to tremble. Bulgaria was turned into a mad house. They continuously called for a "People's Court" for the guilty, but hundreds and thousands were being shot all over the country without a court establishing their guilt or innocence.

A week or so the village crier, beating his drum at his usual corners all over Zelenikovo, announced that the next day there was to be held a "People's Court" for the enemies of the people. All were advised to attend it at the Middle Square. The next evening the square was packed with people. Who dared not to attend? When I arrived there I saw the platform where the Court was to be held: on the steps of the barber shop of Yovcho's father. An old man, Diado Maniu Pashata, was standing up there, with hands tied behind his back. A large sign was hanging over his neck on his chest: "Enemy of the people." He stood there – an old, pale and frightened man – with his big protruding belly prominently exposed. He was not known to have been involved in political activities whatsoever. Maybe he was known to be anti-communist but I knew nothing of that. I knew him as a humble, meek man. He had a small coffee shop, where he sold cigarettes and small essentials for the household. When I would go there to buy something, there were always elderly men, in winter times, sitting around the stove and conversing in their own peaceful manner. As the meeting was opened someone made a speech, sort of an indictment. He accused the old man that when selling something on credit to poor peasants, he was overcharging them. This is all the enemity of the people that they could bring out against the poor man. Someone from the crowd pleaded with them to let the old man alone, that what they had accused him of, was not worth all that attention. Then he was admonished not to do it again, his hands were untied and he hurriedly walked down the steps, tears running down his face. All this was a grotesque performance not worthy the big hulabaloo.

A few days later the village was asked to attend a second "People's Court." This was a more serious affair. Up on the platform for trial were five or six "enemies of the people" – Koicho Manevsky, Diado Spas Tafrov, Marin Papaza, Dimitar Banov, and others whom I do not remember. Speeches against fascism and the fascists filled the menu for the evening. Those arrested were charged as being fascists and enemies of the people. But no specific crimes were brought out for which the captives had to answer. When the speeches were over, those who conducted the trial called on the people to propose what punishments should be imposed. Someone from the crowd suggested that it was enough that they had been humiliated, and finished his suggestion. "Let them go! Forgive them!" The people were asked if they supported this proposition and asked, if they agreed, to raise their hands. The entire village raised their hands. Then the hands of the arrested were untied and they were given a chance to say a few words. I remember that Diado Spas, the perennial clerk in the municipality who at the time was handling the requisition for the armed forces, somehow indignantly said: "I have not committed any crime. I was just doing my job. That job had to be done. You yourselves are already finding that someone had to do this job." Koicho Manevsky began with praises for the new rulers, that he has seen democracy in action and pulled his party card of old times, of the Social Democratic Party, may be in an attempt to curry favour and declare his acceptance of the "Revolution." He was interrupted by the fellow who wanted to spit on the grave of Yovcho. He shouted: "Please do not show party cards." Whatever he meant by that, and whatever his reasons were, I could not tell. The rest of them simply walked down from the platform without saying anything. I watched these proceedings and could not understand what kind of a game was being played after all their shouts of "Death", "Death." All this was happening in a village, Zelenikovo, while the national Radio was bombarding the country with the calls for death penalties. However, the worst was yet to come.

On October 6, in the evening, going uptown, I passed the church square. Up the street I saw an old lady, Baba Ivanitsa, running up and down the street, cursing the communists and wailing. As I learned later, her son had just been arrested. They were our distant relatives. Her brother-in-law was also arrested. She was pulling her hair and screaming. I did not understand what was happening. As I proceeded further up, somehow the crowd past the central square was getting bigger. I was alone but someone joined me. As we were walking I noticed ahead of us two men, with carbines held down, following Marin Papaza, the village bull keeper – a tall man, in his forties. His jacket was over his shoulders, but I clearly saw that his hands were tied behind his back. I realized that something was going on, and it was not for the good. Passing by the Municipal building and glancing at its window, I saw suddenly the lights going on in the main office. Back to the door, with hand on the switch, was the priest. He apparently was watching the passersby and when he saw me coming, turned the lights on and waved to me. He snapped the lights off at once. Was this a "Good Bye?" Did he know what was in store for him? Probably not. The next day, October 7th, around 9.00 a.m. I went to the municipal building to sign up for cutting wood in the forest for winter heating. This was something that happened every year. I walked in, went to the respective room, signed for my father and walked out. While walking in the hallway, I noticed men sleeping on the benches, covered with their overcoats, their boots still on, all covered in mud. I do not remember well, but I think one of them was Parlapansky. I did not see anybody around to talk to, but leaving the premises, I felt that something strange was going on. Further down I met somebody, I do not remember whom, and heard the horrible news.

That night six people were taken out of the village, having been arrested one by one, and had been murdered during the night somewhere in the Yunchal. They were the priest, Stefan Tafrov, Koicho Manevsky, Bano Darzhaliev and Georgi Darzhaliev – the brother-in-law and the son of Baba Ivanitsa, whom I had seen screaming and cursing the communists the previous evening, Marin Papaza, whom I had seen escorted, and Dimitar Manevsky. People whispered one to another the terrible event. I hurried home. Whomever I met and listened to the stories, was stunned and frightened. Much later, in exile, while I was collecting information on victims of communism, I noticed that many people, in different parts of the country, had been executed that same night. This led me to believe that apparently it was done by order coming from some central place, which would have been Sofia, and, ergo, from Communist Party headquarters. It just happened that that day. October 7th, the state Journal had published the Law to prosecute those guilty of war crimes. The special Courts, the "People's Courts," were being instituted for that purpose. My interpretation was that the murders were ordered from above, to murder people who, otherwise, brought before any courts, could not have been accused of any crimes. This way, murdering them without trial, would have been an arbitrary settling of account with leading "fascists." So, this is how the Bulgarian Bartholemy's Night massacres were staged.

Later on I was to find out that when they had been discussing whom to murder in our village, my name had come up for consideration. Who put it on the list and why, I never asked for explanations, but I could only guess, that some of the young bucks with whom I had crossed swords, may have asked for my head. Eventually it became known that it was my younger neighbor, Koliu Pekhlivanov, and my classmate in the middle school – Koliu Bureto, who wanted me dead. But in the Party Council the Secretary, that is the highest authority, Tsaniu Daraka, the father of Ivan and Vidka, otherwise always known as a reasonable and sophisticated man, had objected, arguing that I was just a kid and that murdering me may have negative reaction among the people. There was nothing that I could have been accused of, any way. My guess is that Parlapansky may have supported him and I had escaped the horrible end of my young life. The others had not been so lucky. The priest had been a marked man, any way. He most actively had been working against the communists. He was defending the Church, and they had targeted the Church and the priest as their archenemy. They were trying to win the youth for their cause, while the priest was trying to form a Young Christian Group in the village, as unsuccessful as he was. The bitter struggle between the two camps in the village was not a secret. When they had occupied the village on July 21, they had gone looking for him, but could not find him home. Later I heard that he was home but as soon as he realized what was going on, had hid in the chimney.

Following these events he had sought refuge out of the village, I understood to be some monastery. Having come to power on September 9, the partisans had urged his wife to persuade him to come home, that they would not do anything to him. After the second "Trial" in the village, when they had arrested and "tried" Koicho Manevsky, and the others, and had released them, their promises apparently appeared credible. Sometimes I have wondered if this "trial" by the "people" had not been a deliberately staged deception to entice the priest to return to the village. At one point, in the beginning of October, perhaps 4th or 5th, he had returned to the village, and had at once been arrested and kept in the municipal building as prisoner. On the 6th, in the morning, his wife came to see me home and pleaded with me to go to Plovdiv, to see Metropolitan Kiril, later Patriarch, and ask him to intervene with the authorities to have her husband released. I realized how dangerous this mission would be for me. I told her that if they saw me taking the bus, they would immediately pull me down and nothing would have come out of it. I was trying my best to lay low, so that I do not raise any suspicion that I was looking for an escape. I urged her to go herself, after all, she could fear nothing about herself. I do not know if she followed my advice. She had three small children at home and I did not know who was helping her. Apparently their plans for the murders may already have been set, and if my name was one of those for consideration to be murdered, if I tried to leave the village, and if I was stopped and questioned, I could have probably been threatened with death if I did not tell them the truth. If she had been followed to my home, I most certainly would have fallen in their trap.

As to Koicho Manevsky, he most certainly was an outspoken anti-communist. He was an older man, highly sophisticated, active member of the Social Democratic Party before 1934, and by all means known as anti-communist along these lines. Sometimes in the summer the partisans had taken his ox from the pastures and killed it for food and as a signal to Koicho that he was on their list of marked enemies. Information had reached me that he was one of the members of the "Ratnik" organization, which was outrightly a fascist organization, but I have never heard how he figured in that organization. Much later, I was to learn from Koicho's son, that some years after the murders, Parlapansky had been asked by his mother – apparently they had some distant relationship: "Why Georgi, why did you killed my husband?" To that he had answered: "Kako Vidke, we made a mistake." What a mistake!

As to the others, Bano Dimitrov and Dimitar Manevsky, all I heard was that in the events of 1923, they have been on the side of Tsankov, the extreme rightist, but I never heard of anything reprehensible that they might have done. Most probably in those days, after the September 1923 uprising initiated by the communists, they may have participated in arrests and beating the leftists. Manevsky had served for many years as a police chief in Brezovo, but Darzhaliev had gone into the bakery business and out of politics. After the murders on October 6-7, I heard the story that someone went to Encho Tantikov, father of Spas Tantikov, who was my classmate in the middle school, and two years ahead of me in the Seminary, and had told him: "We have them now in the *Izba* (the basement of the municipal building, which was serving also as a local jail, for a few days any way), come and avenge yourself". It all suggests that in those days of 1923 Encho had been a leftist, had been arrested, beaten, and released. This is what was suggested to him, "to return it to them." Now an old man, he had refused, telling the messenger: "No! I do not want it! I do not want to create hatred between my children and their children. What happened long time ago, is a forgotten story." Years later, when this noble man had died, and the funeral procession was going down to the church, the son of Koicho Manevsky, Georgi, seeing it to pass by, impelled by some force inside of him, without much thinking, had joined in. I have heard this story from him, the latter part, that is.

I never heard what the crimes of Marin Papaza were, if there were any. As to Georgi Darzhaliev, a distant cousin of mine, a young man at the time, may be 24-25 years old, whom I knew, I must say that he was a boisterous fellow, unrestrained in his language, somehow on the adventurous side. It was said about him that he had joined with the military in companies searching for partisans. He had been married to a leftist's dauther, she had left him and if I remember, had tried to shoot her. She was first cousin of my nemesis, Koliu Merazchiysky. The Merazchiysky's were now in power, and if his murder was asked by them, I wouldn't be surprised. But it was never suggested that he had blood on his hands in these events.

As time passed, little by little, some details of the murders themselves have come out. It is said that all of the men arrested for execution, had been taken to the Yunchal, lined up at a ravine and shot or hacked to pieces. The priest had been left to watch from the side. Then they had proceeded to cut his tongue and asked him to perform a funeral service. They had cut his right hand at the elbow and asked him to make his cross. What had happened, is not clear. Such gory details may and may not be true. When it was all over and they were about to cover the bodies with dirt, Georgi Darzhaliev had yelled that he was still alive and begged them to finish him. Then a young man, brother of the Mayor, Mitiu Kraevsky, had gone with a pick and split his skull. What a horrible scene? How any human being could have indulged in such bestiality? After completing their mission the murderers had returned to the village. What I had observed in the hallway of the municipality, the sleeping men with the muddy boots, were the murderers. It was later talked that the bodies of the victims have been washed out by the rain and dogs had been gnawing on their bodies. Someone had been dispatched there to cover the evidence of this brutality. A week after all that, someone found the sweater of the priest, with his identity card in the pocket and had brought it to the municipal authorities. Nothing was heard of that anymore. I cannot help but think that I could have been one of those executed in the dark night of October 6-7.

The murders in my village were a senseless exercise of power. Those who ordered these murders may have deliberately sought to plant hatred in the communities and commit forever those who would have bloodied their hands. Or they may have sought to sow terror in the hearts of the rest of the population, so that no one would ever dream of opposing the communists. It was accomplished, because following these bloody orgies, repeated in thousands of villages, frightened people to their wits end and they would not dare challenge the power of the party. Of all the countries, "freed" by World War II, Bulgaria suffered the worst bloodbaths.

The terrible fate of the six at the hands of the murderers, apparently was avenged by a Higher Judge. Some years later the son of Parlapansky, his only child, has been killed while visiting Turkey on business. Another one of those who were after my head, a boy I remember only by his first name, Ivan, their house opposite the church, at the time one of the better houses in the village, now in ruins, had jumped from a plane, his parachute had not opened and was killed falling to the ground. Rumor has it that he was pushed out of a plane without parachute. The parachute has never been found. A third one, also one of those who was after me, Mincho the Mole, had committed suicide. Koliu Pekhlivanov, komunata, ended mentally disturbed. His brother had chased him out of their house and would not let him in. During my last visit to Zelenikovo, he was pointed to me, sitting on a bench, looking at me from distance of twenty feet and grimacing in a horrible manner. I did not feel sorry for him. He was the one who had been pressing for my inclusion in the group for execution. Parlapansky himself had been demoted. His incompetence and ineptness had been recognized very early by the party leadership of the district, he had been stripped of all his positions of power and had found himself begging for a job from one of the village "reactionaries", Todor Grozev, an enterprising man who had been in and out of jails, but had managed to establish himself as the leading entrepreneur. But on September 9th, and after, these people were waiving the machine guns and sowing terror all around them.

3. Back on the Track of Politics

At that time, in the wake of these murders, I felt dead in my soul, crushed in despair and without any prospects for the future. When I would leave my home, and walked on the street, or ventured up to the chitalishte and to the central square, I lived in terror. I felt that all the eyes were focused on me and that everybody was avoiding to meet me or talk to me. I was a marked man. The young group with which only a few months before I associated, communists and non-communists, formed the Komsomol in the village. The leaders were my enemies. They were inviting everybody to join in, and everybody was joining. They would not invite me. I was in complete isolation. I was ostracized. Even Todor Vassilev had found the way to get in, and started avoiding me. I would go to the central square, I would stand outside, people would be passing me, but I was avoided as a plague. Nobody stopped to talk to me. Everybody was looking for cover, and I was a danger for them.

Once I found myself near Georgi Luchev, a graduate of the Seminary a few years ahead of me. We used to be friends. He turned to me in the very beginning of the conversation with the warning: "Watch it that you do not step on a snake's tooth." What he meant was that I should not turn to some political subject where he anticipated that I would be critical of all that was going on. I immediately realized that he had sold out, that he had passed to their side and that he did not particularly appreciate my talking to him. I understood him to say that if I said anything on the political situation, critical of those who had taken over, he would report me to them. He may have been asked to cooperate. This was the last I saw of him. When I returned to Bulgaria in 1991, he happened to be in a group around me, and asked me if I recognized him. I did not. How could I? I did not recognize anybody. So he introduced himself. I never saw him again, he never came to see me as many others did. I used to see him to stand on the sidewalk, stare at me, and then turn away. I suppose, I could not and I should not judge him. He had accommodated to the regime soon after September 9th, and he had survived, some sort of a clerk in the local cooperative*.

---

*I found in my dossier on May 15, 2000, that the informers of the State Security, working on my case and against my brother had half a dozen pseudonyms. One of them clearly identify himself as a seminarian from my village whom I supposedly trusted.

Sometimes I wonder what I would have done if they had invited me to join in the Young Communist Organizations. I probably would have joined. The ostracism was killing me. If I had to pretend, then, well, I would have pretended, and eventually, if I had been given opportunity to demonstrate my capabilities and make my contribution, I probably would have gotten involved and in the course of time would have integrated in their group. This is where my rehabilitation would have begun. But I was not given this opportunity. I vaguely remember that I wrote to Parlapansky in that sense, hoping that he would help me to join in, but he was too busy with his management of the Party affairs. I never heard from him again.

All I was looking for now was a shelter to save myself physically, and if I could work from inside of the new regime for my inner convictions, then this was the way to go: to join them. And pretend! But it was not going to be. My enemies in the young Communist group were steamed against me, they hated me so much, that even if they accepted anybody else, they would not accept me. It was hate. Pure and simple! There was no place where I could go. I could not leave the village. Where was I going to go? Nowhere!. I had to stick it out there. After a few days, I stopped going out. After work in the fields, with the sheep or with the cows, I would come home, eat what I could and retire to my room, with a vigil light hanging over my bed, and a book in hand, or newspapers, and read. I would go to the *Chitalishte*, I would buy newspapers and go home to devour every bit of information printed there. There were long political discussions, articles by communists or non-Communist writers, much of it political history with pro-Communist interpretations, still information which I had never seen before.

Meanwhile some important events were taking place in Sofia which could not escape my attention. Dr. G.M. Dimitrov, the most prominent Bulgarian Agrarian leader with the Allies, soon after September 9th returned to Bulgaria and became Secretary of the Agrarian Union. The man who all along was representing him in Bulgaria, Nikola Petkov, was now, as of September 9th, Vice Premier. The legitimate, majority Agrarian leaders – Dimitar Gichev, Vergil Dimov and Konstantine Muraviev – were in jail, awaiting trial as "War Criminals." They were members of the government overthrown on September 9th. Behind Dimitrov were the Americans and the British. In the spirit of the war time alliance of the United States and the Soviet Union, the Agrarians led by Dimitrov were political allies of the communists and the "Zveno" people, who for the third time had overthrown an Agrarian government in Bulgaria – in 1923, in 1934 and in 1944. I was watching these developments. On October 12, six days after the mass murders, Dr. G.M. Dimitrov and Traicho Kostov – number 3 man of the Communist party, at the time number one, because the top leaders were still in Moscow – signed a political declaration, as leading men in the Fatherland Front, to the effect that the Communist Party and the Agrarian Union, publicly express their resolve that outside the Fatherland Front there were no other democratic forces in the country, that of the Democrats of Nikola Mushanov and the Agrarians of Dimitar Gichev were fascists. The last one was an obvious lie, a historical distortion contrary to every bit of information which I had on the subject.

During the war, and for many decades before it, Mushanov was leader of the most reputable Bulgarian Democratic Party. He had never compromised himself with the right wing authoritarian or fascist parties. He was Prime-Minister of Bulgaria and Deputy Prime Minister in the cabinet of Muraviev. He was now in jail, even though he had fought against the pro-German policies of the King and his governments. It was a matter of historical record, and no man with his right mind would deny his democratic credentials.

Dimitar Gichev led the moderate majority Agrarians between 1928 and 1944. Fought by the royalist governments, emerging as the leading force against the intervention of Bulgaria in the war, recognized as the most powerful representative of the traditional democratic forces in Bulgaria before and during World War II, he was now, by virtue of the Kostov-Dimitrov declaration of October 12, declared a fascist. G. M. Dimitrov, as I was to learn later, had been the leader of "Pladne", a splinter pro-leftist, pro-Serbian wing of the Agrarian party, foe of Dimitar Gichev, who, incidentally, had been best man at his wedding and god-father of his children. Documented information brought out by his son-in-law, Prof. Charles Moser of George Washington University, Washington D.C., in the biography of his father-in-law, proves, that he, Dimitrov, had been a Serbian and English agent and spy. He had escaped the police in February 1941, a week before the coming of the Germans, in the midst of a plot , discovered by the police, to overthrow the Government in Sofia and join Bulgaria to Yugoslavia. Subsequently he had been under the protection of the British and the Americans, broadcasting to Bulgaria against her government. Formally head of the Agrarian Union, the "Pladne" group that is, he had advised his men in Sofia to join with the Communist Party in the Fatherland Front coalition and on September 9th, the leader of his faction in Bulgaria, Nikola D. Petkov, emerged as Deputy Prime Minister in the Government of Kimon Georgiev. Though that Government of sixteen ministers had only four communists – Traicho Kostov, Deputy Prime Minister, Anton Yugov – Minister of Interrior, i.e. of the police, renamed now People's Militia, Racho Angelov – National Health, and Mincho Neichev, Minister of Justice – the government from its inception on September 9th, in Sofia and in the whole country, was in the hands of the communists.

This Government was a camouflage of the Communist Party, which had taken control of national affairs – from local councils to the central government. Now, this same G. M. Dimitrov, who returned to the country not with the British or American Missions, but by making his own arrangements via rail-road, by way of Istanbul, was thus outlawing, with Traicho Kostov, all political parties and entrenching the coalition of the Fatherland Front, dominated by the communists, to govern the country. So, this was the verdict of this infamous declaration of October 12th, 1944.

In those years I did not have the vaguest idea of the political set up in the country. All that was known, politically, was the pro-German governments of the King, the underground Communist party and the anti-Communist pro-nazi Legionairs. All libraries had been purged of political literature, all political newspapers closed, all political organizations outlawed, and all political discussions critical of the existing regime, punishable. As a young man, I knew nothing of the old politics of the country. The choice was between supporting the regimes of King Boris "without parties", which I could not bring myself to do, or joining the Communist party. I could never forget how the tax collectors came to our house and carried away the buckets of my mother; how the owners of the land my father cultivated came to the threshing field and took half of his work as a rent and all humiliating experiences when I could not pay the school taxes in the Seminary. At the same time I could not join the Communist party. Its philosophical foundations, which clashed with my religious beliefs (In my time they had made this part of their ideology the cornerstone of their policies in relation to the church) and its foreign policies did nor meet my understanding of the world. The third choice was the Legionairs. It proved to be very difficult for me to erase from my mind what Yovcho told me they were: "Bastards of the wealthy." What was left, was to stay neutral. I did stay neutral for four years. When I joined them, out of necessity, rather than out of conviction – to shelter myself against possible reprisals for staying out – I continued to feel that I was independent in my thinking.

Now things had very quickly changed. In a few weeks I was educated in the new politics. I studied the political space around me. There was a forbidden zone in this space, where one could not step without risking to burn himself. It was the zone of fascism – pure and unadulterated. It was being attacked as a plague and no one dared to come even close to it. The other zone was the Fatherland Front parties – the Communists, the Agrarian Union of Nikola Petkov-Dr.G. M. Dimitrov, the Social Democratic Party of Dimitar Neikov-Grigor Cheshmediev, the "Zveno" group of Kimon Georgiev and Damian Velchev, and the "independents". Independents were those who had demonstrated their support for the communist party in the past and had helped the partisans, but were not attached to one or another party. My brother was an Agrarian, and as much as I could deduce – a left wing Agrarian. He became immediately a heyward. He was illiterate, but still he was appointed for some time to serve the new municipal government. In a short time he was dismissed, probably because of me. He had never discussed politics in the house, which indicates that he was not much informed.

In theory all parties were open to everybody to join in. I was caught in a quandary. I could not join the Communist Party. That was made clear in the very beginning. There was no Social Democratic Party established in the village. Koicho Manevsky, its most prominent member in the past, was murdered. All of a sudden, my teacher in the past, and leader of the "Brannik" in the spring of 1944, emerged as a representative of "Zveno" Group. "Zveno", the party, rather the circle of Kimon Georgiev and Damian Velchev, were a handful of people in Sofia, who in 1923 had overthrown the Agrarian regime of Stamboliysky, when the latter was assassinated. In 1934 they, the same group, had engineered a coup d'etat and had overthrown the democratically elected government of the National Bloc – Nikola Mushanov and Dimitar Gichev. They openly advocated a political philosophy, which, if it was not for their left-wing leanings, was not much different from fascism. A circle of military, intellectuals and politicians, they had climbed to the top of the political scene, subsidized by Yugoslavia, enemies of Bulgarian monarchy, and rejecting the democratic system of parliamentary democracy. They did not have a broad social base. If they had not joined with the Communist party, supporting the Soviet Union, they would probably have been taken as fascists. Peter Dimitrov apparently was declared "Zveno" representative and included in the local Fatherland Front group just to have a third political group – with the Communists and the Agrarians.

So, if I had any inclination to re-enter public life and the political debate, I had no other choice, but to join the Agrarians. I was badly hurt by the communists. The events of July 21, September 9th and October 6th, my personal humiliation, the ostracism which was imposed on me, my bitterest enemies running the show, all this turned me off on communism as a possible option of adaptation to the new realities. There were also the Agrarians. There was much that I did not like there. I did not like the fact that they joined with the communists in forming the government on September 9th, but there was G. M. Dimitrov who had just returned from the West, and presumably was committed to stand for western democracy in Bulgaria. Besides, in the middle of October, I somehow came to believe that the Fatherland Front was a political coalition and that this coalition would go to parliamentary elections where the established parties would compete for the national vote. If this was the case, I might find an opening to re-enter public life, with my honor and dignity intact.

I was not far from the Agrarians in ideological terms, and in any circumstances I would side with an Agrarian party. Now that the Agrarian Union was recognized as a political party and open for membership, I saw no reason why I should stay out of it. In fact, if I was to oppose the communists in a democratic way, my only available alternative was to join this Union. It did not occur to me at that point that this communist-agrarian partnership would break up and that agrarians and communists would start fighting each other. I was deceived that the new democracy was here to stay, and it was safe to join the Agrarian Union. I should have recognized the signs of the times, that the communists would not allow the agrarians to practice democracy outside of their control. I was inexperienced in politics. I believed what I was reading and hearing, and I decided to join in. If I could not join the Communist Party, I could join the Agrarian Union, which did not have the philosophical and political trappings of communism repugnant to me. And I proceeded to apply for admission in the local chapter of the Bulgarian Agrarian National Union.

I asked the Secretary of the local chapter, Ivan Merazchiysky, known in the village as a former Gichevist, if they would accept me as a member of their party. He was glad that I expressed this desire and invited me to the next meeting, where he was going to sponsor me for admission. The meeting was held on October 26th, my 22nd birthday. He submitted my name for admission to membership in the BANU. Several other leading agrarians in the village spoke well of me. It appeared that I was not going to meet any opposition. But there was opposition. One man stood up and threatened that if I was admitted in the Union, the group will split. He vehemently opposed my admission. He had sided with the communists long time before the war, supporting the "Pladne" agrarians. He was now one of the leading supporters of the communists in the village. He told the members that I was a fascist, that I had been against the partisans, that I was against the new regime. He was listened to, but when it was proposed to vote my admission, the overwhelming majority raised their hands.

When in 1991 I met the son of the man who opposed my membership – his father already dead – he told me that it was his father who had fought for my admission in the BANU. I was there, I witnessed the whole thing, how his father mercilessly attacked me. I was puzzled by these assertions but who knows, maybe his father had misrepresented the whole affair to him. I had not forgotten what had happened. When it was all over, I had the uneasy feeling that I had put myself again at the barrel of the communist gun. When I was told this story, I did not contradict it. Why should I continue this vendeta? Let evil sleep under the rock, as the Bulgarian saying goes.

So, I became an Agrarian. But there was something there which did not sit well with me. Prior to October 12 I was prepared to take G.M. Dimitrov. After all, he came from exile in the West. Even if he was in the Fatherland Front, he was not a communist and one could follow him without a sense of surrender to the communists. But this allowance for his cooperation with the communists in a coalition government was severely undermined with the declaration of October 12. How in the world anyone would deny the political credentials of the stalwarts of Bulgarian democracy, Gichev and Mushanov and dump them in the pool of Hitler and Mussolini? By that time I was already well versed in Bulgarian politics to appreciate these two leaders. Gichev was the successor of Alexander Stamboliysky, emerging side by side with Nikola Mushanov in the early 1930s as the most respectable political leader in the country. Now G.M. Dimitrov and Traicho Kostov proclaimed them to have been fascists. I had already read about G.M. Dimitrov and his

Agrarian faction, the "Pladne" group, as having been subsidized by the Yugoslav government to overthrow the regime of King Boris. Much later I was going to learn that he had been recruited to serve the British Intelligence Services by the Serbian Agrarians and British spies – Milan Gavrilovich and someone by the name of Tupanianin. In 1941 he was accused of conspiring with a Georgi Vulkov aganst the Governmenr and when he was sought to be arrested he hid in the British Embassy and in a few weeks spirited out of Bulgaria in the baggage of the British diplomats, to start up broadcasting from the Near East anti-German and anti-Bulgarian programs. Much later, thinking of these events, I was to question his influence with the Allies. I was to ask why he did not return to Sofia with the first plane of the British and American representatives, but was left to make his own arrangements for travel by train. For all that there was to it, he was not given an official welcome in Bulgaria, except for the agrarian leaders. Was this done by the Allies, so not to irritate the Russians? Was the cold war about Bulgaria already under way, and Dimitrov looked at by the Allies as a tripwire for Allied-Soviet relations? Was he designated by the Allies to undermine the communists and take Bulgaria to their side? But he definitely was seen by the Soviets as an agent of England and the United States. As far as we were concerned, having experienced the sting of Soviet power, preferred to look at him as a messiah among us, the man who was fighting to save us, against the forces of Stalin. May be he was! May be he was not! The Declaration he signed with Kostov in October made him look like a pseudo-messiah, but we kind of ignored it.

We chose to look at this whole thing in a different perspective. In these difficult times, he was the only hope which we had as a defense against soviet communism. Even the declaration which he signed with Traicho Kostov we chose to take as a maneuver to appease the communists, so that he could, after all, built the BANU as another force in the country, a rival of the communist party. The BANU was still considered to be a formidable political force, which, if organized and compete in democratic elections, could sweep the communist party out of existence. The said declaration could easily be taken as a license for rebuilding the BANU, as an alternative to the communist party. The immediate events following this declaration were to confirm that such an interpretation was right on target. I personally perceived in this set of circumstances a crack through which I could squeeze into the political complex, within the parameter allowed by the communists, without becoming myself a communist, but simply joining the Agrarian union. I had other choices too. The political circle "Zveno" was there too. So was the Social Democratic Party. They were equally respectable alternatives to the communist party, though I never looked to "Zveno" with any sense of approval, but now, among all other political groups, they were attracting the hard core right wingers, which I, in view of my previous associations, were well qualified for joining them. But I was an Agrarian by conviction and surveying the political landscape around me, found the BANU the best choice to re-enter politics. Even more important, as a Legionair, I had perceived the ideology of Agrarianism as the most attractive political philosophy for me.

I was not favorably impressed by the group of Kimon Georgiev and Damian Velchev any way. As I was to study Bulgarian politics in the future, I came to a conclusion that they had operated on the Bulgarian stage paid by, and serving foreign interests. In 1923 they had overthrown the government of Stamboliysky in the interest of foreign powers which were afraid that the Bulgarian maverick Prime Minister, like Kemal Ataturk of Turkey, could easily turn to Lenin and make Bulgaria a Balkan outpost of communism. At that time Lenin had publicly defected from the communist line of total socialism, had proclaimed the New Economic Policy, which was not far away from the ideas of the Bulgarian Agrarian leader. One could speculate, and it has never been suggested, that if Lenin had lived longer, Stalinist socialism might have never been forced on the Russians. Stamboliysky flirted with the communists and this threatened the reparation-collecting Western Allied powers. As Germany had already moved to a policy of rapprochement with Russia at Rapallo – April, 1922 – so also could Bulgaria. Whether this did not enter into the making of June 9, 1923 coup, where Georgiev and Velchev played the leading role, has not been established, but that this coup had this effect, serving the interests of the powers which had ill – treated Bulgaria at Neuilly, is beyond question. In 1934 these same men led another coup, against a government of the Democrats of Mushanov and the Agrarians of Gichev, which again favored the same powers. By that time, the forces of the coup of 1923 had split – one branch leaned towards fascist Italy, another to France and England, with their protégés in the Balkans – Serbia, Greece and Romania. It had become abundantly clear in 1934 that the group "Zveno," which by that time had appeared on the Bulgarian political scene, had been subsidized by Yugoslavia, and then, again at the services of France and England. This put them right against Bulgarian nationalism and independence in international affairs, when Hitler had just risen to power. In 1944 they were at it again. Now they joined the communist party and overthrew the agrarian-democratic government of Muraviev-Gichev-Mushanov, now in favor of the Soviet Union. Sure, they were admittedly a bourgeois political party, but they never appealed to me, though the more conservative and reactionary forces in Bulgaria leaned towards them. The Agrarian Union was a better call for me. If the Dimitrov-Kostov declaration was to secure the freedom of the BANU to rise again as a key party in Bulgaria, this was the way to go. I was not interested in the Social Democratic party. It always had been a non-entity in Bulgarian politics anyway.

Be that as it may, my joining the BANU immediately turned sour. I was assigned the task, together with Matthew Palamuda, to organize a youth group in the village. Few weeks later Dr. G.M. Dimitrov scheduled a mass rally for Decenber 19th to be held in Plovdiv. He had already several such rallies and hundreds of thousands of people had flocked to them, a formidable challenge to communism. His overwhelming success had alarmed the Communist Party leadership. It obviously was not an Agrarian affair. The entire anti-Communist world in Bulgaria was using the occasion to demonstrate against the reds, indicating that if there were free elections, the communists would be thrown out of the government. They were not to let this happen. They lowered the boom on him. Who was assigned the task of organizing the group in Zelenikovo to go to Plovdiv? Yours truly!...

4. In the Jaws of the Beast

Things did not go as expected. On December 15th, about 6.00 p.m. someone knocked on our outside door. My father went to see who was there. He was told by the "visitors" that they wanted me. He called me. I walked out there and saw three armed men, with carbines on their shoulders. They were led by Stahia Pekhlivanov, first cousin of Natcho Pekhlivanov. Our houses were adjacent to each other. The second man was Ivan Patura

(Ivan the Gander) another neighbor, a tall and slim illiterate man who some weeks earlier had mistreated my younger brother and I had written a long letter to Parlapansky appealing for protection under the laws of the country. He ignored it. The third man was Natcho Pekhlivanov, father of my "friend" Georgi Pekhlivanov. He seemed to be reluctant of all that. My younger brother told me when I returned in 1991 that this man, Natcho, a low level official in the Kolkhose in the village, had taken him under his protective wing and had treated him as a son. He tied my hands behind my back. They told me that I was wanted in the municipal building and that they had come to arrest me. They asked me to put my hands behind my back, tied my wrists, and with guns pointed in my back followed me to the Municipal building. As soon as I walked in the hallway, I was ushered into the room of the village cop where a group of unknown to me individuals, dressed in military uniforms, apparently partisans, were waiting for me. As soon as I walked in, one of them, sort of a stocky man, with a vicious expression on his face, asked me unceremoniously: "Come here you S. O. B! So, you have taken it upon yourself to fight the Communist party?" I was looking at the floor and answered him in a low voice. "I have never had any intention to fight the Communist party." Even before I finished he loaded a crushing blow with his hand on my face. My seminary cap flew away. I never saw it again. I lost my balance and was going to drop to the floor, but one of the other partisans caught me and stood me up in front of my torturer. He was staring in my face and screaming: "Tell us the name of the Legionair who had given you orders to join the BANU?" This was a news to me. I had no contacts with any Legionairs since I left the Seminary in October 1943. I told him so. I said that I had had no contact with any Legionairs, that I have been Legionair only for a few weeks, but he did not accept my answer. I did not know that somewhere in the country, some leaders of the Legionair movement, or so it was said, had instructed the Legionairs to join *en mass* the Agrarian Union.

I had no idea that such an order, or instruction, existed. This was the first time I heard of it, if it was ever issued at all. It is possible that the leaders of the Legion, Dr. Ivan Dochev and Co., who had escaped from Sofia only days before September 9th, gone to Germany and speaking by way of radio "Donau", the German-Bulgarian Broadcast station, had issued such a call, but I never had access to a radio and never heard such an instruction. Nor I heard from the man beating me up that there was such an instruction. But my captor was insisting that I must have been in communication with someone and that I had joined the BANU on instructions from above. In vain I tried to tell the man, and those around me that I had joined on my own, but they would not accept my explanations. After every answer I was slapped in the face so bad that sparks began flashing in my eyes. At one point the man who was interrogating me and beating me, kicked me in the left leg above the knee with his boot. The pain was so excruciating that I could not stand any more and collapsed to the floor. The kicking continued, all along my legs, my back, my ribs, blow was followed by blow, until I lost cousciousness. They had then dragged me to the basement and left me unconscious in the corner.

When I regained cousciousness, I looked around and saw that I was not alone. Some thirty or so people were in the basement. But no one appeared to have been beaten. Only Kuncho Tafrov, the brother of the dead priest, appeared to be convulsing in one corner. Later it was said that he had been asked to jump from the window of the cop's room, but had refused, had lied on the floor and had been beaten there. It was said that while he was being beaten and asked to jump out of the window, armed guards were stationed outside to shoot him, for attempting to escape. So I was lying in the corner and recognized my blanket from home. My father had brought it in and was allowed to push it through the narrow basement window. Someone from inside had received it and had covered me with it. I was in excruciating pain. I tried to get up, but my left leg was like a piece of wood. I could not bend it without feeling the unbearable pain. In the morning I was called up again and practically carried by the partisans upstairs. Now they were less brutal with me and tried to tell me that all they wanted to know was who had given me instructions from the Legionairs to join the BANU. I could not tell them. What I could tell them when there was nothing of the kind? They returned me back to the basement. I held to the walls and the banister and managed to go down the stairs. They did not beat me up any more.

All those who were arrested with me were all people known from before as non-communists. There were no Agrarians in there. Among my fellow prisoners was the School Director Peter Videv Zidev – a very kind man, but strict disciplinarian in the school. I never had any contact with him before. He lived in the upper end of the village, he was too much preoccupied with his school work which he was taking seriously. He had created a model of a village school and protected the buildings as if they were his own house. His wife came to visit him the third day after our arrest. They let her inside, at the top of the steps. She had brought him food. I was nearby and she took one apple and gave it to me. I had not eaten since my arrest. Apparently no food was allowed for me, and, as I remember, I did not feel a bit of hunger. My stomach had shrunken out of fear and out of pain that food was the last thing I would think about. The apple which Zideva gave me, while intently observing me with sorrow in her eyes, was so delicious that I never forgot the jesture which she made in this most trying of circumstances. I remember that as I took the apple and took a bite, I suddenly burst into tears. I was inconsolable and it took me several minutes to compose myself. I did not feel humiliated. I felt totally abandoned to the beasts in this world of hatred and naked power. I was morally and spiritually crushed to the ground. Mrs. Zidev extended a loving hand and caressed me behind my neck and my cheek. I have never forgotten her kindness.

There were some disgustingly repellent experiences in this basement. One evening there appeared Georgi Nachev, the former "Ratnik", with a carbine in his hands. He was in officer's military uniform. He came to me and loudly said that that evening Christo Statev (Minister in Hitler's Bulgarian Government in Vienna), was going to speak on radio "Donau". He asked if I wanted to hear him. Sitting on the steps and playing with his gun, sliding his palms over the barrel of the carbine, he was staring in my eyes when uttering his question. This was the same kid with whom we grew up together, played together and studied together for many years, to continue our friendship even after we left the village, studying in Plovdiv, and socializing as teenagers during our vacations in the village. I do not know how he brought himself to that level, to kick an agonizing man, a friend in the past. How many long summer evenings we had sat together and discussed all sorts of issues, agreeing and disagreeing, but never hostile to each other? I turned around and left him gloating behind my back. How did he turn around, and how he concealed his communism all these years, is beyond me. I knew of his leftist leanings, but never realized that he was so deep into it. Or he was just a turncoat, joining with them in the last months before September 9th. During the following years he vanished from the village and I gathered that he was working for the Militia.

I saw him in 1991 on the street, in the village. He came to me and we shook hands. I had forgiven him, or all memories of that dark past had faded. I was glad to see him. He was kind of cold and taken aback. As to his brother, Stancho, who was younger than us,

I was genuinly glad to meet and embrace him. I had nothing against him, I even did not know what he had done all these years, only, I believe, he had been a teacher. The next year I visited Zelenikovo, 1992, they were not so forthcoming. A newspaper "Nova Era" had published my IMPRESSIONS of Bulgaria of the previous year, where I had attacked the Communist regime, and the plight of Bulgaria in such a scathing way, that the village communists had turned bitterly against me. A retired dentist, Georgi Kraevsky, brother of Mitiu Kraevsky, the mayor before and after September 9th, had called the peasants in the village, had read them my *Impressions* and had viciously attacked me. Some people, former prisoners and concentration camp inmates, ordinary farm workers, came to me to advise me not to write such things.

The two Pekhlivanov brothers, had become aware of my articles. One day, when I was taking a walk with my brother down to the collective farm, and had to pass by their house, I saw them from far away standing outside. As they saw us coming, they ran inside the house, not to meet me again. But, for some reason, I had nothing against them, it was all forgotten, but they apparently still feft embarrassed or hostile towards me. After all, they were beneficiaries of the communist system. But my last meeting with Georgi was a different story. It was in 1997. An article written by me had appeared in "Duma," the former Communist party's newspaper where I was critical of the UDF Government's foreign policy. I had taken the position that Bulgarian national interests had always been well served under Russian protection, that our national security was threatened by the Western powers who always had supported the neighbors of Bulgaria at their expense. Georgi, or as we called him way back "Gocheto," told me that they had discussed my article at some reserved officers meeting and was pleased of the positions I had taken. On this occasion I told him that I have been always independent in my thinking, though I have not been always understood and have found myself in hot water. Just like in older times.

But let me go back to the Municipal basement, which I found myself in on account of G. M. Dimitrov's rally appointed for Plovdiv on 19th of December. That day my mother told me later, my father was awakened by a horrible nightmare. He had dreamed of me as decapitated. He had exploded in desperate cries and had cried for a whole hour. My mother told me that she had never seen him crying.I do not remember him crying either. I imagine the ordeal of this old simple man, after having such a dream. It pains me to say, that the day, a year after I escaped from Bulgaria, in 1952, he himself suffered a horrible, similar death, while I was far, far away. But my fate, in utter ignorance what was going to happen to me, had shattered him. As it turned out, that day they let us go. My father had to bring the ox cart to take me home. I could not walk.

As I was laying in the cart, we passed through the central square. Several trucks were lined up there to take those willing to go to Plovdiv for G. M. Dimitrov's rally. Someone approached me in the cart and asked me: "Do you want to go to the meeting? Get on the truck". I said nothing, but turned my head aside. At this time I could no longer think of G. M. Dimitrov, his meeting, and his politics. I do not remember who approached me in this most painful moment in my life with this invitation, but it probably was someone who loved to kick men when they are down. So, I was back home, and I stayed in bed for a week, with all sorts of home remedies applied to my leg, until I was able to walk again. Many a time all these years I would still feel the pain on the spot where the man in the cop's office kicked me with this boot. I have never forgotten these experiences. One thing was sure, my illusions for a blooming democracy in Bulgaria were dashed, they evaporated like a smoke and I had to rethink my future behavior.

5. Victim of My Illusions

I went through all that because I joined the BANU. Thinking back of my decision to take this step, I realize that I was seeking a shelter where I could hide from the political storms raging around me. I was seeking a place, where, without becoming communist, which I could not become in any turn of events after all that had happened to me, to settle down. Somehow, I was convinced that Bulgaria was on the threshold of democracy, that it would become a parliamentary democracy once the war was over, that elections would be held and that the communists would be overwhelmed by the BANU. I was trying to persuade myself that the communist dominated Fatherland Front was a temporary arrangement, that all the talk about democracy was a genuine political prospect. But behind these deceptive ideas, deep in me, I was seeking a way to resist the communists. I felt that the BANU was going to stop the Communists, was going to neutralize them. In this sense I felt a desire, a spark in my mind, which was burning and generating energy, to retaliate for all the humiliations and insults which I had gone through, and all the brutalities which they had committed on the people, to me, and were continuing with their insatiable thirst for blood.

All events that had transpired since July 21, where I was personally touched, made me rebel against the new realities and only a policy which would challenge this reality made sense for me. Under the circumstances, why did I need anybody's orders or suggestions by some Legionairs how to express my resistance to my enemies? I saw the symptoms of this emerging policy in the ranks of the BANU. Even though I had deep doubts of the possibility of such policy succeeding, though I was so impressed of the overwhelming, crushing power of the communists, and that the axe of the guillotine was hanging over the neck of Bulgarian democracy, making out of it an empty illusion, I still saw in it, BANU, the part of my own salvation and redemption. No matter how strong my doubts in the success of the Agrarian resistance to communism was, even if I had seen the guillotine hanging over my head, still I would have gone ahead. At this time I think of Patrick Henry's admonition: "Give me liberty or give me death." So did Levsky write on his banner: "Freedom or death." Some seven years later, I was going to repeat this plunging into the unknown, only that this time I expressed it in a often repeated phrase: "We are taking our heads in our hands." It was a freely chosen path for life or death.

What I did not see in the beginning, when I decided to take the first steps to join the Union of Agrarians, where I became victim of my illusions, was the certainty that this path was not available to me, that in my first attempt I would be hit, and hit hard. Had I recognized this certainty, I would not have gone through my December ordeal.

Much, much later I read somewhere about an incident, which apparently was at the bottom of my arrest and beating. In the days when I was considering my options, by chance, I had met a young man on main street in my neighborhood. He was coming on foot to my village. He stopped with me at the head of my street and started a conversation. His name was Ivan Markov. He did not waste time and told me that he was coming to explore the possibility of founding a Youth Agrarian Union. He invited me to join. I declined. I did not know who he was. Much later I was to learn, that, right or wrong, he had been a Legionair, who was organizer of the Agrarian Youth section in the Plovdiv district villages. Apparently, when I was being beaten by the communists, they may have had been informed of my chance meeting with him on the street. If this information, which I learned later, that he was a Legionair, was true or not true, I could never establish. But, I paid a high price for this chance meeting on the street. Without realizing, I had jumped in the ring where the gladiatorial battle between communists and agrarians was just beginning. In an atmosphere where G.M. Dimitrov was emerging as the Beelzebul on the path of Bulgarian communism, I had joined his forces and found myself under the sword of the communists, seemingly following an order from the Legionairs, on account of the meeting with Markov. Neither he, nor I, in our brief conversation ever mentioned the Legionairs.

On many occasions, much later, I often wondered what would have happened to me, if in those difficult days, some rational communist had come to me and invited me to join with them, had asked me to forget the past, that it was the future that was important and that for this future they wanted every capable young man to dedicate himself. I cannot answer this question now, because the truth might be twisted somewhere. But it did not happen in October 1944. Then they looked at me with despise, as an enemy, and I did not see in them anything less than mortal enemies. It is probable that I would have accepted the invitation – not as a sincere acceptance of their way, of their philosophy, but as a way to avoid persecution. I had no confidence in them. I was a wounded animal and I was seeking a safe refuge, and with all the trappings of pretense and hypocrisy, I might have accepted. In time, given the consolidation of the system, I might have found a niche in their society and would have adapted myself to their ways. But looking at all this from my present perspective of things, I feel that I would have never conscientiously accepted their system. Even, I am inclined to think, that even if the door was wide opened for me, I probably would not have accepted, with my psychological attuning at the times. I would have felt that I was betraying myself.

Perhaps this is what stopped me from looking to them and making an attempt to reconcile with them. There was a deep gap, a precipice between me and them. The events of July 21, September 9th and October 6th, and now the events of December 1944, had dug this precipice deeper and deeper, and any bridge over this precipice was impassable. I might have adapted myself to the conditions, as so many did at that time, but such adaptation and compromise would have caused me infinite pain and a sense of betrayal towards myself. It is because of such feelings, that I did not try to seek them, to apologize for my past, to indulge in hypocrisy that I was accepting their philosophy, that I was ready to become a communist. It is with such feelings that I followed the events after the Plovdiv rally of G.M. Dimitrov. And most important, I was hurt by them and still feel the pain caused to me.

6. The Great Deception

Dr. G.M. Dimitrov was attacked viciously as a traitor, but for every individual who was not a communist, he was hero and was running to his spectaculars to listen to him. After all, he obviously was the favorite of the Americans and British, as opposed to the Soviets. If there was any hope for doing away with communism, it was going to come from America and England. Then it was Dimitrov who was going to lead it. The old fascist gang, or however it is called, had gone with the winds, and there was no hope for their return. The only place to look for salvation were America and England.

As far as I was concerned, G.M. Dimitrov was a hero – not so much for what he had been, but for what he had become. He had thrown the gauntlet to the communists. Nevermind the Declaration of Ocxtober 12! No amount of professions of loyalty to the Fatherland Front – there were many – was to change anything. He acted as if the Fatherland Front was a political coalition which was to eventually conduct free elections where the democratic parties – the ones in the Fatherland Front – were to compete for power. With such understanding of the political scenario, he was the obvious choice of everybody who was not a communist. But this was not how the Communists had seen things in Bulgaria. Later on they were going to come for joint election ballots, where they would in advance nominate all candidates for the Parliament and were not to compete with each other. The Dimitrov formula would pit communists vs. agrarians, and given the circumstances, the Agrarian Union stood to score an overwhelming victory. That would have put an end to Communist rule.

Dimitrov was not the choice of the broad political spectrum in the country. After all, it was he and his party which had put the communists in power, or, rather, had joined them and served as a cover for them before the rest of the world. They would have come to power with or without the support of the agrarians and the "Zveno". The entry of the Soviet army in the country was a factor which would have catapulted them to the ministerial positions any way. BANU and "Zveno" et al. were there only to present the face of the Fatherland Front as a multi-party coalition. In other circumstances he would have been thrown out for cooperating with the communists, but in the given political set up, he would have gotten the votes of the overwhelming majority of the people. In a historical perspective he would not be considered innocent for the coming of communism to Bulgaria. But now, he was the only hope to displace the dreaded communists. A Turkish Sultan having made an alliance with the Russians, justified himself with the adage: "A drowning man will hold to a snake."

All this had the appearance of a great deception. If Dimitrov believed, or if anyone believed that his popularity was a genuine political support by the majority of the Bulgarians, he was fooling himself. In fact, the Declaration of October 12 was a protective shield for him against the "Fascist"Agrarians of Gichev and the "fascists" democrats of Mushanov. Proclaimed as "fascists," they would not be allowed to participate in the elections and then the nation would have no choice, but to vote for him or the communists. Likewise, the said declaration was a protective shield for the communists. Nobody would be allowed to oppose them, and if they held the Agrarians of Dimitrov in a joint ballot, the Fatherland Front would be confirmed in power.

It is in the framework of this configuration of political alternatives, that the Legionairs, as the communists charged, had jumped on the bandwagon of Dimitrov, figuring out, like him, that the next election was to be a battlefield between Communists and Agrarians. The crucial point in this political underground war was whether the communists were going to allow a contest with the Agrarians in a free ballot. I was skeptical that this would happen, but, if I joined the Agrarian Union, it was for the minimal hope, that I would get out of the ghetto of the "fascists" where I have been pushed into. Besides, in October and November it was not yet clear what direction these events were going to take. If Agrarians were to split with the Communists for election contingencies, then I would be fully integrated in the political processes as Agrarian, and I would be with the winning side. This was the premise upon which subsequently Yalta was built. If they did not split, and went into the elections with joint ballot, then I still would be a party in the Communist-Agrarian partnership, integrated in the political process.

But things started falling apart as soon as I joined the BANU. In no time I was arrested and suspected that I had received orders from the Legionairs. By the middle of December, indeed, the battle between Agrarians and Communists was going full steam. It appeared that the Agrarians were rising as a formidable power and in a free election would far outvote the Communists. They were not prepared to allow that. When in the beginning of December Damin Velchev issued an order which exhonerated the military on the front against the Germans of any transgressions they might have had prior to September 9, that is in the campaigns against the partisans, the Communist ministers in the Cabinet resigned en block and a crisis at the top of the government threatened to throw the country into a civil war. The nation was in a state of turmoil. Velchev withdrew his order and the Fatherland Front coalition was restored. This crisis over, the Communists went for the head of Dimitrov. He was accused of undermining the military forces on the front. Charges were made that the Agrarian paper, distributed to the soldiers, was carrying inserts with the slogan Dimitrov had popularized in the country – "Peace, Bread and Freedom". It was interpreted as a call to the soldiers to return home. By the middle of December, there was an outcry all over the country against him. The political pressure was so strong that in the middle of January he was forced to resign as General Secretary of the BANU. By the end of the month he was placed under house arrest. The old fox managed to escape and seek refuge in the residence of the American member of the Allied Control Commission, Meynard Barnes. His "escape" seemed to have been pre-arranged under the pressure of the Allies. Dimitrov was visited in his apartment by two American officers. In the apartment he exchanged clothes with one of them, and walked out, passing the guards with the other one. How convenient! He stayed a long time in the residence of Meynard Barnes, till the Allies prevailed with the Russians. The Bulgarians gave him a passport and he left the country accompanied by his wife. Much later his daughter was given a passport to join them in Washington where she was given a job in the "Voice of America". Now she is back in Bulgaria and leading a faction of the Agrarians, partner in the government of the U.D.F.

This is how this episode with Dimitrov began and ended. The General Secretarship of the BANU passed to Nikola Petkov. Petkov himself resisted the communists on many issues and in May an Agrarian Conference in the general headquarters of the Party, May 8-9, purged the governing councils of followers of Dimitrov and replaced them with loyal to the Communist Party stooges. With the support of the Communist Militia (the new name of the state police), they occupied the BANU headquarters on Stefan Karadja 10. Petkov, in his absence, was designated as the new General Secretary but he refused to accept the decisions of the Communist-staged conference and at the end of July resigned from the government, to lead the opposition parties for the following two years, until he was arrested in the Great National Assembly and executed on September 23, 1947. In the meantime, in February, Roosevelt, Churchill and Stalin had their pow-wow in Yalta, signed the high-sounding Yalta Declaration for establishing freely elected governments in Eastern Europe, and went home. Nothing came out of all that. When it was all over, Stalin had taken firm control of Bulgaria, Poland, Czechoslovakia, Hungary and Romania, and the Iron Curtain descended from Szhezchin to Trieste.

Looking back at these events, one will not fail to conclude that it was all the consequence of a great deception. There months before September 9, the British and the Americans had made a strategic decision, to give Stalin a free hand in the Balkans. Stalin gobbled up Bulgaria. This temporary arrangement proved to be a permanent one. The "Fatherland

Front" proved to be a convenient coalition government, an instrument for taking all power in Bulgaria and prepare its bolshevization. Under the guise as a democratic coalition, it was, from its inception, a political deception. The Yalta agreements proved to be also a deception. Victims of these deceptions were the little people like me, and thousands more, who trusted the Great Western Powers that they would establish a democratic government in Bulgaria. If we were told of the decisions made in the beginning of June, if we were told of the Churchillian barter of percentages in Moscow on October 9, 1944, if Dr. G.M. Dimitrov himself was told of these arrangements, if we were not deceived by the double-edged Yalta declarations, we wouldn't have come out to expose ourselves to the wrath of the Communists. We were deceived! We went out battling for nothing. Nikola Petkov mounted the gallows to become a national hero, but on the gallows, any way. We, the ordinary people, and the country leaders, were pathetically naïve and heroically aggressive. We paid dearly for our trust in the Allies. There is a story that when an English officer was expressing to Churchill his fears that Yugoslavia would fall to the Communists under Tito, Churchill retorted by asking him if he intended to live, after the war, in Yugoslavia or England. So, there was all the philosophy of the war.

7. Back to the Seminary – My Island of Safety

Meanwhile, in the beginning of March, 1945, the Seminary resumed its activity. Since the building in Plovdiv was occupied by the Russian army, we were called to finish our course in the Bachkovo Monastery, some 30 miles south-east of Plovdiv, in the Rodopi Mountains. Following the Russians the Seminary was made into a School of Agriculture and then a School of Foreign Languages. It did not return there until after 1989. The premises in the monastery were too small to accommodate the entire student body, so the classes were staggered into two sessions, each one for five months or so. Since they had to make for time lost, the two and a half years, which, as of November 1943, were still remaining to complete, the program was so adjusted that we had to graduate in five months. This meant missing the most important core of theological studies, generally scheduled for the last three years. This further contributed for our being graduated as illiterate theologians.

I had expected resumption of classes with tense readiness, to leave my virtual house arrest and find myself again with old friends and my academic milieu. In Zelenikovo I had lost all my desire to socialize with anybody and to participate in any activities. So when the appointed day came I flew out as a bird from a cage. It was for the first time since October 1943 that I felt free from all fears that had plagued me all along. There, in the Bachkovo Monastery, I had nobody to fear. We all, students and teachers, were careful to tow the line already firmly fixed by the new authorities over the entire nation, but since all of us were made of the same dough, and there were no communist weeds among us, we found it easy to deal with the situation. We freely shared feelings and experiences we had gone through and our return to the Seminary was like coming back to the safety of our shelter. Sure, there were a few who had accommodated to the new order of things, had joined in with the Communists, but they were so few, and themselves frightened from the excesses they had witnessed, that it would be uncomfortable for them to act out their new faith, and thus cause scandals. I had experienced the poisonous sting of the new power personally and was very careful not to show my horns. Besides, I was already a confirmed Agrarian and could safely confess my convictions. The Legion, which once dominated the Seminary life, was never mentioned again. It had vanished as smoke into the air. It was a bad dream of the past. Later I was to discover that during these few months in Bachkovo, I had read a paper on cooperativism, my old obsession.

Quite a few of our classmates did not return to the Seminary and had gone to public high schools. My very close friend, Genko Keranov, had been tried by a "People's Court" and thrown in jail – as I was to learn later, for some misunderstanding, mixing names or so, but even after discovering this error, he was sent to jail any way. My friend of the Kazanluk district Traiko Nikolov had lost his father, priest, killed even before September 9th, and never came back. So did my friend Yanko Boev of Assenovgrad. So also did my friend Atanas (Nako) p. Ivanov, from Elhovo District. Much later I met him in Sofia, a student in the School of History, an entirely changed man. There were the times when he was spending long time kneeling at his bed and praying. Now he had turned a convinced atheist and Communist. His father, a priest, had in the meantime died. So also did my closest friend Ivan Borudjiev of Muglish, Kazunluck District. Another one of my friends, Christo Borisov, from Plovdiv, did the same. Years later he was to migrate to Canada to join his father, an old immigrant. He was ordained a priest and received a parish in the city of Akron, Ohio. He retired from there a few years ago and is now residing in Arizona. What was left were the veterans, some 29 of us or so, who pushed to the end. We got ever so close, behind the thick walls of the monastery, that the storms raging outside in society did not affect us. We felt protected and we enjoyed the incomparable beauty of nature's environment in this mountainous little paradise. It was spring, we were in the middle of the mountain, there was no militia to meet in our daily activities. We were studying Theology, a discipline so far removed from what was going on outside our walls, that we could not but enjoy the freedom which we found there. We felt like a fish in our own pond and as birds in our vast mountainous surroundings. We also felt close to our teachers, who, in their turn, many of them, had gone through experiences similar to what I had gone through.

We did not have much communication with the monastic community, except for two of the governing superiors – the Igoumen (Abbot), and his assistant. Igoumen at that time was Archimandrite Pimen. In fact he took the chair of literature, of Mr. Lilov, who did not return. Pimen proved to be a good teacher. He taught us Bulgarian literature and as he concentrated on the fundamentals, we did not find him incompetent. As years were to pass he was to become the most controversial figure in the Church. An opportunist first class, he soon got involved with leading political authorities in the country. Rumor had it, that he was entertaining men of the stature of Damian Velchev, and other military officers, and that he had betrayed a military conspiracy to the Militia, true or not true. But it turned out subsequently that he had struck a close friendship with Interior Minister Anton Yugov, the hangman of the country after September 9, 1944. When I left Bulgaria in 1951, my information was that he had become the favorite high clergyman to the Communist Government. In fact he was elevated to the rank of Bishop and appointed Vicar Metropolitan of Sofia with the assistance, or under the pressure of Anton Yugov. Subsequently he was elected Metropolitan of Nevrokop, again as a protégé of Anton Yugov. The resistance of the Holy Synod against confirming him was crushed by the same Anton Yugov. During the decades that followed, as I was reading the church press abroad, he was the mouthpiece of the Synod in support of the Communist regime. In my writings all along I have been denouncing him as Communist number one in the church. After 1989 he continued as member of the Synod to function as Metropolitan, but was careful not to profess his pro-Communist leaning. I was given to understand that the church leadership was united in condemning communism, but delayed taking an open stand against the old regime of Todor Zhivkov on the insistance of Pimen, who believed that the Party was strong, that the street brawls will soon simmer down and it will continue in power. I met him in Hotel Sheraton and treated him with an expensive dinner to discuss church affairs. He acted as a loyal member of the Holy Synod. At the same time he had already turned his back to the other metropolitans and was plotting with the new democratic government for the overthrow of Patriarch Maxim and his own elevation to the position. Indeed, in 1992 it all came out and he led the schism in the Bulgarian Orthodox Church. I could never understand how they had come to take him, being what I knew, and I had denounced him to be. In 1996 he was "elected" as a schismatic patriarch, with the support of the new, democratic government, and died sometimes in 1999 still a "Patriarch."

His assistant was Arkhimandrite or Hieromonakh Kiril. He was Pimen's protégé. I had no personal relations with him but could not miss his presence in the monastic community. Two years after my graduation, he stopped at the School of Theology in Sofia and spent a few days in our midst while awaiting visa to go to Switzerland to study. How he got a scholarship to study abroad these times, during the severest persecution of the church, I could never understand it. I just took it that on account of good connections, he was a lucky one. I asked him what was he going to study, and was puzzled by his answer: "Agronomy." Who knows, may be this was the pretext and the way for his protectors to let him go. He never returned to Bulgaria to practice agronomy or to serve the church. He was sponsored by Metropolitan Andrey to come to America, broke relations with him a year or later, and went to serve a Macedono-Bulgarian Orthodox church parish in Toledo, Ohio, under the auspices of the Macedonian Political Organization which was fighting with the Metropolitan. In 1963 he led the Macedono-Bulgarian Orthodox Churches out of the jurisdiction of Andrey on account of his resumption of relations with Sofia, passed under the jurisdiction of the Russian Synodal Church in exile, and eventually joined the Orthodox Church of America where he was elevated to Metropolitan of Pittsburgh, later acquiring the rank of Archbishop, where he is still in high honor. After 1963 I established close relations with him, for the same reason for which he had left the Bulgarian Orthodox Church. In 1991, while in Sofia, and after discussing it with him, I tried to intervene with the Church in Sofia to lift the condemnation imposed on him. It is for this reason that I met with Pimen, who assured me that by the end of the month, June, it would be done. It was never done. When Pimen led the schismatic clergy against Patriarch Maxim, the Holy Synod suddenly lifted the condemnation and Kiril retained his canonical position in the O.C.A. After that I rarely had any relations with him. But as the result of his exhonoration, he visited Bulgaria on several occasions. On one occasion we were together guests in the home of the then General Secretary of the Holy Synod of Patriarch Maxim, Bishop Neophit, subsequently elected as Metropolitan of Russe, with all the chances to succeed Maxim. On a number of occasions he concelebrated Divine Liturgy in "Alexander Nevsky", with Patriarch Maxim and other bishops.

During the summer, July I believe, a guest at the monastery was Metropolitan Kiril of Plovdiv. Sometimes in the fall, 1944, he had been arrested by the communist authorities, together with the Metropolitan of Vratsa, Paissiy. They have suffered a most humiliating treatment. Their assignment as prisoners had been to clean the toilets and sometimes the guards had asked them to crawl on all fours and they, the guards, would have amused themselves by riding on their backs. They have been held for a trial as war criminals. God knows what war crimes they had committed, but in the ecclesiastical world they have been considered as the most loyal supporters of the governments before 1944. Paissiy was known to have been favorite of the palace people. When King Boris was dying, they called him to administer communion, not Stefan of Sofia, who by rights should have been called but was ignored. During his stay at the monastery Kiril was visiting the nearby Narechen cold mineral springs, ostensibly to recover his shattered nerves. One day I ventured to approach him during his afternoon walks to the adjacent fields. I introduced myself as a Plovdiv diocese resident, from Zelenikovo. He wanted to know more about our parish priest. I had to tell him the story of his murder. He remarked: "We all are martyrs. Look at me! When I entered the prison I had only a few white hairs. Now I am all white."

One important event occurred during this period of my studies. On one occasion, I do not remember why, perhaps it was a semester break, on May 8 I was going to Zelenikovo from Plovdiv. With a few, five or six men, we were traveling on the roof of the bus. In those days when there were no tickets for seats and the bus was packed, one had to accept a place on the roof, among the suitcases and bags, and hold tight to something, not to fall. It must have been about nine o'clock, because it was dark when we were nearing the village. All of a sudden, I heard a fusillade of gunshots. No one knew what was the occasion. The next day I learned. The war had ended that day, the war in Europe that is, and the officials were celebrating. I remember what Koicho Manevsky told me when I was entering the Seminary in September 1939: "The war just began. In a few days the schools are going to be closed..." I had told him that I still would like to register. I did. Now the war had ended and I had only three months to go to graduate.

These three months passed quickly and peacefully. I enjoyed every minute of them. Sure, much of the material which otherwise had to be covered was cut, but for every consideration of time, events and conditions, it was rewarding for me. The rules and procedures for graduation stipulated that the A and B grades on any of the eleven subjects for final examination, excused the student from taking them. Out of the eleven, on eight of them I had a cumulative average A (6). On the other three I had a cumulative of B (5). I wanted to have a perfect score on my graduation – A (6). So I voluntarily appeared for a final test on the three subjects and my goal, a perfect score of A, was achieved. This was the highest score in class and I was proclaimed the Valedictorian. Looking at my Diploma now, I count 20 A' subjects, 13 B' and one single C (4), in Chemistry. Somehow I felt that I was entitled to an A, but the teacher, a good old man, Mikhailov, was quite absent minded and it seems to me he was giving grades without keeping records. So many times in class he would ask questions, I would raise my hand and give him the answers which no other of my classmates could come up with. In such instances he would ask me: "What is your name young man?" I would tell him my name, he would make a note on his cigarette box, then would throw it away, and my grade with it. Well, we had chemistry one year at that time, and this is how I ended with one C.

So, I was proclaimed valedictorian of the class. I was to deliver the final sermon at the last Divine Liturgy on 29th of July. Metropolitan Kiril, later Patriarch Kiril (1953), celebrated the liturgy. He was also the Synodal representative who signed our diplomas. I sketched my sermon with all seriousness, very cautiously discussing the Gospel message and relating it to the events which we were going through. I sprinkled my sermon with all my cherished ideas about a Christian society which had troubled me all along from childhood to this day. I was getting even with society – the society before the Communists came to power and the society which they were now building. I did not spare anything that bothered me. I spoke of the supreme values which had guided me all along – truth, justice and freedom. My delivery was the best I ever remember having made. I was angry, and I was gratified that I had made it through all obstacles which I had been confronted with in my youth. I delivered my sermon and walked out of the church to mix with the crowds. The crowds were big. At one point someone came to me and told me that the Metropolitan wanted to speak to me. I went into the altar. Kiril was taking off his vestments. He congratulated me and admonished me to never let this fire in me be extinguished. Now, remembering all that, I feel that in my sermon I had delivered a crushing attack on communism, and a magnificent tribute to Tolstoy and Gregory Petrov who had impressed me in my youth. I had spoken half an hour, the church was packed. Of course, there was not one person from my family – my father, or my mother, to share with me my triumph. I did not expect it any way. But I was happy for them too. It was they who had brought me to that moment of redemption of all my suffering.

Most of my classmates departed by the end of the day. Only a few, about ten, were left to leave the next day. That evening we went out to celebrate in a restaurant. Needless to say, quite a bit of wine was served and by the time we finished, we were inebriated. Returning to our dorm, we indulged in "free speech". I remember climbing on my bed, and making such a speech. It was a speech of bitter denunciation of communism. Whoever said *In vino Veritas*, must have known human nature very well. Nobody reported to anybody this speech of mine. The next day I pocketed my diploma and headed for Zelenikovo, without the faintest idea of my prospects for the future. Oh, well! As valedictorian I was given a scholarship from the Holy Synod to study Theology at the University of Sofia.

8. Out to Nowhere!

I returned to Zelenikovo when the threshing season was in full steam. I found myself again on the *dikania*, behind the *viyachka*, with a fork, a *graba* or whatever. I had no idea where to turn to. I could not go to the Communists and beg them for a job. They would not allow me even to approach their door. I was trained to become a priest but I was not married and there were no prospects for such a resolution of my problems. I was looked upon in the village as a man with leprosy. The girls I had been keeping company with before, vanished from the periphery of my vision. Besides, I felt that I had outgrown the village mentality and there was no match for me there. Such events are not settled in a matter of days or months. I could not go for the priesthood in Zelenikovo. It was now being taken by another priest, an opportunist and accommodationist. So I would not compete with him. At one point I was struck with the idea of becoming a diocesan preacher and went to see Metropolitan Kiril. I told him of my predicament and inquired if I could apply for such a position. He looked away from me, and stroking his beard told me, as if he was talking to himself: "It is very good, but this job is not for you. See what happened to Mincho Mandulov? He was our diocesan preacher. He was arrested and vanished somewhere in the concentration camps. He was accused that in performing his duties he was engaging in politics. He was not involved in politics, but this was their way to hit the church." I knew Mandulov. He was the son of a prominent priest in Plovdiv and had graduated a few years before me. So my idea of becoming a diocesan preacher was popped like a soap bubble. I found myself again in Zelenikovo.

The threshing season was over. We plunged in other farm activities. I found myself again behind the cows in the fields, reading all the time and not knowing what to do next. Very soon I befriended the son of Koicho Manevsky, some five years younger than me, *Chodjuma*, as he was known by his nickname, meaning "The Kid", in Turkish. He had gone through traumatic experiences after the murder of his father. He did not have company in the village either. He had been going to High School in Plovdiv, but that had come to an end. Like me, he was treated in the village as if he had leprosy. So we were a good company for each other, though it did not help either of us very much. He was mechanically inclined and as he had inherited from his father a machine for scraping the corn ears, was busy with it during the days of September. He was a great talent in playing chess, so he taught me. He had a radio at home, and since Radio London was broadcasting on short waves, and we thirsted for news, we began going to his house in the evening "to play chess." He would put a blanket on the window and we would glue our ears to the radio, which, the cursed thing, was having too much static and we hardly heard something, except one word here and one word there. It did not cross my mind that they could have followed me – I tried to reach his house by side streets – and we always had the chessmen on the table, just in case. The door of course was locked and when we could structure one or two sentences of the news, which was devastatingly anti-Communist, we were deeply pleased. This is how my days without prospests for the future began in Zelenikovo.

At this time, August-September, the political tensions in Bulgaria were reaching a climax. Parliamentary elections had been scheduled for the end of August. The Fatherland Front, or whatever was left of it, a Communist – controlled coalition of non-entities – Nikola Petkov and Kosta Lulchev, the BANU and the Social Democratic parties, having withdrawn at the end of July – was going with a joint ballot. The opposition parties were not yet organized and demanded postponement of the election. The Communists, backed by the Soviets, refused to yield. The opposition asked the Allied Control Commission to intervene, but since the Soviet representative, General Biryuzov, was the permanent chairman, and since he had a veto power on all decisions, the Communists held to their plans. The tensions were growing by the day until the morning of the elections were to be held, and at the last minute, yielding to allied pressure, they were postponed. What deals were made behind the doors between the Allies, or what threats were exchanged, is still a secret, but the postponement was interpreted as a crushing victory for the opposition and defeat for the Communists. The opposition from the very beginning had decided to boycott the elections and never offered candidates. What may have been the deal is the promise by the Allies, that if the elections were rescheduled for a later date, as they were for November 19th, announced sometimes later, even if the opposition boycotted them again, the Allies were not going to object. Indeed, this is exactly what happened. It made the postponement a pyrrhic victory for the opposition.

All this changed nothing for me. It only diverted my attention for a while from my personal problems. It was a moral satisfaction for me, but they, the Communists, still held all power and were rampaging physically and ideologically in the country. My days were passing in boredom and desperation, without any prospect for a change. What was I going to do? As a valedictorian of the class I was awarded a scholarship to study Theology. But what was I going to do with this Theology when both, Church and religion, were virtually outlawed? It meant only that I would be jumping from the pan into the fire. The other alternative was to follow the cows and the sheep, the plow and the hoe. But how long was this to continue? My only entertainment was reading opposition newspapers. Only a few papers were brought to the village every day, and in order to get them I had to be first in line. I was there every day. To cover myself, and claim that I was after broader information, I purchased also the Communist newspapers – "Otechestven Front", "Zemedelsko zname" and "Rabotnichesko delo." I did not discuss anything with anybody * .

By the middle of September I had made up my mind what to do. Since I had no other choice, but to study Theology, it was not too bad, after all. I would take the Scholarship of the Synod, and come what may. I hoped that the Theological Schools, the seminaries, would continue functioning one way or the other, and if so, I could try to secure a position as a teacher. It happened that about that time *Chodjuma* was going to Sofia to visit his uncle. I stuffed my diploma from the Seminary in an envelope, wrote a letter to the Dean of the School of Theology, and asked him to deliver them at the office at 19 "Sv. Nedelya" Square. This decision proved to be a turning point in my life and I never regretted it. At first it appeared that it was giving me just four years of study. And after that? After that I was going to see what I could do further.

---

* In my dossier I recently found an informer's report that I have been reading the opposition papers and discussing them publicly. The last was not true, and the man omitted saying that I also was reading the F.F. papers.

VI

AN OASIS OF SAFETY IN A GAME OF HIDE AND SEEK

"... The Bulgarian Orthodox Church presently is one of the major reactionary nests ... upon which our internal and external enemies are placing great hopes ... The Synod had become a place where fascist elements, enemies of the government, have found a shelter in the Church ..."

General Yonko Panov, Deputy Minister of Internal Affairs, in charge of State Security. Report to Anton Yugov, Minister of Internal Affairs. August, 1949.

1. Under the Roof of the School of Theology

A week or so after that I received a letter from the Dean of the School. I was accepted to study there. I was going to live in the dorms at 19 "Sv. Nedelia Sq." All I needed was a note from the municipality that I was being stricken out of the list for ration cards in Zelenikovo. I was overjoyed. I was going to leave the village which was getting on my nerves. The problem was that little note. If I went there, in the municipality, I would be turned down and that would have been the end of my hopes and expectations. All officials there were fanatic Communists and hated me. But .. I had Drazha Parlapanska there, the sister of Georgi Parlapansky and my aunt. I confided to her my predicament. She told me when to go and look for her. When I appeared there, she came out of her office, discretely pull the precious piece of paper from her pocket, shook my hand while hiding it in her palm, I took it, she wished me good luck, and I was free to go. She had written the note, had signed someone's name, had affixed a seal and in full secrecy had prepared my – to put it this way – passport for the future. My long friendship with her, her appreciation of my struggles to get an education and her disregard of political conventions when it was a matter of advancing someone's career in the field of scholarship, opened the way for my studies if Sofia. I have told this story before but it was worth repeating it.[*]

By the end of September I packed my two suitcases with whatever books and clothes I had and flew out of the village, never to return there, except for short visits. After the bus to Plovdiv, I took the train for Sofia and by 6.00 p.m. arrived at Poduene, the first station in the capital. My money was short, so I did not take a taxi, asked for directions to "Sv. Nedelia" which were not hard to follow, and started my journey to the future. I did not have any idea how far it was. This was my first trip to Sofia. The heavy suitcases were killing me. Here and there I stopped and asked how far it was and the answers I was given were not very pleasing. But I was determined, and finally arrived at the high, solid stone building facing the cathedral church. Built in the beginning of the century it was architecturally and in coloring very similar to the Seminary. As I was to discover later, it was also very similar to the Holy Synod at "St. Alexander Nevsky" Square and the surrounding walls of the Sofia Seminary. At this time I had little interest to study the architecture. Just for information, the benefactor who had paid for the building had designed it for that purpose – to serve as a theological school when it would be opened as a part of Sofia University. The Theological school was authorized by the government of Alexander Stamboliysky and opened shortly after he was toppled from power and killed in 1923.

I entered through the heavy doors and met my first challenge. The caretaker, an old man, Diado Misho, serving as doorman, would not let me go further. I had the Dean's letter with me, but it was not enough. He wanted a note from the Professor in charge of the dorms, Dimitar Dulgerov. Diado Misho was categorical and treated me as an intruder. Fortunately a couple of students were watching from the side and offered to help me. They were going to take me to Dulgerov's home. It was not far away from the School. We climbed the steps to the third floor and knocked at the door. An old man, with severe look and penetrating eyes studied me from top to bottom, asked us to wait outside and disappeared in the apartment. He knew the students who had taken me there. I showed him the Dean's Letter and in a few minutes he was out with the note to Diado Misho. We went

---

[*] When the Bulgarian version of this volume was published Drazha, in her 90s, was still alive, and I was glad to present her witn a copy.

213

back and I was admitted to the dorm. The dorm consisted of three large rooms connected with wide open doors. All along the walls were lined up, lengthwise, pairs of iron spring beds, very few occupied. A matress was found somewhere, stored in a corner, Diado Misho gave me a few blankets, one or two. They were old raggs and I settled in my new quarters for the next four years. The accommodations were a hundred times worse than I had in the Seminary. But still, all that was a hundred times better than Zelenikovo.

I do not know, whether it was fate, whether it was luck, or I was just propelled from place to place by my hopeless situation and my initiatives fighting the odds of life. I was to hear later the comments of my neighbor Koliu Pekhlivanov. Hearing that I had gone to study in Sofia, he had Acidly retorted: "He should have gone with those in the Yunchal, and now he is going to study at the university. We do not know what we are doing." Too bad. I was away from their reach. In a few days I was in the office of the Chief of the department of Social and Cultural Affairs in the office of the Holly Synod to straighten the problems with my scholarship. At this time the Holy Synod was settled on the third floor of the same building. The Synodal headquarters near "Alexnder Nevsky" were a ruin since the bombing of the Capital and were under repair. It was to take long time. The School of Theology was also damaged, but it was a minor problem on the top floor and they were working to fix it.

All around the cathedral, once impressive buildings, were ruins, burned down or destroyed, with high pillars sticking here and there, with buildings standing up here and there without windows, without doors, blackened by smoke all over, high piles of debris swept back from the streets, big chunks of cement laying all over. Now this waste land houses the modern Sheraton Hotel, the Presidency, the Council of Ministers and the TSUM Department store, the Bulgarian Macy's.

The only obstacle which could have prevented me to go to school now was the system which was introduced the next year: special notes from the municipal authorities – "O.F. Belezhki" (Fatherland Front Notes) issued by the local authorities, testifying that the candidate student was politically in good standing. Had there been such requirements in 1945, I could have never come to Sofia to study anything, least of all Theology. But the government caught up with this omission in 1945. In February of 1949 some five thousands students were expelled from the university on political grounds – all those who had entered the university without being politically scrutinized. I will discuss this later and indicate how it almost brought my studies to an abrupt end.

The building of the Theological school was very soon turned into an all purpose church headquarters. Besides the Holy Synod, it had to accommodate the Church-Archeological museum, which previously had been located in the Holy Synod building. The Russian army had occupied the building of the Sofia Theological Seminary, so some place was going to be used on the upper floor, whatever the accommodations were, to continue a semblance of a Seminary. Sometimes later the wing of the Bachkovo Monastery where the Plovdiv Seminary was housed, burned down and what was left of the seminarians, were also brought to our building. Later both Seminaries were merged into one, were transferred to the Cherepish Monastery and conditions in our school were a little relieved. The biggest problem very soon appeared to be the lavatories on our floor. First the plumbing system was old fashion, and was broken most of the time and the floors were perpetually flooded. Sometimes all sorts of things were to find their way to the floor in the hallway.

There was a caretaker living in a small apartment between the lavatories, the Church Museum and our dorms. He was a nasty middle-age man with his family. He happened to be a brother of the Cathedral's sacristan of "Sv. Nedelya" who in 1925 had been part of conspiracy which organized the bombing of the church. Some 150 members of the government elite were killed while attending the funeral of a general assassinated by the Communists a few days earlier. The intended victim was the King himself, but for some reason he was delayed in coming and survived. The Metropolitan of Sofia had emerged badly shaken, covered with plaster but alive. The sacristan was subsequently hanged, and apparently his brother – I do not remember his name – was given the job and the place to live as a compensation. He was a fanatic Communist and we tried never to be close to him or have any relations with him. So, he was supposed to take care of things. He never did. The conditions under which we had to live, were never improved. Still, this was the best that we could have in those post-war days.

So, this is how my studies in the Theological School of Sofia University began. The lectures were conducted in the Aula – the vast ceremonial hall which could accommodate over three hundred people. Sometimes we would be no more than three students listening to the professor, huddling in the front seats. But this was the only room left intact from the bombing and not occupied by anyone, except the library at the end of the hallway and the two offices of the school. At the front end of the Aula were three long separated daises, the middle one a speaker's lectern, all mounted on an elevation from end to end of the room, some one or one and a half feet. We felt lost in this room, but soon we became used to it. These were bad times and we just had to grin and bear it. Our professors delivered their lectures as well as they could under the circumstances. Sometimes I felt embarrassed for them, to lecture to three or four students. They never took attendance, they received their salaries and that was good enough for the times we were living in. Two or three times during the semester they had to sign our student books as a verification that we had attended the course, but there was a way of bypassing this rule too. So many students, after registration, would return home and leave their student books to us who were permanently there, we would hand the book to the professor, he would sign it and this was the end of it. Such absentee students would just appear at the end of the semester, borrow our notes, study, pass the tests and go back home. I held the books of several such students, one of them was Father Dimitar Dimitrov of Varna whom I was going to look for help in my most critical hour of need in May 1951.

I already mentioned the conditions in our dorms. Let me say a little more. The beds were iron spring beds, over which we would place a mattress, God knows how old and dirty it was. We were supplied with a few blankets of no better shape. As to sheets and pillows, they were commodities to be rarely found here. It was every man for himself and most of us went out without them. We were lucky that in those days we had place to come to as our "home" away from home. We stored our suitcases and possessions under the beds. No broom or anything else went there to clean up the floors and probably decades of dust had accumulated. Mice and rats had a free run all over the place and it was quite often that in our sleep, if we would not cover our faces, they would run over us. Disgusting! Sometimes someone would complain that the cursed rodents had tried to pick their ears. When we were graduating and someone gave us a banquet in the halls of the Sofia Mitropolia, (Diocesan headquarters), presided by Metropolitan Filaret of Lovech, and I happened to be again the Valedictorian, I did not miss the occasion to gently suggest and express my hope that such things would not have to plague future students, that the Holy Synod would take a closer look at the physical conditions under which the future theologians were being trained. He apparently took offense, but promised that they would take care of it as soon as times improved.

The professors in the School of Theology proved to be highly competent in their respective fields. In Biblical studies Prof. Ivan Markovsky had published an impressive amount of articles and several volumes of textbooks which were the last word of biblical scholarship. When I appeared for a test in one of his subjects he asked me for recent literature in the field. I had just read his latest publication. When I mentioned it and made some complimentary comments, pointing to specific subjects, he was enormously pleased and did not ask me for other works. Prof. Gancho Pashev, teaching Moral Theology, also a prolific writer, but a poor lecturer, was very boring, though listening to his analysis, one noticed a profound understanding of Christian ethics. But he was difficult to follow because of his monotonous delivery. The way I was, my thoughts rushing at high speed, very soon would lose interest in his lectures and find something else to occupy myself, most of the times writing on the back of my notebook whatever my mind was set upon at the time. In one occasion he apparently suspected that I was not following his lecture and asked me in class if I was listening to him. I told him that I was keeping notes. He asked to see my notebook. I showed it to him, the front page where in large letters I had put the name of the course and his name. He could not go further and sent me back to my seat. His Assistant Professor Dr. Ivan Panchovsky was a close friend of mine. He had been subjected to political pressures, to say the least, and once we knew who had suffered what, we kept in touch. An incident occurred between us, where I, foolishly, repeated the same mistake as in the case of Salgandjiev in the Seminary, though not under the cover of anonymity. He had assigned to as to read two chapters of Pashev's textbook on Moral Theology, and discussed them in the seminar. He called on me to make my comments. With all my frankness, not thinking of the consequences, I said that it all was a pile of straw, where may have been some grains, but I could not find them. He hurriedly changed the subject and my remarks passed unnoticed. Later on, privately, he warned me to be careful. After all, he was Pashev's assistant and if this thing reached him, he would be called on the carpet. Fortunately, nothing of that happened and from then on I was very careful. On the other hand Pashev was very willing to help us when we needed a speaker for our Student Society activities, boring as he was. I will say more on that. I concluded the above described incident by making positive observations on the works of Pashev and his scholarship at the next session of the seminar.

The most remarkable personality among the faculty was, unquestionably, Prof. Dulgerov, teaching Dogmatic Theology, but better known as a public speaker on theological subjects. Though his course was scheduled for junior students, I established close relations with him during my sophomore year. I already mentioned my first meeting with him at the door of his apartment. Soon after that, actually in my second year, I began to play an active role in the Student "Patriarch Evtimiy" Society, organizing weekly public lectures in the Aula of the Theological School. He was our best speaker and always at our disposal when we invited him. In a way, we became partners in a common cause. But my favorite Professor was Ivan Snegarov, an elderly man, born in Ochrid, Macedonia, old fashion in his dress – crumpled suits, everlasting tie, an age-long fedora hat, somehow out of size shoes, and quite often his socks at the heels with large holes. For one thing, he could not help it, his wife had been bed-ridden for long time, I believe his son had committed suicide, and he was dedicated to his scholarly work in Church history. Once he confided to me that he had made the mistake of getting married. A scholar's work needs the full attention of man, without any distractions. A married man had to split his time between the family and his work, which not always worked well – either the one or the other will suffer, he counseled me. I never missed any of his lectures, absorbing every little detail which he would mention. When he was lecturing on Sts. Cyril and Methodious he listed a number of arguments in support of his view, that they were Bulgarian Slavs. I noted all his arguments. When the time for the exam came, it happened that he asked me exactly this question and I repeated every point which he had made. Many years later, when I read a letter in *The New York Times* by someone who referred to them as Greeks, I responded and the *Times* published my letter. I repeated the arguments of Snegarov. Then, at one of the Conferences of the AAASS (American Association for Advancement of Slavic Studies) a professor wondered how I could have said so much in such a short letter.

But my most important achievement under Snegarov was a seminar report on "Bogomilism," a medieval heresy originating in Bulgaria, and spreading to the Western Balkans, Italy and France. I had chosen this topic at the beginning of the semester. When my turn came to report, it took me three seminar sessions, two hours each, to present my paper. Usually these reports lasted half an hour or so, but Snegarov never interrupted me and gave me all the time which I needed. I did take advantage of this opportunity and showed all my abilities for scholarly work. Of course, he congratulated me. In my Junior year, when I had to select a subject where I would like to specialize and write a thesis, I expressed desire to work under Prof. Stefan Tsankov in Canon Law. Snegarov stopped me in the hallway and somehow regretted that I did not choose Church History and turn my seminar report into a thesis. I thought it over and changed my mind. I chose Church History under him, and "Bogomilism" as my topic. I started work on the subject. Just about that time a young historian, professor in the School of History at the Sofia University, Dimitar Angelov, had published a book on "Bogomilism", a Marxist interpretation. I noticed so many loopholes in his discussions and undertook to give my version of understanding this heresy from a theological point of view, with all inferences on social, political, cultural and historical issues connected with it. I visited every library that I could think of, I tracked down every book or article that had been written on the subject, I examined every issue that my study led me to and when I finished it, it was a masterful work. I have never forgotten how I would sit in the library, with a dozen books spread open before me, how I studiously copied church Slavonic, Greek and Latin texts of the Middle Ages and what a pleasure it all gave me. I was snug as a bug in a rug. This, indeed, was me. Later on I had to hear from Radko Pop Todorov – I will say more on him next – Snegarov's wondering how courageous I have been to challenge Angelov.

I finished my study of "Bogomilism," my thesis for graduation, in the summer of 1949, all typed, some 300 pages, or so. I made five carbon copies, and had them bound professionally, before submitting it. Snegarov approved it and I graduated. In 1965 I wrote to the Rector of the Theological Academy, Nikolay, asking him to send me the copy of the Thesis, so I could xerox it and would return it to them [*]. I intended to work more on it and publish it. They had refered my request to the Office of Religious Affairs of the Ministry of Foreign Affairs which had permitted it and in a few months I received it. After xeroxing it, I returned it to Sofia, as promised. But I never had a chance to sit on it. At that time I was working hard to develop, researching the courses which I was teaching, about half a dozen, and my project was thus shelved. Much later Dimitar Angelov

---

[*] In 1951 the School of Theology was dropped from the university, but continued to function as Theological Academy under the Holy Synod.

published a far bigger study on the subject. I wrote him and he was very kind in sending it to me.

And one last word on the subject: I had four copies of this work left with my aunt Plina when I left Zelenikovo for safe keeping. She had later turned them over to my brother Stoiu. And since the copy paper I had used was very good for rolling cigarettes of home produced tobacco, he started helping himself of them, piece by piece. Once my neighbor Marin, whom I have already mentoned, had seen his cigarette paper and asked to see the paper he was cutting from. My brother showed him a copy. He had exclaimed that this was a doctoral dissertation. Indeed, I had done far more than it was required for graduation. But I never regretted my hard work in preparing it. I experienced the greatest of pleasure doing it. This is how I plunged in the study of history which was to become my profession in America. I owe it to Snegarov.

Among all the professors, I must confess, my favorite was Boris Marinov, teaching Apologetics. Short, with a paralyzed neck, a big head, brisk walk, always carrying a big briefcase bulging with books and papers, he was always ready to expose the fallacies of atheism – the other word for Communism. We became politically acquainted very early and freely discussed the topics which were taboo in Bulgaria. His alter ego, Alexander Velichkov, his assistant, was our very close friend. I certainly would name all the professors, Tsonevsky, Piperov, Goshev, Tsankov et al. They were all friends and we related as if we were a family. The world outside of our fortress was hostile to us, but we all held together in the School as an oasis of peace.

The majority of the students were clergymen – priests and monks. Some of them were in the same situation as I was: staying in Sofia to escape persecution by the Communists in their localities. If anyone would have used the name "a nest of reactionaries" in Sofia, the way that word is used for anti-Communists, we fully deserved it. Luckily, there were no Communist infiltrations among us and we felt free to discuss our problems and share any information from the outside world which would bring us some satisfaction[*]. Dean of our student body – by status, if not by appointment – was Monk Sofroniy of the Trojan Monastery. I have mentioned him already. He probably was about forty years old, small in built, black beard, a little graying, not too long hair, down to earth man. He had set his bed and other paraphernalia in the middle of the dormitories and his corner had become sort of a social club where all of us gravitated for reason or for no reason, just for friendly talk. After every vacation he would bring a trunk of food supplies from the monastery – lentils, beans, potatoes, oils and whatever. He had there his electrical hot plate and every day something was being cooked for his lunch and supper. Quite often, when he saw me there, and knew that I was starving anyway, he invited me to join him. I shared in his goodies. Sometimes after our graduation, I must say many years after my escape from Bulgaria, he had become Metropolitan of Russe. When I was in Bulgaria in 1991 I heard about him from Pimen and others. Pimen told me that he was refusing to retire, on account that he could no longer walk, and insisted that he was conducting the affairs of his diocese with his head, not with his legs. He was good, very good, very human soul and I counted him for one of my best friends. Another monk from the Trojan Monastery was a young fellow, Filaret, very handsome, and very proud and somehow very cocky. When I was planning to escape with Radko Pop Todorov, he took me on the side and offered to give me a revolver, if it came to that. Years later, he had become

[*] Recently, reading my Dossier in State Security, I discovered that they had an informer among us, a priest from the town of Pirdop, and I guessed who he was. His clowning imitations of the authorities were apparently an act of deception.

Metropolitan of Vidin. In 1978 or 79 he visited the United States with a delegation led by Patriarch Maxim. I wished very much to go and meet him, but given the circumstances, I did not want to give the impression that I was patronizing the communist emissaries, as they, indeed, were. I never asked myself the question how he had qualified to be given any such consideration and allowed abroad. I had too much trust in him in order to accuse him of collaboration with the communists. Apparently they all had made their peace with the regime.

And there was Pencho Donchev, a younger student from Plovdiv Seminary, one year behind me. He and his classmates joined us in February, 1946. Pencho was from Muglish, Kazanluk district, the village of my personal friend Ivan Borodjiev, and of Nikola Zagarov, one of the students once expelled from the Seminary with the Gashtev group for leftist politics. I knew that Pencho was sick, so the word went out in the Seminary, of tuberculosis. He had a Synodal scholarship. We were close friends in the Seminary. We all belonged to the group around Hieromonakh Gerasim, a teacher-supervisor, who was always trying to persuade young seminarians to become monks. In a combination of circumstances, Pencho became the reason for my surviving physically in Bulgaria. If it was not for him, my bones might have been for long rotting in some Bulgarian slave labor camp, concentration camp, where they fed dead prisoners to the pigs. I will come back to this story later, when I come to discuss his tonsuring as monk by the name of Pankraty.

Years after that Pankraty had become Metropolitan of Stara Zagora and in charge of Foreign relations of the Synod, for many years. I read numerous of his statements in the press as dedicated defender of the Communist government, and I was sure he meant it. He was known back in the Seminary as a leftist, under the aegis of Nikola Zagarov. It turned out that in 1992 he joined with the schismatics and I attacked him in my writings as a Communist fellow traveler. I had done that long before 1989. However, a few years ago he returned to the Holy Synod. I was able to meet him briefly in the Holy Synod, before he passed away from heart attack while attending a session of the Synod.

Among the other clergy in our class I must mention a monk, Ioan, from the Sofia Seminary. After our graduation and years after I left Bulgaria, he had been elevated to the position of Bishop and become Abbot of the Rila Monastery. For some reason he had never become a Metropolitan. Probably he did not play politics with the people who mattered. For Sofroniy, Filaret and Pankratiy, I suspect, it was Metropolitan Nikodim of Sliven, who had promoted them. He had strong ties with the Troyan monastery. He had studied at Oxford, and in 1947 or 1948 became Metropolitan of Sliven – as I understood – with the support of the Communist Government. Much later I read an article by him in the Journal of the Church where he called all American senators sons of American slave owners. I met him in Toronto, sometimes in the 1970s. He did not fail his reputation of lover of the spirits. The Abbot of the Troyan Monastery, Clement (that monastery produced the famous Trojanska rakia – Trojan plum brandy) had become the godfather of Nikodim, who, once a rector of the Plovdiv Seminary in my time, had become Metropolitan of Sliven and in his turn – sponsor of Filaret, Sofrony and Pankraty to become metropolitans, respectively, Vidin, Russe and Stara Zagora. Yoan was not in this click and probably this is the reason why he did not rise to the coveted rank of Metropolitan.

There were a number of priests in our class. Of them best remembered and friendly, that is politically of my persuasion, were Assen Petkov of Pleven District, Petko Petkov of the Vratsa Diocese, Jordan Novakov of Svishtov, Georgi Chakurov of Borisovgrad, and Trajan Georgiev of Pirdop.

Of the laymen my closest friends were Boris Petrov and Christo Prodanov of Turnovo District. The three of us were inseparable for all the time that we were in the School, but after graduation I lost track of them. In 1991 when I visited Sofia Radko PopTodorov attempted to contact Boris, who was ostensibly dying of cancer, but failed to find him. So I could not talk to him. I believe Prodanov became secretary of the Turnovo Diocese, as I read his name in the papers, but we lost contact. The one who made a big name for himself was Prof. Todor Subev. He and Pankratiy were inseparable – as students in the Plovdiv Seminary and as students in the Theology School. Subev had come under the auspieces of Prof. Snegarov, had become professor of Church History, which, if I had a better luck, might have been. But Todor Subev was another protégé of Nikodim of Sliven. Apparently, he made out very well with the Communists and with their help, and the help of Nikodim, was sent to Geneva to represent the Bulgarian Orthodox Church at the World Council of Churches and eventually became its Secretary General, the highest position of that international organization. He accompanied Patriarch Maxim during their visit to America in late 1970s. We were very close friends and I wanted to go to see him, but due to the circumstances which I mentioned before, I did not go to New York, as I was urged by my once very close friend Kalin Koichev.

In those days I felt that I had to explain my position on the church collaboration with the Communists, and I did it by writing a long letter to my friend in the Ecumenical Institute Ilse Friedeberg, where Subev was at that time. She wrote me back that she had used a chance to read my letter to Subev while he was driving from Bossey to Geneva. This letter, dated March 7, 1980, I have published in the third volume of my *Politicheski Problemi*, Vol. 3, pp. 237-249, under the title "A Personal Confession." Whether Subev had informed somebody of my letter and my "confession" I never heard of it. But, somehow I was pleased to have done as much as I did.

2. Radko Pop Todorov

I will take more time to remember my other closest friend in the Theological School – Radko Pop Todorov. Radko was from the village of Gradets, District of Vidin. His father was a priest. He had graduated from the Sofia Theological Seminary, had served his military duty and in the spring of 1946 joined our company with Christo and Boris. In 1995, I believe, I visited him in the Holy Synod Building, which they, the schismatics, still controlled. There, in the presence of the "General Secretary" of their "Holy Synod" Prof. Anatoly Balachev, and their supporter, a priest Yanko Dimov, biographer of Patriarch Kiril and Patriarch Maxim, told the story how we – Christo, Boris and I – had turned him to the Church. After serving his military duty, confused and not knowing which way to go in life, he had registered to study Theology, just to kill time. Christo, Boris and I had taken him under our arms (this was the usual way of walking together in those days, which in America today is viewed as a scandalous manner, suggesting indecent moral depravity, but in our times just a close friendship) and led him for a Vesper service at "St. Alexander Nevsky". He said that while attending the service, something had shined in his mind and he had decided to dedicate himself to serve the Church. So, having entered the School of Theology out of having nothing else to do, he had taken seriously his work. After my escape from Bulgaria, so I heard from friends, he had established good relations with Metropolitan Paissiy of Vratsa, who asked Prof. Stefan Tsankov to do him a favor and appoint Radko as his assistant, which eventually led to his appointment as Professor of Canon Law in the Theological Academy.

We remained very close friends with Radko to the end of my studies. We were so close, that in the spring of 1947 and 1948 considered defecting from Bulgaria to Yugoslavia. He knew a man in his village who could take us across the border. We planned to place ourselves at the disposal of Metropolitan Andrey in America. For this purpose we needed more information about him, and the only person who could give us this information was Arkhimandrite Gorazd, at the time Teacher Supervisor, I believe, in the Sofia Theological Seminary, which was housed in the School of Theology building. Gorazd had served in Constantinople where Metropolitan Andrey was during the war. We could never understand why he had returned to Bulgaria, but trusted him, any way. He had a colorful past. Before the Communists took over he had been assistant to Metropolitan Kiril of Plovdiv. When Kiril was arrested and kept in prison, he reportedly was plotting to take his place, trying Kiril's Mitres and looking at himself in the mirror. Upon his release Kiril, having learned of that, had thrown him out and had vowed to never allow him to become bishop. Besides, Gorazd had some sort of feminine mannerisms, so rumors had ascribed him all kinds of character weaknesses. When we visited him in his apartment, all kinds of lavender curtains were being used for separations and this confirmed all rumors about him in our minds. Well, we confided in him our plans, he was rather vague about Metropolitan Andrey and we left it at that. But, to his credit, he never reported our intentions to the authorities. Sometimes in 1992 or 1993, I am not sure, he was found murdered in his apartment. No one investigated the murder and it still remains a mystery. Our plans to escape to Yugoslavia fizzled. The man Radko talked about had vanished and our plans collapsed. My last attempt to involve him in my schemes to defect was made in 1951, when I wrote him from Balchik to find me a compass. I was then in the Military Construction Troups. I mailed the letter in Balchik. He answered to me that he could not find "the book" I was asking for. Had the officers read my letter, and his reply as they most of the time did, I could have cooked my goose.

This brief forey in the "revolutionary" field with Radko brought us very close and I came to consider him my most trusted friend. When I was layed off by the Military Construction Troups in the winter of 1950-1951, I went to Sofia to look for him and discuss my firm decision to run away. The reader will find later how serious my intention to defect was. I found him in the Theological School surrounded by friends. He was involved with a girl student. She was there with another girl, with her boyfriend – I do not remember who he was. I could not talk to him in that company. He was in a giggly mood. His girl was a brunette. The other girl was a blond. I regretted that I met him in these circumstances. For one thing I could not confide in him my decision to defect. For another thing I made a faux pas which bothered me a long time. In our ordinary, small talk conversation, I inadvertently used the language I had acquired in the service and blurted a fat swear fit only for a locker room. I realized immediately how inappropriate it was for a mixed company, blushed and separated with them in a matter of minutes, carrying with me my embarrassment. I realized my falling out of the class I had been with for so many years, and the class I had acquired. When I met him in 1991 in Sofia he told me that he had married the "black" one. He invited me for dinner and I found the "black" one to be a charming lady. This meeting with Radko, in 1991, was the beginning of our parting of ways. He came to me in hotel Park Moskva. We talked there for over two hours. It was mostly he talking, and I listening. We discussed church and personal affairs. He had thought that I had escaped by boat via the Black Sea. I told him my story. It is found at the end of this Memoire. They had heard that I have been captured and killed.

Radko spent a long time talking about his misfortunes in Bulgaria all these years. But there was nothing that impressed me. All his complaints revolved around how he had been neglected by the higher-ups in the Church. He had written a book on History of Bulgarian Culture and was encouraged by some German professor to publish it. He had tried to interest Patriarch Kiril in his project. But Kiril would have helped him if he would have agreed that it is published as a collective work, and if Radko accepted that his, Kiril's name, appeared first, which to him appeared like stealing his work. He never finished the project, that is, publishing it. He went into a long tyrade as to how, on many occasions, he had been persecuted, rather, how his academic activities had been sabotaged by people in the Academy, how the Secret Services had been after him. He described numerous incidents, where members of the Faculty of the Academy, having connections with these Services, had prevented many of his attempts to come out of the getto where he was. His greatest regret was that he was never allowed to go abroad and learn a language, was left "without language". There was nothing that I heard to suggest that there was some political entanglement. He was portraying himself as a victim of personal rivalries for benefits bestowed by the powers of the days. His biggest complaints were against Todor Subev and Metropolitan Pankratiy who reaped all the benefits with the support of the Communists. He was swearing to me that they were Communists and agents of the KGB. He was assuring me that he personally had seen their membership RMS cards (RMS – the Komsomol in Bulgaria). I was rather skeptical of that. As to their KGB connections, it was not difficult for me to believe it. Subev in Geneva and Pankratiy in Sofia, after the Moscow attitude towards the World Council of churches in early 1960s had changed, were strategically well placed to serve the KGB, which was using clergymen and theologians for their international propaganda. In exile I was well acquainted with the cases of Prof. Hromadka of Czechoslovakia, Mitropolit Nikolay Krutitsky of Russia and the future, post revolutionary leader of Hungary, Janosh Kadar. I had written so much how these high clergymen were used as tools of communist propaganda, so that a Subev-Pankratiy KGB connection was not outside of the loop.

Far more important were my discussions with Radko of the current problems of the Bulgarian Orthodox Church. I had not been impressed by the performances of the Church leaders after the fall of Todor Zhivkov. They were going very slow in disassociating themselves from the communist authorities in the past, as I was anticipating a blast on behalf of the Holy Synod against their tormentors. Nothing of the kind had happened. But I was prepared to live with it and hope for the best. A little pressure from outside might produce some results. We discussed possible reforms in the Church. Our discussions on the subject continued at his home where we were invited with my wife for dinner. Over the dinner table he continued a line which I had already heard from him: Patriarch Maxim would be persuaded to resign, Pimen would take over as a *pro tem* President of the Holy Synod, and then, following the By-Laws of the Church – which were in effect dictated by the Communists to the church in 1950-1951 – things would be put on the right track. I listened carefully. In general I agreed. In the middle of the discussion he gave me to read a rough draft of a letter to be addressed to the Director of the Office of Religious Affairs, Metody Spassov. I gave a perfunctory reading of this letter – two sheets torn from some notebook, but typed, as a draft, not a final version. I noticed that on several occasions the authors were making references to Canon Law and suggested that all these references should be dropped, because the Director of ORA was not a competent authority to interpret Canon Law. Never during these conversations did I grasp the real story behind all that, which was somehow concealed from me: that a plot was already been hatched to have the leadership of the Church dismissed by the authorities, by the government, and in its place install a new Holy Synod with Pimen as the *pro tem* president. I was to realize all that much later, but in these conversations, Radko was giving me the outline of the Schism in the making. What I have never been able to understand was how he, Radko, had accepted Pimen, the most active supporter of the Communist regime in the country for all that I had known, written and published extensively in exile. To that, to accept also Pankratiy for the new "Synod" who was presented to me in 1991 as a KGB agent, and also Kalinik of Vratsa, a younger man whom I had not known, but had learned from the Bulgarian church press of his fanatic support of the regime. Later it was reported to me that he wanted to place the pictures of Lenin and Stalin in the Synodal chapel in the Theological Academy. When I published my *Impressions of Bulgaria,* in 1993, following my visit there in 1991, and had made some critical remarks about the Church leadership they had xeroxed the respective texts and had them circulated widely. I was not aware of the whole scheme of splitting the Church, if they could not take it entirely. I would have never agreed to such a plot, and when they did it, May 25, 1992, I was stunned.

When I arrived in Sofia on 10th or 11 September 1992, the first thing I saw on a Newspaper stand, was a paper, "Reporter 7". There was an highlighted headline on the first page stating: "The Schismatics Debunked. Spas T. Raikin, American professor of the Pennsylvania University says that Maxim is the Canonical Patriarch." I was one step from the bus when I saw and bought the paper. I had written a letter to President Zheliu Zhelev in August and sent it to several Sofia papers. I have formulated there a series of arguments contesting the case of the Schismatics. It was a shattering piece of analysis of the schism. At that same time visiting Sofia were two Bulgarian church activists from New York, Puio Puiev and Damian Georgiev, who at the times of Todor Zhivkov were on the front line of cooperating with the Communists, but now, apparently seeking exhonoration, had passed to the other side and were supporting the Schismatics. Very soon after that they were to return to the Canonical Church. But before that, it was reported that Puio had commented: "It was all over for Maxim and the Synod, but Spas Raikin ruined everything." The next day I was given an article of "Duma", published the day before the Article of "Reporter 7", in a prominent place with a large headline, summarizing my letter to Zhelev. So, I had thrown my hat in the ring, and from now on, to this day, had to write and publish numerous letters and articles attacking the schism, including a long open letter to Radko Pop Todorov.

My last confrontation with Radko took place two days after my arrival in Sofia. All along, since May 25 – I learned about it all in the beginning of June – I felt that I was betrayed by Radko. I was on the forefront in fighting the Church leadership during the Communist regime in Sofia, I do not know of anybody to have been more consistent, and more vociferous (I would say) in denouncing the collaboration with the Communists, but now I had switched to the side of that leadership, to defend the church. The last thing I would have backed now would have been a schism. I was for immediate departure from the old pro-Communist policies and immediate but gradual reforms. I would have expected the other metropolitans to retire and new men to take over. It did not happen, but this did not warrant destroying the church. The Church was not a secular organization, to throw one set of officials and replace them with another one. The Church hierarchy was sacramentally confirmed and no amount of politics would take away the grace vested in Bishops and Metropolitans, to undo the sacramental acts. It is with such an understanding of the crisis that I approached it. I have never believed that the rank and file of the clergy had sold out to the Communists. I was always convinced that they were accommodating in order to survive, and only a few of them were traitors, among them Pimen, Pankratiy and Kalinik. Having arrived in Sofia with such understanding of the crisis in the church, I plunged in the struggles for its defense. Of all things, I had to defend it from my once most trusted friend, Radko Pop Todorov.

A number of events, following the Government's Decree of May 25, 1992, by which Patriarch Maxim and the Holy Synod were declared illegitimate and a new Synod, under the presidency of Pimen, was appointed, had taken place. These events included the elevation of an adventurous monk, Christopher Subev, to the rank of Arkhimandrite and Bishop by Pimen, with all intents to make him a Patriarch. This entire episode is very similar to the case of Cola di Rienzo in 14th century Rome, during the "Babylonian captivity" of the Western Church. Next was the occupation of the Synodal building by a gang of ruffians led by the same Subev, and a scandalous attempt of Pimen to take the patriarchal Cathedral "St. Alexander Nevsky". On this occasion the cathedral had become a scene of a stormy confrontation, amidst a pandemonium caused by the contesting forces. While Pimen stood in his vestments in the center ready to start the liturgy, Maxim passed by him, locked himself in the Altar with his bishops and celebrated the Divine Liturgy while the nave was resounding in shouts and curses. Pimen, unable to enter the Altar, retreated. Maxim won the confrontation. Had the schismatic won it, Maxim's reign would have been over. They failed. The next thing was the Seminary episode, where Pimen appointed Radko as the new rector, and Radko had no difficulty occupying it. He himself had led a band of armed men, had scaled the walls and had taken over. Radko had advised telegraphically all students that the school year was postponed to begin on October 1, instead of on September 13th. The Rector of the Seminary, Bishop Gregory, at the time attending a conference in Prague, having heard of what had happened, return immediately to Sofia and countermanded the telegrams of Radko, instructing students and parents to return to the school as scheduled on September 13th.

I decided to go there and observe what was going to happen that day. Gregory was a fighter. He singlehandedly had taken over the Seminary from the Communists after the change in 1989. He had hired trucks and busses, piled all that they could from the Cherepish Monastery, students, teachers, furniture or whatever, headed to Sofia, appeared at the Seminary, entered it, walked to the offices of the "Pioneers", the Communist Children Organization, told them to take their coats and leave the building. After forty or so years the Seminary was back under church control. Now, this same Gregory had organized the recapturing the Seminary from the schismatics. When I arrived there, a large mob of people, students, and their parents, had gathered in front of the iron gates. I was interviewed by a group of journalists. At 2.00 p.m. the old rector, Bishop Gregory, arrived at the gates accompanied by the teachers. I and my wife were standing right there. Gregory started an emotional speech. All attention was focused on him. I understood later, that it was for diversion. In the meantime I managed to yell through the gate to Radko, who was inside, that what he was doing was a shameful performance for a clergyman. He had not known that I had arrived in Sofia. He had heard me. Hired guards with guns were stationed at the gates. It appeared to me that the whole thing was an exercise in futility.

All of a sudden, from the east side of the building inside the park, I saw some fifteen to twenty young boys running to the front of the central building where two or three guards were stationed. The boys jumped on the guards, had them on the ground and disabled the large fire department hose stretching to the iron gates to be used against invaders from outside. At the same time we were pushing at the gates, while some monk was climbing over, next to me. In a minute the guard was overcome and the iron gates were wide open. The mob surged in to the front of the main building. While Gregory was delivering his speech my wife had been standing in the corner between the gate and the cement pillar. A guard had stretched his arm around her to reach the gate. My wife had turned to him in Bulgarian, the little that she knew of the language, with the words "ne e te sram", ("shame on you") and he had jumped back. Later I was to learn that Gregory had organized the recapturing of the seminary by arranging with the young seminarians, when in his speech he would have uttered some sentence, they were to climb over the wall and attack the guards. It worked. The whole operation lasted some five minutes. The Seminary was taken over. All of a sudden the park and the long amphitheater of steps were covered with people. It was an exhilarating moment. At one point all seminarians, standing as candles scattered over the steps, started singing the Troparion of St. John of Rila , Patron of the Seminary. It was a magnificent moment celebrating the victory. It was a victory for the church over the state. I was in the middle of it all, and proud of the seminarians. Bishop Gregory then invoked a prayer for the police detachment which was there, but did not interfere in the confrontation between the schismatic forces represented by the guards, with Radko inside the building, and the seminarians led by Gregory. They stood and watched the dramatic scene. The entire crowd sang to them "Mnogaya leta" ("God grant them many years").

At one point I noticed that Radko had come out of his hideout, was surrounded by an irate group of seminarians, ready to lynch him. I made my way through them, I reprimanded him and asked him to tell the guards to go home, which was pointless, any way, because they by that time were disabled to do anything. If they had guns, they did not use them. I doubt it if they had guns and if they had any authority to use them. Radko was shaken by the events, besides, he was in danger in this crowd. By that time I was already recognized as the professor from America. I asked him if he wanted me to stay with him. He was glad to tell me to stand by him. I suggested that it was time for him to leave the Seminary. I was going to escort him out. He told me that he had some things in the office. I took him to the office, he picked his cassock and we went to a make-shift bedroom where his pajamas were. I took him under his arms, with his clothes on my arm and led him to the gates. Somewhere along the way, Bishop Gregory joined us. Outside the gate I threw his cassock and pajamas at him, he entered a taxi and was driven away. It pleased my heart that at this critical moment I could join in this action in defense of the Church.

So, this was the end of my friendship with Radko, which had begun in such an inauspicious moment in 1946. Still, even to this day, I have never lost my feelings for him. I just regret it that he fell so deeply in this schism. My last meeting with him was in 1997, on my way to the National Historical Museum. A monk was taking me to introduce me to the Director of the Museum, Dr. Bozhidar Dimitrov, his uncle. We met Radko on the sidewalk behind the school of Theology and could not avoid a brief conversation. He told me that he wished he could deliver a crushing blow in my face. I told him that I felt the same. Then we were left alone, and in a friendly chat for a few minutes tried to persuade each other who was right and who was wrong. He did not yield. Neither did I. But we separated as friends. I am writing these things three days before Christmas, 1999. A few weeks ago I sent him a card with a wish that he would return to the Church. I must add that most of the schismatics have done just that. Only a few are left, and one among them is Radko. I hope he will see how wrong he was to engage in this destructive struggle against the Church.

3. "Render unto Caesar what is Caesar's and to God What is God's!"

So, these were my friends in the school of Theology – Christo, Boris and Radko, who joined us in February 1946. When I came to Sofia, besides the fact that psychologically, under the circumstances which accompanied me, I was not attuned to studies, I was completely disoriented in the big city. There was so much to be seen and experienced there which I was not familiar with, that I too often did not attend classes, and even less reading anything on the courses which I was taking. I did more exploring Sofia than attending lectures. In the meantime I was receiving my Synodal scholarship, a meager, but sufficient amount to cover my most essential needs. Every month I went to the Director of the Cultural and Educational Department in the Synodal office and received it in cash. I do not remember who was in the beginning in that office, but may be the second year of my studies it was taken by Arkhimandrite Joseph, later Bishop.

Joseph (Dikov) had been tried by a "People's Court" for war crimes. He had been assigned by the Holy Synod to represent the Church in the international Red Cross Commission invited by Hitler to see the evidence of the Katin Forest Massacres of the 15,000 Polish officers. That Commission had issued a report consistent with the Nazi charge that the murders were committed by the Soviet Union. Joseph had been sentenced to two years jail, but was released ahead of time. I am going through this story, because in the 1970s he was designated by the Holy Synod to be Metropolitan of New York and I had a long relationship with him. I will discuss it later. Recently, last year, I received my Dossier from the State Security in Sofia. There was a notation there that Metropolitan Joseph had submitted in 1979 a nine page report against me to the Communist authorities. Life, indeed, takes sharp turns. How he deserved the privilege of serving the church abroad, with his background, and report to the Bulgarian KGB in Sofia, is not difficult to grasp.

So, my first semester ended without me accomplishing anything. So began the second semester. Sometime, in March, I do not remember how it all happened, I was sitting somewhere, rethinking my position and it hit me. Why was I wasting my time? It was Theology that I was studying, there were no prospects for a future in it, but this was the only thing I had, why don't I make the best out of it? I took it seriously, sat on the books, went to lectures, and by June I was ready to take the tests which I ought to have taken in January. During the summer I cracked the books again, and in September I took all the tests which I ought to have taken in June. From then on, I took my studies seriously.

226

But somehow, politics injected itself again in my life. Sure, I was very careful. I had experienced its sting and I was not going to expose my back again. Sofia, unlike Zelenikovo and the Seminary, was a place where it was easy to maintain anonymity and avoid the enemy. At this time, after July, 1945, the new opposition parties were appearing one after another and their papers blasted the Communist regime to high heaven. We had to go out the first thing in the morning and grab these papers before they were sold. We devoured every word which was printed there and which unsparingly devastated our tormentors. What we could not say openly, it was all there. If your soul was bathed in pain and frustration, these papers were the balsam for it. New elections were scheduled for November 19th and the capital was in a state of bedlam – speeches, meetings, bill-plastered walls… November was a cold month and I remember how a government truck was racing through the streets loaded with firewood. Of course, the opposition again boycotted this exercise in futility. It was obvious the Communists were not going to conduct a free election, and the opposition was still not well organized. Since 1934 the political parties were not in existence and their local activities virtually eliminated and forgotten. As far as I was concerned, and my friends, though we bitterly opposed the government, we could not risk not going to the election precincts. Not voting was a signal for joining the opposition parties. So we did our "civic" duty. Early that morning someone brought from somewhere a batch of slogans, typewritten, with a message upholding the opposition leaders. So, we voted, our votes were not "real", containing the names of Gichev, Mushanov, Nikola Petkov, Kosta Lulchev, Pastukhov et al. We could not be identified with the opposition, having voted, and we still had the chance to rub the noses of the Communists. But we, in the Theological school, were cautious not to join or form an opposition political party. Personally, I had already paid my dues with my joining the Agrarian Union in 1944. But unofficially, we loved and acted when we could as a political group.

It happened that very soon among us appeared a medical student, who made our library and our company his daily habitat. He was a short, likable fellow, from somewhere in the Burgas or Aitos towns. Daniel Zheliaskov was the son of some leading Agrarian in the past in that region, who was by now deceased. But he used the connections of his father to get close to high level opposition Agrarian leaders and became a source of information for us which we could not obtain anywhere. It further happened that he had studied in the Seminary in Cherepish Monastery and our company appeared congenial for him. He had graduated from some Italian school, spoke perfect Italian and was in contact with the Italian Intelligence Service. On occasions we visited the Italian library in Sofia. He would take a book from the shelves, would read for sometime, sneak a written message in it and then return it to its place. Then, sometimes later, an Italian agent would pick the book and get the message. What was he reporting to the Italians I did not know, but I was getting a kick out of all that. He was trying to persuade us to establish a formal opposition Agrarian group in the school. I argued against it. First, it would be a catastrophic for us personally, and then for the whole school. We were better off if we stayed with the pro-government Agrarian Union of Alexander Obbov and Dragnev, than with the Nikola Petkov party. We still could work for the opposition secretly, while formally associated with the BANU collaborating with the Communists. This was not enough, but it was safe and better than joining openly the opposition. Well, in no time, Obbov and Dragnev, who had taken over the Stefan Karadja Office where Petkov and G.M. Dimitrov were prior to May 8-9, were denounced by the Communists and on their way to be expelled from the Cabinet of Ministers, as it really happened. But we gave them our support. We attended a meeting at the Ministry of Public Works on Slaveykov Square where they appeared to tell us that they were in full understanding with the Communists, but we felt that they were less than truthful and gave them a rousing applause, anyway. The next day they were ousted from the government. I do not recall exactly the dates of these events.

On another occasion, still informally members of the BANU in the government, when the communists were about to throw the School of Theology out of the University, we went to see Prof. Mikhail Genovsky at "Stefan Karadja 10" and plead with him to prevent this blow to our school. We sat there in his office working on the text of a petition, under his guidance. At one point we were stuck for the right word, when I proposed one which appeared to be most appropriate. He stared at me with approval and appreciation and followed my suggestion. I believe I impressed him, but I would not go further to cozy up to him after that. The School of Theology was saved this time. Apparently his words carried weight on these matters. In 1949 he was the man who reported to the Great National Assembly the Law of Confessions and carried it through, on February 24. However, when the Opposition needed our support, and we could offer it still preserving our anonymity, we were there in full strength, applauding the speakers. I will never forget such a rally at the club of the Social Democratic Party on Preslav Street, when the speaker, a leading Agrarian, Stefan Tsanov, mercilessly attacked the Agrarians in the Government. "You," he said, "are cemetery bitches who loiter over our graves to shlurp the oil which our mothers pour in our vigil lights." The hall exploded in frenetic, long applause, so much so that it felt that the roof was going to fly off. On another occasion, we listened to the soft spoken, powerful, devastating criticism of the Communist regime of Yordan Kovachev of Plovdiv. His populism as an Agrarian and a Protestant background of Christian vision of the world, was like shattering lightening followed by ear-breaking thunders, as if it was a biblical prophet who was lambasting the Communists. These were really unforgettable treats of our injured souls which we experienced. So was the case when Nikola Petkov was addressing the pre-election rally at the National Bank in October 1946. I had screamed so much that it took me days to recover my voice. The pain that had been accumulating for years in my soul could not be restrained from expressing itself in irrepressible anger. It is a terrible thing to give freedom to a heart rotting in pain for so many years. Nothing can curb the passionate feelings from explosive expression in fierce bitterness. Only a person who has lived in such a miasma of suppression of all the pain deeply into one's heart can understand the relief experienced in such a moment of unjaming the whirlwind of accumulated psychotic energy. This is how we participated in the national struggle against Communism in those days, in full anonymity, hiding in the crowds. But back in the School of Theology we would not create even the slightest impression that it was a nest of the most virulent reactionary opposition.

We had to make many compromises in order to preserve our existence. One of these compromises was to join the National Student Union – OSNS (Obsht Studentsky Naroden Saiuz – to be referred hereinafter as NSU). The NSU was a Fatherland Front Organization, under the auspices of the Government. It was politically committed to the cause of Communism, parading as a non-party formation. It conducted public activities in support of the government. Its mission was to unite all students behind the Fatherland Front. Besides its political mission, it was the authorized agency to distribute ration cards and perform all sorts of social services for all students of the university. Membership in that organization, if not required, was a necessity. Thus, they were taking us into the Fatherland Front through the back door. We, in the School of Theology, had to form such an organization. As it turned out, there already was an organized student theological society in our school under the name of "Patriarch Evtimiy". Its president was Gancho Velev, a very close friend of mine. His sister, Rosa, was my teacher in the Middle School in Zelenikovo. He passed away a few months ago. When I came on the scene he had already been contacted by the NSU and had arranged that the "Patriarch Evtimiy" was going to be a collective member of the Union. I, and in February my friends from Plovdiv Christo and Boris, were coopted by Gancho in the governing Committee of the Society. Again, I and my friends were drafted in the NSU, entering into the scheme of the organized Communist association, from the back door.

Sometimes later a representative of the NSU came to address our society along Fatherland Front lines*. Such a "renewal" of the Church had already been urged by Georgi Dimitrov in a speech in the Rila Monastery where he called the bishops "old men with ossified brains". Only one student responded positively on his suggestions, a priest, from some village in Plovdiv District. I believe he was one of those who had been in Gashtev's group, expelled from the Seminary. In the course of the speech of our guest, at one point he asked for the death of Al. Tsankov, known as the bloodsucker, Prime Minister following the coup of 1923, and at this time in exile whom the Bulgarian Communist government wanted extradited, to be tried. Years later he died in Argentina. But the voice of this priest, who, incidentally, graduated in a few days after this incident, was met silently by all those attending the meeting. In the discussion which followed, the NSU representative was given to understand that we, as theology students, and as servants of the church, could not follow political creeds of any kind and we would be better off if we were left to conduct our affairs in the context of our obligations to the church. Subsequently, having taken over the leadership of "Patriarch Evtimiy" I attended several NSU conferences and on many occasions were lectured how to turn around the church into the tracks of Fatherland Front politics, where I had to repeat this argument. So, it appeared that we, in the School of Theology, had to live a double life – openly to support the Fatherland Front Communist government, and privately, in anonymity, to support the opposition. Our faces were with the Fatherland Front, while our hearts were with the opposition. This double life was to continue in the future. Our student life was to be conducted according to the biblical adage: "Render to Caesar what is Caesar's, and to God, what is God's." If this was the rule of the new game, we were prepared to play according to this rule... until we, especially me, were caught cheating Caesar, the Communists.

4. Behind the Mask of "Patriarch Evtimiy"

So, formally we were members of the NSU, but among ourselves we let things go our ways. As soon as we, the newcomers, took over the leadership of "Patriarch Evtimiy" we gave it a new image. I was in charge of a Committee which was to organize public lectures in the Aula of the school. I had to find speakers, had to go and invite them and make

---

* He tried to persuade us to work for "renewal" of the church along Fatherland Front lines and take part in the campaign for extradition of Alexander Tsankov.

all the arrangements for the events. The lectures were given every Wednesday, 7.00 p.m. and were open to the general public. Every Monday I, Christo and Boris rounded the editorial offices of all papers, so they could put it in their publication, announcing the speaker, the topic, the place and the time – always under the name of our society, not under NSU. The opposition papers always accommodated us. We were careful not to omit the government papers, which, sometimes, published our notes. But there we introduced ourselves as NSU students. I introduced the speakers, I concluded the meetings. There were no question and answer periods. If we had allowed it, we would have found ourselves in hot water. One never knows what provocations may be staged, and how irrational sometimes people are. The topics were oriented in the realm of religion and specifically Orthodoxy, but they always touched on issues closely related to the daily experiences of the people and this included social and political issues.

The speakers were not advised what to say and what not to say and quite often the border between what was permissible under the circumstances and what was not permissible was crossed. The public was carefully following the nuances of the speakers and when they stepped out of line of the permissible things, and touched upon issues closely reflecting and expressing views contrary to what the Communist authorities would consider an attack on the government, the audience exploded with applause for the speaker. Very often it was all covered under symbols and concepts that had double meaning. So, the speakers had to be experienced in dodging the bullets of the current politics. Our lectures became very popular in Sofia and many students from other schools were our regular visitors. Years later, when I was a refugee in the camp of Lavrion, Greece, I met a law student at the time, B.I., who told me he was one of our regular guests and could not believe that I, the scroony little guy who week after week stood there, was the same man. We remained very close friends for life, he became assistant to Ivan Mikhailov of IMRO of past glory, and arriving in the United States under my sponsorship, became editor of *Macedonian Tribune* for a few years. Our friendship still continues.

My first assistants in this work were Christo Prodanov and Boris Petrov. Most of the time we invited the speakers together, did all the preparatory work and served as ushers when we did not have other help. At one point we realized we needed a microphone and loudspeakers. The Society had no money to purchase these things and we grabbed the tin cans and rounded the businesses on the main streets around "Sv. Nedelia" Square. I had never done such a work, but somehow I prevailed over my inhibitions, entered the premises, delivered our pleas for help, and surprisingly many of the people responded. This encouraged me. Christo and Boris wondered how did I muster so much courage to do that. Well, we were successful and managed to purchase the necessary equipment. More often than not our speakers were Theology professors. Ever ready to speak was Prof. Gancho Pashev. He was not the best speaker, he was quite boring, but when we did not have anybody to fill the spot, we always could count on him. If one listened to his talks though, carefully, one could see how profound and correct his analysis were. Our most popular and most effective speaker was Prof. Dimitar Dulgerov. I have mentioned him already. With his humble appearance, a frail small man, but a deep baritone voice, piercing eyes which nailed in the space ahead of him, he captured the audience with his oratorical skills. One of his lectures, on the subject of "Science and Religion", was a masterpiece of oratory, diction and persuasive power and has left a lasting impression on me. There he stood on the dais, behind the speakers podium, immovable, focused into space and throwing thunderous lightening bolts. His concluding remarks threw the audience of the packed Aula, hallways, and the entire two wide stairs embracing the well, down to the street, in a frenzy of applause which lasted for minutes. He recited an old poem for children, from the old primers, I do not remember the author: "Mother, My dear Mother, My golden little shade, Let God give you mom, to live long and be healthy." The Communists had deleted the last two strophes, the reference to God, and had allowed only the first two lines. Let me give it in its original Bulgarian text where it rhymes:

Mamo mila Maichitse	Mom, my dear little mom,
Zlatna moia senchitse	My golden shade
Gospod mamo da ti dava,	Let the Lord grant to you mom
Da mi budesh zhiva zdrava.	To be alive and healthy for me.

And the little man with strong bariton, standing like a prophet there, cried out his iron clad words: "You may erase the name of God from the children's book, but you will never erase it from the heart of man." The audience went wild while he was descending from his mountain top and making humbly his way out through the crowds. The roof could have crashed over our heads. I was satisfied! I was triumphant! All my labors were not in vain. I had provided this spiritual feast to so many of the tormented souls in the capital. I had challenged the communists here, in the heart of Sofia, and there was nothing they could do to me. Standing on the dais and watching this sea of people, standing on their feet and applauding the professor, the prophet amidst us, I felt at the top of my power to inflict such a powerful blow over the heads of my persecutors. Oh, had they been here and seen it all, I probably would not have lasted a minute in my oasis behind the stone wall of the school...

Well, there were setbacks. At some point I was struck by the idea to invite a Protestant speaker to address our public, a sort of ecumenism in our times. Intuitively I felt that we had to demonstrate our common cause against Communism, where both Orthodox and Protestants were at gun point. Who would have been better than the editor of the Protestant paper "Zornitsa", a pastor by the name of Trifon Dimitrov? He was later tried with the fifteen Protestant clergymen as a traitor, spy or something. I went to his office alone. We had a short friendly chat and I invited him to be our speaker. He readily accepted. I do not remember the topic of his discourse, but it was well received. The problem for us was the speaker scheduled for the following week – Arkhimandrite Parteniy. If I remember correctly he was either General Secratary of the Holy Synod or Protosyngel of the Sofia Diocese. In the 1960s he was assistant to Metropolitan Andrey of New York Diocese. So I introduced him to the audience and took my seat at the right side of him. His first words while looking at me were: "I regret to be speaking from this podium, after it was desecrated last week by a heretic." I was stunned, smiled and let it go. The audience was unresponsive, as if nothing had happen. I believe they did not understand what had happened. Parteniy was scheduled as speaker for that Wednesday long before I had invited Dimitrov. He, Parteniy, belonged to the ultraconservative, say fundamentalist, wing of our clergy – if his remarks have to be explained, but his discourtesy was inexcusable, and I did not even bother concluding the session with my customary remarks of appreciation for his contribution to our program.

I met Parteniy again in America, in New York. He had asked me through others to meet him. I agreed and invited him for lunch at the New Yorker hotel. I picked him at the Diocesan Quarters with a taxi. My wife was with me. Out of courtesy I asked him if he cared for a drink. I expected him to decline, knowing his reputation in the past as strict monk, a man given to ascetism. I was surprised when he accepted and asked me to order whatever I liked. I ordered three manhattans. We watched him how he finished it in a few minutes. I ordered him another one. He enjoyed them. I do not recall if there was a third one, but during the lunch I began wondering about his monastic sincerity, or may be, he had changed during the decades. Who knows? But I never reminded him of our encounter in Sofia and the scandalous scene which he caused, where I was concerned. His idea of meeting me was to explore the possibility of me getting ordained as a priest in his diocese. I believed he was considering me as a parish priest in Detroit. I told him that I would think it over and will write him. At that time I was well into my profession as professor of history and I was not about to leave it and go to became a parish priest. In addition to that there were some matters of political consideration which weighed heavily on my balance of values.

I was bitterly opposed to our clergy abroad submitting to orders from, and serving the Communist government. I suspected that they all were being sent to America as agents of State Security. I was not far from the truth. In 1970 or before, probably long before, I had met a priest coming from Bulgaria. I was a social worker at that time and it was in that capacity that I was at the pier to meet him. Not suspecting my views on the subject, he confided to me that when they were processing his passport papers, he had been questioned by State Security agents. They had asked him if he knew me or had heard of me. He denied that he had ever heard my name, when, in fact, he had read the periodical which I was publishing at that time, "Borba". Most recently, when I received my dossier from State Security in Sofia, I learned that the late Metropolitan Joseph had written reports on me. But let me finish this story with Parteniy. Returning to my University at Potsdam, I wrote him a letter, later published in my collection of articles *Politicheski Problemi* where I lashed at the synodal clergy with a vehement merciless attack. This was the end of my dealings with Parteniy. As it turned out at the end he had involved himself with political good-for-nothing individuals, and was recalled to Bulgaria. Before returning there he had gone to see King Simeon in Madrid. He died in Sofia in the 1980s. Poor man! I learned later that his two brothers had been killed by the Communists. But why he did not stay abroad and condemn them? Well, it was up to him to make such a decision.

So, I was playing a double game in Sofia. A member of a society, admitted as collective member of the NSU, I was carrying on surreptitiously activities which were squarely anti-Communist in conception and in execution. This double game had to be played with compromises and in 1946 I found myself playing their game. At that time the Government was recruiting young men and women to work on State construction projects on a grand scale, without pay, as brigadirs. One of these much heralded projects was the Pernik-Voluyak Line, building a bed for railroad tracks through the mountains near Sofia – requiring dynamiting rocks and clearing the bed. I was assigned to recruit Theology students to join with other students from the other schools, for a brigade. The quota assigned to our school was ten individuals – considering that most of our students were priests and that our number was very small, perhaps one hundred, more or less. When it was all over I had recruited twenty five or more, which was considered great success. At that time going *brigadir* was considered to be a demonstration of support for the regime. For people who did not enjoy the confidence of the authorities as politically reliable individuals, such voluntary work as brigadir was the first step for political rehabilitation. So all those, who one way or the other were in that category saw in this the opportunity to better their chances in the Communist society by sacrificing one month of their time for such purpose. Indeed, I was first on the line for such rehabilitation. I registered myself in the first place. But by the time the scheduled shift in July came, I developed some pain in my hip joint. I was afraid that I was getting the same thing that my mother suffered. She was already bed ridden. I saw a doctor, told him my fears, he advised me not to do heavy work and gave me a note to that effect. But I wanted to go brigadir. I hoped this would improve my political standing. I presented my case to the organizers of the brigade. They suggested that I should go, and that something could be found for me there not to go to the quarries. I did go and settled as a kitchen help. My friends who worked with the blowing up the rocks and moving them out of the projected tracks were in a terrible shape after the first few days.

Thousands and thousands of students joined the brigadir movement. Those whom I recruited did not need persuasion. They flocked to me to be included. In those days everybody was looking for cover and "rehabilitation." I remember a situation which had nothing to do with the brigadir movement, but illustrates the state of mind which moved people to seek ways to make their peace with the red barons who had taken over the country. One evening the residents of the blocks where our building was, members of the Fatherland Front organization, were holding their monthly meeting in the Aula, next to the library. The Aula and the library were connected with a big mahogany door, and between the two wings was a rather wide crack, some quarter of an inch. So whatever was going in the Aula could be heard through this crack. We listened to the speeches there. That evening they were accepting new members. At one point the name of a woman was called. She stated her reason to join the organization: "to work for the people's republic of Bulgaria, together with all patriotic citizens," and all that nonsense. When she finished, someone from the public got up and said that she should not be accepted, because her husband had been a military officer, an "enemy of the people", and had been thrown out of the army. The woman started crying, insisting that he was not an enemy of the people and that she wanted to join and atone for whatever was held against him. She was rejected and asked to leave the meeting. As she was leaving she was crying and begging to reconsider their vote, but no one listened and she was thrown out. The case of the brigadir movement was of the same nature. Sure, there were those who were supporters of the regime, but also there were so many who were seeking rehabilitation, or in anticipation of little privileges which they were seeking. Acceptance in the University was one such privilege, as they had already introduced the system of "Fatherland Front Notes" of political "reliability."

But my double game could not continue indefinitely without being noticed. I have been using the NSU to carry on activities which were, if not in form, at least in content, anti-Communist. I was not joining the opposition parties, I was for all practical purposes claiming that I was supporting the BANU in the government coalition. But this was not enough to dissipate the impression that I had made the Aula pulpit a platform for speakers of anti-Communist persuasion. In the fall of 1947 a number of students had signed in the school who had been already involved in Komsomol activities, and, sincere or insincere, were pushing for taking over the leadership of "Patriarch Evtimiy." They were joined by disgrantled graduates of the Sofia Seminary, who were agitated because our leadership was made mostly of Plovdiv graduates. When the time for elections came one of those on the left came to me and advised me to withdraw from running for the presidency, which by now was held by an Agrarian from the upper classes. He suggested that

I just should not attend the meeting and make things easy for them. I read the signs of the times, that if I did not follow his advice, a coup was to be executed any way and things could become unpleasant for me. So I figured out that a wiser move on my behalf was to step down. In addition to all that my studies were beginning to suffer from my giving too much time to public activities. If I went out fighting, my political double game might be exposed and the consequences could be much worse for me. My friends, too, were not prepared to take a fight, and so we gave up the leadership to our opponents – a combination of leftists, led by a Plovdiv graduate, son of a Russian, White Guard priest, but for some reason well entrenched with the Fatherland Front youth organization, DSNM (Dimitrovsky Saiuz na Narodnata Mladezh, Dimitrov's Union of the People's Youth). I did not show at the meeting. They elected some leftist priest as President. One benefit for me was that when they were distributing cards for free meals at the Student Mensa at the University, I was included in the list of the needy students. I suspect one of the professors may have advised them to do that. It might have been Dulgerov, though he reprimanded me in a friendly way for having given up the leadership. Well, it was easy for him to give me this advice. If I had chosen to fight, I would have been placed at the barrel of the gun, and I could not afford it. In any event they continued the weekly lectures, which was the only activity of "Patriarch Evtimiy" to show its existence, but after a while they gave it up. I focused on my studies.

5. Behind the Iron Bars

In the meantime, after my days as brigadir, something happened which changed everything. On way home from Pernik-Voluyak I stopped in Sofia for a few days. My friend Daniel briefed me on the latest political developments. One of the big news was that Damian Velchev had been placed under house arrest. All kinds of rumors about the shaky state of the government and exaggerated expectations that the regime was falling, were flying around. What a treat that was! We in Bulgaria have a saying: "a hungry chicken dreamed of millet." From there I proceeded to Zelenikovo. Only a few days later I met my old friend Diado Marin Darzhaliev, some seventy years old. He asked me about news in Sofia and I had the imprudence to tell him of the rumors about Velchev. He lost no time to share it with his friends and in two or three hours this hot news was all over the village. Before the day was over I was called to the municipal building and Parlapansky locked me in the basement where I had spent some days and nights before. Of course he asked me about the "news". I admitted that I had brought it from Sofia, that it was a rumor, but he was furious. He accused me that I was purposely trying to undermine the confidence of the people in the government. So, I ended in the cooller. What was I going to do? I tried to use my having been brigadir, but it did not go far. I was told that I had done it to hide and cover my subversive activities. The whole thing was taken as a vicious propaganda against the Fatherland Front. I spent the night in the village "Jail." There were all sorts of threats to frighten me, but they did not touch me, to beat me up. The next morning I had a visitor: Koliu Mirazchiysky, my nemesis. He stood at the top of on the steps up there, looking down on me while holding the two banisters on the side, ready to spit in my face. Hatefully, he told me that I have not changed, that I have not understood the spirit of the times, that I am working with the enemies of the people and that for all that I was going to pay very dearly. "We gave you freedom, so that you should show your horns," he said, "and then we will smash you like bed bugs." This is the same Koliu who was to be exposed as a police agent of the "fascists.' As of now, he had not been discovered, and for a long time he was going to hide, until his dossier was found and he was thrown out of the Party and his job. As about the rumor that Damian Velchev was under house arrest, it proved to be true. A few weeks after my arrest he was appointed Ambassador to Switzerland (probably on account of Kimon Georgiev, the Prime Minister, his friend in so many mischievous adventures in Bulgarian politics). In no time he defected and much later died in Switzerland.

But this incident disturbed me to no end. I was not beaten this time, but it was a signal of worse things to come, if I continued to hang around there. I discussed all that with my father. He was frightened too, that something worse could happen to me if I did not leave Zelenikovo. There were still rumors of people vanishing into the unknown. Who knows what they could set up for me in the village? As bad as things were in the house, we decided that if I could find some place to go, and disappear from their sight, I should go. I packed a little suitcase, took the bus to Plovdiv and went to Metropolitan Kiril. I told him what had happened to me and that I was looking for a place where I could stay until classes start in the fall. He pulled out his card, wrote a message on it, gave it to me and told me to go to the Rila Monastery and give it to the Igoumen, Arkhimandrite Varlaam. *
In Sofia I picked a ration card release, to present it wherever I was going to stay, and the next day I arrived at the Monastery. I read the message on the card: "Diado (Grand Father) Igoumen, take care of this young man." Varlaam was a very gentle soul, with erect posture, bribing smile under his clear-frame glasses, beautiful beard, pleasant to no end. He sent me with somebody to the monastic room where I was going to stay. It was a newly remodeled cell, with bathroom and kitchen inside, separate from the bedroom. I settled there and felt relieved of my fears.

A few days after my arrival, a new man arrived from Sofia, a monk, Father Theodore. Father Theodore was a nephew of Metropolitan Kiril and student of Medicine. He was sent to share the room with me. Theodore proved to be a street-wise Plovdiv boy. He brought to the monastery a large suitcase filled with small pamphlets on the life of St. John of Rila. His intention was to sell them and make some money for his studies. He asked me to become his partner: I was going to sell the booklets, twenty levs each, and split it between us: ten levs for him and ten levs for me. It was not quite to the liking of other monastery residents, but since he was the Metropolitan's nephew, nobody could say anything. It just happened that that year, 1946, was the millennium history jubilee of the Monastery and pilgrimages were organized from all over the country. It was estimated that about a million people visited it from early spring to late fall. This made our business partnership work well. I was making money and Father Theodore was making money. I proved to be a good salesman. Besides, so many people coming to the monastery, they all wanted to read something about the patron saint. The sales were going briskly. One day a tall, slim clergyman, with dignity radiating from his aristocratic pose, stopped me in the middle of the monastery yard and asked me: "Young man, what are you selling?" I repeated my sale's pitch. He asked me if he could have one and gave me a

---

* Years later he became Metropolitan of Plovdiv.

banknote bill of 200 levs. I did not have enough to return the change and felt a little nervous, reaching in all my pockets. He looked at me with compassion and told me: "Keep the change for yourself." I had already told him that I was a student of Theology in Sofia. It turned out that he was Metropolitan Boris of Nevrokop, now Gotse Delchev. Two years later he was shot dead in front of a village church in his diocese by a disgruntled defrocked priest. The murderer did not stay more that a few months in jail. The general consensus was that the murder had been ordered. Boris was known as one of the most conservative leaders of the Church.

We were working well with Father Theodore. He was a down-to-earth young clergyman. It was generally believed that after obtaining his doctoral degree he would shed his cassock and go into civilian life. I believe this is what happened. But while in the monastery, he behaved appropriately, read a lot his textbooks and studied for his tests. At one point he asked me to help him with something. Did not tell me what exactly he had in mind. He took two pillow cases and led me to the ossuary of the Monastery. He wanted me to help him pick human bones and skulls in one of the pillow cases while he would fill the other one. I could not do it. These were human bones which sometimes in the past had bodies and souls. I could not do it. He tried to explain to me that they were nothing different from baked bricks. I still could not do it. He selected his bones himself. He needed them for his studies, and to take some to his friends. I was watching how some shin-bones were sticking out of the pillow cases. I carried one of them, but looked up and to the side. In our room he washed them in the sink and lined up the three skulls on the counter. What a sight? I dared not touch any one of these "Bricks."

I felt safe in the monastery. There was nobody there with whom I could or I would discuss politics, and this was a respite. There was a Militia post on the ground floor in one of the side wings, but they were somehow inconspicuous and we kept away from them. But I was dying to know what was happening in the world of politics and could hardly wait to return to Sofia. In the beginning of September the government had appointed a referendum for Monarchy or Republic. There was not much electioneering in the monastery, but when the votes were counted, the entire monastic community, including me, had voted for Republic – some 177 votes. Who dared to vote otherwise in this small community? The Communists had no reason to suspect anybody in opposition activities, and even if they did have their eyes on somebody, they had no reason to take any action. So peace reigned in these majestic mountains of Rila. In the middle of September I flew out of this cage and arrived in Sofia, to prepare for the new semester.

VII

STRANGER IN MY OWN
LAND

"I have loved righteousness, and hated
iniquity; therefore I die in exile."

Pope Gregory Hildebrand, at Salerno,
1085.

1. Monasteries, Political Rallies and Conspiracies

When I arrived in Sofia the referendum for the republic had just finished and the government was ready to schedule or had already scheduled elections for the Grand National Assembly, to be held on October 25. The country was soon thrown again in election fever. The bitter rhetorical battles fought in rallies and wall posters, and in the papers, were now bloodied with the bodies of many opposition activists, murdered in one region of the country or another. The *Narodno Zemedelsko Zname* of Nikola Petkov was displaying at the bottom of its first page Agrarians killed by the Communists. The bloodier the battles became, the more vicious the attacks from one side or the other were becoming. The articles of Nikola Petkov, Kosta Lulchev and Trifon Kunev were devoured by us in our dorms. I believe I have already mentioned my participation in the grand rally before the National Bank where on account of my screaming I lost my voice for several days after that. Sofia was bristling of hate and one could feel in the air the imminence of a civil war. One day some of our students had ripped off a large poster with the picture of Georgi Dimitrov, pasted at the entrance of our school. Some taxi drivers had seen him, jumped from their cars and pursued him with open knives into the building. He managed to get in the basement, while they ran up to the attic. Had they found him, God knows what might have happened. I never learned if the man lived in our dorms or was an outsider. He probably was an outsider who knew the building well. Otherwise, we would have known his identity. But the sight of these vicious looking pursuers with their knives made us shake in our boots. It turned out that on several occasions, when people were looking in despair for some place to hide, our school was attracting them as a magnet. When the government had conducted a census of the population, numerous individuals fled Pirin Macedonia, not to register themselves as Macedonians, fearing that if they uphold their Bulgarian ethnicity they would be subjected to persecution. They inundated the school and we housed them the best way we could. I do not remember how the election day, October 25, passed, and whether I voted or not, probably not, but since the various parties in the pro-Communist coalition voted with colored ballots, the Communist party alone secured for itself 57% of the votes, an absolute majority in the parliament. Its allies, remnants of the Agrarian Union and the Social Democrats, not to mention the "Zveno" people, added to its parliamentary strength and gave it the appearance of a coalition government. This was important for the Western Allies – so the Communists thought – to argue that the elections were free and democratic. The opposition was allotted some 100 to 110 seats in the parliament. So, the government could use the argument for free elections. It was later charged that the elections were stolen. I believe that to have been true, but who was going to change things? After the elections the tensions were relaxed. The new Assembly undertook to prepare a new republican constitution and pass new laws. It all was controlled by the Communist majority.

By the end of December I had to figure out where to spend the winter. I could not return to Zelenikovo. So, I ended back in the Rila Monastery. The Igoumen, the amiable Diado Varlaam, put us, five students, in one big cell. A big stove in the middle, burning wood, which was in abundance there, kept us warm. It turned out that all my room mates were of the same mind where politics was concerned. Though everybody was reserved to reveal his story, we all may have had the same reason for being there. We felt free and safe in that room. As long as we had free bed and board, all else was going to find its place. But there was also some other student, living there. He was a Communist and often visited us for lively discussions. I remember one day I got involved with him on the subject of money. He argued that the day would come when money will be no longer necessary. When the perfect Communist society would be built, there would be no need for money. I argued that the money was a certificate that an individual had accumulated so many credits by his work, and would go and exchange this certificate for goods. There was no way to establish that one deserved the goods he was getting from the common store. Human nature, being what it is, some people will easily go out and gorge themselves with goods, while others would be producing them. All else is a fantasy and illusion, which will never come to materialize, unless men turn into angels. After so many decades of Communist practices, it appeared that my position was right and his wrong. But who is going to keep a score of these illusions with which the young were fed in those days?

In February I was back in Sofia, but had little taste to re-enter politics. It all appeared to be settling down. The Communists were fully in control. In the Grand National Assembly the controversies and the battles between opposition and Communists were reaching for a climax. For us there was nothing to do, except to watch from the sides and be careful not to say the wrong words at the wrong places. But on one occasion I somehow lost control of my better judgment. Sometimes in the Spring of 1948 I was invited by my neighbor-friend Todor Vessilev, who at that time was studying Forestry, to join him at a secret meeting. It was held near the Seminary, in a rather large ground floor room. He did not explain to me who was sponsoring this meeting. It was political, it was clandestine, it was against the regime. It dawned on me, in the course of the discussions, that these were former Legionairs. There were between twenty-five to thirty people. The meeting was delayed. They were waiting for somebody from the "Center" to come and speak. Finally he, with some two or three companions, showed up. I vaguely remember his speech, only that it was against the regime, and that liberation was to come, that the Americans and British were to wrest Bulgaria from the claws of the Soviet Union. On a question when all that was to happen, he gave a disappointing answer. It was to come about 1953. 1953? From 1948 to 1953? Was I going to survive? How was I going to last for so long? During the following years I kept looking for this white barnswallow to show up, but it was not coming. In 1949 I graduated, the Communists were still there. In 1951 I escaped to Greece. They were still there. Came 1952, I was in Switzerland. Nothing happened. In 1953 I was in England and close to the beginning of 1954 I returned to Switzerland. So, the prediction that by 1953 the English and Americans were going to liberate us, never came true. When this guy uttered this prophecy I was examining the faces of all those present. There we were, too many of us. I did not know them. But what if among them there was someone who was a Communist plant? As agent provocateur? What if this whole thing was a set up for politicians with green ears like us? Then and there I decided that this whole thing was stupid and dangerous. Leaving the meeting I told Todor not to look for me again for this thing. This was the end of it. I do not know if he also dropped out, or it fell apart by itself. It crossed my mind that some of these young men might have been a Legionair, might have been in a concentration camp or something, may have been promised something to ensnare people like us, old Legionairs like Todor and his friends. Much later I was to learn that such traps were layed out by the secret services and many young men have been caught in them. In 1951, while a fugitive, I almost fell in such a trap.

Meanwhile, the parliamentary debates were getting to the point where one could easily characterize them as political assassinations. If it was not that the Communists were the only armed force in the country, the tensions would have exploded into a civil war. But, knowing that they were armed to their teeth, and remembering their wanton killings in 1944 of their "enemies," no one dared to confront them with a rebellion. The nation was cowered in a fear beyond words – in the parliament and in the newspapers. Nobody dared to take action. In this atmosphere in the beginning of June, Nikola Petkov was arrested in the parliament, after being deprived of his immunity. In August he was tried as a traitor, conspirator and all that was "appropriate" in the case, was sentenced to death and executed on September 23rd. The "Allies" made some representations to the Bulgarian Government, but did not lift a finger to save the man whom they were instigating to fight their fight against Communism. My conclusion on this occasion was that if one had the Americans and the English for friends, he did not need enemies. From this point on the opposition skidded down to the precipice of its destruction. The Agrarian Union of Nikola Petkov was outlawed, to be followed in the same path by the Social Democratic Party et al.

I do not remember where I spent my summer in 1947. But sometimes after the hanging of Nikola Petkov I read an article, an editorial in a Communist newspaper, where they had made a very strong and believable point that they should extend a hand to all those who had followed Petkov, and offer them a chance for joining the Party, or outside of the party to build socialism in Bulgaria. I must admit, this intrigued me. I was yearning for a chance to put my life in order and felt that may be this was a genuine call for accommodating to the regime. It is with such hopes and expectations that I wrote a letter to my neighbor and "friend" Marin Pekhlivanov, to which he had responded. I, in my turn, wrote him a long "confession" of my thoughts.

2. Tempting the Devil

When I visited Zelenikovo in 1991, Marin came to see me at my brother's home. He acted as if nothing had happened between us some forty five years earlier. I was also well disposed to forget all of the past. After all, there has been so much water gone under the bridge. He did not hurt me physically. Neither did I do anything to him. So, let bygones be bygones. In the middle of our conversation he pulled from his pocket my letter of November 24, 1947. I had completely forgotten about it and did not have the vaguest idea as to what I had written. Who knows, sometimes things are written in one set of circumstances which may be used in a different set of circumstances for blackmail. So, I was afraid that may be there was some indiscretion there which was not appropriate to be publicized in 1991. I grabbed the letter from his hand and put it in my inner pocket. He wanted it back. I told him that I will need it when I am writing my memoirs. At that time I did not have any intention to write memoires. When I started writing this part of my reminiscences and reached the events I am discussing now, I remembered that letter. It took me a while to find it in my papers. I read it for the first time since I had taken possession of it. Reading it I was impressed of its contents and saw it as some sort of a revelation of myself, made some forty five years ago. I saw it as a very important document of the past, a testimony of my thinking in the times when I was under such a stress and at a crossroad, not knowing which way to go. I discovered that with all the pains and sufferings which I had been going through, I had not lost my sense of honor and dignity, had not fallen on my knees to beg for mercy, to whimper and whine for forgiveness, that I had stood firmly in defense of my positions. I am printing this letter in translation, without making any changes in it or additions, adding also the brief comments made by Marin in the margins. Here is the letter.

Dear Marin:

I am quite late in answering your letter. The reason for this delay is that I am over-loaded with work, have very little free time, and my response, in view of the questions raised by you, could not be limited to a few lines.

I was expecting your letter, and exactly such as it is. I have been for sufficiently long time, so to speak, outside of the view of the activists in public life in my native village. This weighed heavily on my heart, but I could not think of a way to state my views, to relate them to the views of those with whom I have been associated in the past. I have been isolated from the dynamics of the cultural life of my native region – not only ideo-logically, but also physically. On numerous occasions I had sat down, had composed a few lines of the character of this letter which you will be reading now on the following few pages, to restore our relations, but, let me frankly admit, my attempts always turned unsuccessful. Sometimes I would begin writing, would finish my letter, but at the end would tear it up. I could not see to whom I could address myself, who would understand me, who would resonate to my views. I was afraid I would not be understood correctly, that my words could be interpreted one way or the other, or that they would be taken not as a personal confession, but as a concocted hypocrisy.

I felt this way on account of the well known opinions of me, and whatever I would say it would be looked at with suspicion. I was being told that I could work in the mass organizations, to be observed there, and if it appeared that I could, or that I had radically changed, then I could be taken into confidence. This placed me in a most inconvenient position. Every minute I had to be aware of the burden of non-confidence hanging over me, and in the sense of this awareness, to watch for every word I said and every act I was engaged in, so that they would not give reasons for negative interpretation. I had to tune in the waves of my existence carefully, so that I do not land in a disagreeable situation. By limiting the logic of my thinking this way, in view of the stated objective, I had to experience every moment the unpleasant feeling that I am a second class man. This may have been justified from the point where you stand, but it had its adverse effects on me. Excluding its subjective reflections in my conscience, I would call your attention to the fact that it would create a wrong impression of me in the eyes of society. On one hand there were people who would like to present themselves as better catholics than the Pope, so with or without reason to throw rocks at me. On the other hand, the really reactionary people would have concluded that I properly belong to their circles. The result of all this, as it appeared, was what you say, that I had become "a silent observer of the socio-economic life of the country". However, for an active individual this is an intolerable suffering. I suffered but I could not see anybody to whom I could turn. (Now, fifty three years later, I could sum it all up in the words: "I had to walk on egg shells.)

I felt that the mind of the intellectual could be correctly understood and interpreted only by an akin soul, the soul of other intellectual of real standing. However, I had no connections to such an individual in my village. I had to wait for someone to approach me. This waiting already continues for three years, and I had to come to the sad conclusion that nobody cared about me.

How justified are my conclusions, I find confirmation in the very Communist press of the Fatherland Front. In this respect, I was specially impressed by an article in "Trud", entitled "To go to them". It was published immediately after the arrest of Nikola Petkov. I had to think over it for a long time. I felt that I, myself, had gone to the editor of "Trud", had related to him my predicament, and he, like a good friend, wishing to help me, had published the article. The article indeed treated this question. Accounting for the error of the Fatherland Front to leave some people outside its wings, hesitant to get in, etc. it recommended that they should be approached, to be attracted and persuaded to join the organization. I preserved this issue of the paper, so that I could show it to somebody in a convenient moment. This paper is now home and if I come to the village for Christmas, I will show it to you.

Half a year has passed since that time. Nothing happened. Simply, everybody is looking after his own affairs. During the summer I was in the village, but nobody, except you, showed any interest in me, I noticed. If sometimes something was thrown my way, it was done in such a way that no results could be possible. It was only in your attitude that I observed that I was not treated as second category individual. Irrespective of the ideological differences which you had supposed to exist between us, you at least have found something of value in me. On every side I met the attitude that I was nothing, and yet, I could be something. You gave me to understand, that I, after all, was something, but could be even something more. As it may seem immodest on my side, I came out with this conclusion about your attitude, as I did also from your letter, so that I could think of some words and clarify the questions which you have raised.

I thank you for your letter. For three years I have been waiting for such an opportunity, so that I could clearly, concretely and unambiguously, explain my views on the problems arising in our times. You are the first who brought to my attention clearly and concretely the problems which excite the conscience of every intellectual in our times. In my reply I will try also to be clear, concrete, nambiguous, and most important, frank and sincere. Without any, whatsoever, hidden thoughts, I would be guided by the principle which I have scrupulously observed always and everywhere, namely, following the iron logic of my own reasoning, without relying on authorities, I will try to give you my personal view of the events. So to speak, I will make my own personal confession – not to try to ingratiate myself with you, for one or another purpose, but simply to clarify my views on the problems of our time.

In this instance I beg your indulgence to allow me to express my perception, my deep conviction, that there is no man in our society, especially one who claims to have some of the culture of our times, who would have no opinion on the questions of life confronting us. It does not matter whether he expresses his opinion publicly or not. The important thing is that everyone has his own views. This is the criteria for me when I judge the intelligence of an individual – to what extent he could form his own convictions in the whirlwind of the social struggles and bring into accord, or juxstapose, or oppose the generally accepted public norms. I would not give much value to an individual for intelligence and culture who settles for a sorry existence, who drag himself after the opinions of authorities, incapable of giving personal evaluation of historical dynamics.

Even if it is considered immodesty, I dare affirm that I have always formed my own opinions on all questions, not to accept completely the opinions of anybody, any time and any place, always seeking to form my own opinions, independently from others. [*] I hope that this has been noted. If it has not been understood by now, the future, sure, will confirm it. And anybody who has not understood it, he does not know me and it is natural that he may have one or more erroneous opinion about me.

It is possible that on some occasions my opinions may have been wrong. But this does not mean at all that I am not entitled to them. On the contrary, it means that my life has made sense when I have been coming closer and closer, step by step, to the truth, and to experience the pleasure of the man struggling in life when he scored his victories over the inequities and overwhelming its secrets and its storms.

You are reprimanding me for my erroneous views prior to September 9th. But I am unaware to have done anything, except that I had held one or another kind of views, regardless of the fact whether they have been wrong or right. If I have been wrong, if I have been right, it was my right to define them. I had to correct myself alone. It would have been too bad if I did not do that. The historical events were going to correct me, were going to refute me, and if I did not take a note of their refutation, I could have been justly judged as incorrigible.

But I noticed something else. I noticed that I was seen in such a way, from your side, that I was a sort of par excellence second Tsankov (Prof. Alexander Tsankov was the leader of the Bulgarian fascists). My natural psychological reaction to that was to experience unpleasant feelings, which I had to suppress with great difficulty. [**] I feel that I was not so black before September 9th, so that I had to be insulted in the most inner fibers of my soul, to be fingered as enemy [***] of the people, to the Fatherland Front, as a flag bearer of the reaction, of the exploiting classes, of the social, economic and political slavery.

Believe me, if I was told that I was a bandit, that I was a brutal murderer, thief etc., I would not have felt so hurt, it would not have been a burden enough so that I would not be able to carry, rather than to be told the fearful, unbearable painful insult that I had been against the people, that I had been enemy of the people where I directly come from.

How much sorrow, how much joy, how many flying hopes and dreams I had lived through, how many worries and what exalting feelings have shaken my soul over the fate of this people?

I will not permit to anyone to speak of its suffering! I will not permit to anyone to speculate with its tragedy, because – this I have to be allowed to assert – nobody knows the suffering of the people, the tragic ordeal of this people in the way that I have come to know them. [****] I know all that not through sentimental fantasies of immature poets, not through the dimmed conscienceness of moral capitulants and bankrupt pessimists, not also through the glasses of shallow observers, but through my own suffering, through my own tragedy, as a son of this people.

I know the people's suffering at its worst and if I have to express my opinion of it, it would be negative to the extreme, nothing less. I am convinced that in my understanding of this problem I have never taken a second seat to anybody. It is not a question, of

---

[*] Marin's comment on the margin: "Like the speech at the Branniks' meeting."
[**] Marin's comment: "Right!"
[***] Marin's comment: "It was so because you could, indeed, with one vigorous swing to organize, you had all the possibilities, you were a force and not an ordinary hand, ordinary link in a mechanism."
[****] Marin's comment: "Why then you became a tool of the bourgeoisie against the people (Legionairs). Perhaps you have understood wrongly the Legion!"

course, whether the positions from which I have attacked the suffering of the people have always been the best ones. I would not deny it that one could fall on such positions, especially in times of war, where it often happens that the artillery may be battering its own infantry. I am not denying that before September 9th I had fallen on such a position. But I will protest from the bottom of my soul to have done that with some forethought. No! I have been, I am, and will be consistent defender of a populist outlook. It is another question how well was this outlook articulated. If before September 9th it had not been sufficiently articulated, in other words, containing inevitable factually wrong positions; and if it is now in the stage of its finalized form, to be expressed tomorrow in a sound realistic world view as a realistic system which would meet the criteria of my own logical reasoning, to legitimize itself to my moral conscience where the central place is occupied by my love for the people; then, in as much as I have caused any pain to anybody, or in as much as I am not causing any pain to anybody, I would beg that I should not be judged so harshly.

What is my view of the present historical reality?

You are seriously concerned about that. Your impression is that I am silent observer.[*] You have come to this conclusion on the ground that I have not been involved in any activity in the village. It is true. In the village I have done absolutely nothing which would show exactly what I think on the issues of the day. Why? I think I answered this question. There I was subject to a special treatment. Every demonstration on my behalf in favor of the Fatherland Front, by all means, would have been interpreted on all sides in a negative way. I was afraid of such a contingency. Some would see it as lack of character[**]. Others would have seen it as weakness, still others as hypocrisy, careerism, cowardice ... I am sure that nobody could have penetrated in the essence of my personal political creed, nobody would have looked for the strong continuity in my views, the consistency in my way of looking at things in life. Everybody would have seen it as a turn around of 180'. This is so offensive for me, that I am prepared to suffer endless physical consequences, rather than allow to be publicly branded as "traitor".[***] I have put the word in quotation marks, because, in reality, there was no treason.

Indeed, the events took such a course, that I deserted from the village. But, is it only there that a man could find a job? No! The world is wide. Prospects for work and success will be found everywhere. I found myself among new people, unknown, and, consequently, could show myself such as I really am, without reasons to think that I have fallen in one or another false position.

Having found myself in a really conservative circle of people with reactionary inclinations where all undertakings of the Fatherland Front are subjected to scathing attacks, I could not remain indifferent, could not stand out there without an opinion on the issues. So, one after another, in the focus of all criticisms, the National Assembly passed the laws for T.P.S. against the illegally enriched, for the T.K.Z.S.s, etc. What could I think on these questions? To defend the big landowners? No! You may remember what I thought

---

[*] Marin's observation: "I thought that observing the life, you weigh everything on a balance, exert yourself to juxtapose your old views of given realities and the new way of looking at them today, and thus to make some new conclusion."

[**] Marim's observation: "I would have understood you, as After September 9th."

[***] "This is true. A man like you (If I was on your place) with reputation in his own eyes, would not crawl at the feet, would not have shouted and made noise as many such people did, who before September 9th beheaded the Communists, and after September 9th began massacring the fascists."

on this subject before the 9th. I support the thesis for 80 decars per family. To defend the capitalist exploitation? I am certain that if there is one who has experienced misery in its worst form, it is me. For me to defend a life of misery, it is absurd. To come out against cooperativism? No! I think my views on the subject were known to everybody long before September 9th. In 1945 I read a Report on cooperativism in the Seminary. If you had heard the thesis I was defending, I am sure you would have shaken hands with me long time ago.

Such questions are debated every day in my circles. We carry vigorous debates where I – it is quite immodest and inconvenient to write about myself but the question is now put in such a way that I have to do it – defended positions for which I have been called often, more or less, what do you think? A communist! Tragedy! When I would explain how the communists treat me in my village, the puzzlement of my interlocutors would be followed by doubts that I have been telling them the truth.

This is nothing, you would say. Allow me to add two more words: there are facts. For two years already, I am active member of the Joint Students Peoples Union (O.S.N.S.). And I have worked so hard that no one in the village have any idea. Last year I was in charge of recruiting Brigadiers (Voluntary workers on State projects). In my school I surpassed the quota by 150%. My activity was not passed unnoticed. This year I was rewarded for that by the O.S.N.S.*

I feel that all that I have given to the Fatherland Front, in the limit of my possibilities, is not negligible. How it will be seen by you? I do not know. What is important for me is that I have fulfilled my duty for my people and the Fatherland Front.

You are interested to know how I could juxtapose my activities as described, with the philosophy which I believe in.. You even try to explain the supposed "silent contemplation", my standing on the side, outside of the Fatherland Front and R.M.S. (The Communist Youth Union), with this philosophy, with my philosophical orientation, which supposedly determines my political outlook. This concept of yours is clearly in accord with the basic materialist thesis that conscienceness is determined by material reality. If I do not clarify this issue, naturally, my response would be inadequate. I must stress that I see philosophy as a tool for penetrating in the mysteries of reality. For me it is more gnoseology (Epistomology?), logic and metaphisics** rather than a social and economic platform. To come to a full understanding***, to conquer gnoseologically the nature, I think, does not mean establishment of one or another system of social organization, one or another blueprint for social reformation. When I take up philosophy I seek to understand things as they are in themselves, the world in itself, without one or another reference to me. (I am using here terminology of the philosophy of existentialism, which I was unaware of at that time – STR)

When a philosophical system satisfy my inquiries, regardless of the name it has – dialectical materialism, idealism or naïve realism – I would accept it. And if I do not find any philosophical system satisfactory for me, I will continue searching for principles corresponding to my own reasoning and will settle for my own philosophy. This may

---

* All these things are twisting of the facts to appease him and those in the village. When I come to explain all this "activity" in the O.S.N.S. it will be shown that it all was a cover for our anti-communist work in the School of Theology.

** Today I would use the words Epistemology, Metaphysics and Axiology.

*** Marin's observation: "One without the other does not go. Philosophy is a method, outlook, glasses through which you see everything - nature, arts, culture…"

sound like a paradox, self-exaggeration, but it is the fundamental principle of my existence. Whatever I settle for, I want it to be legitimate itself with its own arguments to me.
I do not recognize other authorities. I do not accept anything on faith. This is why I read
any book – be it espousing materialism, be it espousing idealism – it does not matter to
me [*] – whichever reveals to me even the smallest particle of the world truth. This is how
I look to philosophy – and I want to stress: I am seeking only gnoseology, logic and
methaphysics, not an ideology for some kind of a social and economic restructuring of
the world. [**]

When I approach this subject, the last one, I leave all philosophy on the side and
consider the issues as they are presented in front of me concretely, as reality in history, in
my presence. And to come to conclusion on one or another question individually, or to
the entire concrete historical socio-economic complexity of problems – mutually condi-
tioning themselves and so conditioned, I follow one – allow me to call it this way –
golden rule, namely: THE HUMAN INDIVIDUAL IS AN ABSOLUTE VALUE. Starting with this principle, one could devise a system of social, economic and political relationship, the social, economic and political system should be arranged, should be re-
formed in such a manner that the golden rule cited above become a living reality, a real
fact.

What all this means? This means that the life of every individual, regardless of his
racial or ethnic background, his religious or philosophical convictions, become a harmo-
nious song and music, in happiness, in wellbeing. If I have to search for a name of this
view of mine, I, without hesitation would call it philosophy of life. I would doubt it if you
have anything against this philosophy. Don't you yourself serve such a philosophy? Isn't
Communism, in its deepest essence, following such a philosophy? So, then, what divides
us? (Today I would call it Christian humanism).

It may be that you know of something that divides us! Perhaps your friends in the
village [***] think that there is something which divides us ... but I see no such thing. I
should frankly tell you how much pain I have experienced, how many sleepless nights I
have had, tortured by the thought that those, with whom I should be marching hand in
hand, on account of not knowing me, push away my extended hand. The class out of
which I have come, to which I belong, which was the starting point of my cruel struggle
against the currents of life, against the hellish conditions of this life which reduced to
dust the joy of the so much praised young life [****], but which I had to vanquish, and I did
vanquish; the class which planted in me hatred for social slavery and unlimited love for
the people-slave, to consider me an enemy?

This caused me endless suffering. Further up I referred to my personal tragedy. I am
mentioning it here only to stress, that if there is something which is like unbreakable tie

---

[*] Marin's observation: "G. Dimitrov says: One should read that which will build him up. There is not much time
for reading. You will not become a specialist. We are loaded with work. We should follow the real life. How
Logic will help to free the poor peasant?"
[**] Marin's observation: "Why then we live and are struggling? How are you going to look at historical reality
without knowing the laws of its evolution? These laws are the laws of philosophy?"
[***] Marin's observation: "I have always tried to judge people not by their title. If some individual with great
potential, in my judgment, takes a wrong turn, I would go and help him. In as much as you are concerned, I have
the unshakable conviction that it could be changed only by you and nobody else."
[****] Marin's comments: "Be like Katrova, like karata et al. This is what I want you to be. You want something
more. This does not make you better. E.G.Monica is not concerned about the past." (I do not know of these
characters)

247

between me and you, between me and your group, the RMS, the RP (c), it is our common view of the social tragedy of our nation. I fully support your platform for social-economic reforms. I do not find it necessary here to go over it point by point. I think you have had a chance to recognize that. So, if our subjective views on gnoseological and philosophical issues are different in some respects, the philosophy of life can unite us. In the name of such unification, you may have my right hand. Will you accept it? I could not answer this myself, but if your response is positive, my tragedy will be over. At long last. I would have found *my place* * .

Before I finish, I want to clarify some questions about the Young Agrarian Union. You write to me that I have attended the meeting of its organization. This is absolutely untrue. It was by accident that I met the organizer Ivan Markov. And because it came out that he knew some of my friends in Sofia, Agrarians, he invited me to attend that evening the meeting ** . I found the way to decline the invitation and not to attend. This is all. I cannot understand how this legend has appeared. Be that as it may, for several years now such events are happening around me with unpleasant consequences. I hope that after the organization of the Union of Democratic Youth (C.D.M.) and the single political anti-fascist union in Bulgaria, my difficulties, arising from the suspicions about me, will cease. In any event let me stress that the question of establishing a regenerated Socialism in our country is being advanced. The Party rivalries, divisions and sectionalism will come to an end. I hope that the situation will be normalized, so that, at long last, our nation, united and unified, could peacefully and creatively do its business.

This is all that I wanted to write you, Marin. I am expecting to hear that you have understood me. In any event, I was sincere, as you were with me. I will be waiting for your reply.

Cordial greetings to the Zelenikovo colleagues in Plovdiv.

Sofia, November 24, 1947.

Yours: (Signature)

Received November 29.

After I typed this letter, rethinking its contents, I was deeply impressed of the passionate defense, as I have noted elsewhere, of my independence. All my life, from before the Communists took over, and after they came to power, in all configurations of public life and circumstances I had found myself in, I have stood on my feet and have not bent under the pressure of one or another force. I most certainly could have joined to save my skin. But I did not do it. I could have sworn loyalty to Marxism, to Communism and to atheism, but I did not do it. Instead, I have stood there with all my mental power in defense of my independence and not bow to any authorities. Here and there I have resorted to slipping concepts and generalizations difficult to define in order to evade concrete commitments. In those ambiguous remarks I have succeeded to pull myself out intact, though insinuating reconciling with Communism and the people's government. And how I have dressed my anti-Communist activities in Communist vestments, is another matter. I have discussed it at another place. I never received an answer to this letter. In all probability Marin had written to me his preceding letter under Party instructions, in the spirit of the

---

* Marin's comments: "I would like to see you a member of the Party like Botiu Botev (philosopher, economist...")

** Marin's comment: unreadable …

248

policy of relaxing tensions to attract people like me. But their attempt had hit a rock, judging from my letter. I continued to be the man I was before, after everything that had happened to me after July 21, 1944 – enemy No. One of the "People's government".

As I am reading this letter now, I am trying to figure out what was I trying to say? It seems I was defending my honor and my dignity, I was not capitulating, I was arguing for my independence to be what I wanted to be, I am accusing them of ostracizing me, I was resisting their demand to accept their philosophy, I was asking them to let me be what they did not want me to be. What they wanted people to be was to be Marxists, to be atheists, to be intellectual robots, to be disciplined Party members, and with my capabilities and frank attitudes, to be fanatic Communists. On the other hand I was offering to do in public life what they wanted me to be (I had no other choice) and out of their presence I was doing it, going to Pernik-Voluyak, membership in O.S.N.S. etc. I was adjusting to their society. If they wanted me to participate in their reform of the society, which they were forcing the whole people to do any way, so I would do it. It all came down to that: They wanted not only my body, they wanted my soul. I was prepared to give them my body, but I would not surrender my soul. In my soul I was against them, and I secretly used my body against them. After all that had happened, I had not joined them. By November 1947 it was clear to me that no liberation was in sight. I did not trust the Western Allies to go to war. They had surrendered the country to the Soviets, and all that was left for us was to go along with their, the Communist, social and economic system, to submit hypocritically to their authority. But they could not have my soul!

The truth was that I had always been skeptical – before 1947 and after 1947 – that England and the United States were going to war for our sake. It was this skepticism that kept me mostly from jumping openly into the struggles against Communism. But this hope was openly preached by the opposition leaders, that America was to liberate us. At public meetings opposition leaders openly, by insinuations or by whispers, sought to assure the crowds that this was going to happen. My skepticism turned into certainty that it would never happen, that I had to try to live as much as I could with the system, not to expose myself to danger, and not to betray my soul. Inside of me I was bitter, I was devastated, I was humiliated, and I could not find for even a moment a spark of acceptance of the situation. I wanted the destruction of Communism, the gap between me and the Communists could not be bridged. I felt that physically I was a slave, I was in chains, I could not escape it, but inside of me I was boiling in fear and hatred. My letter to Marin was mental acrobatics, which, judging from his comments, he did not understand and accepted at face value. Or may be he was not baked sufficiently in the Communist oven. My letter was accusatory, it was not an apology and more than anything else, it was a defense of my right to have my own independent opinions.

3. Keep Quiet to Save Your Soul!

My studies were going well. Never again did I fall in the state of confusion as I was when coming to Sofia in 1945. I was now focused, looking forward to graduation. At least here, in the dorms of the school, we were all young men and priests of the same mind, we had become a family, we felt that if one of us fell, we all would be falling together, and if we could survive, we would survive by holding tightly to each other. This is how we kept our balance in this unbearable atmosphere in the society around us. Meanwhile, the political storm in the country was raging like a wild beast. In the beginning of December the Communists adopted a new constitution, to replace the old one of 1879, which was a good fundamental law, but was trampled by Koburg kings and Bulgarian politicians whenever and however it was convenient to them. As one looks back on our history of the XX century, one will discover that not the constitution, but the will of those governing the country was the law of the land. Three weeks after adopting the new constitution the Communists moved to nationalize all industry and trade. On December 23, like thieves in the night, the government had sent military and police detachments to guard the gates of the business establishments and turn away the owners. Nobody of the "patriotic industrialists and businessmen" lifted a finger to protest. During the next four decades plus, they never raised a voice, or gave their support, moral if not material, to any of the fighting men who were rotting in jails and concentration camps. Those of them who were abroad, never gave a penny to the struggling exile groups in support of their cause. It is amazing how, after the fall of Communism, they crawled out of their caves and sought to recover all that they had lost in December 1947. And they got it.

Socialism was marching with big leaps, and we were watching it in consternation from our small windows of the school. In agriculture, from the very beginning, 1944, they started organizing Agricultural Cooperatives. It was done on "voluntary" basis. A few people, Communists, would form the cooperative as a division of the Consumer cooperative. They would expropriate the best land around the village for which owners were given in exchange waste lands, never cultivated before, with poor soil, rocks and brush, so that many of them, instead of taking what was given to them, would elect to join in the cooperative. Then they would hold public meetings, argue for the benefits to be derived from the cooperative farms, where all work would be done by machines. In our region Brezovo was the first to try the new system. In the winter the peasants who had been forced into it would come to our village and beg for a piece of bread. They planted fear in the hearts of our people. But eventually they started it in Zelenikovo too. I personally observed how two Communists, members of the farm, were dragging the goat of an old woman to the Cooperative farm holding it by the horns, and the woman, screaming and crying behind, was pulling the animal back. It was an ugly picture which one could not forget. When some opposition was clearly taking steam, they started writing on the walls of the houses of the people who refused to join in derogatory and threatening slogans as "Fascists", "Enemy of the people" etc. Eventually most of the people gave in and joined. It happened that at that time I had left the village and did not see the process further developing. But information was reaching me in Sofia what was happening. Socialism was marching full speed.

All this time, I was hiding in my little oasis: the School of Theology. I did not dare step in the village. At that time, the worse that happened to people was to be sent to concentration camps – the much heralded Work-Education Camps. Many priests who had been sent there, advised us to be careful of our actions and our talk. I did not speak against the regime, except among well trusted friends. Left on our own, we all very carefully looked around us and would not utter a word which could be overheard by dirty ears. Any hope that Americans or English would come to free us, was as dead as a door nail. Those who dared speak of such hopes were usually very close to the leaders of the opposition parties, from whom they were getting their courage.

In 1948, when the political storms seemed to be simmering down, the opposition crushed, the threat of concentration camps hanging over everybody's head, freezing any desire for politicking, I settled down for academic work. I started the research for my thesis for graduation. I expected to graduate in June 1949. I looked for materials on Bogomilism in every library where I could. The more books I read, the deeper I sank in my research. I was writing, I was re-writing, I was copying Greek, Latin and Church Slavonic texts. I became a familiar figure in all places of scholarly interest. My friends often joked about me that I had become a book worm. I was genuinely absorbed in my work. May be that was my way of escaping what was really bothering me for long periods of time. I could not go to my village to see my mother who was bed-ridden and was being taken care of by my younger brother and my father. If I would dare visit, they would grab me and God knows where I would have ended. Ii seems to me that I have taken some chances to visit the village, but have not stayed there.

In the dormitory life was poor. I have already described the miserable conditions. Rats and mice were still contesting the ownership of the premises with us. The toilets in the rest rooms continued to be flooding the floor, and quite often the water was reaching the hallway. The food was never enough. In the morning we would go and eat *shkembe chorba* – intestines, liver, pig's knucles, chicken stomachs etc., all in one, with garlic. It was delicious for breakfast, for lunch and for supper. It was relatively inexpensive. For supper we would have pilaf of beans and splurge with a salad – turnips, thinly grated, with a little vinegar.

There were days when we would allow ourselves the luxory to dine in Hotel "Bulgaria", at that time the Waldorf-Astoria in Sofia. There we would order Green-Beans plaky, the cheapest, but very tasty dish, just to feel that we were still human beings of class. As to clothes and shoes, socks and shirts, this was a subject of what you find and what you had. When the first winter came, and I had no overcoat, a student, Ivan Kirchev, sold me his Seminarian uniform-overcoat, a military type overcoat, *shinella*, which lasted me for four years. Who cared? My shoes were always with holes in the soles. It was virtually impossible to find shoes. I will not speak of underwear, socks and bed sheets. I do not remember how we supplied ourselves with these necessities. For shirts, turtleneck covered all that was under them and old sweaters from home our most prized possessions. From time to time we would hear that somewhere in Sofia some of these items were available with ration cards. We would run all over the city, and sometimes we were successful. So, life in the dorms and the school was going on. This was the period after the war and we learned to cope with all miseries.

4. Dark Clouds Over the Church

I have already mentioned it that the entire church leadership and schools were gathered at 19 "Sv. Nedelia Square" – Synod, Seminaries, Theological school professors,

Seminary teachers, seminarians and theology students. In these conditions there was much interaction between those on the top and those on the lower floors. So, whatever of significance happened in the church became soon known to all. Sometimes in 1949, after the resignation and internment of Exarch Stefan, the government had asked the Holy Synod to close the synodal bookstore at the corner of "Sv. Nedelia" and "Vitoshka", the same building where all church institutions mentioned above were. The pro tem President of the Synod, Metropolitan Mikhail of Russe, an arch-conservative, had replied that if someone wanted the corner store, he may throw the icons and the crosses on the street, but he would not order them collected. The request was withdrawn. There were numerous events of this nature. In 1951, when I escaped to Greece, I was asked by the representatives of World Council of Churches to write a report on all that had happened to the Church since the Communist take over. I wrote this report, some eleven type-written pages, single space, and submitted it to them. I wrote in Bulgarian. A refugee, graduate of some English school in Bulgaria, Todor Sharankov, translated it and typed it. I have never written a piece of anything which was to have such a decisive effect on my future. If I was taken out of the refugee camp in October 1951, and given a scholarship to study in Athens in the Theological school, and the next year be sent to Switzerland to the Ecumenical Institute, then to England and eventually to come to the U.S., it was my reward for this report. In the Ecumenical Institute I met the General Secretary of the World Council of Churches, Dr. Visser't Hooft, at lunch, and he told me that my report was the first detailed information they had received as to what happened to the Bulgarian Orthodox Church and to his friends, Prof. Dr. Stefan Tsankov and Exarch Stefan of Sofia.

There was so much going on in the Church, which the general public never heard of, but became known to us immediately. For the general public the bells of the churches were still ringing, but the church and the theological community was living in constant pain. As early as 1944, there appeared a Union of the Priests. In fact, the old Union of the Priests was now taken over by a new, pro-Communist leadership. It was headed by a priest of the Sofia Diocese, the notorious Georgi Bogdanov Georgiev. I may have already mentioned him. He reportedly was with the Communist partisans and on September 9, 1944, had led a heavily armed battalion to the capital, with submachine gun in hands. He had initiated a movement for reforms in the church along Communist lines, had taken the Priests Union and formed a Committee for that purpose. One of the leading men of that Committee was our already mentioned friend Atanas Gashtev, who soon was tonsured a monk and became assistant to Metropolitan, later Exarch, Stefan. The Organ of the Union, known before as *Pastirsko Delo*, *Pastor's Cause*, was now changed to *People's Pastor* under the editorship of Nikola Zagarov, one of the Gashtev's group. The paper and its new editors waged a vigorous campaign to abolish the traditional street dress of the priests and wear civilian clothes, to allow a second marriage for widowed priests, permitting the priests to become members of the Fatherland Front organizations, et al. They met a stern resistance from the Holly Synod, and, indeed, did not have the support of the rank and file priests, except that they controlled the Union and paraded if as if they had that support. They had the support of the director of the Office of Religious Affairs at the Ministry, Mr. Dimitar Iliev, and exerted heavy pressure on the Synod with his backing. The Union was holding its conferences in the Aula of the Theological school, and since the meetings were open, I never missed them and listened to all debates. I was getting a pretty good picture of the church.

By the end of the second year after I arrived in Sofia, Georgiev was confronted as leader of the Union by a strong opposition among the delegates at the Conference. A group of moderates, mostly from the Plovdiv Diocese, led by Pavel Grozev and Ivan Yuliev, were successful of pushing out the pro-Communist leadership and gained control of the Union. At one point, loosing the elections, Georgiev bluntly stated: "This year we are twenty, next year we will be two hundred (delegates)". It turned out, that even before the time for the annual conference for next year, Georgiev had the Union leadership removed and himself back on the saddle. In later years, however, he was accused of pilfering the treasury of the Union, and thrown out of it. Meanwhile, before all that was revealed, he was riding high with the support of the government and played a decisive role in shaping the future of the Church. In him the government had their agent inside the Church and when the new By-Laws were being prepared, he sided with the Director of the Office of Religious Affairs, and at the end they had their way. A Law of Religious Denominations was passed on 24 February 1949 which obligated the Confessions to submit their By-Laws to the Office of Religious Affairs in three months. Georgiev was one of the members of the Committee which the government accepted to prepare the new By-Laws. Dimitar Iliev, behind their back, was exerting his influence through Georgiev to have a By-Laws convenient for the atheistic authorities to manipulate Church affairs to their advantage. Metropolitan Paissiy of Vratsa, and Prof. Dr. Tsankov represented the Synod – Georgiev the Union. Instead of three months, the new draft was discussed for twenty two months, and at the end a text was promulgated by the government in circumstances which suggest that it was dictated to the Synod. Georgiev did not have all his ideas inserted in it, but the government had secured for its organs the power to rule the church with surrogates. We, in the school, were informed that the government wanted to create institutions and procedures in the church, which would guarantee them the ways and means to intervene any time it was necessary for them, and thus control the Church. The Holy Synod had resisted all that, but at the end Paisiy and Tsankov have capitulated. Paisiy was immediately replaced by Kiril of Plovdiv, who undertook to implement the new By-Laws and was elected Patriarch – five hundred fifty years after this institution had been abolished by the Turkish conquerors of the country. This By-Law is in force to this day. When the Communists established their control over the church government, they left the new leaders to manage church affairs without following strictly the By-Laws as cited above, to cause a schism in the church, because the Holy Synod had not bothered to implement the Communist imposed procedures. We, in the Theological school, followed closely the conflicts between Synod and Government in the preparation of these By-Laws.

I have mentioned already the suspension of the state subsidy for the Church as a means of pressure by the government when it wanted to obtain concessions. This happened several times in the course of my studies. There was an occasion when the government requested that the Church submit a plan for its self-support. Kiril of Plovdiv was assigned by the Synod to draft such a plan. As we learned about it, he proposed that the Holy Synod set up a bank, pulling all resources of the church in it. The Synod approved it. The plan further included a provision that the Synod print special stamps to be offered to the public as a form of anonymous donations. The government rejected it. It saw in it an attempt of the church to free itself from the tutelage of the State. The introduction of anonymous donations by way of selling such stamps would open the possibilities of the church tapping sources which the State could not control. All this became known to us, though the public at large was never informed. Thus the Church was left with only one source of income – selling candles. The church leaders, including the Union of priests, panicked. Zagarov was practically crying from the pages of *People's Pastor*, writing that the church has to support itself with the "little candle." The Holy Synod assigned quotas of candles to be sold by every church, and how much money was to be sent to the central synodal treasurer. The Synod was to divide the pie, all funds thus collected, as salaries for the clergy. When this situation developed, the government intervened with promises to continue the subsidy. Then, when the Communists wanted some concessions from the church, they would suspend the subsidy, and then, having obtained the concessions, they would restore it.

On one occasion, when the government wanted to force the church to discontinue its attempts to introduce Sunday School for children, and also when it wanted the church to stop organizing pilgrimages to the Rila Monastery, it suspended the subsidy and the church bowed down. On the same occasion, the Office of Religious Affairs pressured the Synod to allow the priests to become members of the local chapters of the Fatherland Front, but the church stood firm against this demand and the request was withdrawn. I never forgot this instance when, on October 20, 1948, Dimitar Iliev, the Director of the Office of Religious Affairs, appeared at the Conference of the Priests' Union with the following declaration: "After the resignation of Exarch Stefan the relations between Church and State are rapidly improving. I just received a statement from the Holy Synod that they have decided to stop religious education for children, group visits to the Rila Monastery, and to encourage the priests to support the politics of the government... The priests should not worry about their salaries, the government will take care of supporting the church." These words of Iliev impressed me as a revelation from above. Here was the answer to all my questions and I recorded it in my notes. Prior to that declaration of Iliev, the priests had not received their salaries. I was also affected, because I had not received my scholarship. I quoted this declaration when I wrote my report to the World Council of Churches. Later I saw it published by Robert Tobias in his Book *Christian-Communist Encounter in Eastern Europe* (Indianapolis, 1956). He had taken it from my report, which I had already read in one of the confidential files of the WCC in Geneva.

Dimitar Iliev was charged by the Government to suppress the Church. In May, 1949, he made a most flagrant attempt to eliminate the authority of the Holy Synod. He issued a circular letter to all church leaders – Synod, school faculty, students in the Theological School and the Priests' Union by which he was requesting their cooperation to obtain a number of political commitments on behalf of the church and the clergy: to come out for the nationalization of industry, to accept the proposition that the State had authority over the church and to show full support for all government initiatives in every field. In response to this request of Iliev, Bishop Pimen, then Vicar of the Sofia Diocese, wrote an article in *Tsurkoven vestnik*, (Church Journal), urging the priests to include in their sermons advises to the peasants as to how important was the deep plowing of the land. It was silly, but it was there. Iliev also urged the church leaders to educate the seminarians and the theology students in a spirit of loyalty to the Fatherland Front. He suggested that the programs of the theological schools be modified to include subjects in support of the social policies of the government, to urge the priests to join the Fatherland Front local chapters, to attack anti-Communism from the church pulpits, to attack the russophobia, wherever it was noticed and to display the portraits of the national Communist leaders and praise them in the churches. He further requested that the priests should attack the leaders and the rank and file of the opposition parties – those who were still in existence. Likewise, the church was asked to recognize the already organized children's and women's organizations established by the government. These were sweeping demands. If they were met by the church, it would amount to its politicization in favor of the established Communist regime, when at the same time the church was forbidden to participate in any political activities.

This letter provoked a storm of indignation on all levels of church leadership. Iliev then attempted personal contacts to exert pressure for acceptance of his demands. He started calling to his office individuals and groups to explain his request. He also called a group of theology students. I was among them, with some twenty or more of us. He delivered a long lecture as why he had made these demands, which was not different from what he had already written. At the end he opened the meeting for discussions. Some students asked some inconsequential questions, which gave him opportunity to repeat many of his points. I felt that prudence required that I keep silent. It was pointless. It was widely debated in church circles. The Holy Synod sent a special delegation of metropolitans to the Prime Minister to protest Iliev's attempt to dictate to the church. The whole affair was concluded by an arrangement between the Synod and the government where it was decided that the letter was never written, was never received by the Synod, and then there was no need to answer it. The Synod was told that this was just Iliev's personal initiative and did not represent government policy. All that was published in the *Church Journal*.

But not all tensions between church and state were resolved this way. There were occasions when we felt painfully how things stood between church and state. At some point we learned that the Igoumen, Abbot of the Rila monastery, had been arrested and held for trial. This was not published in any papers. Kalistrat was tried for espionage, or God knows for what. The story was that he had received in the monastery some western diplomats. How this was construed as espionage, it was never made clear. There was a fear that he might be sentenced to death. He was given a life sentence. After a few years we heard that he died in jail. There was also the case of Arkhimandrite Myron, Protosyngel of the Sofia Diocese. Myron had taken the initiative to restore the once very popular Christian Brotherhoods at the churches. It so happened that a few of us in the theological school found ourselves involved in this activity and had become his assistants. In a few months a number of churches in Sofia had their brotherhoods resuming their activities. It was then arranged that several Sofia Brotherhoods would hold Sunday afternoon joint meeting and celebrate vespers. When we had these joint meetings, several hundred Christians from all over Sofia joined at one or another church. It was all going well, until Arkhimandrite Myron suddenly disappeared. We did not know what had happened to him and where was he. Later it was rumored that he had been arrested in the middle of the night and taken somewhere by the police. It was also rumored that Bishop Pimen, Vicar of the Sofia Diocese, to whom Myron was a subordinate, was instrumental in the disappearance of the Arkhimandrite. I have never been able to establish the truth. In vain I read the Church Journal for decades, when I could put a hand on it, searching for my friend Myron. I never saw his name ... until late in the 1980s where I learned somehow that he had died in some obscure monastery. When I was in Sofia in 1991 one of his relatives came to me to ask for information about him. I could not tell her anything more than what I am writing here.

Other middle level clergy of the church suffered too. There was Arkhimandrite Vassiliy. He was a chaplain of Alexander Nevsky Cathedral. He was a great preacher. Every Sunday, while he was in charge, his sermons lasted sometimes for one hour or longer. This was not considered appropriate, but he was doing it and apparently someone in the government did not like it. It was reported that he had been quietly dismissed or just arrested and sent to concentration camps, as I heard it rumored . Much later it came to my attention that he had become Protosyngel of Russe Diocese. This was still better, because not everybody was so lucky. We had already learned of the tragic fate of Arkhimandrite Irinay. He was chaplain in the Plovdiv Theological Seminary in my time. He had translated and published a book from a Romanian theologian, *Christianity and Communism*. During the turbulent days of September, 1944, he had disappeared, and murdered. And so was also the Protosyngel of the Vidin Diocese, Archimandrite Paladiy, about whom it was reported that he had been cut piece by piece while still alive. There was another Arkhimandrite of that group, Stefan or Sofroniy – I do not know his office in the church – who was also killed.

The upper clergy of the church was thrown into turmoil. Some of the Metropolitans apparently looked for an opportunity to make a fast career. The Metropolitan of Varna, Yosiff, may have had his eyes on some position and in 1948 or 1949 attempted to play politics by coming out publicly for a leftist program for reforms in the church. He made a play for the support of the leftist Priest's Union. He published a series of articles in their paper where more or less he endorsed their program. It did not go far, because the Holy Synod came out with stern rebuttal in *Church Journal* and this ended his flirtation with the Communists. The Holy Synod was a fortress of conservatism and stood opposite anyone in the church who would try to change things by taking advantage of political opportunities. In August 1949 a Deputy Minister of the Interior, General Yonko Panov, a leading communist from our region, Brezovo, who had spent long time in jails, prepared a report for Interior Minister Anton Yugov where he correctly assessed the Holy Synod as "a nest of the most reactionary forces." I did not see this report until 1995, but reading it I recognized the correctness of his conclusions. It is published further down for the first time.

By far the most important event in the life of the church during these years, where the heavy hand of the Communists fell upon the church, was the "resignation" of Metropolitan Stefan of Sofia. This affair had never been adequately explained. Even now, ten years after the fall of Todor Zhivkov, nobody, as far as I know, had dug out the true story. The official version of these events was published in *Tsurkoven vestnik – The Church Journal*. According to the Synodal version, Stefan caused some sort of a scandal in the meeting of the metropolitans, over some minor issue, and threatened to resign. Later during the day he sent his written resignation. He had done that many tines before, but never confirmed it in writing. As we learned the story, the Secretary General of the Holy Synod, having received the paper, reported it to Metropolitan Boris of Nevrokop, who hurriedly ordered that it is registered in the official records. As it became known, Stefan had meant to resign only the Exarchal office, but the Synod interpreted the rules to the effect that having resigned the office of exarch, he was automatically resigning also the office of Metropolitan of Sofia. The rational behind this interpretation was that the Exarch was automatically Metropolitan of Sofia. So, if Stefan retained his Metropolitan's seat, they could not elect a new Exarch. As it turn out, after removing Stefan, they never elected an Exarch, and his place as Metropolitan of Sofia remained vacant for five years. The Synod published the protocols of its meetings and giving additional explanations, offered a sweeping condemnation of Stefan portraying him as despot in the church who had intimidated all metropolitans, using the Communist authorities. It is a document which is unique in its kind in the annals of the Bulgarian church.

Indeed, Stefan was closely associated with Kimon Georgiev, the Prime Minister of September 9, 1944, and for several years thereafter, he was not quite the most popular among the Metropolitans. He was not called to the royal palace when King Boris was dying, to administer to him the holy communion. He was disliked by all royalty in Bulgaria. Rumor had it that on one occasion Prince Kiril had jumped on him with his sword ready to hit him, but had been restrained. When he would be celebrating Te Deum services for the military and the king would be present, he would dump his bundle of flowers in the Holy Water before blessing the king and then slap him in the face with force, with so much water that the King would have to go and change. During the war there were rumors that he, Stefan, had plotted with a Lady in Attendance at the palace to kidnap Prince Simeon and deliver him to the British. He publicly opposed the policies of the government of making things difficult for the Bulgarian Jews and was baptizing them into the Christian faith to save them from deportation to Germany. Prof. Dulgerov wrote an article condemning such baptisms. For all these things, one of the royal palace sycophants, Stefan Grouev, in his biography of the King, did not abstain from telling derogatory stories of the headstrong hierarch. It is understandable then how strong he was when the old regime fell in 1944, when he immediately became pro tem President of the Holy Synod, and then manipulated things to become exarch. But his ties with the new government soon cooled off. He had too many connections with the political world in Bulgaria, especially with the democratic leaders in the country. When the communists lowered the boom on them, Stefan found himself exposed into the crossfire.

It is my belief that Stefan was probably considered and plotted with the opposition to become a a pro tem premier while the political crisis in Bulgaria was resolved. But the Communists were not going to allow such a course of events where their future would be jeopardized in some political combination which they could not control. It is probably because his involvement in such political schemes that the Communists went for his head. There was enough bad blood between him and the other Metropolitans, so that they readily plotted with the Communists to depose him. The truth about these events still has to be revealed. One thing is sure, he did not voluntarily give up his Sofia See. A group of theology students, priests of his diocese, went to visit him. They reported back to us his vowing that he would be taken out of Sofia only on a stretcher. He was! In the middle of December, 1948, a police wagon stopped at his apartment, he was taken in the middle of the night as prisoner and driven to a royal villa near the town of Karlovo, Banya. There he stayed incommunicado until his death in 1957.

There was a very dark cloud over the church while we were studying theology. How all this thing could inspire us to go ahead? Well, our tenacity was surpassed only by the absence of any avenue which we could have taken. There was much more that we had to put up with and fight against. Above all, there was a constant threat that the School of Theology was going to be dropped from the university. The university was an umbrella which gave us a sense of security, a false security as it was. We attended classes on general education subjects in the other school – History, Philosophy, languages et al. Sometimes in 1947 or 1948 they introduced the subject of Dialectical Materialism, which every student from every school and major had to take. So we, the theologians, had to take Dialectical Materialism which was a philosophy of denial of our *Raison d'etrte*. Not only we had to attend lectures, we had to take the tests. The tests on all subjects were not written, but oral. We had to appear before the professor and he would twist things any way he wanted and then decide to let us go passing it, or recommending that we appear a second or a third time. I had no difficulty. I listened to the lectures, I read the literature and digested it all. By the end of the course I had mastered this philosophy well enough to pass any test. I did pass under Assen Kisselinchev. Now I only remember that while he was asking the questions and I was answering them, he was eating feta cheese and bread, and between bites was asking me his questions. But this subject was not so easy for the priests, who were at the mercy of the professors. There was the case of a priest, who had taken the course under Prof. Zhivko Oshavkov. In the 1960s Oshavkov had conducted a study of religion in the country and had published a report which could have been anything but a science work. So, the priest, Father Boris I believe was his name, answered perfectly all questions which Oshavkov asked him. The last question was: "Do you believe all that you said?" The priest answered him: "This is what the Dialectical Materialism teaches." The professor was offended and told the priest that he would have to appear a second time for the test.

Quite often these professors were speaking obvious stupidities. I remember attending a public lecture by a professor, I think his name was Atanasov, but do not remember from which school. He stated that "The Germans have not contributed anything to world culture. Some student had to put him in his place by mentioning Kant, Hegel, Luther, Max Planc, Beethoven, Mozart, Durer and ... Karl Marx. I do not remember how the professor dealt with this reminder. There were many instances when the Communists were galloping all over the fields of science and humanities. When the biological theories of Michurin-Lisenko were propounded against the theories of heredity by Weissmann, the most prominent Bulgarian scholars in the field had to fall in line. The most important, the leading Biology Science professor in Sofia at that time was Prof. Metody Popov. He had just published an Introduction to Biology where he had presented the traditional view of that science on heredity. He had to appear at a symposium organized by the Party scientists and publicly tear off two chapters of his book, which later had to be republished with the new orthodoxy ordered by the Party and its Guru – Stalin. The same incident had occurred with Prof. Spiridon Kazanjiev in the field of Psychology. I personally attended a symposium of historians where the veterans of this science in Bulgaria had to march to the podium and swear loyalty to Historical Materialism. My professor of Church History, Ivan Snegarov, likewise, accepted the method, though observing that there are situations which Dialectical or Historical Materialism could not explain. He pointed to the case of St. John of Rila. While all that was going on, some five feet from the speaker's podium was ... Vulko Chervenkov, brother in Law of Georgi Dimitrov. Chervenkov was then Minister of Education, and subsequently prime minister. He stood there with a notebook in hand and a pencil, ... keeping notes. Once, in my class in Old Bulgarian Language I asked Prof. Stoyan Romansky what was the reason for eliminating the letter Double-er from the Alphabet (This letter was read and spoken in Eastern Bulgarian as Ya, and in Western Bulgarian as E). the old man indignantly replied: "Because it is too high and it was poking them in the eyes." He was against this change.

My life in the dorms was going more or less uneventfully. It is better to say that the difficulties became a routine and we did not dwell on them. Sometimes, things happened in such a way that our lives were not much different from that of animals – not in any way derogatory, but simply in a sense that one animal lived for himself and in himself, as if the others did not exist. There was the case of a young boy, from the village of Vrabevo, somewhere North of the Balkan, in the districts of Gabrovo, or Turnovo. He fell sick one day and stayed in his bed a day or two, may be three. People were passing by, some would go to him, talk to him, and leave him to tend to their own affairs. I had no relations with him. Then he was taken to some hospital and … died in a day or two. It was reported he had meningitis. A handsome boy, full of zest, friendly as he was, but somehow nobody looked after him.

I had a similar experience. One late November day I was returning from the university when a heavy cold rain hit me. I did not stop to protect myself, had no hat or umbrella and by the time I reached the school I was all wet and my head was freezing. The next day I was in bed with high fever, and piercing stabs in the chest. I could hardly breath, I was shaking from cold and no amount of blankets would help me. Friends around me tried to put over me whatever they found. I remember having gone through such fevers when I was in the village in 1938. It was malaria. This was not malaria, but the fever was the same. I was agonizing in my bed, friends would stop, say a few words, and would go after their business. What could they do? So, this is how things were. If you were hit by illness, you were on your own. I was on my own for two or three days, when it all subsided and I felt better enough to get up and slowly resume my activities. However the pains in my chest were to stay for a long time with me. It was a bad attack of pleurisy. So when a man has days left to live, which apparently was my case, he survives like a dog. My attack of malaria some ten years earlier was the same thing. After two or three days, when I was in such a bad shape that I felt that my end was near, someone urged my father to take me to the doctor in Brezovo. There I was given some injection and pills, and a warning that if I had not been there for a few more hours, I might not have lived. The biology of that desease, malaria, was simple. The bacteria, or whatever, resulting from mosquito bites, were invading the red corpuscles. When the breaking begins, that is when the attack of high fever and experiencing of cold starts and lasts for some fifteen to thirty or forty five minutes. This is the hour of agony. I have heard somewhere, that for long time the malaria bacteria survives in the body in latent form, and in favorable conditions are reactivated again, and a man goes through a new crisis similar to the first. Apparently that happened to me. Well, in a few days I was back in circulation and little by little recovered … like a dog.

In the meantime the storms of 1947 appeared to be simmering down in 1948. Nikola Petkov was hanged, the opposition parties were turned on the run, everybody was looking for a cover and things seemed to be returning to normal – if anyone could say in those days what was normal. People were frightened to their wits end of the government. The thought or the threat of going to a concentration camp was very real and this was enough to silence anybody and freeze him on the spot. I focused on my research and writing of my thesis on Bogomilism. I had to be ready with and hurry it up. But I bogged down in so much material, expanding my research in so many directions, as the information I was putting my hands on, was becoming almost unmanageable that I did not see how I was going to be ready. I was running from library to library, collected information, kept notes, copied texts, though, I must say, experiencing infinite pleasure of it. In all that I found inner satisfaction, better than my experiences in politics. I did not mind. I felt that this was my true vocation. I discovered my own spiritual and mental peace, bringing me to forget my nest in Zelenikovo, where I could not ever come near in those days. There were moments, in the midst of my studies, when I would drop everything, my thoughts would wander around and I painfully remembered my sick mother and young kid brother, yearning to embrace them and kiss them, to show them my deep love. And then, as if to forget it all, I would plunge again in my scholarly work, as if to escape the miseries of life.

5. Deafening Thunders from Blue Skies

But, there was more to come. In the middle of January, late in the evening, a friend approached my bed, I was just falling asleep, tapped me slightly, not to awake me if I was already sleeping, and opening my eyes I read in the expression on his face that something was deadly serious. It was Vasil. I do not remember his last name now. A year ago my representative in Sofia Slav Peev, a journalist, told me that he had met Vasil, he was his neighbor, and he had told him the story which I am about to mention. Vasil leaned close to me and said: "You have not heard the news, have you?" "What news?" I asked. He continued.: "The University has told Tsonevsky (Dean of the School) that twenty students of Theology are being considered for expulsion. Do you know who is on the top of the list?" I did not know a thing. I had heard that such a move was under way for students who had entered in 1945, but I knew nothing definite. As I heard him, I felt that the earth under me was splitting and I was falling in the crack. But I was not surprised of the news. I was, sort of, expecting such a development. I asked him: "Who?" and he answered looking straight into my eyes: "You!" He added that Tsonevsky will do all that he could to save the theology students. He would call those on the list for interview, in trying to build his case for the defense.

I thanked Vasil for the information and he left me. I do not know why, what I heard did not upset me. I felt no fear. I was not surprised. As if I was anticipating that to come around, sooner of later. Or may be I was already used to such unforeseen perils. My sense of anticipation of such things was somehow blunted to such news. I had lived through so many such events. I had lived through all of them, and somehow had managed to weather them and survive. I had passed through the rain-drops so many times that some instinct was telling me that I would survive that too. But I had very little hope that this was going to go away without me finally being caught, without falling in the pit of the beast. I felt that this was the last precipice over which I was going to jump and was not going to reach the opposite end. How was I going to defend myself from charges which are made against me, made by people who were away from me, away from the scene, knowing that they had accused me of things that were not as offensive as they would appear on paper? Somehow, I saw no room for maneuvering, for hiding, for running. I did not loose any sleep on account of this "news". I fell asleep, engrossed in a blind fatalism, rather than any certainty that this too was going to pass.

Two days later Prof. Ilia Tsonevsky called me for the interview in his cabinet. Thin and small as he was, humble like a shy monk, he appeared to me a beaten man, sinking in fright, his lower lip trembling when he invited me to sit down. I was unusually calm and tried to calm him down. I told him that I knew why I was being called, and he should not worry about our conversation. I assured him that I was ready to take everything that was coming to me, and if it could not be avoided, then let it come, whatever it is. Still shaking, he was trying to convince me that he would do everything in his power to prevent the expulsion of any of the theology students and wanted to know if I had some political activities that might be held against me. I assured him that I had nothing that warranted such a drastic step, that if I have been charged by my local enemies, it is not for any valid reason, but because I was a seminary and theology student. I told him that I have stayed out of the opposition parties during these years, which he knew, because whatever we were doing was not politics. He would have known if I had been involved in opposition politics. I told him that if it is a question of political orientation I was supporting the

BANU in the government coalition, mentioned my NSU membership, my having gone as a brigadir, having been in charge of signing others as brigadirs, having fulfilled my quota 150 per cent. The accusations against me were coming from my village, but if I had any activity for which I would deserve punishment, they would have taken care of it long time ago. He was keeping notes and assured me that he would bring all that to the Academic Council of the University. His defense, he said, was to be that all Theology students are cited for expulsion not for any valid reason, but because they are theologians, involved in religion, and that locally such involvement is tantamount to negative politics, identification of the church with fascism. He was successful in preventing any of the twenty students being expelled from the University. When on February 5 or 6, the lists of the five thousand students being expelled were plastered all over the walls of the university, my name was not there, neither were the names of the other eighteen of us for whose expulsion the university had asked. Only one of the twenty could not be saved – Hristo P. Ivanov, a military officer of war times, who, it was reported, had registered in violation of some procedure. Later on he had become a priest and after 1989 wrote me a letter. The poor man had gone through so much tragedy, persecuted, jailed and ravaged by illness – himself and his family.

In 1988, when the Sofia University was celebrating its one hundredth anniversary and the Bulgarian Academy of Science had proposed a panel, for reading of scholarly papers on that occasion at the annual conference of the AAASS – American Association for Advancement of Slavic Studies – sending to Washington half a dozen of professors, I made a comment following the papers to the effect that in 1949, when I was a student there, a violent storm had hit the University and had taken the life of five thousand students, the attending scholars were stunned of the news. They had never heard of such a disaster. I proceeded to explain that the storm was the purge of the students for political reasons. I had a few hot disputes with some of the Sofia professors there on the matter whether there was academic freedom in the university in Bulgaria or not under the regime of Todor Zhivkov.

So, I was not expelled from the University. The storm passed me over again. But this was not the end of my temptations. It was only a new page in my struggles with destiny. Those who were expelled were given three days to leave Sofia, otherwise they would be arrested. Saturday morning, the third day, I and a few of my closest friends left for Bachkovo Monastery, led by Prof. Dulgerov. One of our close friends, Pencho Donchev, from Muglizh, was scheduled to be tonsured as a monk the next day. Donchev, Todor Subev and I belonged to the circle of Father Gerasim, a teacher supervisor in the Plovdiv Seminary. I could not miss this occasion honoring him. Sunday morning he was tonsured and given the name Pankratiy. Decades later he had become Metropolitan of Stara Zagora and the Holy Synod had put him in charge of the office of foreign relations of the Church. In 1992 he joined two other Metropolitans, Pimen of Nevrokop and Kalinik of Vratsa, and effected the Schism in the Church, instigated by the government. During the 1950s he emerged as one of the leading collaborationists with the Communist government and I had on numerous occasions attacked his stance. In 1991 Radko Pop Todorov was telling me that he, Pankratiy, and Prof. Todor Subev, had been members of the RMS, the Communist Youth Association, that he had seen their membership cards, that in addition to being agents of the Bulgarian State Security, they both have been agents of the KGB. I could never understand how Radko accepted them as leaders of the Schism, to which he himself, of all things, was the initial advocate, with his article in "Otechestvo", of January 1990. As it turned out last year, 1998, Pankratiy abandoned the schismatics and returned to the Holy Synod. Sometimes this year, 1999, he suffered a heart attack during a session of the Synod, and died in the hospital.

So, half an hour or one hour after we had left, the Militia of the district precinct had sent three of their officers to the School of Theology, looking for me. They were told that we were going to Bachkovo. They established that I had not left Sofia as I should have as a student expelled from the University and that I was going to return. They had then asked and received my address card registration from the doorman, Diado Misho, which he could not refuse to surrender. He was told that I am expelled student and that if I returned back I would be arrested. The Dean, Ilia Tsonevsky, told of what had happened, had gone to the Precinct to argue that I was still student in good standing, hoping to get my address card. But it was Saturday, the Chief of Precinct was not there and the matter was to be considered next Monday. Monday morning Tsonevsky was back there, but after one hour, he was told to come back at five o'clock. He had gone again at four, and after some checking was done, the matter was resolved, he was given back my card and he took it to Diado Misho. Half an hour later we returned from Bachkovo. My friends met me with an air of disbelief, after all that had transpired in my absence. It all had had a happy ending, but had left a bitter taste in the mouths of all those who had participated in it or had witnessed it. I was looked at as if I had returned from my inferno and as if I have been a leper. I listened to the story of the events in my absence with a sense of amusement, even with indifference. After all, it was all over. One of those things that had happened to me before in different contexts.

I was grateful that it all had happened in my absence. It crossed my mind that if I was in Sofia, if they had found me, they would have arrested me on the spot, and this would have been the end of my studies. Nobody could have found me where I was and no one could have saved me from sinking in some concentration camp, and from there ... God knows!

Apparently, of the entire group of twenty theologians, I was the only one who was singled out for such an action. Had they dropped the charges for all of us, and then taken a second look at my case and had decided to expel me? Had this second information reached the police and the army, but the decision not to expel me had never been communicated to them? I am asking these question now. At the time I was more concerned to save myself, rather than seek answers to questions which I would not have received any way.

So, it was the tonsuring of Pankratiy which saved me this time. Later someone told me the comment of Dulgerov: "I cannot understand this Raikin, what self-control he is showing in all this, as if nothing has happened." His impression was correct. Not for one moment was I alarmed by all that. I was seized by a fatalistic sense of inevitability of all that was happening to me, that I felt powerless to change anything, and then I was accepting it as a blind destiny. So, why was I going to fight it or be disturbed by it? Or maybe I had lost my sense of reality and was moving around as a somnambul amidst the weeds of my evil fate?

If all that taught me a lesson, it was that I should mobilize all my strength, take all my tests in June, submit as soon as I can my thesis to Snegarov and graduate before something else happened. I sat over the books again, behind the typewriter, and resolved not to let my attention be diverted to anything else. But hardly had I recovered from the shock of the past week, on February 17th I received a telegram from my family in Zelenikovo.

My mother had passed away, and that day was the funeral. Even if I could go to the village, by the time I would even take the train, the funeral would have been over. Besides, the hatred against me of the authorities and the communists was still so bad, and still alive, that if I showed my face in the village, for whatever reason, they would have grabbed me, and then, God knows what would happen. I had no place where I could turn to. I was hit in the most vulnerable fiber in my soul. I had not seen her for long time, I do not remember since when. In my soul she was shining as a brilliant star. If there was a fire burning in my soul of uncorrupted love, this was the love for my mother. She was a martyr, she died a martyr without being able to see her son once more in her life. She had raised me when I was a small feeble child, she had lighted in me the flame of knowledge, in a primitive and simple manner, but a flame, nevertheless. I was her pride, though she could never express her feelings. I have never forgotten her sweet smile when rising from her sick bed, resting on her elbows, she admired me dressing for going out and combing my hair in front of a small mirror. I saw in her dearest smile her pleasure of having such son. I was chased from my home when she needed me mostly. The news of her passing away crushed me.

For a moment I was stunned, not knowing what to do, to whom to turn to share my pain. I could not run to the village for the reason already stated. In despair and frozen in disbelief, I could not even shed a tear … And then, all of a sudden, my soul was seized by such an uncontrollable eruption of a volkano of pain, that I cried out with all my force. I threw myself in my bed, covered my head with a blanket and continued sobbing for hours. Sure, my friends saw me in this state of agony and tried to find out what had happened. I could not say a thing. At some point father Sofroniy came to my bed, sat next to me, did not say anything. I showed him the telegram. He sat their silent, in prayer and compassion may be for hours. I told my friends that my mother had died and begged them to leave me alone. They did. What could they do? What anybody could do? I just cried unconsolably. But I felt Father Sofroniy next to me and in this moment of personal tragedy, he was my only angel of mercy that I needed for spiritual support. Late in the evening, I had composed myself. When I felt somehow relieved of the pressure coming from my heart, I got up, walked out in the city, wandering aimlessly in the streets for two hours, in loneliness and sorrow, until I returned to the dorm, a little at peace. I never stopped castigating in my mind all those who separated me from my mother when she needed me so much, and I suffered so much pain for my poor father and my young brother in this hour of pain and sorrow.

Again, this was not going to be the end of my trials and tribulations. About the end of February I received a letter from home with a summons to appear for induction in the military service sometimes in March. All these years I had been excused as a student and kept postponing my military service. I was still entitled to stay out to the end of the semester. But here was the summons, and I could not ignore it. It was a military affair and noncompliance could lead to jail. Something had to be done to rescind it. I had very little hope that I could do that. So, I packed all my things, papers and clothes, in a wooden box, obtained all documents which I could use to ask for one more postponement – from the Office of the University and the Office of the School of Theology, certifying that I was regularly registered student, and headed for Plovdiv. If I succeeded to gain a postponement, I would return to Sofia. If not, I could go to Zelenikovo and in a week or so, report to the draft commission. I arrived in Plovdiv, left my things at the Station and went to look for the office of the military draft. I entered the office, and reported to an officer

263

who was serving as a receptionist. I showed him the summons and explained my predicament. I gave him my certificates from Sofia. He took all papers and disappeared in the offices. He returned and told me to wait. After some fifteen minutes there appeared another officer. He told me that according to information they had, I was expelled from the university and had to go and serve my military duty. I kindly pointed out to him that I was not expelled from the university, that I was still a regular student and entitled to postponement. I mentioned it that the documents which he was holding in his hands would testify to that. He was somehow upset and retorted to my explanations with the suggestion that we ourselves write these letters. "You yourself write these letters", were his exact words. I objected that the military service was not a joke, it was a serious business and that I would not dare fabricate false information. He stepped back, looked at me again, and told me to wait. Sometimes later he came back and told me that they are postponing my service for now, but as soon as I finish my studies, to report back. "By all means," I said and flew out with the summons – a resolution written across it: "Postponed". On the street I felt like a big burden had fallen from my back. Now the question was to go or not to go to Zelenikovo. I figured that they did not know about my postponement, they would be anticipating that I was going to serve my military service, and would not touch me. If someone asked me why I was coming to the village, I would say that I was going to do my military duty. I took a taxi, went to pick my box and things and back to the bus station. At about 7.00 p.m. I arrived in the village.

A group of people were standing about fifty feet from the bus stop at the square and staring at me. I recognized several leading communists among them. I noticed some of them had scornful smiles on their faces. Nobody approached me. Nobody asked me why I was returning to the village. They looked at me struggling with my box, but nobody offered to help me. I went home and my father had to come with the cart to take it home. Apparently they guessed why I so suddenly reapeared home. But someone of the peasants that was next to the bus asked why I am coming back. I told him that I was going to serve my military duty. I explained to my father what had happened and told him not to tell anybody the truth. If someone asked, to tell them that I was going to serve in the military. I told him that Monday morning I was returning to Sofia. The next day I went to the cemetery to visit the grave of my mother. On my way back I met my cousin Goriu (Gregory). He guessed: "You went to visit my aunt, did you." I confirmed it. We chitchatted briefly generalities and I left him. Otherwise I did not see or speak to anybody during this visit to Zelenikovo. Anyone speaking with me was taking a chance, a risk for himself.

As I understood, they were already gloating in the village over my defeat, the termination of my studies, thinking that, after all, I have been expelled from the university. This is what they had requested from the village. I was on their black list. Monday morning, after collecting all my notes for my thesis, my books, and my clothes, not that I had very much, all in one case, I was on my way to Plovdiv. My father insisted on coming with me, to make sure that nothing happened to me on my way to Sofia. Who knows? They might stop me and arrest me. There was nobody at the bus stop. I looked around the station and when I did not see anybody who knew me, I jumped down, my father following me, and soon we headed down along what was then known as Kurshum Han, going to the Plovdiv Museum. In a few minutes we were out of the danger zone. I was hurrying for the railway station. My father was keeping up with my brisk walk. I was taking the short cuts. I felt more secure with him along with me. He was marching with his raggedy Turkish style *poturi* and his oily old fashion hat, staring at the ground as we were walking. I now remember how depressed he was. As if the whole world was weighing on his back. I felt his concern for me, his love, his suffering and his fears that something could happen to me. I was tormented by the thought that he did not deserve all that. From time to time he asked me where were we and I was trying to explain to him. From time to time he would utter with grunting some swearing addressed to my persecutors, as if he was talking to himself. He was swearing in Turkish. This had been his habit, learned in his youth. We stopped here and there to rest a little bit. The distance was considerable and the suitcase, held by both of us, quite heavy. He would role a cigarette, strike his steel piece at the little flint stone, with a small piece of touchwood, to light his cigarette and we would resume our sad journey. I felt like I was running. I wanted to get out of Plovdiv a minute sooner, to catch the train.

When we arrived at the station, the train was already there. I ran to the ticket counter, bought my ticket and ran to the train where my father was waiting for me. I jumped at the first step, he gave me the suitcase, so I pushed it in, stepped back, embraced him, he kissed me on the forehead and with tears in his eyes wished me a farewell. I hurried in the train, took a seat at the window and the train instantly pulled forward. I waived, my father responded, he looked at me and I looked at him until we could no longer see each other. It was a sad separation. I still remember how depressed he was, what a pain was I causing him, as if, again, he was carrying the entire world on his shoulders. Later I learned that when he returned to the bus station he had met Slav Gidikov*, top partisan from Zelenikovo. Gidikov asked him again about me, if I was expelled from the school. To him he told the truth – that I was not expelled and that I had just left for Sofia. He had pleaded with the partisan that they leave me alone, that I was not a danger for them, but Gidikov had replied threateningly. "Sooner or later he will fall in our hands, he will not escape us, and then…" he had not finished his threats. So, this entire new peril for me missed me again.

When I all of a sudden reappeared in the school, my friends were happy to see me, but they were also stunned of my "resurrection". I have been given up for gone. I explained to them how I got out again. Life began as if it had never been interrupted. I hit the books again, I was writing and typing the final copy of my thesis on Bogomilism furiously. I had to hurry it up, and to watch that no black cat would cross my path again and spoil everything. In June I passed all my tests and finished my thesis, had it bound in red hard cover and submitted it to Prof. Snegarov. In August he called me to his office, praised my diligence and my research and congratulated me as if I was already out of the woods. Well, there was no more to be done. I had to fill out a long questionaire for the central university office. I had to give my personal data and write a summary of my experiences as a student, specifically my contribution to the people's government while I had been a student. Apparently this was the most important part, and I did a wonderful job in ascribing myself all the goods which I had done. I wrote there that I was from a poor family, that I was able to study Theology thanks to the scholarship which I had received, wrote about my having been valedictorian in the Seminary, my joining the BANU of the government faction, my activities as a member of NSU, my brigadir experience and anything that I could think of to build my case, including my hopes to do more for the people's government when I graduate. It worked. In the beginning of September I was awarded

---

* I have mentioned this incident before.

the certificate of having graduated from the university with a major in Theology. But now, at once, I found myself back in square one.

Where was I going to go with this diploma, what was I going to do? "Fortunately" for me, an old teacher in the Sofia Theology Seminary had just suffered a mild heart attack and was going to recuperate for a year. I submitted my papers, I was highly recommended by the Dean of the School and Professor Dulgerov, and in a few days I was appointed as a substitute teacher to lecture to the upperclassmen New Testament Theology About that time, or about some months before, the Seminary had returned to their home building, temporary, and I had the blessing of spending a few months there – until the next March, 1950. I had a great time with the students and the teachers. I had room and board, I was also teacher-supervisor and also entitled to be certified as Seminary teacher. I enjoyed my job immeasurably. I was receiving the minimum salary for starters, but for me it was an enormous amount of money. For the first time I could send some help to my father. All this came to an end in March, 1950, when I received my summons to go and serve my military duty, to report to the same office where I had been before. There was no escape now. I returned home, stayed there a week and then ... to my military service. That day came. It was a turning point in my life, a new page in my life, far, far different from everything that I had experienced so far.

VIII

THE ULTIMATUM AND THE
GREAT COMPROMISE

"...They will lay hands on you and
persecute you and deliver you to
synagogues and the prisons; and
they will bring you before kings and
governors for the sake of my name...
for these are the days of vengeance... "
St. Luke 21: 12, 22.

1. State Security (Durzhavna Sigurnost) and the Church

I will interrupt my personal story here in order to interject the government's views and plans for the Church, which, in a way, affected my decisions, consciously or subconsciously, what course of action I should follow.

Plato explained his theory of ideas with the famous example of the cave. He suggested that we are creatures tied before the entrance of a cave, facing the wall further in, while big flames of a fire are looming behind us. Since we could not look back, we took our shadows cast on the wall as a reality.

The following three documents illustrate much the same deception. The first document is a Report of General Yonko Panov, Deputy Minister of the Interior in charge of the State Security Office (Durzhavna Sigurnost) in 1949. In this Report, addressed to Anton Yugov, Minister of Interior, Panov discusses the political state of affairs in the Bulgarian Orthodox Church, specifically the attitudes of the Holy Synod towards the Communist government. The second document is an abbreviated version of the Report, and is addressed to the Secretariat of the Central Committee of the Bulgarian Communist party, with proposals as to how to deal with the situation.

The existence of these documents was never known to the general public. They became public knowledge only after the fall of Todor Zhivkov's dictatorship on November 10, 1989, and then only in brief quotations in the press. Apparently their full text was given to the small group of dissidents who soon after the palace revolution in the Party, and the take over of the Office of Religions Affairs by the opposition, plotted to overthrow the Holy Synod of the Bulgarian Orthodox Church. The leader of that group, Prof. Radko p. Todorov, once my best friend as theology students, trying to persuade me in the justice of their cause, gave me a copy of them. When I read them I felt that the chains which Plato envisioned at the entrance of the cave fell from my hands and I was able to turn back and see the fire burning behind me, staging the dramatic shadows on the wall opposite me where the tragedy of the BOC was being played.

I will leave it to the reader to make his conclusions. I must give credit to Panov that he was perfectly well informed of the resistance of the Holy Synod to the government of the Fatherland Front. All the information which he has put in his report is correct to the last word, except for the Communist paranoia that behind all this was some big international espionage affair. His report apparently was written by the section of the State Security overseeing and using the Office of Religions Affairs (ORA) for their work on religious organizations. Much of what he has written there was known to me and it has found its place in my reminiscences as a Theology student. The reader is already familiar with my observations.

But this Report contains information which was not known to the public, or whatever was known, or suspected, was now confirmed, directly or by inferences, by Yonko Panov. There is the case of the fall of Exarch Stefan. The public was told that he was deposed by the Holy Synod after he had submitted his resignation. This may really have been the case but Panov's report indicates that the Synodal Minutes had been edited and published in such a manner that to indicate that it was the Government who staged it. Thinking over these events, it seems to me that Stefan was the Bulgarian Cardinals Mindsenty and Stepinats. Only that the Bulgarian Communists handled the matter differently. Such is also the case of Metropolitan Andrey in the U.S. where the respective documentation points to the government as having requested it. This was the truth. Panov charged that the Synod deliberately edited these documents, so that to blame the government for the dismissal of Andrey.

By far the most important revelation in this Report is the plan which its author proposes to effect radical changes in the Church – to purge the church leadership by reducing it first by three, then by more metropolitans, and eventually, out of eleven, to retain only three, thus dividing Bulgaria into three dioceses. They would accomplish that by way of bureaucratic methods – budgetary considerations, age limitations, and, or, the easiest one, to force them to resign by blackmail. However, it never came to that. No metropolitan was removed, no purges in the church took place. The personnel of the church remained intact. But some important changes at the top took place. Soon after these Reports were written, without anybody knowing what happened to them, Metropolitan Mikhail of Russe, *pro tem* President of the Holy Synod, resigned and was replaced by Metropolitan Paissiy of Vratsa. Yonkov's Report indicates that Paissiy proved uncooperative, and by December 1950 he was also out and his place was taken by the Metropolitan of Plovdiv Kiril.

The rise of Kiril was not surprising. Back in 1946, on August 28, he had addressed thousands of worshippers at the Bachkovo Monastery, urging them to vote for the Republic, against the monarchy, on September 8th. He had written several articles in *"Tsurkoven vestnik"* about the Soviet Union, in favorable terms. He had stayed out of open confrontations with the communists. After September 9, 1944 he was arrested and put in jail for several months with Paissiy. It was during this period – 1949-1950 – that a new By-Laws of the Church were being prepared, and, as it appeared, it was Paissiy who was opposing the demands of the government. On December 31, 1950 the Holy Synod "received approved" by the government the new By-Laws of the church. I am quoting the words used in *"Tsurkoven vestnik"*. Paissiy was immediately replaced by Kiril who accepted them and went out to implement them. When all the procedures were completed, he was elected patriarch on the 10th of May, 1953. It should be noted that in the Report of Panov one could find many references to the new By-Laws, and will understand what it was all about and what the government wanted from the Holy Synod.

Another very important change in the church, required by the government as per Panov's report, was the appointment of a new General Secretary of the Holy Synod. The ORA tried to force the Synod to accept a priest, by the name of Ivan Karadjov, one of the expelled Gashtev's group seminarians in 1943, but the Synod declined to accept him. The next candidate of ORA was even less acceptable. It was the former teacher of my seminary days Bogomil Bossev. He had defrocked himself after September 9th, was fired from the office of Principal of the High School in Brezovo for immoral behavior and transferred to Svoge, near Sofia. Much later he was to author some anti-religious pamphlets. The Synod turned him down. Then the ORA came up with the right man, my teacher in the Seminary, Arkhimandrite Yona. Yona was known for his drinking, he had sided with the Communists and took the position. The Synod could not object. In no time we started reading in "Tsurkoven vestnik" articles in praise of the Communist revolution, on Lenin, on the October Revolution et al. over Yona's name. Likewise such articles were written by Bishop Pimen, Vicar of the Sofia Diocese. In Panov's report Pimen is mentioned in not so complimentary terms, as if he was collecting information for foreign intelligence services. But, something happened to him and soon he was going to emerge as a leading favorite of Yugov. In the report of Panov it is strongly recommended that the Party takes seriously the task of "persuading" some of the clergy at the top level to work for them. He uses the word "obraboti" or "podraboti", meaning a workout of the individuals, so they become their agents or their slaves, whom they would help in rising in the church hierarchy. That thing may have happened to Pimen, as he was to show in the future that he was prone to such "persuasions". The charge against him by Panov may have been a half-baked scheme, because his ties to Yugov seems to have dated from much earlier.

One way or the other, by the middle of 1950, having cut my ties with the theological circle and having lost my contact with church leaders, moreover, reading here and there tid bits of pro-Communist statements of high church leaders, I came out with the impression that the Synod and the high leadership of the Church had sold out to the Communist Party. The third document I am publishing here is an essay which I wrote in Drama, Greece, two weeks after I had crossed the border. I was bitterly disappointed of the course which church affairs were taking and to that, led by the Holy Synod. It was not the same church and the same Synod which I had known before. I could not digest this new course, and my experience as Trudovak and as Gorianin, deepened my bitterness, to last me for forty years during my exile.

Now, going back over these documents, the Panov and Yugov Reports, I suddenly asked myself: What made Kiril and the rest of the Synodal bishops to make such a radical turn, to join in the hymnology of the Communist Party? To my surprise, I found the answer in these documents. It crossed my mind that somewhere along the lines these Reports were shown to a few of the Metropolitans, even only to one of them, and that might have been Kiril. The more I concentrated on their contents, the more they sounded to me as an ultimatum to the church and its leaders: Either accept to be a slave of the government, or it is a goodbye for the church. The purges would take place, the dioceses would be reduced to three, there will be other leaders who would be "persuaded" to serve the Party. It became clear to me that this ultimatum led the Church, the Synod and Kiril, to make the great compromise. For Kiril it was either he takes over or it would be somebody else, of the kind of Yona, Pimen, and the gang in the Priests Union, the disgusting Georgi Georgiev. In addition to that Kiril may have come to the conclusion that the Communists were here to stay, that the English and the Americans would never come, their hot propaganda appeals being a pure deception. He and they, under these circumstances, had made the right choice. But this was not obvious until I read these documents. But even if it had become obvious to me, though I always suspected that this was the case, I could not have changed my mind as expressed in my essay.

The good part of all this is, that even without the evidence revealed in these documents, I had always assumed that the reality was exactly what Panov described it to be. Panov simply confirmed my assumptions that all along the Church was a slave to the Communist party, without its consent. I do not know whether the communists ever trusted the clergy, including the Synod, and how much the church sincerely served the Communists. Or it was all an inescapable *modus vivendi*. One thing is sure, this *modus vivendi*, and this great compromise, prevented the implementation of the Panov-Yugov plan and the church, injured and humiliated, survived the ordeal of the Communist yoke. For Kiril it all came down to this: if anyone had to cooperate with the Communists, it was better for the church to be him and the Synod, rather than the lackeys of the Priests' Union. For the Communists it was a better alternative to have a strong, submissive church leadership, as Kiril and the Synod certainly were, rather than a bunch of nobodies seeking ingratiation with the Party. This settled the question of church-state relations to the end of the Communists era November, 1989. And as things turned out it proved beneficial both for the church and the government. In no time the church accepted the status of a vassal to the state, and the state turned the church into a tool of its international propaganda, under Synodal leadership. By 1955 Kiril disbanded the Priests' Union with the help of the government and the Synod was left to manage the affairs of the church, as much as was left of it. What was left for me, and the exile community abroad, was to relentlessly attack this symbiosis of the cross and the hammer and the sickle.

2. Yonko Panov's Report

Strictly-Confidential

To Comrade Minister of the Interior

REPORT

of General Yonko Panov – Deputy-Minister of the Interior

RELATIVE TO:

the political state of affairs of the Bulgarian Orthodox Church. Its attitude towards the people's government of the Fatherland Front and measures to be taken for the creation of a people's democratic church which would be engaged in service to the people.

Comrade Minister:

During the investigation of the Protestant pastors and Gavriil Tsvetanov, presently under investigation, information was developed which indicates that for many years, continuing to this day, contacts have been established between some of the leaders of our church and the Western Intelligence agencies along the religious lines of the imperialists. This information, matched with the intelligence received by our services since September 9, 1944, enables us to fully expose the state of affairs of the Bulgarian Orthodox Church, which at this time appears as the principal reactionary nest upon which our internal and external enemies have placed their big hopes in their struggle against the government of the Fatherland Front.

All this demands that the Orthodox Church should immediately be paid most serious attention, so that its destructive influence and its transformation into a people's democratic church, which would be in a position to attract in support of the government and for the building of socialism in our country that part of the population which holds religious views.

HERE IS SOME OF THE INFORMATION ON THE
ANTI-FATHERLAND FRONT ACTIVITIES OF THE CHURCH

1. The composition of the supreme government of the Bulgarian Orthodox Church (BOC), the Holy Synod, from September 9, 1944, to this day, in spite of the several changes in its Executive Body (Maluk sustav) and the resignation of the former Exarch

Stefan, is reactionary and in opposition towards all initiatives of the Holy Synod, stated on several occasions, have been, as a matter of fact, undermined by subversive and open activities of individual metropolitans and some of their priests.

This is confirmed by enormous material evidence collected by our services, which indicates that metropolitans and priests appear to be active disseminators of all kinds of hostile rumors, campaigns for war, sowing fear and insecurity among the population, slandering the government, campaigning against the TKZS (the Bulgarian version of the Soviet Kolkhoses), against the Brigadir movement ("voluntary" labor battalions for construction of State projects), against the education policies, the economic plans, etc. – approaching a most acerbic struggle against the government – banditism and espionage.

2. The Synod has become a place of concentration and shelter for fascist elements. The Synod and the Metropolitans have appointed in their departments many military officers fired from their offices, disbarred lawyers and other hostile enemies of the government.

As for instance: Alexander Ankov, an Officer of the Reserve is appointed by the Holy Synod as Consultant. A month ago we found concealed weapons belonging to him, with the knowledge of Paissiy (Metropolitan Paissiy of Vratsa) in the monastery of "Saint Ilia". Assen pop Iliev, fired from his military office, has been appointed a choir master ot the Church of "St. Nedelya". As a bookkeeper of the Candle factory of the Holy Synod is appointed officer of the Reserve Konstantin Popov. On the staff of the Synod are Nikola Vassilev Ilchovski, a former member of the Military Police Force (Gandarmerie); Panayot Stefanov Kiradjiev and Ivan Ivanov, fired from their jobs in the Government Printing Office and many others. The disbarred lawyer Tikhomir Naslednikov is appointed as chief of the monasteries in Bulgaria. The disbarred lawyer Doncho Dochev, former notorious director of the Food-Export Agency, is appointed as chief of the real estate properties of the monasteries in the Sofia Diocese. The former wholesale merchant in Sofia, president of the Merchants' Union and Executive Director of the Agency for Supplies, Boris Dimitrov, has been appointed as chief of the Department of Economics of the Synod. Ivan Zagorski, fired from the former Ministry of Propaganda as a fierce fascist, is appointed as a caretaker of the Synod.

The situation in the diocesan offices is the same. If we start mentioning all the fascists and enemies of the government having found shelter in the church, our listing them will have no end. It should be noted, however, that in the Sofia Diocesan office, in the Synod, in the Theological Seminaries and in the rest of the diocesan offices no Communist or individual favorably disposed to the government has been appointed to any office, except in a few instances.

3. The decisions of the All-Orthodox Conference in Moscow of 1948 for breaking ties with the World Council of Churches in Geneva and the Vatican have not been implemented by the Synod. The latter had agreed, after a long insistence of the Office of Religious Affairs (Direktsia na veroizpovedaniata), to publish these decisions in their church press, and nothing more. However, it (The Synod) continued to maintain contacts with the Ecumenical Council through Prof. Protopresbyter Tsankov, and probably through others too, and receive all kinds of assistance.

Along these lines, in May this year, the Holy Synod received two shipments of woollen cloth and silk, all together 293 kilograms, and not too long after that an announcement was received that some ten tons of grain supplies were on the way. In exchange they asked for information on economic subjects. It is interesting that the Holy Synod was ready to give out this information, even though they are a national state secret, if the Office of Religious Affairs would have permitted them. Yet, we are not sure that this information was not given out in some other way.

4. Following the resignation of the former Exarch Stefan, the present Holy Synod, after long struggle, has agreed to concede on some important questions over which there has been a dispute with the National Council of the FF (The question of education of the youth, the participation of the priests in the FF organizations, in the government bodies of some cooperatives, etc.). But this consent of the Synod was not translated in deeds.

Thus, at the same time, the Synod sent to the congress of the priests last year one of its most reactionary metropolitans, Clement of Stara Zagora, to address the delegates, where the same delivered a strictly dogmatic reactionary message, without mentioning the above concessions or the Fatherland Front (FF). Presently too, the metropolitans are seeking the opportunity to dismiss those priests who act as active and good FF supporters. The case with the firing of priest Dimitar Yordanov * of the village of Pobit Kamuk, district Razgrad, who takes an active part in all FF initiatives, is a typical one. The intervention of the District Committee of the FF and the outrage of the local people who wanted to call a protest meeting notwithstanding, Metropolitan Mikhail of Russe, refused to reinstate him. The question now is referred to the Office of Religious Affairs.

The impudence of the Holy Synod lately goes as far as to ask the Office of Religious Affairs with a letter No. 2346 of May 11th to intercede with the Ministry of Interior and the National Council of the FF that in the future, when it becomes necessary to call the priests to meetings where they would be informed of government initiatives and their cooperation is sought by the local FF authorities and Committees, the requests should be addressed to the diocesan superiors, who, in turn, would give the respective orders.

5. The Holy Synod does not demonstrate any particularly favorable disposition towards the Russian Church and avoids contacts with her. They have refused to permit a permanent concelebration of Bulgarian and Russian priests in Bucharest, as it has been desired by the Russian Church and has declined to allow the coming of the Russian patriarchal choir to Bulgaria.

6. The new Ecumenical Patriarch sent from America, Athenogoras, a tool in the hands of the Americans, who arrived with the private plane of Truman, was congratulated by the Synod without the knowledge of the government. The same Patriarch was not congratulated neither by the Russian, nor the Romanian churches.

7. The Holy Synod hands over materials to international reactionaries with the intent to harm the government and provoke intervention of the Anglo-American imperialists in our internal affairs.

Such was the case with the question raised in the United Nations "For the Freedom of Religion in Bulgaria". It was established that the Holy Synod had sent to Bishop Andrey in America all Minutes of their meetings in connection with his removal as administrator of the Bulgarian diocese there, together with the resolution of the Prime Minister Georgi Dimitrov and all letters of the Office of Religious Affairs on the same question. The Minutes of the Holy Synod are written in such a way that it is emphasized that the removal of Andrey was made on insistence from the government, and not as a wish of the Synod. These materials, consequently, on account of the current metropolitans, are in possession of our enemies, and are sent there not by way of our Ministry of Foreign

---

* I suspect that this might have been my friend in Varna, mentioned earlier. But I had no idea of that.

Affairs, but through their own channels. This way of proceeding by the Synod and Andrey can be characterized as 100% espionage for the benefit of the imperialists with the intent to discredit our government. On 28th of April our office received information that the Synod was preparing to send still other documents abroad with the same intent and that these documents are in the house of Prof. Markovsky, who was copying them. On this tip the house of Markovsky was searched and, indeed, the confidential archive of the Holy Synod, in connection with the resignation of Exarch Stefan and his transfer to residency in the village of Banya, were found there. The Minutes and the files are written in such a way, as if the Exarch had been removed from his office not on account of his resignation, but by the government.

8. The campaign for peace was not publicized in the church press, and it, the Church, looked to it with considerable coldness. Only in the 17-18 issue of May 9th, of *"Tsurkoven vestnik"*, (Church Journal) an article was published by the editor, Prof. Markovsky, on the subject of peace, but it was in a conservative spirit and contained some dangerous allegations, for which it was justly criticized in No. 1454, 22 May, of the "Fatherland Front" newspaper. Not one single metropolitan or bishop came out to say or write something on this campaign for peace. Such silence is kept also over other important actions, as for instance about the Atlantic pact, et al.

9. The organ of the Synod, *"Tsurkoven vestnik"*, is strictly a church and reactionary paper. It refuses to publish items with progressive contents. Such is the refusal to publish the speech of Kosmodemianskaya at the Peace Congress in Paris, and also to honor the memory of Diado (grand father) Blagoev, under the pretext that they were political statements. They have forgotten that in fascist times they have published political articles and even printing the portrait of Hitler. Likewise, in 1946, when the reaction was raising its head in our country, they have published such articles. Putting this question to Prof. Markovsky in this manner, he responded that it was possible in those days, but presently the paper is "apolitical". It is clear what is being concealed behind this claim of "apolitical" stand.

10. After Easter this year all Metropolitans were visiting their dioceses and even though this was close to the elections for local People's Councils, none of them, in their speeches, has ever mentioned it. Some of them issued circular letters, but in their speeches they said nothing. Even more! There are materials in our office indicating that some metropolitans and priests have campaigned against the elections.

11. There are instances, where the Synod uses insignificant misunderstandings with the government or other agencies – administrative – and tries to turn them into big "incidents", with the intent to use them to carry out hostile activities. Such is the case of the refusal of Paissiy to appear at the dais on May 1st, because he had not been invited to the central line up, and at the same time had scheduled to serve in the "St. Alexander Nevsky", as some sort of a demonstration. So also on the day of Kiril and Methody, in his capacity as Pro Tem President of the Synod, he has not appeared to the place at the line up of officials and had gone to serve liturgy in some church in Sofia. In addition to that the Holy Synod has issued several circular letters by which they use the name of the Director of the Office of Religious Affairs (distorting the facts - STR) as is the case with Circular Letter No. 2069 of April 20, on the subject of continuing the service of old pensioner priests, among whom it appears a most virulent campaign against the government is carried on from the church pulpits, as is the case with the church "St. Petka" in Sofia. Some priests of that church have been recently interned for their hostile campaign against the government among the worshipers.

12. Lately the Holy Synod tenaciously, with all its actions, demonstrates its independence from the state, so that it could mount opposition to the government initiatives on the one hand, and on the other not to commit themselves with the policies of the FF, trying to show its independence of the state. This was particularly obvious after the signing of the Atlantic Pact and after the question of "The Religious Freedom in Bulgaria" was introduced in the United Nations. This has become apparent from the following: Until now all circular letters of the Ministry of Education were sent to the Theological Seminaries for their information and execution. Lately the Synod, with letter No ...... has let the Office of Religious Affairs know that such circular letters should not be sent there, because they, the Synod, are the supreme authority over the Seminaries.

The Office of Religious Affairs last year, on the grounds of Art. 57 of the By-Laws of the Exarchate, which entitles it to confirm the District chief priests (The Arkhiereiski namestnitsi), had replaced a large part of them, selecting priests known as good FF supporters. Having lost these District Chief Priests, in spite of the stipulations of the present By-Laws, the Synod, with circular letter 1171 of May 18th, took away all functions of these priests, thus eliminating them for all practical purposes. This reform is substantiated with the argument that after the separation of the Church from the State, the financial means of the Church had been reduced. This way they have inflicted a double blow to the government's policies: the responsibility for cutting down the service in question is layed on the back of the government, and the progressive apparatus of the District Priests is taken out of the hands of the Office of Religious Affairs. This is a bold and defiant struggle with the government, which is conducted in secret, even in violation of the By-Laws.

As a result of this heightened selfconfidence, there appeared a new draft By-Laws of the Church by which they raise claims which the Church has not had and has not asked for.

13. The Holy Synod avoided raising the question of electing a new exarch, even though there is no exarch now. They have done that because now nobody bears responsibility for the actions of the Synod, and this way it is easier for them to escape commitment for FF policies, hiding behind each other.

These, even though very briefly presented facts, show that the Holy Synod and the individual Metropolitans, in general, have taken a stand on enemy positions. On all fundamental questions they are in solidarity and at the service of international reaction by uniting around them in the country the enemy elements, sabotage the Party and government initiatives and make attempts to make the Church much more independent from the state, in comparison with what had been the case in the past, in spite of the new law (The Law of Confessions of February 24, 1949. STR, translator). This way the Church, instead of helping to attract that part of the population, which are the believers, in support of the government and the building of Socialism, continues remaining and with every passing day becoming an ever greater stronghold of the external and internal enemies in their struggle against the government of the Fatherland Front.

Side by side with the open and subversive activity, the leadership of the present Orthodox Church resort to other surreptitious forms of underground activities. It aims at locking in its own shell the priesthood and officially declares that "consistent with the spirit and the traditions of the Church", the priesthood should stay out of the political life. At the same time they take all possible steps against priests manifesting themselves as supporters of the FF, to neutralize their influence in the Church, using, depending on the case, bribery, flattery (the case with Ikonom Georgi Georgiev), blackmail, threats, even engaging in persecutions.

Because, on one hand, they are not able always to act following the mentioned methods, and on the other hand it is dangerous to resort to open opposition, they carry on their line under cover, as for instance, writing some circular letter from the Holy Synod on the economic planning, on the brigadir movement, on the elections, on peace, on harvesting the crops et al., without, of course, engaging in active work. They are written in a manner which they deem sufficient that they could not be accused tomorrow for having given up expectations for restoration (of the old older – STR) that they had committed themselves to the FF. They extend this argument as much as it is possible, to serve as a cover for all categories of "people of the past" who have suffered from the initiatives of the people's government and close their eyes to openly reactionary acts of the priests.

They expand the greater controversial issues as a dilatory tactics and usually prefer to postpone *ad infinitum* their solutions.

HERE ARE SOME OF THE REASONS FOR THIS ATTITUDE OF THE CHURCH LEADERSHIP TOWARDS THE PEOPLE'S GOVERNMENT

As it is known, long before September 9, 1944, the Church, represented by the Metropolitans, had created a monastic oligarchy, which was suffocating every living and free thought, nailing it in some canons, which, in fact, nobody has observed, least of all the bishops. Likewise, the personal and public life of these "holy men" who all their life have practiced unheard off pharisaic hypocrisy, deception, slander, intrigue, gluttony, treachery, drunkenness, fornication, sexual depravity, and what else in addition, and always at the beck and call of the immoral life – now in service to Boris (the King – STR), now in service to the English or the Germans, but always invariable enemies of the USSR and BCP (Bulgarian Communist Party – STR).

This situation in the Church did not change after September 9, 1944. In the beginning part of the progressive priesthood took a stand. A Committee for Reforms and Controls in the BOC was formed and became known to the public. It worked two months in preparing the By-Laws for democratization of the Church. These By-Laws have been submitted for approval to the Ministry for Foreign Affairs, but the Exarch and the Minister for Foreign Affairs, Prof. Petko Stainov, have suppressed everything. The Exarch, in a short period of time, has succeeded in taking everything in his hands. To all key positions in the Church, and as his assistant, he appointed the greatest reactionaries. He succeeded in twisting the neck of the progressive priesthood under the old harness. As of now, only single and small groups of priests fight for a people's progressive church, but they are not united, and do not have reputable representatives. There is nobody to encourage and embolden them.

In the third place, as a most serious cause, is the very Holy Synod, the Metropolitans and those individuals, who, though formally outside the leadership of the BOC, play an enormous role in its leadership and policies. These hardened political enemies of ours engage in desperate efforts to defeat the Law of Confessions and to this day not only have they failed to draft new democratic By-Laws in the spirit of the new Law, but are proposing an arch-reactionary one.

It is clear, that, given this state of affairs and composition of the synod, if no measures are taken towards the Church, it would not be able to change its spirit, and only hypocritically will hide itself, preserving its feelings for restoration (of the old order – STR) and its inclinations to work actively against the people's democratic government.

PROPOSED MEASURES TOWARDS THE CHURCH

In view of this state of the Church, we propose that the following measures should be taken in regard to same, so that its hostile attitude towards the government should be neutralized:

1. In the first place some changes in the Holy Synod should be made. There should be a minimal purge from the top. It will alarm the rest of the metropolitans, and they may be asked to make commitments to the FF government and the progressive priesthood will be encouraged.

As a beginning the Metropolitan of Russe Mikhail, the Metropolitan of Turnovo Sofroniy and the Metropolitan of Vidin, Neophit, should be removed – as leading exponents of reaction in the Church. This could be done in three painless ways:

a. By abolishing their dioceses for budgetary reasons.

b. To implement a terminal age for service of the bishops, as it is for all citizens of the Republic.

c. Or to be led to the realization that they should resign their office and go into retirement – with medical certification. This seems to be most convenient way. Enough compromising materials may be found for them to do that.

2. To enforce the Law of Confessions by firing five to six active reactionary clergymen from each diocese. We should not be stopped by the fact that the international reactionaries will use that to claim that there is no religious freedom in Bulgaria. We are witnessing that whether the law is implemented or not, the international reactionaries are raising noise that there is no freedom of religion in Bulgaria. The church reactionaries in Bulgaria, seeing that the law is not being implemented, arrive at the conclusion that the government is afraid of international pressure if it implements it, and are encouraged.

3. Loyal to the FF clergymen and laymen to take key positions in the Church. Such key positions are: General Secretary of the Synod, Director of the Cultural and Educational Department, Director of the Economic Department, Director of the Publications Department, Editor of the "*Tsurkoven vestnik*", Rectors of the Seminaries and the Theological Institute and the Abbots of the Rila, Bachkovo and Troyan Monasteries.

4. After these changes, to suggest to the Holy Synod to prepare a new, short democratic By-Laws and to proceed with the election of a Patriarch. As probable candidates for Patriarch appear to be Paissiy of Vratsa and Kiril of Plovdiv. Out of the two Kiril appears to be more acceptable, but in any event, we have to settle for the Metropolitan, who would make greater commitments that he will democratize the Church.

5. If the Holy Synod opposes drafting of a new democratic By-Laws, the Congress of the priests may be used to raise the question of democratization of the Church.

6. By all means to reduce the dioceses. Now they are eleven. They should be reduced to six, and later to three. This could be accomplished for budgetary reasons.

7. To pay attention to the available cadres of the Church. Because certain considerations of foreign policy demand from us not to carry out a wider purge in the Church, some reactionary clergymen who are prone to "persuasion" may be encouraged with advancement in the church hierarchy, while others may be stopped in their rise to higher church positions. (This question may be developed in concrete terms).

8. The progressive priesthood can be encouraged. Special attention should be paid to the President of the Priests' Union, now Ikonom Georgi Georgiev, who is a member of the Bulgarian Communist Party, but has bourgeoisie inclinations and there is a danger to be "conquered", i.e. to fall under the influence of the episcopal reactionaries and become

a conduit of their suggestions as was the case with the draft By-Laws. He should be charged with progressive tasks and to expect that he carries them out. If he does not perform, to be replaced by another man, better suited to carry them out.

9. To pay attention to the preparation of a truly FF clergy cadres. This question has not been discussed. For this purpose it is necessary to seek a knowledgeable and tried man which cannot be found now. For this purpose we should create a special department in the Office of Religious Affairs. A suitable subject for this purpose is Arkhimandrite Damaskin, a former General Secretary of the Holy Synod. (Damaskin, of the monastic clergy, was found to have been secretly married and father of a child. He was removed from the position of General Secretary of the Holy Synod and lost his status as an ordained clergyman in good standing. Nevertheless, he continued to wear his priestly cassocks, while employed in the Ministry of Foreign Affairs, probably in the Office of Religious Affairs – STR). A communist could be appointed as his superior. When admitting new students for the Seminaries and the Institute, a special selection should be made by asking the candidates to present certificates from the local People's Council that they have not shown themselves to be involved in reactionary and fascist activities.

10. To review the study programs of the church institutions, especially a thorough review of the education of competent organs. All this could be carried out by the Office of Religious Affairs with some help from us.

11. On our side – special attention should be paid on winning and speeding the cultivation of the persons identified as foreign agents who have great influence in the church and their unmasking and rendering harmless.

Date: Illegible. Probably the second half of July, Preceding the Proposals of Anton Yugov, based on this report, dated August 3, 1949.

Signature: illegible.

3. Anton Yugov's Report

TO THE SECRETARIAT OF THE CENTRAL COMMITTEE
OF THE BULGARIAN COMMUNIST PARTY

PROPOSAL

RELATIVE TO: TAKING MEASURES FOR DEMOCRATIZATION
OF THE BULGARIAN ORTHODOX CHURCH

COMRADES:

The materials which are in possession of our office and come to our attention every day relative to the state and the activities of the Bulgarian Orthodox Church indicate that after the blow inflicted on the reactionaries with the processes against Nikola Petkov, the Legionairs, the Lulchevists, the Protestant pastors and the depriving of the capitalist elements of legal possibilities, the said church emerges as one of the principal reactionary nests, upon which, with justification, our internal and external enemies place great hopes in their struggle against the people's government.

As it is known, in spite of the deep political, economic and social transformations which took place in our country since September 9, 1944, no changes whatsoever have taken place in the state of the BOC for its democratization.

The church reactionaries at the top and lower level, who waged active struggle against the Soviet Union and the Bulgarian Communist Party, have succeeded by way of a variety of machinations to preserve even to this day the leadership of the church in their hands. The attempts of some individuals or small groups of priests to initiate a struggle for democratization of the church have been in different ways quickly countermanded and suffocated. In this respect the metropolitans of the Synod have always demonstrated their unanimity.

One single glance over the activity of the church these days will establish that these hardened enemies of USSR and the BCP are increasing their hostile activities. On all fundamental questions they are in solidarity and in service to the international reactionaries with whom they have not yet cut their ties officially, unite around them hostile elements, sabotage Party and government initiatives and attempt to make the church much more independent from the state than has been the case in the past. They are not ashamed to declare that their good will is present, that they could work in full harmony with the government if there is a "mutual confidence and consultation", while at the same time do everything possible to bypass and sabotage the Law of Confessions. They use every means to lock into their shell the Bulgarian priesthood and officially say that according to "the spirit and the traditions" of the Church, it should be kept out of the political life, while at the same time apply all repressive measures against the FF priests and neutralize their influence in the church. They delay and extend *ad infinitum* decisions on the big issues, shifting responsibility for that from one to another in order to gain time.

One typical activity of the Synod lately is the attempt made through Bishop Pimen to take census of the population in Sofia as per parishes and divide it into categories by collecting information about the number of the believers and the non-believers, their attitude towards government and religion, their economic status et al. to make a full card-index, which could not be interpreted otherwise, except as a hostile act of the Synod, with the intent to collect materials which are of interest for the Intelligence Centers abroad, and could obtain them easily.

The leaders of the Church think that the pressure which is exercized from outside, in connection with "violations of the terms of the peace treaty" and "the religious freedom in Bulgaria" has led to the decisions of the June Plenum "for concession to the peasants". From that they arrive at the conclusions that the Party is forced to make concessions, and that they could exert pressure, in order to receive the same, especially after the death of Comrade G. Dimitrov. They indeed increase their hostile activity.

It is clear, that given this state of affairs in the Orthodox Church, with this composition of its leadership, it will continue to remain and to transform itself day after day into an ever greater and solid mainstay of the foreign and domestic reactionaries with their serious hope for restoration. It is necessary to take timely measures, so not to permit that, and, in the country, to create conditions for democratization of the church. For this reason,

WE PROPOSE:

1. To effect changes in the Synod. To carry out a minimal purge at the top, which will alarm the rest of the Metropolitans, some of whom to make commitments to the FF government and this way to encourage the people's priesthood. To remove at first the Metropolitan of Russe Mikhail, the Metropolitan of Turnovo Sofroniy and Metropolitan Neophit of Vidin who are devoted representatives of the reaction in the church.

This could be accomplished in three painless ways:

a. By reducing the dioceses for budgetary reasons.

b. By applying the terminal age for service of the bishops, as it is for all citizens of the Republic.

c. Or to be brought to a situation to submit their resignation and go into retirement – with medical certificate. This manner of proceeding appears to be most convenient. Enough compromising materials may be found for those mentioned in order to do that.

2. To begin applying the Law of Confessions. To dismiss four-to-five active clergy-reactionaries from every diocese, in conformity with the same Law.

3. Loyal to the FF clergymen and laymen to take key positions in the Church. Such key positions are: General Secretary of the Synod, Director of the Cultural and Educational Department, the Director of the Economic Department, the Director of the Synodal Publishing House, Editor of the *"Tsurkoven vestnik"*, Rectors of the Seminaries and the Institute of Theology, the Abbots of the Rila, Bachkovo and Troyan Monasteries.

4. Following these changes, to suggest to the Synod to draft a new, short, democratic By-Laws and to proceed with the elections of an exarch or patriarch if this is necessary. As probable candidates to head the church are indicated to be Paissiy of Vratsa and Kiril of Plovdiv. To prefer the one who makes greater commitments for democratization of the church.

5. If the Holy Synod opposes drafting of new democratic By-Laws, to make use of the Congress of the priests to raise the question of democratization of the church and submit a draft By-Laws.

6. To pay attention to the available cadres of the Church because certain considerations of foreign policy demand not to undertake a broad purge in the church. In this respect, some reactionary-active clergymen, prone to "persuasion," may be encouraged with advancement in the church hierarchy, and others to be stopped in the rise to higher church positions. (This question may be worked out concretely).

7. To encourage the people's priesthood. Special attention should be paid to the President of the Priests' Union Ikonom Georgi Georgiev who is a member of the BCP, but has bourgeois inclinations and there is a danger to fall under the influence of the reactionary bishops and become a conduite of their suggestions as was the case with the draft By-Laws. To be given progressive assignments and to expect that he carries them out. In a contrary situation to replace him with other, more suitable man to carry them out.

8. To pay attention to the preparation of a truly FF clergy cadres. This question has been done in this respect by now. For this purpose we should seek a competent and tried man. In the process of admitting to the Seminary and the Institutes of new students, to make a certain selection by requiring of the candidates to submit a certificate from the local People's Council that they have not shown themselves to have been involved in reactionary and fascist activity.

9. To review the study programs of the theological educational institutions and especially a thorough review of the education of competent organs.

10. The Office of Religious Affairs should be shown a greater attention. The man at the top of the said Office to be a comrade answering to the Central Committee. This is imperative, because the Office of Religious Affairs exercizes the role of a guide not only to the clergy, which includes also all foreign churches and sects, but, likewise, the minority groups in the country.

11. On behalf of State Security (The DS) to pay special attention to strengthening and "persuasion" of the individuals identified as foreign agents, who have big influence in the church, for their unmasking and rendering harmless.

3 August, 1949
Sofia

Minister:
(A. Yugov)

Deputy Minister for D.S. (Signature)
(R. Khristozov)

4. The Bulgarian Orthodox Church in the Conditions of the Red Terror [*]

In over a thousand years long historical life of our nation, the Bulgarian Orthodox Church (BOC) has played an extraordinary important civilizing role. The political establishment and the acquiring of cultural identity of the Bulgarian people had become possible and have survived on the stage of history thanks to the inexhaustible energy and the rich creative activities of the Orthodox clergy during the Middle Ages. Their names are a testimony for the enormous contribution of the BOC to the treasure of our national culture.

The red evildoers know the power and the influence of our national church over the broad masses of people, know the creative potential of the Bulgarian clergy, the unshakable strength of the Christian spirit and with Sisyphean efforts try to suppress the church as an institution, and the religious spirit of the Bulgarian people as a powerful force of resistance. However they dare not come out to confront openly the church. They fear that such a confrontation will boomerang against them. They liquidated with great ease the incorruptible conscience of the honest and honorable Bulgarian intellectuals in the professions – teachers, lawyers, physicians, writers and scholars. They liquidated with one blow the legal political opposition, but when they had to liquidate the church – they pulled back. They had the examples of Nero and Diocletian which conflagration and bloody hecatombs, instead of putting an end to Christianity, scattered it all over the world, and adopted a new tactics of their anti-religious war.

All of a sudden they appeared to be the best friends of the church, proclaiming with their constitutions, that the freedom of conscience and religious confessions are sacred principles of the Communist dictators. Even more, the bloody evildoer Georgi Dimitrov showed up on several occasions in "St. Alexander Nevsky" and in the Rila Monastery.

---

[*] This essay was written in the middle of July, 1951, in Drama, Greece.

282

One might think that it was a case of miraculous conversion similar to that of St. Paul on the way to Damascus meeting Christ, or perhaps a suddenly flashed love for Christianity. Nothing of the kind! Behind their empty statements the Communists conceal their perfidious plans to liquidate the church, but here they have chosen to play chameleons, and under the masque of benefactors, treacherously plunge the dagger in the back of their chosen victim. The leaders of the BOC experienced this tragic truth. If they still keep silent in telling the truth to the Bulgarian people and world public opinion, they themselves have to be blamed, not the BOC. Neither the telegrams of the Holy Synod to the United Nations, nor the articles and the statements of the fatal Arkhimandrite Yona – General Secretary of the Holy Synod – published in the political and church press, can deceive the Bulgarian people and world public opinion, to persuade them that the Communist satraps have abandoned their satanic intentions as to what to do with the church.

The facts speak much better than the General Secretary of the Holy Synod with his telegrams to the United Nations and the Communist pseudo-constitutions. Even more! They not only refute the top leadership of the BOC, but they do expose them as open liars. We remember that Someone had said: "Do not lie!" it is in *His name* that the Holy Synod and the General Secretary have occupied such high and responsible positions before the Bulgarian people, before Bulgarian history and before world public opinion. This already suffices to suggest that nobody has a license to lie and sow deceptions from such a high and responsible place.

It could not be said that the BOC is free when its leaders are being transformed into a voiceless tool in the hands of the government's powerful officials, as the one serving as Director of the Office of Religious Affairs (ORA). These civil servants have no difficulty to frequently replace the *pro tem* Presidents of the Holy Synod in order to carry out the government and Party policies aiming at liquidation of the Church. Typical case in this instance is the demotion of the former Exarch Stefan I.

This Bulgarian hierarch, well known to the Bulgarian people, as well as to the Christian community in the world, instead of raising his voice and protest the cruel murder, without a trial, of hundreds upon hundreds of priests, and in the presence of two metropolitans (Kiril and Paissiy – STR) with handcuffs on their hands, approached the pagan shrine of the Bolshevik beast to accept with trembling fingers the white veil from the bloody hands of the red hangmen. He hoped to serve as exarch until old age. Alas! His day came soon enough. A tragic and comic scene in the session of the Holy Synod on September 4, 1948, threw dust in the eyes of the church community by creating the impression, that the fall of Stefan was a purely internal church question, but the declaration of the then Director of ORA, Dimitar Iliev, at the Priest's Congress in October, effaced this deception and the whole world understood that the fall of Stefan was orchestrated by the government. But Stefan refused to submit. Even to this day he has not accepted this act of the government and considers himself as a legal guardian of the Holy see of Sofia. A river of tears in his eyes before a delegation of theology students visiting him soon after these events testifies for his deep conviction of his innocence. He was not willing to quit Sofia and on account of that those who had handed him over the exarchal crown, one morning forced him into a police car and took him to internment in the village of Banya, district of Karlovo, where to this day he laments his old age and atone for his not too few sins.

After the fall of Stefan it was the Metropolitan of Russe, Mikhail, who became *pro tem* President of the Holy Synod, but very soon he was deposed too, because he proved to be a hard nut for the iron teeth of the Director of ORA. He was ordered to vacate in one day the bookstore of the Holy Synod at the corner of "Sv. Nedelya" square and "Queen Yoana" street. This hierarch advised those who gave him the order, that if they wanted to take the premises they were to take out all religious books and icons to the street and he would not order them picked. The order was not implemented but Mikhail was deposed.

After Mikhail came the turn of the political prisoners – Metropolitan Paissiy of Vratsa, who had played a central role in bringing down Stefan, and Kiril of Plovdiv. Paissity stayed a longer time, but when he too became inconvenient, was replaced by the most active, though otherwise serious and balanced Metropolitan Kiril of Plovdiv.

This game with the government of the BOC convinced the church community that the Bulgarian metropolitans had become puppets, a toy in the hands of the Director of ORA, being also leaders of the Orthodox Bulgarian people. Even more, their human dignity and personal honor were publicly attacked by the evildoer Georgi Dimitrov. In a moment of drunken posturing, on the day of the 1000th anniversary of St. John of Rila, in the presence of the Patriarch of Moscow, in the Rila Monastery, he called the Metropolitans "old men with ossified brains". They, holding doctoral degrees of world renown universities, the supreme government of the BOC, the Church of Sts. Cyril and Methodius, of Paissiy of Khilendar and Sofroniy of Vratsa, of Metody Kusevich and Exarch Yossif, had to wink, to keep silent in a slave-like obedience to the cruel master who was admonishing them as disobedient children or as senile old little men... What a humiliation! This is what they call freedom of religion. They know very well what this freedom is all about, because they, more than anyone else, know how many priests had been killed without trial, how many priests are rotting in jails and how many are in concentration camps for no other reason, but for their following the synodal instructions for organizing Sunday schools, how many church and monastic welfare institutions had been closed – orphanages, old age homes et al. – have been taken away from the church and turned into Communist and anti-religious centers; by what a brutal manner the study of religion in and out of the schools has been terminated; how all religious-oriented literature in public libraries was destroyed; how many churches had been desecrated; now many clergymen had been defamed, how many church festivals had been prohibited or subjected to provocation by Communist gangs, etc., etc...

They, the members of the Holy Synod, know very well the tragic fate of the former abbot of the Holy Rila Monastery, the good-spirited Arkhimandrite Kalistrat, tried behind closed doors, innocent as he was, by a Sofia Regional Court, and sentenced to ten years imprisonment, soon after that done away with in his dark jail cell. Arkhimandrite Yona may sing hymns for the freedom of the BOC in the era of the red Communist terror, if the soul of Arkhimandrite Kalistrat, a martyr for the faith of the Bulgarian people, does not bother him.

The Metropolitan of Plovdiv, who at this time is getting ready to take the Patriarchal scepter from the hands of the men who only five or six years ago were plucking his beard and were riding him like a mule, who murdered in a beastly way tens of priests of his diocese, knows very well, that his former diocesan preacher, Nencho Mandulov, went to jail for nothing more than a dedicated performance, with zeal, the duties of his office, assigned to him personally by the metropolitan, spreading with passion the light of the Christian faith from the pulpit.

Dr. Paissiy Raikov, Metropolitan of Vratsa, knows very well, that his Protosyngel in Sofia during 1949, Arkhimandrite Myron, was lifted from the capital in a most brutal way and taken no one knows where on account of his revival of the Brotherhood movement in the diocese and because he had the courage to found a Youth Association of the Sofia diocese.

The Holy Synod and Arkhimandrite Yona know very well that the government is preparing to raze to the ground the Cathedral of "Sv. Nedelya" in the center of Sofia, in order to errect in its place a monument of Lenin. The first step in the direction of this Herostratian act is the closing of all accesses to this central cathedral in the capital, except the most dangerous crossing over the tramway lines.

The Holy Synod and Arkhimandrite Yona know, that the Communist evildoers are implementing an irreversible policy for the liquidation of the Orthodox Church as an institution and as a spiritual bulwark of the Bulgarian people, and that the first step in that direction is the closing of the church educational institutions.

Immediately after September 9, 1944, they occupied the building of the Plovdiv Theological Seminary. All kinds of promises before 1950, when that Seminary was closed, that it was going to be returned, made to obtain concessions from the Holy Synod, remained unfulfilled. In 1950 that Seminary, which had been first opened in Constantinople long before the 20th century began, where many outstanding Bulgarian patriots and church leaders were educated, was closed without anybody raising his voice to defend it. At the same time they closed the Theological Institute and the School for priests at the Cherepish monastery. The Holy Synod swallowed these bitter pills and sent another telegram to the United Nations giving assurances that there was freedom for the Bulgarian Orthodox Church in their country. The Holy father hoped that the church still could go on. These self-assurances lasted till September 11, 1950, when the education institutions of the church were administered the last and most devastating blow.

A government decree, issued at the same time, ordered that the students of the Seminary in Sofia, having just begun the fall semester, were to be sent home and if anyone of them desired to register in a high school, to be admitted in the respective class where he had been in the Seminary. The teachers were warned, threatened with sever punishments, if they would try to dissuade the students to remain in the Seminary. All school furniture and implements were to be transported in two to three days to the Cherepish Monastery, where those electing to continue, could return. The Holy Fathers silently watched the tragedy of this half a century old center of theological studies, where hundreds and thousands of prominent cultural leaders of Bulgaria had graduated or had taught – beginning with the world renown professor of linguistics at the Sofia university, Stefan Mladenov, and to end with the Communist professor Evgeniy Mateev. The Sofia Theological Seminary, the dearest jewel of the BOC, which location in times past had been a wasteland, but under the immediate cares of the church has been made a beautiful park, had to be turned into, and after September 11, indeed it was turned into, a center for offering anti-religious education, home of the Communist children's organization "Little September Kid". The Sofia Theological Seminary died under the accompaniment of the tasteless empty declarations of the Holy Synod for the freedom of the church in Bulgaria.

It is impossible that the Holy Synod does not know, indeed, it would not have forgotten the slap in their face administered by Dr. Tagarov (ORA Director in 1949) on the occasion of a student strike organized by the Communist in the Sofia Theological Seminary during the month of November, 1949. On this occasion the entire teaching staff, with the exception of Dr. Tutekov and Arkhimandrite Gorazd, unanimously voted to expel eight students, ring leaders, a Communist 5th column in the Seminary. The Synod approved unconditionally the decision of the Council of Teachers. The actions of the students were of such a nature that it was impossible to justify them on moral or legal grounds, even on political grounds along Party lines. But Tagarov decided to use the occasion to show the Synod and the teaching staff, indeed to the entire church community, in a manner bordering to absurdity, that the freedom and the independence of the church, even in the area of its internal insignificant affairs, do not exist. The Holy Synod was advised to rescind its decisions and council the students to recant for their actions and appeal to the teachers to accept them back as innocent virgins.

The teachers came out of this duel with the Communist-inspired gang with its honor intact, but the entire church community was left with the impression that the Holy Synod of the BOC is not a free and independent governing institution over church affairs in Bulgaria.

All these things are known to the Holy Synod and to the entire church. Yet, no one dares raise a voice of protest and lift the curtain a little. Does not dare or does not want to do that Metropolitan Kiril, though he knows very well the actions of one of his predecessors whose biography has been written and published by him. Does not dare to do that Paissiy of Vratsa who occupies the See of Sofroniy of Vratsa... They don't dare! It is a pity that our younger clergy follow the line of the least resistance and either get fat in the monasteries or get lost in trivial literary pursuits, romantically sighing over tender flowers of field weeds visiting the monasteries around Sofia to taste the yogurt prepared by monks, and then return to the Sodom and Gomorah, to stretch on the plush chairs in the Holy Synod and dream of the good old times, to write empty, meaningless stanzas for the weeds, for the stars, not able to contemplate with their fattened by idle brains the tragedy of the spiritually and politically enslaved Bulgarian people. Poor souls! To hold them responsible for their idleness would mean to bestow upon them a great honor.

But there are exceptions. While Arkhimandrite Yona, Archimandrite Gorazd, Bishop Pimen et al. are engaged in betrayal, hundreds, thousands of Bulgarian priests, simple laborers in the field of national life, silently wage a struggle against the oppressors. Thus, the refusal of Arkhimandrite Gerassim, Abbot of the Bachkovo monastery to slay lambs for a group of Russian delegates during the lenten season, is a showing of great courage by him and an honor for the Bulgarian church community – nevermind, that soon after this incident he was to loose his position. Arkhimandrite Yona took it upon himself to advise his "friend" to withdraw from Bachkovo and go to the Preobrazhenski Monastery, a dishonorable demotion.

Such is the "religious freedom" in Bulgaria. It is this "religious freedom" which the Holy Synod dare not communicate to the United Nations, while Arkhimandrite Yona exerts himself to invent proofs that a real freedom and independence of the church exists, indeed. But neither the Holy Synod, nor Arkhimandrite Yona would ever mention how many times the editors of "*Tsurkoven vestnik*" and "*Dukhovna kultura*", Professors Markovsky and Marinov, are called to the ORA, to be threatened and instructed to write about peace, against the Yugoslav traitors, in favor of the USSR or for the spring and fall sowing.

The Holy fathers dare not tell the truth. And why they dare not tell the truth? Where the Bulgarian people and the world public opinion are concerned, the excuses of the Holy Synod are indeed trivial and humiliating. They keep silent out of fear that they may irritate their red master, he may suspend the subsidy for paying the salaries of the Orthodox clergy. When a few years ago the church was separated from the state and the question of its sustenance came up for consideration, the Holy Synod drafted a plan which among other things envisaged printing of special charity stamps which would be offered to the public for contributions in support of the church and the clergy. The government vetoed the plan, because it saw in its implementation the danger of losing the most important lever in its hands for controlling the church. In exchange for that it promised to provide the church with a subsidy to pay the clergy. After that, every year, in fact every month, in releasing this subsidy one could see a number of tragic and comic exchanges taking place. Some humiliating bargains begin where the Holy Synod is asked to surrender one or another position of the church.

This is why the Holy Synod is silent. This is why they lie to the Bulgarian Orthodox people and to world public opinion, and in both cases they just deceive themselves because the Bulgarian people and world public opinion are aware of all this. In the best of cases the Holy Synod is not deceiving themselves, they are simply frightened to come out and tell the truth of the state of the church under the red terror. In this case we would permit ourselves to ask an obvious question: How the Holy Synod of the BOC perceive their obligations as spiritual leaders; how the Holy Synod understand their responsibility before History, before the church, before the One Who came into this world to show us the Way to Follow, as a Truth to be confessed and as a life to be realized? Is the Holy Synod following His ways, does the Holy Synod confess His Truth, does the Synod show themselves to be living His life?

However, the good part in all this, strengthening our hope, is the reality, that if the high leaders of the BOC have abdicated from their positions of responsibility, the rest of the Bulgarian people, loyal to their faith, led by their shepherd-martyrs, keep deeply in their souls the traditional faith and do not miss a chance to show their devotion to the church. The clearest example for that is that on big religious holidays they attend *en mass* the religious service and that on other occasions the attendance not only has not decreased, but in some places has increased. However, there is the impression that the youth, in spite of their religious convictions, abstain from going to church, and this may have fatal consequences, especially where the educated youth is concerned. This may be explained with the fact that Communist functionaries carefully watch who enters and who leaves the churches, and they, the young, for one or another reason, would sooner or later have to seek the party recommendations to get a job. Attending church service is the equivalent of hostile, enemy acts. Thus, a president of the Communist youth association in a Plovdiv district village, on Great Thursday, having defiantly mocked the church service, viciously threatens: "Let us now see who is going to the pop!" (the priest). In spite of all this the attendance of church services has not diminished. Especially in the cities, where the Communist agents are less able to follow their victims. There, in the cities, they spy on the priests, and it is being done by people of high education, able to grasp fast and clearly the meaning of the sermons. It has been noticed how in the Cathedral of "St. Nedelya" an assistant professor from the School of History and Philology, to be attentively listening to every sermon of the priests and to take notes.

It is not difficult to discover the reason why the Bulgarian people have preserved their religious conscience even under such extraordinary circumstances. The life of our people is becoming more and more unbearable. Families in distress are multiplying very fast. Where the mothers and fathers, the brothers and sisters, the wives and children of the tens of thousands missing or rotting in jails Bulgarian patriots are going to seek solace and compassion, if not in the churches and the monasteries? If one takes a look at the faces of those entering or coming out of the churches, quite often he will notice the traces of heavily shed tears, will observe the seal of a unknown to anybody, but obvious to all tragedy. There, before the throne of God, the people shed their tears, reveal their suffering soul and with broken hearts pray for liberation as fast as it could be done. And then, when the dedicated to his mission priest immerses himself in this personal and national tragedy, he cannot but share the pains and the hopes of the entire nation for freedom. Our priesthood, in their overwhelming majority – humiliated, reviled, spat at and abused by the Communist mob – to their honor, have not failed their flock and the church. The faithful who seek them before the altar of God, leave the holy place with renewed overpowering faith and energy, and with confidence that the end of the Communist tyranny would soon come.

It is for this reason and in such cases that the faithful support morally and materially their shepherd. This shepherd is not concerned for his material support, remembering the word of our Saviour: "The worker deserves his reward", which have not lost and will never lose their meaning. Only the hired men are concerned for their reward. Regrettably, there are plenty of them in the BOC. Such are the people of the drunkard Georgi Georgiev Bogdanov, president of the pseudo-priests' Union in Bulgaria. After he was rejected by the priests at their congress in 1947, this bandit used his party card and joined by several of his confederates, morally and physically corrupt, of the kind of Balev, Zagarov, Kotsaliev et al., managed to re-take the presidency of the Union. It is only recently that this gang felt a bit inconvenient on account of not quite decent actions of their chief and called on the Director of the ORA, Dr. Tagarov, behind his back, to secure his downfall. This way, these innocent virgins try to divest themselves from responsibility for three million of levs, Union funds, spent by their chief for banquets, drinking parties et al. They are the hired men in the BOC.

The government pays special attention to the sermons of the priests. Under the pressure of the ORA, the Holy Synod and the Priests' Union issue very often circular letters, ordering the priests when and what to speak from the pulpit on political subjects. They are ordered to persuade their flock to join the TKZS (Bulgarian version of the Kolkhos – STR), to build socialism – which is said to be the equivalent of the Kingdom of God on earth; to teach the people from the pulpits when, what and how much to sow of their crops and to deliver to the government their quotas of produce, etc., etc.

The believers who attend church services – it should be mentioned that the Communists do not go to church – cannot suffer with indifference such profanation of the pulpit. In such instances, as soon as the preacher touches on such themes, "the worshippers demonstratively head for the doors and the priest has to either end up his sermon or talk to the walls." (Quoted from an article of Prof. Markovsky in *Tsurkoven vestnik*. STR). Such reaction could be observed in "St. Alexander Nevsky" and in some obscure villages in the Turnovo and Vratsa dioceses, as well as in churches all over the country.

This, of course, troubles the priests devoted to the church and the people and they find the ways to bypass the synodal and union instructions with the circular letters. But also there are others, who try to curry favor with the Communists and desecrate God's temple with their tasteless, impropre for the place political campaign. Thus, for instance, a priest from the district of Plovdiv, who is never sober and keeps company to every Party official in the village tavern, on the eve of Christmas had been drinking the whole night with a Communist company and taken off his cassock, thus allowing his drunken friends to mock the dignity and the honor of the priesthood. Standing with difficulty on his feet, he goes to the church to serve at the Holy table, and instead of delivering a sermon on the event, considered the holiest day of the year, he engages the worshipers with a talk on socialism, TKZS, et al. This man had taken the place of a priest who had been murdered in a horrible manner by the Communists on account of having been a worthy servant of Christ. This is how his successor wastes the rich spiritual fields which he had inherited, together with the negligible means of the church entrusted to him. He engages in morally reprehensible behavior and at every turn exposes the church to mockery. All this pleases the Communists and they tap him on the shoulder in approval. With such a priest, they would not need to carry on anti-religious campaigns.

The people are scandalized by such behavior of the priest and do not follow such "shepherds". When people saw the danger of attending worship in the churches, or realized that the churches are taken over by lackeys of the Communist government, they turned their own homes into a place for prayer, thus preserving their faith. The following case illustrates this point. On one occasion a teacher visits the home of one of his pupils, sort of an inspection, to find out if the child is taught religion. This takes place in Plovdiv. The teacher notices in the corner of the room a vigil light and an icon of the Mother of God. He turns to the eight-year old child, while her parents are flabbergasted, anticipating a true answer to the question:

"Do you know this woman?" He asks the kid pointing to the icon.

"This is a portrait of a peasant woman!" Answers quietly the girl, having understood already the mission of the education inspector.

The parents are relieved. The inspector concludes that the child had not been taught religious "deceptions". It all ends well. This same child had been kneeling every evening with her mother to say her prayers.

The people expressed their religious beliefs with faithfully observing their national customs and traditions closely related to their life in the church. The Communist cultural leaders are trying to kill the faith of the people and replace the national customs with a cult for the Bolshevik bloodsuckers, but the Orthodox faith gain strength every other day and emerges as a powerful force of resistance against the tyranny of the reds. It is regrettable that the leaders of the BOC are still irresolute to come out of the red terror. Years ago, an old teacher told this writer: "At least one metropolitan ought to have been hanged, so that the Bulgarian people know, that they too, have a church, and that this church is in the hands of worthy successors of our spiritual leaders of renaissance times."

We subscribe to the words of this old man. Not that we want to see somebody hanged, not that we want to trouble the peace of whoever he is, but because we think that if it necessary to offer sacrifices at the altar of freedom, the BOC should be in the first lines of the struggle, and that if there should be sacrifices, the first sacrifice should be the national shepherd, as it is sad: "The good shepherd gives his life for his sheep." The truth about the state of the church in Bulgaria under the bloody Bolshevik slavery should be told by the Holy Synod of the BOC. But this should be done while there is still time. It is possible that the 12th hour will ring soon. Is this ring going to find the BOC with its present attitude? It will be regrettable, it will be a crime before History and before the Bulgarian people if their spiritual leaders continue to keep silent – be it out of fear, be it for any other reasons. The days the civilized world is living through are crucial. The days the Bulgarian people are living through are tragic. And if the Orthodox Church does not raise its voice today in defense of the people, tomorrow its leaders may find themselves on trial, if not in a Court of Law, then on the bench of history's judgement and the latter does not forgive anybody. We hope that things will not go so far, that in not so distant future there will be an awakening, and the reputation of the BOC, greatly lost in the eyes of the nation and world public opinion, will be restored. We hope so! We, the Bulgarian political exiles, loyal sons of our mother Orthodox Church, are eagerly waiting to hear its voice in defense of *the truth, the justice and the freedom*. Let out hopes not be deceived. *

5. Excerpts of the Bulgarian Church Press

The following notes were taken from *Tsurkoven vestik* and *Naroden pastir*, respectively, publications of the Holy Synod and the Priests' Union – the issues available in the Library of the British Museum in November 1953. I have lost my Bulgarian notes, but sometimes, before that, I had translated them into English, which text I am using here. They clearly indicate what course the Bulgarian Orthodox Church has taken after the Panov-Yugov Reports and proposals, and what I personally, with all my previous experiences, had to cope with and define my position for the decades to come, beginning with my first essay, published on the preceding pages.

Tsurkoven vestnik, <u>No. 4-5, 1951.</u> Church and Society Chronicle. New By-Laws of our Holy Church: "On 31 December 1950 in the Holy Synod was received, approved by the government, the new By-Laws for the organizational structure of the Bulgarian Orthodox Church. Church life in our country now follows the organizational principles underlying the new By-Laws and in conformity with the needs and the demands of the times

New pro tem President of the Holy Synod. Following the stipulation of Art. 17 of the newly approved By-Laws for the organizational structure of the Bulgarian Orthodox Church, the Holy Synod, in plenary session, on the third day this month held an election of a *pro tem* President. His Eminence, the Metropolitan of Plovdiv Kiril, was unanimously elected.

<u>No. 6-7.</u> Announcement. The Union of Church Employees has held a meeting on occasion of the anniversary of Lenin's death where a Ms. Ilina Korakova, my classmate in the School of Theology, fired from her job in the public school system, known to me as a bitter anti-communist, has delivered a report on Lenin's life and work.

<u>No. 6-7.</u> Article by L.Ch. (Lazar Cholakov – STR) dismised from his job in some government agency, (at the time of writing this article a high level official in the Holy Synod) under the title of "The People's Economic Planning and the Church", writes: "The upper level government of the church (the Holy Synod – STR), with their orders and fatherly recommendations, has always been in support for a quick subscription of the loan intended for the development of our economy…"

<u>No. 6-7.</u> Article by Bishop Pimen (Vicar of the Sofia Diocese – STR): "After so long, it is now, under our People's Government, during the times of the socialist people's republic, that our Orthodox Church is given its Statutes which provide and permit a historical act to take place: the people's Orthodox church to change from Exarchate to Patriarchate. The principles of these Statutes are fully democratic. The second chapter stipu-

---

* This essay reveals my state of mind, my injured feelings at the time of my escape from Bulgaria – June-July 1951.

lates for some new institutions – the Patriarch, the People's Church Council, the Supreme Church Council, Diocesan Councils et al. The people's government of our republic played a positive and creative role.

No. 15-16. Article by M. (Prof. Ivan Markovsky, Editor – STR), under title: "Why they do not wait for the end of the Divine Liturgy", he notes the increase in attendance of the services, the good sermons which are sometimes delivered, but is critical of the length of them, and he adds in italics: "It is with reluctance that the worshippers listens to political speeches from the pulpit … We see how they walk out of the church as soon as the preacher appears…"

No. 15-16. The Holy Synod has issued a circular letter in support of the Berlin Peace Manifesto, and in general terms urges the priests to explain it to the people, so that everybody signs it.

No. 15-16. At a meeting of the Priest's Union chapter in Sofia, Metropolitan Kiril, *pro tem* President of the Holy Synod, discusses the same Manifesto and urges the priests to include themselves in the capmaign for collecting signatures under it.

No. 19-20. Arkhimandrite Yona, General Secretary of the Holy Synod, has addressed a meeting of the same Union on the subject of "The Meaning of the May Day as a Feast of Labor and the International Solidarity of the Working Classes."

No. 27-28. The Holy Synod has issued a circular letter on occasion of the harvest times, urging the peasants to gather it as quickly as possible.

No. 27-28. Article by M. (Markovsky). He goes on to say: "… How regrettable it is to see that some churches are seldom, and for a short time, open. The doors and the windows are not open widely. Fresh air and sun's rays do not penetrate there. Inside it smells of mould, the church courts are covered with wild weeds and not enclosed, some animals wander and graze around the church. There is no care for the buildings… It seems that this is not a church but wilderness… The last elections for parish councils shamed us. They proved that in many places, even in the capital, the connection between the spiritual leader and the believers is quite weak, and that the parish is not living, has lost its former meaning… Even *Naroden pastir* has written: "The 27th of May brought us disappointment. It proved that we are Generals without an army." …

No. 40-41. Arkhimandrite Stefan, publishes a poem under title "October":

> … Russia, … Mighty, great,
> Your sons stood up…
> They knew that Lenin is in Stalin
> That Stalin is calling them
> For the last battle
> Then October became a Spring time
> October became work and freedom
> The first month of the happy state
> Where the people is lord
> Because Comrade Stalin is in Kremlin
> And Stalin means Peace and Freedom.

No. 40-41. Metropolitan Kiril makes a speech profusely filled with Russophilia.

No. 44-45. Bishop Pimen writes an article on occasion of Stalin's birthday, under title: "72 Years".

"… The greatest genius and the greatest son of our epoch, rightfully named after him, is Joseph Stalin… Dealing with Marxist literature and revolutionary activity, he was imbued with the most profound faith in the revolutionary genius of Lenin and followed in his steps. Joseph Stalin gave to the world, in a clear and concise form, the fundamentals of dialectical and historical materialism, the work of a genius, developing them on the ground of the latest data of science and revolutionary experience… Joseph Stalin is great leader, irreplaceable teacher, great father, valuable friend, and a genius of a military leader."

Naroden Pastir. No. 19-29. "The elections for church councils before September 9, 1944, were a pure formality, automatic voting for the "proposed" ticket…"

No. 26-27. Rev. Nikola Zagarov, Editor of the paper, comments on the financial situation of the church, after the state subsidy is cut, and it was to depend only on the sale of candles: "If last year's sales of candles is used to determine this year's quota, while there has been a massovization of the TKZS (Collectivization of the peasant lands – STR), probably the sale would fall to half of that of the last year…"

Another priest writes in the same paper: "Let all Christians who consider themselves members of the church pay a membership fee… In such case, there will be no need for the existence of the "Christian brotherhoods" who do not exist in 18% of the churches, anyway…"

No. 26-27. A poem by a D. Hadjiysky:

> Truman's greedy, cruel, dark hords
> Are making their business
> With the blood of numberless victims
> But the fighters of new great regime
> Are mightier, because they are protected by Stalin
> The loved teacher and father.

No. 28. A priest writes: "The contemporary priest's clothes are a mockery for non-believers and confuse the believers. They are heavy burden for the priests… If someone disputes this he is from the past… His own children are ashamed and some abusers are openly "transmitting the pop" (a gesture of mocking the priest on the street, where a man would touch his private parts while smiling to other passers-by – STR). Why is it necessary for the Bulgarian priests to be exposed to such a shame?"

No. 28. Comments on the forthcoming elections of parish delegates to diocesan conventions: "It is absolutely necessary that the organized priesthood show particular vigilance and go out to work. It is of great importance as to WHO and with what kind of views will be elected. In no case should we allow the election of persons … alien to the spirit of our times or secret enemies of our people's government."

No. 28. Metropolitan Kiril on the lifting of the Schism of 1872, a statement: "The lifting of the schism was due to the intermediary of the Russian church, with its great love for us."

No. 29. Comments on the new Statutes by Stefan Landjev, one of the influential members of the Priest's Union: "The preservation of the Bishop's right to punish the priest with docking his salary for 15-days (Art.18), or binding the priest, as with chains, not to leave the parish without permission from the bishop (Art. 155) are incompatible with our times… Generally speaking, the priesthood is under the impression that what they have been fighting for in the past, has not found place in the new statutes. However, without trying to create disorders or anarchy in the church, it will continue the fight… (This is part of the Report of the Union's President).

No. 30. Praising the British fellow-traveler Hewlett Johnson: "… The workers in England, who have nothing to do with the faith, call him "Our Dean"".

No. 31. Comments: "The church press is completely free and is not censored in advance or after that. The Holy Synod has its own two weekly and monthly periodicals… The state supplies the church regularly with paper…"

No. 34-35. Comments on the forthcoming elections for delegates to diocesan conventions: "Let people, who are tempted by the thought (The Holy Synod – STR) that they could cancel some of the elections by some new amendment, so that they could arrange for a second round, keep in mind that such a procedure will only mobilize the organized priesthood, that in the future they will not include in the ticket those people who have now somehow managed to get in."

No. 36. Article under title: "The Awakening Role of the Church: The idea of the nonpolitical essence of the church has never been popular."

1952

Tsurkoven vestnik, No. 6-7. The Holy Synod has issued an appeal for collecting aid for North Korea. The appeal is in moderate language, but the article of Kiril of Plovdiv and Sofia speaks of "Imperialistic Invaders."

No. 10-12. Article by Bishop Pimen: "In this campaign for spring sowing, our parish priesthood has the best opportunity to show once more in the most effective way their desire to help the policies of our people's government not only in words, but also with deeds, and so demonstrate their full cooperation. The believers – in the churches and in private conversations – must be told how urgent and how great is the benefit for our nation from the full implementation of the plan for the spring sowing. Every showing in this sense on behalf of the priests will be rightly appreciated and will be useful for the general work of our church."

No. 15. Announcement that in the May Day demonstrations large number of Metropolitans, priests and church officials have participated, clergy and laymen.

No. 16. On occasion of the spring sowing, the Holy Synod has issued a circular letter, saying, among other things: "In the name of our Holy Orthodox Church and following the decision of the Holy Synod, we are fatherly inviting you to demonstrate in these difficult days of spring sowing all your zeal. … All your sense of good citizen of our fatherland and all your obedience as faithful children of our church should be shown, so that the plans for spring sowing be fulfilled and over-fulfilled."

No. 17-18. On April 3rd, Arkhimandrite Yona has delivered a speech on occasion of the execution of the Greek Communist leader in Athens, Beloyanis.

Naroden pastir, No. 6, 1952. Stara Zagora Priests' Union chapter in conference has considered: "…How much it was necessary for the people's priest to join in the spring deep plowing, sowing, gathering of the harvest and how important it was to deliver to the state as quickly as possible the grain."

No. 7-9. Comments on church elections: "… Many priests showed no interest as to how many voters would participate, and others, with criminal laziness, did not even hold elections… There are no other ways to explain this, but with lack of political or church information and ignorance of the political and church situation in our country … (also) as cleverly taking advantage and tactically using the international situation. The full freedom which the people's government assured them to prepare and conduct the elections, the reactionary priests, many of whom occupy important positions in the church, demonstrated unbelievable audacity and extraordinary impudence in their desire to be included at any rate in the ballots."

No. 7-9. Editorial by Zagarov. "... And now, the pseudo-Christians from the other side of the ocean, will meet the true crusaders – the servants of the Orthodox church, standing at vigilant guard for peace and good will among the people..."

No. 7-9. Comments on the new By-Laws by an influential member of the Priests' Union, Stefan Landjev: "The elections of 27th May for parish councils and of 30th September, have been carried under the provisions of statutes, as conservative and reactionary as they are."

No. 38-39. Editorial: "The Dimitrov's Constitution and the Church." "After the separation of our church from the state, it was completely liberated from its dependence on the government. But this separation does not mean that the church becomes a stranger to the interests and the ideals of the state, because in the people's democracy state, people and government are completely identical. The church, being a free and independent institution, in fact, is an excellent collaborator of the government."

IX

TRUDOVAK I
BEZMER – THE FORGOTTEN
MANUSCRIPT

Vae Victis!

ТРѸДОВАК

Въведение.   23.I.1951 г.

Като започвам да описвам стра-
данията си в бѣлянското каторга, аз се
страхувам, че нѣ ще мога да доведа до
край започнатото дѣло. Освен туй Несигурния
ми да разчиташ и опасностьта тази тетрадка
всѣки моментъ могла да не попадне да
всѣки моментъ могла да не попадне да

забравя мислѣнте. Все пакъ, за собствено у-
покоение ще се помъча да разкажа нѣщо.
Ако нѣкога времената се променятъ и стане
възможно такива нѣща да видятъ бѣлия свѣтъ
то мнозина ще се чудятъ на дързостьта ми
Нека времената се променятъ, нека новите
хора отъ далечната бѫдеще никога да не
преживеятъ туй което преживѣхме ние, нека

1. A Message from Way Back. The Forgotten Manuscript

In the beginning of December, 1989, one month after the collapse of the Communist regime in Bulgaria, I received a package from Sweden. I was a little puzzled. I expected nothing from Sweden. I knew nobody in that country. The package was sent by Katherine Ismail. The Moslem surname further intrigued me. I opened the package and saw inside an old, grey notebook of some 100 sheets. It still did not tell me anything and for a few seconds I was at a loss as to what was this all about. The mystery was solved instantly when I turned the hard cover to the first page. I immediately recognized my handwriting. On the top of the page was the date: January 29, 1951. I at once remembered what was there. I had completely forgotten that at the end of January, and during the month of February, 1951, I began writing about my sufferings as a Trudovak in the military camp at the village of Bezmer, near Yambol. I had filled some 177 pages, and had stopped, before completing the story. What follows in the next pages is an exact copy of these notes. They reflected my state of mind and my experiences in Bezmer. Even though here and there I do not like the style and the expressions, in some places it has much to be desired, I changed nothing. It is a document of my martyrdon during this difficult period of my life. It explains my further actions to escape from Bulgaria and I will reproduce it exactly as I find it.

January 29, 1951

As I begin to describe my sufferings in the Bezmer penal servitude prison, I am afraid that I will not be able to finish my notes. I am also aware that my style of writing is not up to the necessary standards and there is always the danger that this notebook may fall in some enemy hands, so that all that may cause me to abandon the project at any time. Still, for my own relief, letting some steam off, I will make a few remarks. If times change and my notes become known, some people may wonder of my daring to put these things on paper. But, let it be! Let the times change, let the new people in the distant future know how we have lived through this ordeal and let them wonder at our daring and our naiveté. We are people destined to live an unhappy life. Our sufferings are so big, that we have by now lost our ideas of a true, of a real human life.

We have become robots, machines, nothing more than speechless animals which suffer but are unable to express the pain of their suffering. One could read it only in their eyes. Only, that the animals could find sympathy in us when dying under the burdens forced on them, but we, the inmates of this penal institution, of this hell at Bezmer, not only that we find no sympathy, but, it appears, are deliberately being pushed to our death.

...

Friends whispered to me during my residence in the Bezmer camp, that a record should be left for the inhumanity of our suffering, that the future generations should know of the ordeal of their predecessors, which had led to their liberation, to their freedom. Let them know that their freedom has grown out of our blood and our bones scattered all over the face of our beautiful land. Let their freedom be our reward, for our suffering.

...

The story which I am going to tell in the following pages is a tale of suffering of the Trudovaks at Camp Bezmer, near Yambol. The Communists often refered to them as "gads" (loathsome creature). I was a "gad". I should note here, that I was not offended by this appellation. I consider it to be a badge of honor. I am proud of it. I am not ashamed of it. Not that I have been who knows what, some glorious resistance leader or fighter. I was an ordinary man and have nothing specific to claim as political credit. But I was branded as a gad!...

...

2. Facing the Draft Board

It was not for the first time that I was appearing before the draft board. The procedure was boring and annoying me. Not because I had to stand there nude, in full view of the officers and expose my skinny body before too many people, who, if one had to judge them according to their ranks, their smoothly shaven faces and well ironed uniforms, were all men with education, i.e. men of the intelligentsia. For many draftees this was the first time that they stood in the nude before such a select public, and in confusion covered their private parts with hands, to which the officers laughed. What annoyed me was that I had many times already passed through this procedure since I had come to be of draft age, and I felt that as if it was yesterday that I was here, and now I am going into it again. But there was some difference between the earlier sessions, and this one. The fascists of before were not too much concerned about me, being a seminarian. They did not hold me in the chalk's circle. Now that I was appearing here with the certainty that I had to be drafted, knowing that my dossier was very colorful, I felt uncomfortable. Not that I was afraid. When a man knows his fate, he is not scared, but he is overwhelmed by a feeling of annoyance. He would want to see this charade finished as soon as possible, and then, whatever will be, will be.

So, my turn came and I stood in the chalked circle before the Board. There was a Captain chairing the proceedings, and some lieutenants and sub-lieutenants. I noticed how a small man, black like a devil, was staring at me with evil eyes. I learned later that he was the political officer. He looked like he wanted to penetrate in my soul with his small, sparkling beady eyes. Next to him was an evil small man, sent there from the communist authorities in my village. It seems to me, it was this little man who played the most important role in the determination of where I was going to serve. He had blue suit and a red tie, all quite crumpled. His face, yellow, made him look like a true Quasimodo with his twisted evil mouth, long bone-like ears and sparse reddish beard. In height he was practically a dwarf, a hunchback at that. On a few occasions I had met him in debating public issues where our views clashed – he was a leftist, though he had come from a family of reactionaries, and I considered him a rightist. Somehow this man emerged as the closest advisor to the Communist authorities. Now, as I stepped into the chalk's circle, he leaned to the black lieutenant, and without moving his eyes away from me whispered in his ears for quite a while.

I realized that my fate was being decided now. The black lieutenant noded his head from time to time, without saying anything. The captain was entering the physical data marked by the doctor on my chest. For a few minutes I stood motionless. It seemed to me as an eternity. I sensed how under my left armpit a stream of cold sweat was running down over my ribs. The political officer ended the long pause of silence. He asked me:

"How much land do you have?"

"Four acres", I responded with trembling voice.

"Are you a member of the C.N.M. (Union of the People's Youth)?"

"Yes, I am a member!" I answered quickly, struggling to maintain my composure and the firmness of my voice.

"Since when are you a member there?" he continued the interrogation.

"Since 1945."

"There was no CNM at that time," shot he back.

"There was O.S.N.S." (National Union of Students), I answered him and added: "Subsequently O.S.N.S. became part of CNM."

At this time the little man from my village leaned to the political officer, whispered something in his ear, and he stopped asking me questions. The Captain, Chairman of the Board, turned then to me:

"What school you have graduated from?"

"The School of Theology," I answered his question.

"Theology?" wondered he, "Such a progressive region, and it had produced people who study to be pops (priests)."

I kept silent and looked him in the eyes. He thought for a minute and then turned to the political officer. The latter hurriedly dismissed me and I left the hall. After that I learned from my friend Todor Vassilev, who was behind me, the Captain had just said:

"The same, and" dismissed him.

3. Gathering of the Damned

After all draftees were seen by the Board, an officer came out and distributed the draft assignments – not including me. Apparently, they were to be considered further. When I received the draft summons, it was just designated: "Command 101". Nothing more. Soon I was to learn that "Command 101" was the assignment for trudovaks. After all that, I rushed to Sofia, to wait for orders. The order was sent to my village. My relatives informed me and on March 31, I returned to Zelenikovo. On April 5th, I had to leave for a designated address in Plovdiv – in the Jewish cultural center. When I entered the big hall half of the chairs were already occupied. All new draftees of the Plovdiv district had to gather there. So, quite a few had arrived before me. I thought that they were anxious to go and serve. Up front a young Captain was explaining things about the service, stressing the importance of discipline. The eyes of all were fixed on him so hard that a weak man could hardly stand it. The young man felt that each word was a nail on their skin. The new draftees were meeting a military chief for the first time and looked as if they wanted to guess instantly what their future fate was, as if he held it in his hands. No one dared move. A buzz of a fly would have been heard. The tall captain was walking back and forth from one end to the other and back in front of the lined up chairs. Trying to give to his movements greater importance and poise, as well as greater firmness and commanding tone. At times he seemed to display some human intimacy to be liked by the draftees. At the same time they were making every effort to understand what was hiding behind every word uttered by him, I felt despise for him, thinking that all this was an act, and at times even felt sorry for him, that he had to go through all that in performing his boring task.

I watched the young men, felt that they were as unhappy as I was, and saw themselves as prisoners, once they had entered the hall. It was difficult to read anything in their faces immediately, but somehow I saw them as victims of misfortune. In their eyes, on their faces and their movements, as well as their clothes, I saw human misery. It seemed to me that most of them had not been away from their families before, had not been separated from mothers, fathers, brothers and sisters, and, in general, had not been out of their villages. Most of them followed closely their more sophisticated friends who have been around. There were some who obviously despised the captain, ignored those around them, and preferred to isolate themselves from all that was going on around them and live with their own memories. They were preoccupied mostly with the thought of continuing to their destination, impatiently waiting to get it all on the road, rather than the speech of the Captain. For some reason the departure was being postponed, some village contingents of draftees were not yet in, and we had to wait for all to report to the hall.

I was observing the appearance of the new trudovaks, my brothers in fate, and I somehow experienced moments of fright and disappointment. It seemed to me that they all had intentionally chosen to wear clothes that have been worn out, even raggedy, so the picture sometimes was quite amusing. In an overall view, the group looked like a mob of bums who until this morning had been sleeping under bridges or had been loading railroad cars, and today had gathered here for some strange decision of some fate. Here and there intelligent faces were hiding under their rags. Unlike this mob, I had dared to wear my best clothes – not conspicuous, which I could not afford, but decent enough to make the difference. I had put on a tie, over a nylon shirt, and a sports short windbreaker, even if it looked quite worn out. In my small pocket on the chest I had stuck my pen. So the main attributes of an educated person were becoming immediately visible. The Captain apparently noticed that, because during a break he came closer and sat next to me, apparently hoping for a conversation. But I was not feeling disposed to talk to him.

My whole being was revolting against all that. I felt that my life was being raped and a storm of indignation was boiling in me. I felt like a fish which was being pulled out of the water and thrown on the sand. My yesterdays were in such a contrast with my today, I thought of my years spent over books, in libraries and auditoriums, and I felt humiliated by what I was watching in front of me, and contemplating what was ahead of me. I felt that I was going to burst out of my skin... This captain, who sat next to me, my conclusion was, was a full ignoramus, who was to decide, or already had decided my fate. I felt that that fate was something frightening and all the reasons which had brought me here, were now represented by this captain. I knew why I was assigned to serve as a trudovak. I thought that he too knew the reasons. He must have known them, in every detail. Then why I had to keep him company when both of us knew that we were divided by a precipice? I despised him. My feelings were boiling inside of me and from instant to instant were becoming intolerable. I got up and moved away.

I was trying to penetrate behind the faces of the draftees, to discover some face in this mob which would inspire in me confidence and I could share with him my mood, my disgust and my forebodings. But nobody seemed to be ready to allow himself the luxury of revealing his soul, and the storm of indignation raging in his mind. No one could be sure that the person he would venture to pour his pain to, does not have some hidden intentions to betray you. The suspicion which had frozen society outside of the military barracks, that the authorities had planted stool pigeons all over to spy on everybody, was not absent here, among those who were doomed to slave labor camps, on account of their opposition to the Communist power.

All these men were being herded to slave labor camps on account of their participation in activities in defense of freedom. They have been honest in expressing openly their convictions and shown determination in standing for their ideas before their torturers. They all despised the reigning tyranny, slave philosophy and policies of the regime. It was this secret message which was imprinted deep on the minds of all those surrounding me. These unfortunate men, including me, were not prepared to discuss their fate with anybody they did not know. They were guided by the general principle of the time: when you close your door behind you, you should close your mouth. But this could not last forever, and sooner or later, it was all going to break down. Little by little all these men were going to learn about each other, and were to share their thoughts. But now? Now they all were with closed mouths. It was difficult to breath in this heavy atmosphere, filled with poisonous gases. The waiting was becoming more and more boring. Noon time passed, and there was no sign that we were going to move. Every half hour the Captain was checking the presence of all and marked those who had not arrived yet. All the time someone was absent. In some cases entire village contingents had not yet arrived.

Around noon some draftees started opening their suitcases and taking out the packages of food their mothers had stuffed in. All others followed, and small groups of young men from the same villages were devouring their supplies. It appeared that everyone had found some chicken to bring along. Gypsy boys were munching on their broiled geese drum-stricks. I found it more amusing to watch this mass of unfortunate kids than to sit and eat. I had no appetite. Still I managed to have a few bites and closed my suitcase. Somehow the lunch hour was passing without noise or talks. From time to time someone would ask a question, would get a short answer and the munching continued. One had the impression that this was a memorial dinner for a dead relative rather than a draftee's banquet. Many were still sporting carefully arranged flowers on their lapels. Still, there was no longer the joy and the excitement of their parting away from their loved ones sending them to the service.

For all practical purposes, we were already prisoners. The draftees were admitted in the hall upon their arrival, but were not allowed to leave. The door of the Jewish *chitalishte* was guarded by sub-lieutenant Angel. He was not an Angel. On our way out of Plovdiv we recognized him to be a true devil. During the day Angel stood at the door and would admit all arrivals in, but would not let anybody out. Very soon he won his title: "the Cerber". But it appeared that he was not beyond human frailties, and if someone of the draftees showed him discretely a bottle, he would walk outside, would take a decent gulp, or two, and let the generous young men go for a walk. After having his initiation of the bottle, Angel would return to the room and take a nasty look at the rest of the crowd, as if to say: "I will teach you, S.O. Bs. what the service is all about." Then he would sneak close to some group where some delicacies, like chicken legs or breasts were still visible, would help himself without invitation, revealing them the mysteries of the military service, and by the evening when we were to depart, he was the devil himself. He was a drunk cossack. Short, reddish, with a naïve, quite stupid look in his eyes, he was a sorry picture, a mockery of the service and a bit of a joke. In the beginning he appeared to be trying to create a positive impression for the service which he was performing, but later on his figure evoked only disgust, and, in some instances, compassion. This Angel was playing the role of a clown, but in his own mind he was an upright representative of the service, he was proud of it and trying to demonstrate that he knew it well, only pretending for those who he thought knew nothing of it.

About four o'clock something began happening. A call was issued that everybody had to take a place in the hall. The Captain started making a last check up of those who had reported for service. He was not alone. Next to him was the black lieutenant whom I had met at the Draft Board, and a small sub-lieutenant who was holding in his hands the list of the draftees and was following the calls, marking those who did not answer. It was established that all contingents from the villages were here, except here and there some individuals being absent. After the checking was over, the Captain let the draftees ask questions. Some inconsequential questions were asked and answered. The Captain was followed on the podium by the black lieutenant. His figure radiated evil and despise. His skinny face, his piercing look in the eyes, his motionless body – all this made us feel deep resentment of him and fear his threats. He started his speech. Every word, every phrase, every syllable were emphasized in a manner which made us feel as if he wanted to nail us on the wall. His entire speech seemed to be reduced to one sentence: "Do not bring into the service the baggage which you carry in your heads, leave it outside the door, outside the threshold. Otherwise, this service will be very difficult for you." He chewed this phrase for quarter of an hour. His intelligence was revealed with all its shallowness, and instead of evoking some serious attitude with this speech, he provoked a hidden laugh in many of us. Throughout our travel during the night, someone, somewhere in the darkness would imitate him loudly, for the delight of all the rest.

After the black lieutenant spoke, the captain told us that he was leaving us to the small subleutenant who was going to take us to our assigned place of service. The more this little man tried to inspire in us respect and fearful submission, the more laughable he was. A little over five feet and two inches tall, on the skinny side, large blue eyes, arrogant look and half open mouth he could any moment, it appeared to me, erupt in imposing penalty or scolding. Before arriving in Plovdiv we thought of the military officers as a sort of deities. Now, after meeting them, we saw them to be small little men, with not much intelligence in their brains, or elementary human dignity, to win our respect. If it was not for this ironclad fixed idea, rather an idiotic imagination that the standing of the military leader does not depend on his personality, but on the office which he holds, then the discipline in the armed forces would have evaporated instantly.

Subleutenant Yotov – this was the name of our new chief – attempted to introduce us with some incoherent words to what was ahead of us on the way to our assignment and to stress the importance of discipline; to utter some threats addressed to nobody specifically and to impress on us that we were already formally in the military service, consequently, subject to penalties stipulated by the laws of the armed forces for violation of orders. Following this bland ceremony, we were asked to pick our suitcases and go out on the street. We were then lined three by three, and, led by a military band, marched to the Central Railroad Station.

I could not get out of my mind the word: "Poor young men". This word was stuck under my tongue, looking at life going on outside of our ranks, on the street, where pedestrians were turning and looking at us as some sort of exotic animals. I could not also get out of my mind some dark premonitions of my future fate in this herd of wild animals. We arrived at the railroad station too early and had to wait for a while for a special train. To kill the boredom, the band started playing folks dances. Few however felt like jumping out there for a *horo*. The officers had to give an example. The captain was there and seemingly wanted to display great excitement with his dancing, to encourage the ordinary men to join. He could not realize, or if he had he skillfully avoided showing it, that all this was a farce. He was acting the role of the Party official, who, in tune with the whole communist perception of reality, had to perform an act to serve the interests of the party, that the party and its interests came first, while his individual dispositions were to be concealed. Men look first to their own interests, their own wishes and desires, their own pleasures and displeasures, and only then to the social commitments and requirements.

This captain was clowning there for the benefits of his office, to distract the attention of the bewildered young men, who were wondering where were they going and what expected them there. He was concealing his own feelings when trying to drag his captive audience, which he took to be a peasant group of ignoramuses to his *horo*. He ultimately failed, because, besides a few officers and men from the services managing the transfer, no trudovaks joined in. His efforts were in vain. The mass of the young men just looked blandly, observing the comedy, and stayed out.

I should observe here that neither in our group of new draftees, nor in any of the camps where I served – Bezmer and Balchik, each one of over two thousands trudovaks – I ever met young men from the big cities: Sofia, Plovdiv or any of the district capitals, with very few exceptions. All trudovaks seemed to be coming from the villages and small provincial towns.

4. A Journey to Hell

About 6.00 p.m. we were lined up along three freight cars. What an illusion had we entertained expecting to travel in passenger trains! Our disappointment set in. It became clear that we were reduced from human beings to ordinary cattle. Orders were given for us to climb in these wagons, gave us last orders, and prepared us for the night journey. The captain and the black lieutenants left the station and our new chief, sub-lieutenant Yotov, took over. Yotov's right hand for the journey was our Cerber from the Jewish *chitalishte*, Angel. But they could not be everywhere, and for the first time during this long day, we felt a little freer. The boredom of the day was behind us. The travel always creates a better mood. Little by little the draftees were coming to know each other and from little groups, clearing some basics – where each one was from, whom they knew in that or that village, what was their occupation, families, education and all this small talk stuff, turned the draftees into a friendly crowd.

In no time bottles of liqueur were pulled out, *rakia* and *mastica*, wine and appetizers and all that makes for a good company. By the time the train was miles out of Plovdiv, the ice was broken and the general mood was picking up steam. In time, the group of Goliamo

Konare[*], which appeared to be the largest contingent, occupying the front part of the cattle wagon, led in the gay mood. In the back part of the car were more people from Dulgo Pole. Their leader was a young man, who had studied or graduated the School of Agronomy in Sofia. I had managed to get acquainted with him during the day. He was first giving impression of a very serious, no-nonsense man, but it soon became clear that he was an entirely different character. Very soon he began to mock everything that the military officers during the day had said, giving the impression that he was in full agreement with them, but then turning to their clichés, he evoked loud laughter in the entire wagon. His skill to repeat all that the officers had said with all seriousness, and then turn it into a farce, could not but please all young men, who had been subjected to stupidities, but suppressed their reactions during the day.

When we passed Philipovo Railroad station, he fell on his knees, grabbed his face in his palms and screamed hysterically, wailing as if it was his mother who was crying for him. The whole car was shaking in uncontrolable laughs. Then he started singing, and sang with gusto, with unconcealed irony and mockery, all communist songs, giving special attention to songs that the opposition had used to mock the communists, like "Ot kak doide ofeto, ogolia mi dupeto" (Since the Fatherland Front came in, my derriere lost its pants,) or "Otishla e Tsola, pri Gemeto Gola" – (Tsola – Top leader of the Communist Party, a woman – had gone to Gemeto, the exile leader of the Agrarians, in the nude). He went through his whole repertory and turned our car into a theater of comedians. All this was followed by reopening the suitcases, another attack on the food provisions and the liqueur, and low voice conversations where little by little one was coming to know his companions.

Very soon, songs were sounding on all sides. The Goliamo Konare group was trying to outdo the Dulgo Pole Group, led by our agronom-comedian Chakandrakov. That group was joined by Angel, our angel-guardian, who apparently enjoyed the food and the drinks, and did not much mind the irony and the mockery which Chakandrakov was making of the officers of the previous day. He never stopped assuring his generous hosts that for as long as he was with them, they did not have anything to fear from the service, he was going to "fix it" for them, i.e. arrange for them lighter assignments. Some of them apparently believed him. Everybody thought of getting out of the ringer. Those who were further from Angel, kind of envied those who had befriended him. At one point the good mood had reached a climactic point. As the train was crossing the Thracian plane in full speed, the cattle wagons were filled with deafening screams, songs, speeches. We were able to glance instantly through the open door the fleeting sings of small rail stations along villages in glittering electric lights.

But our good mood did not escape the attention of sub-lieutenant Yotov. Having heard the big commotion at one of the stations where we had stopped briefly, he climbed in our car, helped by a few draftees. He began to make a speech, demanding greater decency and discipline, even threatening that he could fix us in such a way, that we would remember him all our life. He allowed himself to appear mad at us, losing sight of it that the more he acted this way, the more laughable he became in his inability to control his tongue. We were dead silent. But after he finished his admonitions and tried to get out of the car, our Angel, well loaded on liqueur and stumbling, approached him with a bottle of *rakia* (plum brandy). There still was a quarter of the bottle. He could hardly say, his words blurred by drunkenness:

[*] Town of *Saedinenie*.

"Comrade sub-lieutenant …"

Yotov turned to him in a fearful pose.

"Comrade Sub-lieutenant," continued Angel, "Wouldn't you like to take a little drink? The young are treating…"

The sub-lieutenant said nothing. He grabbed the bottle and poured it on the ground reaching out of the door. Then, turning to Angel, he screamed:

"Pig!"

At this time the blow of the station master's whistle made the sub-leutenant jump and go to the other car, which too, was radiating exuberant sounds of gaiety. Angel returned to his companions, murmuring something under his nose, and then exploded in his own speech:

"Pig! I am a pig? He is a pig! If I am a pig, why is he taking me with him? If I am a pig, he is a liliputian… No, he is an idiot. Hey, you listen, don't anyone dare tell him what I think of him! He who would do that should think of what may happen to himself later. He will remember who is this old man Angel, the trudovak… I know… nobody will report me… I know, nobody."

The draftees, as soon as the sub-lieutenant had disappeared, exploded in laughter - not at Angel, but at the officer. Someone told the story how a bridegroom, after taking his bride home, picked the cat and tore it apart just to show to his future wife who is the boss. Yotov, apparently, wanted to make such an impression. Naïve simpleton! He did not realize that the effect was the opposite. Well, some wisecracks were biting their lips and instantly started imitating the officer, and all had a hearty laugh. At one Railway station our car was visited by a candyman, who continued with us trying to sell his goods. As the train pulled out, and nobody was buying his candy, someone of the Goliamo Konare group grabbed his hat and ceremoniously threw it out through the open door. The candyman hysterically screamed:

"Who took my hat?"

As the draftees watched and laughed, following the hat which landed in the fields, the candyman was falling in despair and threatening that this was a country of laws, and that every stupid man will fall in his place. Someone yelled from behind:

"If this is a country of laws and every stupid man will fall in his place, the Communists would have been out for long time, and I would not be going now to serve as a trudovak". Soon everybody got into the act, some reached for his candies and in no time nothing was left for him to sell. The candyman realized his helplessness and sat down in a corner waiting for the next station. As soon as the train stopped, he jumped out, and there came in a minute the sub-lieutenant, green in madness. But it was not noticed in the darkness. There followed a canonade of insults, of dirty words, of swearing, of shouts, and we froze in our corners. He asked for the individual who had taken the hat of the candyman. Did he really expect an answer? I doubt it. He asked also who had taken part in appropriating the candies. No one answered. When he saw that nothing was happening he appeared to be begging, spoke of morality, of human compassion, only to flare up again and start with his abuses and vituperation with frenzy threats for applying penalties the next day. No one dared move. At one point Angel felt that he had to enter the stage as man responsible for order in the car and murmured something. Now the sub-lieutenant discovered whom to blame. He turned to Angel with a vicious attack. He regretted that he had selected him to take along to help him, when, in fact, he was obstructing his command duties. At this point Yotov blew out the little pebble in his mouth: he called Angel with a name, which was reserved for the worst political offenders: "Gad" (loathsome creature).

"Gad and gad! Saboteur!" He went on. "I will report you immediately tomorrow. I will fix you well! Animal! Pig!"

Meanwhile the train had started again for some time, the candyman was left behind and we appeared to be approaching our final place of destination. At the next station the sub-lieutenant, after having pored all his venom on us left our car. Then Angel rose in his turn. He walked over to the group of Goliamo Konary and tried to get information who had picked the hat of the candyman and his candies. But soon the old mood returned, the trudovaks started laughing at him and he was left to talk to himself.

"Comrades", with slurring speech he went on, "This is not a joke. He will report me. I am finished! Listen, if you stand by me, I will stand by you. You look to me with one eye, so I would look for you with two. Mountain does not go to mountain, but man and man meet again. It all depends on you. You stand by me, so I could stand by you."

I do not remember how many times he repeated this phrase. As I recall, he was forced to sit down, because he had began crying, drunk as he was. Some were laughing off the whole incident. At some point we all were very tired and everybody tried to snuggle in some corner or in the middle on the floor and try to catch a little sleep. Little by little, it was all quiet, except for some snoring here and there. At the same time the train was flying across the plain with full speed. Out of the car one could notice the small stations left behind. Some of us were fighting the sleep and shared their impressions with each other. Some were absorbed in troubling thoughts about what was coming to them in the immediate future. For some time I was fighting the sleep, but nature took its course and eventually I fell asleep too.

Sometimes after midnight we were awakened by the shouts of Yotov. He was ordering all to get out of the cars as quickly as we could. Half asleep we managed to get our things and jump down through the door. Standing for a minute while the rest of draftees were coming down, I looked at the small Railstation and noticed a blue glazed sign, where in white letters was inscribed: *BEZMER*!

"Bezmer! Bezmer! A funeral chamber of a thousand dreams!"

5. Arriving at the Hell Called "Bezmer"

An order was given that we lined up three by three. Led by Angel, the column moved across the rails and, as I could determine by watching the stars, was leading to the South. I was under the impression that we were walking down an endless plain, or better, through uncultivated fields. Later I learned that this was an unused road, but I could not recognize it in the darkness of night. There were too many pot holes, too much mud and puddles, and I could not see where I was stepping. We had hardly walked some five hundred feet, and the suitcase handle of my neighbor from Zelenikovo, Todor Vasilev, broke, and so did one of the hinges. My suitcase was tied with an extra rope for greater security. So we had to stop and fix it with my rope. The column was not moving very fast and we managed to rejoin the march. Though we could not see much around us, it did not look like an attractive place. Dark forebodings crossed my mind. As if to make our feelings worse, our friend mentioned before, Chakandrakov, was wailing with his already recognizable voice:

"Olele male (My dear mother), this is where I am going to die... Mari bozhke (My dear God), who has made this Diarbekir" (A place in Turkey where Bulgarian revolutionary leaders were exiled in the 19th century). He carried on and on like woman wailing over our heads. He was not being taken seriously by the officers. By now they were familiar with his style and his sense of humor. Terrible! Deadly terrible! Such were our first impression even before arriving in the camp, having just stepped foot in the land of Bezmer.

One way or the other, we dragged ourselves to the camp. In a matter of minutes we were pushed into three or four half finished barracks where we jumped on the long plank-beds, covered with straw over which they had spread sack-cloth. Soon we were sleeping, tired of the arduous day experiences, with foggy thoughts about our future. When we awoke, day break had long brought light over our camp. Day! Day First! Here and there someone would murmur something meant to be funny, but everybody was somehow keeping silent, waiting for orders. Such orders were given and we had to get out of the barracks. What we saw around us and in the distance in every direction, was not at all pleasing. One could not focus on anything that was something. We were in the middle of nowhere, of a plain, extending from East to West. We were at its Eastern end. To the west it was merging with the blue skies over a long stretch of meadow, in the morning mist. We did not see any tree, or structures, or whatever of note in all directions.

The plain was covered with fresh growing grass over the muddy ground. The morning chill was piercing our light clothes and at one point I started shaking, gnawing my teeth and jumping up and down to warm up. There was no immediate call for any activities, and we just mulled around, looked at the endless plain, a *polyana*, and wondered about it. It was a field as long as one could see. And as we were walking around with heavy accumulation of mud sticking to our shoes, heavy as a ton, somebody commented: "Can you imagine how hot it may be here during the summer?" About 150 feet from the barracks, there was some sort of a ravine, as if it was a dry little stream. It looked like a long row of deep holes and seemingly connected in a long chain, as it had been an uneven aqueduct. It was inclining to the East. In a few weeks, a Sunday, we were led to it with picks and shovels to turn it into a canal. During the summer, running in the dark to the water pump one night, I did not see the holes there, and fell to the bottom, some five or six feet. Somehow I fell on my feet, but the experience for a moment shook me to no end. I was so frustrated, almost resigning from life and cursed the service and all those who had sent me here with all the venom which I could muster. Behind these holes, we had turned into a canal, in a large ravine which seemed to have been specially designed as a mud pool, few days after our arrival, some old trudovaks started putting together some sort of a structure. Then they started pulling water from an old well, suspending some two-gallon aluminum pots, attached on a rope, then pulling them back and filling the big cauldron where the food was prepared.

It was a chaotic operation which we followed from the side. Dirty water was poured in these enormous vessels, beans were emptied straight from big sacks to be boiled and, under them, resting on big stones, wood was pushed as a fuel. In the beginning, actually the first day of our arrival there, we noticed this primitive arrangement of our food preparation. All along, the trudovaks who were in charge of this culinary operation slopped around in mud, yelled, screamed and cursed each other. I watched this whole thing, and felt that I had fallen into a pit of some hell. When we were thirsty, we had to slop through the mud to reach the same well and help ourselves the best way we could. Fortunately, a few weeks after our arrival the well was covered and some sort of an old fashion pump was installed, the kind where you push a handle down and water would come up the pipe. Until this pump was installed, we had to see the dirt running down the well from outside, remnants of food often falling in, and mud sliding down. It was disgusting, but we were reduced to animals any way. In a few weeks things changed, the "Kitchen" was finished, the mud was reduced and a semblence of order was established – except for the pump, which used to break down all the time and we had to patiently wait until it was fixed. Otherwise, the mud around it never disappeared and going there was always an unpleasant experience.

As we arrived here, after a brief sleep, we looked the place over and became reconciled, rather we saw no other way, but to accept this to be the place where we were going to spend the summer. We silently remembered our beautiful Thracian plain around Plovdiv, between the Rodopi and Sriadna Gora mountains, luciously green, cultivated and loaded with fruit, vegetable, vegetation, people, stately poplar trees, large flocks of sheep, in general, bubbling with life. There was nothing of the kind in this God forsaken land. Only the railroad, some one kilometer to the North, where trains were running one after another in both directions, showed us that we were still living in the civilized world. On June 25, when the war in Korea started, and trains loaded with guns were passing, one after another, our clown Chakandrakov made the comment: "The train of peace is passing". It was moving from East, Burgas, to the West, God knows where!

The first day of our arrival in this wilderness, the sun did not show up. It was a cloudy and murky day – to make our forebodings even darker. One hour after we had been messing with the mud fronting the barracks, an order was given to line up. A table was set in at one end, some officer or an old trudovak sat on a chair, another next to him to help him, and we had to march there – to have our names signed in and to receive a loaf of bread – I believe 750 grams. This was to be our daily ration. After that, another order was given to us to grab all our suitcases, to line up and follow the leader. We were led to some temporary structure, some half kilometer westward, which, we were told, was the command headquarters of the camp. We had to go through a medical check up. I was pulled aside by an officer – he had noticed my pen clipped in my shirt pocket, and led to a better looking barrack There were some five or six trudovaks like me. We were instructed to help those illiterate to fill out their health forms. And then followed the same procedure that I described at the Draft Board: undressing completely, measuring, weighing, questioning and probing. We, who were to help the illiterates, passed first. Then followed the rest. A kid from my village, nicknamed Docho Uzunov, named after a prominent criminal at large, but half blind, slightly retarded, and illiterate, was brought in the circle and I overheard the conversation between him and the Captain:

"Why they call you Docho Uzunov, and not by your name Ivan Kurtev?"

"Because my Grand Father is a fascist. He gave me only 159 levs before leaving". I looked at him. His body was small, somehow misshapen, dirty as if he had not seen water since his mother bathed him as a baby, squinting with his small eyes, slightly curved legs… "Why had they taken him here?" I asked myself. But they had no mercy. His family was one of the prominent people in the village, with the biggest house on Central square, but he and his father and mother lived in a small room, in squalor. Once, I do not remember what was the occasion, I had seen it. It was never erased from my mind. Now

I was to share with Ivancho – this is how I called him, diminutive of Ivan – my service. On occasions he asked me to write his letters to his family in the village, which I, with the deepest sympathy for his misfortune, readily did. We served together the next year in Balchick. I will have more to say there about him.

When I had appeared before the Commission, I was asked about my profession. I answered: "Teacher". One lieutenant commented: "We will need teachers here". They gave me a form with my name, marked on it "Company 309" and let me go. Now, I was formally inducted into the Trudovak Service, "Company 309". There were five of us from Zelenikovo. The others were my neighbor and old budy Todor Vassilev, who sometimes in 1945 began avoiding my company, to demonstrate that he was cutting his ties from the reactionary past, Atanas Dimitrov Manevsky – his father was one of the six people murdered on October 6, 1944, Ivancho, a boy Ivan Gurdev, and me. I alone was in company 309, which subsequently proved to be assigned the worst work – the transportation company – loading and unloading materials – cement, sand, gravel, beams and whatever would come to camp – by rail or by trucks. I will have to talk about it all later.

6. An Army of Ragamuffins

Our assignment did not end with the clearing of our Plovdiv contingent. Large groups of new draftees had arrived from Haskovo, Svilengrad, Harmanly and Northern Bulgaria. Those of us who were assigned to help, were held to the end of the day, until it was all finished. When I returned to the camp, I was stunned by the scene which I saw. Half or more of the trudovaks had already changed from their civilian clothes into the military uniforms which were given to them. What a pittyful picture I saw around me! The pants of the military uniforms were just rags, ripped here and there, crumpled, patched; the tunics were not different, the berets were dirty, shapeless, most of them without cockades, the boots where old, used, colorless, misshapen, of different sizes for the same men, and the bags where one had to place and keep his bread were shining of black accumulation of dirt. This was a disgusting scene. I looked and did not believe my eyes. The trudovaks who had already dressed, looked confused, incredulous of what had happened to them. All this stuff was a scrap, condemned for destruction, but was passed to the SLT to be used as uniforms of the trudovaks. Many of those whom I met looked at me from under their eyes as if they were ashamed or wondering of the blows which had hit them.

I lined up to receive my "uniform". The sergeant looked at me, and went back searching for something more decent to give me. He found it … the pants were ripped at the front, had no buttons where the fly was (Zippers were not used in Bulgaria in those days), the tunic was a rag and could serve only for a garden scarecrow. The beret was so big that it covered my eyes and I exchanged it at once with somebody with bigger head. I got a new one. I got better overcoat, so I could cover my rags. The shoes were way out of my size and I exchanged them instantly.

I grabbed the stuff and went to the barracks to change. I took off my windbreaker jacket and my nylon shirt. I took off my tie … and looked at it with regrets. "Miserable man!" I thought of myself, looking at the uniform waiting for me to put it on. I put it on, I looked at myself and with difficulty took control of my feelings humiliation . A big lump was pressing in my throat and I had to let it dissipate little by little. I folded my civilian clothes and put them in my suitcase. "Vae Victis", I thought of myself. This was my conclusion. (Woe to the conquered!). I went out in my new uniform. The trudovaks were loitering back and forth, lonely and in groups, looked at each other timidly, often laughed at themselves, and also to others, made snide remarks, and were making Herculean efforts to adapt to the situation. These young men were not new rookies, this was not a military service, this was more a camp of prisoners of war. This was enough to kill the selfrespect even of the most courageous young men. I do not remember what feelings moved me at that time. Was I mad at somebody? Did I hate somebody? Was I mad at myself? Was I hoping for something better? I only remember that I was frightfully bewildered. I was rattled by the realization that my worst expectations were coming true. What was I to hope for? Indeed, a black trudovak's fate. I felt as if this fate was weighing over me with the sword of Democles. It seemed to me there was nothing I could change, even a comma, that some unknown evil hand had charted it for me, that whoever did that to me, did not want me to come out of it alive, someone was seeking my mental, spiritual and physical destruction.

I was beginning to boil inside me, some bitter hate was setting down in me, but very soon it was overcome by a steely determination that I would suffer all that and I would survive it. I would not give in to them, those who sent me here, the satisfaction to see me defeated. The very first day, when I was confronted with humiliation, intended to exasperate me, I swore in my own mind that I would patiently go through all this, I will gnaw on my teeth, but I will keep myself ready for the day when I would again grab my suitcase and walk out free, will put my cream-colored, carefully folded shirt of nylon, will put my tie and leave this prisoners-of-war camp. In the worst moments of my service, this personal vow was back in my mind and gave me strength to ignore everything and to suffer everything, but not to be defeated. I had to manage to the very last moment!

But this was only the beginning, not the end of my days of misery. This beginning was a murky, cloudy, gray and muddy April day. It happened to be Great Wednesday – three days before the Day of Resurrection, the Easter Day. If not here, these were the days when I would be most of the time in Church, praying and listening to the Gospel readings. Now I could participate only mentally, in the hidden source of my hope and my energy – my faith. This day was over and I was now designated Trudovak No. 241, from Company 309. By the end of the day we were assigned the barracks where to settle permanently. We were given instructions about the order of things in the barracks, how to behave and how not to behave, and let us go and sleep. I turned over in my mind all that had transpired, and for long could not fall asleep. Finally the tiredness prevailed and I like all others passed to slumber land.

This is how my service in Bezmer began.

7. The Starvation Diet

During the following few days we did not do anything. New contingents of trudovaks were coming and were being processed as we were. New barracks were being built by special units. By the time this part of the operation was over there must have been over 2,000 men met, processed and housed. The most important part of our life these days was the food. We would line up in the morning to be given our loaf of bread. They called our names and we had to answer: "I, comrade commissar!" Then the one called had to approach the table of the record keeping man, receive his loaf, and shout with all his voice: "I thank you with all my heart comrade commissar." There were boys who were somehow confused or shy, they would simply grab the loaf and run back on the line. The officers got a kick out of that and laughed it out, but did not let the man alone. They would call him back several times until he mastered the phrase. The boys were turning red from shame, the officers were laughing, and those still waiting intensely listened for their names to be called, and practiced the answers. This would continue for an hour-hour and a half. Then we would be ordered to line up again, split us to groups of fifteen, picked up among them the men to serve as food-committee, which would receive the cooked stuff in special copper buckets – baki – and distribute it to their group. When these bakari, would bring the food, we would line up, place each one our cans on the ground, two feet in front of us, and wait for the designated men to pour in each man's can one big serving spoon.

While all this was taking place, there were shouts, there were cat calls, there was pushing and pulling, there were protests that the distribution of the stuff was not fair, that the bakari showed preference for themselves. It was a chaotic situation. When it appeared that it would go out of control an officer would blow his whistle, everyone would freeze and the pushing and pulling would soon after that resume. After the food was distributed, everyone would pick his piece, and sit on the ground in the middle of the yard and quickly devour whatever was cooked for the day. This was the picture at lunch and so also the supper and the breakfast when a piece of cheese and marmalade were provided. It was a pathetic human experience and more often than not I stayed on the side, said nothing to protest and whatever came my way, that was just that.

It was humiliating and degrading to argue for a bit here and a bit there. After a few weeks a dining "Hall" was constructed – plank tables, rough and ugly wood, with benches along the tables, so we could sit down, patiently wait for the bakari to do their job and eat more comfortably. The arbitrary distribution of the food could never be regulated, and often some of us were left without a piece of meat when it was served, or saltsa from the top, which was of some sort of an oil or lard. There was no sense of comradship, or for justice and fairness. It was every man for himself. So, always there were some who ate well, and always some who were taken advantage of, and made no fuss about it. Still, always, at time of supper or lunch, sharp insults were exchanged, small quarrels and all sorts of arguments over the food distribution were the rule of the day.

Needless to say, there would have been no such arguments if the food was satisfactory in quantity and quality. In reality it was far from satisfactory. Even if the distribution was not affected by such subjective factors as personal friendships, its quantity at least was a sure way for starvation. The bread was not enough to satisfy the hunger of these peasant kids who were used to eating much bread, and what they were receiving was hardly half of what they were used to. I felt sorry for these poor souls who could not save some bread for supper and ate all their ration with their lunch. At least three quarters of the trudovaks ate their supper without bread. Some of them, apparently from well to do families, received packages from home, but those who could not afford it simply starved. I watched with regrets those who enviously looked at better supplied trudovaks when the latter pulled home made bread to finish their meals. Silently, they just attacked the cooked food without bread.

My bread was enough for me, and sometimes I had some left over. I had become used to it from my Sofia days, when, as student, I had to settle for the little which I had. I often offered a few bites to some of these devils and they appreciated it. When my turn would come to be a *bakar*, they volunteered to replace me. This was not a very pleasant duty. The soup which was mostly served for lunch and supper was not enough either. Who knows, perhaps the State was very poor and could not afford it to feed those whom it had called to do the dirty and heavy work, having pulled them by force from their homes. The trudovaks often heard about the food served to the soldiers of the regular army and that made their blood boil. That was an indication that the State looked to the trudovaks as a second class servicemen. The work was heavy as if designed for prisoners in punitive camps. It required heroic efforts to complete the required norms. Enormous physical energy had to be expended so that the officers could be satisfied and stop hustling their men. But how do you reconcile your screaming stomach with the heavy picks and shovels? Who knows, it may be that together with the punitive work they had also planned the poor food and scheduled the hunger in their program. Could it have been that the goal of the whole institution was to send the young men in its service after their term was over to the hospital if not to the cemetery? A rumor was circulating among the trudovaks that up there in the Government offices, they had a report to the effect that after serving their term, 75 per cent of the trudovaks were sexually impotent and sterile.

It is impossible, so it was thought, that the officers in the service were unaware of that. Even the commander of the Trudovak military service, the top officer in the service, visiting our camp, had the temerity to joke with us, shamelessly saying that he knew, that the bread was insufficient. Colonel Angel Tsanev, that was his name, excused it with the fact that the whole nation was short of bread.

Scoundrel!

Every high officer of the Service, visiting the camp, sought to inform himself about our bread. Colonel Neikov wanted to know how we fared with the quantity of bread, was told when he asked that it is insufficient, he promised to look over the subject, but after he left we never heard more of him and our bread ration remained the same.

Our breakfast was the perennial tea with feta cheese. One could hardly taste a grain of sugar in the tea, and at some point, for months, the sugar was replaced by prune marmalade. You try to sweeten a cup of tea with sour plum marmalade and see how it tastes. The piece of cheese in the beginning was said to be 40 gr. for individual and later reduced to 30 grams. It was a joke. Was it meant to be eaten or to be licked? For lunch we had some mesh of vegetables, or beans, or potatoes in the form of soup. For desert we were happy sometimes to get a spoonful of rice or makarony, sweetened with sugar. It was a delicacy. For supper, our desert was forever and ever a tea spoon of marmalade. Sometimes we were served meat, but the *bakkar* could not find a piece for everyone, because, sometimes "by mistake", he had served to somebody a double portion ... to his friends. So most of the times we sipped the soup and at least there was some taste and smell of meat.

In working days, lunch was served at the place of work. The *bakkari* had the responsibility to go to the camp, fetch it and bring it to the men waiting hungry for them. The service of the *bakkar*, where we all had to take our turn, was not one of the cherished obligations. The most unpleasant part of it was when all had finished their supper or dinner, they had to look for water to wash the bakki and run back to work, or if it was supper, when everybody had a little time to relax, they had to continue their chores as such. They had to find water to finish the job, and when *bakkari* from all companies and all subdivisions of the companies had to use the same pump, it was becoming an impossible task. The area around the pump would turn into a pool of mud, the pushing, and pulling, and bumping into each other, the arguments, the chaos, the selfishness of all involved, made the situation ugly. When I could, I gladly avoided this privilege, and retired for rest at the end of my day. Still, there were those who loved to serve as *bakkari*, and if you let them have your turn, they would be ever so appreciative. Some had even monopolized the service and we dubbed them *chorbari* (the men who were in charge of the *chorba* (the soup). Some would do anything to hold the *cherpak* (the big serving spoon). They served themselves the best part in the *bakka* – the meat, the saltza, the beans or whatever. They also helped their friends.

Our company had two boys who were perpetually serving, among the others, as *bakkari*. One of them was Ivan Gurdev, whom I have forgotten. I met him in Zelenikovo in 1992 and he reminded me. He was a shepherd now, with some one hundred or more sheep. He loved to be all the time in charge of the *bakka* and the *cherpak*, and sometimes I found my can with better than usual chorba, but I was not quite happy with the situation. He loved to argue with everybody, and to come out a winner at the end. I did not like his manners and when I could, I avoided his patronage by sitting at a different table. The other one was Mitteto. He was from the Haskovo area and performed this service, as he often claimed, not because someone had to do it but because there were no takers. In exchange his can was always taken good care of. If somebody challenged his fairness, he would reach back to the *bakka*, take a little extra to fill the protester's can, and if something was left, he would finish it in his own can. He was especially generous with the company leaders, with the corporal and the rest of the privileged individuals to gain and keep their disposition. They were happy with him, and since they were free to pick the *bakkari*, he always ended in charge. There were some who often protested the inequalities practiced around the *bakka*, but since they were not too ready to serve as *bakkari*, nothing came out of it. When something is repeated twice a day, and every day, it soon become a routine and one learns to live with it. The food was served under open skies. There was no roof over the table to protect you from the elements of nature – rain, sand storms, or whatever, which was quite often the case. By the time the food was served, the cans were well exposed to all sorts of things which were blown in or rained into. Still you had to eat the food, or starve.

8. Water in the Desert

The water supplies were another matter. We experienced this torturous incidence the very second day we had arrived. In the evening the officers blocked the well, and would not allow anyone to drink its water. It had turned into mud and all kinds of diseases could explode on campus. And we were thirsty. Our mouths were turning dry. A patrol was placed there and he would not let anyone near. The entire camp population was up and protesting. There was no other place to find relief. They did not send someone to bring drinking water from outside. We were flabergasted. Was this thing going to continue in the summer? It was still cool now, the beginning of summer, but how were we going to survive when the summer heat hit us in this vast plain? We had to go to bed thirsty. What we could do? Military discipline required us to submit to orders. We were still just new draftees, and watched helpless, and silently, the officers marching across the camp. Well, soon the well was covered by a pump, it often broke down, but the access to it was not blocked. But our first experience taught us that water was to be a continuing problem.

It indeed was. Since the places of work were some distance from the barracks we had to be supplied by a special contraption – a *buchonka* (a wine-butt like wooden container, may be some 15-20 gallons, mounted over two two-by-four long handles, with a peg in the back. Two watermen would drag it on wheels, like a Chinese "taxi" from place to place and then run back to refill it. It often happened that the men in charge of the *buchonka*, if it was a muddy road they had to travel, or were ordered to take water to some unscheduled place, took a long time to come. We had no choice but to suffer. When the pump was not working, a water railroad car was brought in, they had to first supply the kitchen. We had to wait helplessly in burning thirst. What a picture this was? Some two thousand men, scattered along the project at work, depending on a primitive system of water supply! Sometimes the situation during the summer was intolerable. We would be standing up on the mountains of sand or on the runway in construction, and look helplessly in the distance to see if the *buchonka* was coming. And when finally it would show up, a swarm of men, with cans in hand ,would run to intercept it, to be first there and get the precious liquid.

Soon the bedlam breaking out there was turned into a hell. The lucky ones would fill up their flasks, soon the *buchonka* would be empty, the mass of trudovaks helplessly watched how it was being pulled back, and wait for the second chance. No one was there to introduce order, the officers did not interfere and it was a free for all. Sure, the officers blew their whistles, demanded order, but no one listened to them and they had to give up. And none dared to accuse the management of the camp for this tragedy and human suffering. There was again pushing and pulling, the physically strong simply bumped the weaker ones to the back. The man of the *Buchonka* helplessly argued that he had to reach the project's working place, but it never worked. Only a man with a gun could put order in this anarchy, but guns were not noticed anywhere around here, except on the waists of the upper officers. So, there was the only alternative: wait for another *buchonka* to show up far in distance in the sand storm, to make its way within three hundred feet close to the work place where it would be besieged by another group of thirsty trudovaks, until it is emptied in a few minutes. The lucky ones happily gulped the cold water, and those who had failed again, distraught, waited for another turn. Water! So much water in the world, and how little of it reaches these trudovaks where it is mostly needed! How little attention is paid to it when it is in abundance, and how precious it is when it is not available? Couldn't they have taken care of the problem on time, so the trudovaks be spared this torture under the burning sun? Of course, they could have organized a better supply of water, but they did not care. There always was enough for the officers.

This was a life in a desert. There was this endless field where no tree could be seen, while horrible winds very often blowing from West to East with such a force that sometimes the roofs of the barracks were torn away, sometimes would blow the whole kitchen, and carried clouds of dry sand, all kinds of trash, dust and anything that is movable. The sand penetrated through our clothes, you felt like it is reached your bones. Desert! It was a practice that even in the worst cases work would not be stopped, and we had to be blown by the winds sometimes to the ground with our wheelbarrows. I would never forget the day when the winds were so strong, and they were so cold – it was October – that one could feel how the sand was piercing the skin on my face and was hurting as deep as it could be. It had no mercy! Eyes, noses, ears, mouths were exposed to merciless attack, but our officer, a sub-lieutenants from Sliven, Sotirov, screamed to us not to stop. He himself was agonizing like a wounded animal, jumping from one foot to the other to keep warm, and viciously screaming at some trudovaks who would stop for a moment to clear their eyes, their ears, their mouths of the sand. His heartlessness was bordering on misanthropy. Was this some sort of an order to torture the trudovaks? We helplessly clenching our fists, biting our lips, so that no inappropriate word was uttered in desperation. We were silent, we pushed the wheelbarrows, and pushed and pushed, always remembering that this too, will pass.

There were days in this desert when, very rarely, it would rain. We were pitifully longing for a day when we could have a little rest. Some times we were lucky to have this blessing from above bestowed upon us, and work was to be stopped for an afternoon or a morning, for half a day. The time lost had to be made up Saturday afternoon, which usually was non-working time, and Sunday, as the case could be. Still we were so pleased to have a little rain and grab a little respite from work, because, if it was not for that, they always found excuses to put us to work Saturdays, any way. It was a supreme pleasure to stretch on our beds in the barracks and listen to the rain falling on the roof. Otherwise, the summer heat was killing us. There was no shade anywhere near where we could hide even for a minute. During the lunch break we would stand up the wheelbarrows and rest, eating our lunch under its shade, even if part of our nude bodies was still exposed to the burning sun. It was a pity to watch these thousands of young men baked to bricks in this fire-swept desert. Was this the place for us to be? Is this the place where we had to spend the best years of our lives? Once I met a trudovak, I did not know him and he did not know me. He looked me in the eyes and with tears running down his face, told me:

"Friend, we have been burned here! It is the end of our youth! We are doomed to death!"

"Don't get discouraged!" I tried to appease him. "This too shall pass. We are going to take everything patiently. Who knows, may be all this is for our good."

"Do you believe that there will be a good day left for us?" he continued the conversation.

"The sun will shine on our street too.", I told him. I left him behind me.

We did not know each other, but we understood each other. All of us, all those who were planted in this desert, shared one spirit, one faith and one hope: to survive!

9. The Barracks

The barracks where we slept during the night and where we kept all our belongings, were not anything to talk about. Divided by a pathway of two feet, on both sides were the long sleeping plank-beds, covered with straw, some two-three inches, and above the straw, stretching from one end to the other, perhaps thirty feet, was a sack-cloth. The straw easily stuck out to pinch us when lying down. Twenty people were accommodated on each side and during the night we had to stretch like candles, without room to curl and rest our muscles. If it was not that we were dead tired, there could be no sleep there. Covered with some rag of a blanket, we just closed our eyes and for a minute would be gone. None felt like talking. Besides, there were officers patrolling the place to make sure that none would talk. Our suitcases were directly under our designated places to sleep. So, when one had to open his suitcase, he was blocking the way for others to pass. And there always was someone who would be rummaging in his possessions.

When the morning call for awaking was sounded, our time for getting ready and out was short. The barrack in such instances was turning into a bedlam with yelling, screaming, complaining, and cursing left and right. The worst part was when we were ordered to bring out all our possessions, suitcases and all, and line up outside. For some reason quite often we were asked to open our suitcases and sergeants, or corporals, searched them. What they were looking for? God knows! May be to make sure that none had any weapons! They never found anything, but continued this practice. This was especially done during the first weeks after our arrival. Then they would search inside the barracks, and at end send us to occupy some other barrack. I would not be wrong if I suggest that it was all done to make life difficult for us. Most serious offense was considered not to make your own bed as specified. The Blanket was to be folded lengthwise, from top to bottom, not to leave one single fold sticking somewhere, whatever was used for a pillow was to be carefully folded on the top. The sheet was supposed to show at bottom two centimetres and the top – show 20 centimetres. In the beginning I attached the sheet to my blanket with safety pins, as was the custom in the Seminary. Once this was discovered I was ordered to abandon the practice. Convenience had to yield to "aesthetics".

There was no comfort to sleep in these barracks. We slept body to body, breathing the poisonous gases which the abundant bean soups produced, and no one even tried to be discrete about relieving himself publicly. Unable to turn over we would awake in the morning more tired than we were when we went to sleep. I soon discovered that the best place to sleep was at the very end of the plank bed and when allowed to move in I would position myself in such a place that I could be the first to enter. I had more privacy there, and could raise my knees at least leaning on the wall. This was not always possible, but sometimes I was just lucky. In the beginning, somehow, I got stuck with an individual, sleeping next to me whose vulgarity and cynicism were unequalled by anybody through my service. He had a vulgar name for everybody and every second word in his vocabulary was in the area of erotica and sexual organs.

Culture? Civilization? It was a life of shame. I was very careful in my language and considered it a matter of dignity to preserve my good manners. But it could not be helped. As time went on, inadvertently I adopted the language of the mass, dirty words, and swearing became part of my vocabulary. When a month ago I was in Sofia and met a group of friends – two girls, Radko p. Todorov, and I do not remember who else was there – in the course of the conversation, unwittingly, I used one of these words which in polite company are considered to be the ultimate of indecency. I realized immediately my faux pax and apologized, but the damage was already done. (When I met Radko during my visit in Sofia in 1991, and he invited me for dinner to his home, he introduced his wife as one of the girls which I had met – "the black one", he said. I apologized again for my impertinence of those years, far back in 1951).

The officers were much concerned that we keep the barracks in perfect order. Our platoon officer was obsessed on the subject. He would not let anything to appear in disorder. We had to sweep the plank-beds, under them, the pathway, and let no items of anything showing from under there. Quite often he himself would inspect the place and whatever he found not in its place he would throw it outside and impose penalty. Any time he inspected the premises we would see shoes, blankets, baskets, rags and bags flying out of the door. Special attention and care was taken to fill every corner of empty space with communist slogans. At the end of the barrack's pathway the wall was to be decorated with such slogans in praise of Stalin, or Dimitrov, or peace, the Party, stakhanovism, etc. A trudovak, supposedly a good artist, was freed from work to take care of these political education chores. He was a good guy otherwise, friendly, good humored, but his orthography left much to be desired. Once I pointed to him an error, and sensed that he did not much appreciate it. I never tried to correct him again. And, after all, who was paying attention to these things besides the officers? Usually, on these special corners of the barracks, the central space was occupied by the portrait of some of the big wigs of the Communist Party.

The first few weeks we had hard time to find our place in the barracks. There was no electricity and we had to feel our way in complete darkness. Later on they managed to put a weak light bulb which made things a little easier, but nothing more than just to find our way. When the first light shined in the barrack we had to hear all about the progress made by our Communist government. Much care was being taken of the cleanliness of the camp. If not every day, every second day the entire trudovak body would be lined up at one end, then proceed at a slow pace and each one had to watch for trash, including cigarette buts, picked them with bear hands and drop them in a bag carried by some one from behind. This was called *obborka*. It was all done under the supervision of our platoon commander. He was the personification of evil. The slightest error was punished. He would walk behind the line and if someone missed the smallest piece of paper, he would call him back with screams and have him pick it up. To further beautify the camp, in no time it was arranged that at the front wall of the barracks, some two-three feet deep, a flower bed was to be prepared and planted. Our clown, Chakandrakov, as an agronom, was assigned this task and freed from work. He was given all materials and showed great interest, especially when the officers were around. On his own he planted in every patch of flower bed a sunflower, which in the summer, as much as they could grow in a dry bed, showed their small yellow heads. Someone of the officers also thought of the idea of setting up something like a park near the campus - with flower beds and alleys. It was bigger project and we all had to give extra time to build it, before the May Day.

We had to transport soil with wheelbarrows some three hundred feet. I thought the whole thing was a joke and told Chakandrakov, but he was very enthusiastic about it. After all this was pleasing the officers. I told him that it will be beautiful garden, but he had to plant flowers growing in dry climate. He wanted to know if I knew any. Yes, I told him, *Magareshki truni* (Donkey's thorns) and *Buzuniak* (I have not seen the name of this in any dictionary and would not know how to translate it). But both of these plants were the ugliest thing to see in open fields. He did not mind, he knew that the whole thing will fizzle, but he would lead the naive officers, who pretended or were really believing in the justification of the project. So, this is what really happened. Soon the whole idea was abandoned and forgotten. When we passed by this park in the summer all we saw was wild weeds, thorns, amaranth, goose-foot, quiche and all sorts of ugly vegetation. How much work was wasted there, how many of our free hours for rest were taken away from us for this stupidity... But who cares?

Quite often the officers were thrown in hysterical outbursts of energy. They would order us to take out all our suitcases, then to sweep under the plank beds, dust the rafters, straighten our beds in a perfect form, shave ourselves, clean our clothes, leave no sloppy looking dress uniforms, clean up our tools, arrange them in perfect order, pick every little piece of paper or cigarette butts on the entire camp, fix the little patches of flower beds and memorize and practice thousand times the responses to a higher up of the officers. When this operation was ordered we knew what was the case. Colonel Zmiarov was coming for an inspection of the camp. Colonel Zmiarov never visited our camp. He supposedly was an officer of the central command of the Construction Troops, his glory was that he was the holy terror for the officers. He was not after the trudovaks. He was after the officers, as to how strict they were with trudovaks and what care they took for their men. Perhaps five to ten times, he was expected to visit our camp. Somehow our officers were being tipped by somebody of these visits, and if they could, they would ask us to clean the camp with our tongues. The yelling and the screaming, the threats and the penalties for the slightest omissions were ominous. One thing that we felt somehow improving in those days, was the food. It was better and meat was the delicacy we enjoyed, while mocking and ridiculing among ourselves the fright of the officers.

10. Roll-Call, Night Patrol and Day-Duty Assignemnets

One of the regular – 9.00 o'clock in the evening – events was the roll call. The Campus Chief officer of the day, or his designee, would blow his whistle, we would line in front of our barracks, he would verify our presence, and note the absent, who were usually *bakkari* or in some other service, and let us go to bed. This thing was not taken too seriously. More often than not the officers would delegate this chore to an elderly sergeant, Diado Rusko, about 60 years old, who otherwise was in charge of the uniforms stock room where we would go for another pair of pants or shoes, or blankets, bringing back the rags which were no longer good for us. We called him "Diado" ("Grand Father") and he would get mad and snap back: "Do you have a baba ("Grand Mother"), is she young and pretty? "He would take a few minutes to lecture to us how to protect our property signed off to us by his office, where to go for our natural needs - the camp did not have special place for rest rooms, the open fields were the place to relieve ourselves, usually being some embankments where no toilet paper was used. The primitivism and the practice of this indignity was part of our "culture". Night time was the preferred hours for such performances, if one did not particularly care to expose his privacy publicly. We had taken him for a joke, but his fatherly councils were taken in good humor and when he would address us with his official greetings "Zdraveite Yunatsi" ("Greetings Boys"), we heartily shouted back: "Health we desire, Sir!" and to his letting us go – "Svobodni ste!" (You are free!), we answered "I am listening, Sir!" and then everybody would wander around a little and head for the barrack. The next day serving trudovaks for the various chores: *bakkari*, barrack-chiefs, night patrols, kitchen patrols, etc. were designated at these roll calls.

318

My first Night-Patrol service was quite an adventure. I was apprehensive when my name was called, together with a young man, tall and handsome, from the village of Uzundjovo. We did not know each other. He was as apprehensive of the appointment as I was. We had no idea what we were supposed to do. We were briefly told that we had to check on the night guards at various places. The sergeant in charge of these posts instructed us how to approach the positions, guarded by regular soldiers with guns. The pass word was given to us, and late in the night we had to go to these places. It was so dark that we could not see five paces ahead of us. So we had to first find the patrol of the gasoline storage. We walked in its general direction, without seeing where we were stepping. It looked like we were stepping in bowls, tripping, falling in the holes and wondering where we were. It took us half an hour until we managed to recognize the place and realised that the patrol was somewhere there. Our hearts were filled with fear. We were not armed, but the soldiers we were checking on were, and who could say what may happen in the darkness of the night? To alert the soldier, we started speaking loudly, proceeding slowly, expecting the soldier to hear us, ready to run back if something went wrong. All of a sudden a harping voice, which seemed to us to have gathered all the terror of hell, roared near us:

"Stop! Who is there?"

"Night patrol", both of us shouted at the same time.

"Give the pass word!" shouted the soldier now with higher voice.

"Which letter do you want?" I asked him.

"The fifth!", he said.

"E", I said.

"True!", answered he. "Come closer!"

Who knew anything of these things? I was afraid that anything may happen in the darkness of this night. I hoped that he would meet us friendly and then we would continue. Instead of initiating an ordinary conversation, he jumped a few paces back and warned us:

"Stay there: whisper low the whole password.

"Konche!" I said. (In Bulgarian the sound Ch is written in one letter). The word meant "Pony".

"Go ahead with your duty!" he ordered us, still holding his gun pointed at us.

"Animal!" I thought of him. He is either very scared, or this is how he understands his service.

We moved on. Soon we reached the entrance to the Camp. We saw from distance the patrol standing in the middle of the road. He was looking at us with gun pointed in our direction. When we came some fifteen paces near him he softly asked: "Are you the night patrol?"

"Yes", we answered.

He did not ask for password, nor engaged in any rituals. Asked us to sit, offered us a cigarette, which I accepted as a sign of friendship and managed to smoke it. I was not a smoker. We chatted a little and continued our walk in the direction of the big machinery. The patrol there had seen us coming and had decided not to meet us. We stopped at the road and whistled lightly. Then he called:

"Password, please."

In a minute it was all cleared and we were on our way back to campus. It turned out we had fallen in a pool of mud and had a hard time getting out. There were more posts to be checked, but we gave up. We cruised across the camp, as if someone was going to steal the big treasures of the place, and we were in charge of protecting it. Much, much later, in Greece, I realized how much the intelligence services of Greeks and Americans were interested in the smallest details. I realized that what we were doing was really important. This was my first night patrol. Rarely, afterwards, I was designated for this service. I had told a sergeant that I did not particularly liked this assignment. When I was called again, having already learned the service, things became quite different. Once we happened to be with "Suzi" again. This was the name of my first companion, and as tired as we were, soon we discovered a field of wheat, found a comfortable place, lied down and fell asleep. At one point we were awaken by some noise and mumbling. A drunk officer was returning to the camp, hardly keeping his balance. He did not see us. We did not move to stop him, and that was all. Another time, we found a good company in the kitchen and when the cooks left we filled our pockets with sugar. In the fall, for three or four times, I did not even leave my bed.

Quite a few times I was appointed to be on platoon Day-Duty, sort of a supervisor that all assignments were running smoothly, or Kitchen Day-Duty. I liked both of these assignments, and as a man of education, was frequently selected for this service. It was in fact a day when I would not be sent on the project to work. In the morning, after the bread, the tea and the cheese were distributed, there was nothing much to be done and a brief sleep was a blessing. One could also write a letter or read something. But it was a responsible assignment and for possible errors and omissions the penalties were strong. Many of the trudovaks stayed away from it. Sometimes I had unpleasant experiences. My first kitchen supervision ended with a scandal. A group of men, twelve, were sent to some special assignment and were not going to return in time for supper. I was to make sure that their food is saved. It happened that when they came and picked the *bakka*, the cooks had not put one single piece of meat in it. They exploded in vocal protestations.

An arrogant simpleton, later mocked for his eccentricities, educated, known as Liuben the Mane, picked the *Bakka* and took it to the company's chief officer. I was called there for explanations. I told the officer that it was not my fault, that I did not check all the *bakkas* how much meat the cooks put in each one, since there were tens of them, and could not count the meat morsels they were distributing. At this point my platoon officer intervened. He scolded the complaints in such a manner, arguing that such things inevitably happen sometimes and that it was not possible to see everything. This was one of these occasions, and I, a man with reputation and education, could not be held responsible. They bowed their heads and walked away. Much later, somehow we became close friends with this same Liuben the Mane, having understood that neither I was to blame, nor he was to be accused where I was concerned, that this all was a part of the service and we all could sometimes find ourselves in such predicaments.

Besides a rest for a day, these duties offered possibilities to satisfy better my needs for food. For sometime I had felt terribly hungry. There were some trudovaks, serving such duties, often to get confused, to have less of the necessary bread for distribution, exposing them to penalties, so they would rather work on the field than be on kitchen duty or Platoon care-takers. Such was my friend G.R. He was glad to delegate his turn for such duty to me, when called to serve. Kitchen duty was more demanding that the Platoon care-taker. It required constant presence in the kitchen to watch that the cooks did not appropriate goods for themselves. All kinds of abuse and thievery was possible.

320

11. "Military Training with a Pick and a Shovel."

During the early days of our service at Bezmer, we had to practice some of the military routine – marching, turning around, saluting officers, etc. But we were never given any weapons to learn their operation. Our weapons were the pick, the shovel, the fork, the wheelbarrow and our own hands and backs. We had hardly arrived, the third day, when we were called to the vast field and lined up in a square, one arm distance from each other. A corporal stood in the middle, Peiu, whose figure and expression on his face revealed a very limited intelligence, mental ossification, yet some sort of a good soul radiating from his kind smile. Much later we learned that in the past he had some fascist background, but as of now, we could not judge anything except what we were seeing. At the time I was predisposed to look to any officer, from bottom to top, as a Communist Party functionary and deserved to be despised by me. So, I looked at Peiu through the glass of my prejudices. Peiu began to explain the military rule for standing erect, for standing at ease, and ordered every one of us to practice it, by shouting the orders himself. So, what do you do? You follow orders and one by one the trudovaks began:

"Boets, mirno!" (Fighter, atten-tion!)

"Boets, svobodno!" (Fighter, at ease!)

The first calls sounded uncertain, some sort of embarrassingly, and even mockingly. I had to shout these silly things. I tried it first. Not much voice came out of me. It all looked so silly, so undignifying, and I felt like laughing at it, not at myself. I could not say it the third try. Then I noticed that Peiu was looking at me. I felt like I was shaking. So I started again:

"Boets, mirno!"

"Boets, svobodno!"

Peiu came closer to me. His stare was fixed on me. I tried my best. I let my voice go and shouted the commands more naturally and in a freer manner, but I could not conceal my laughter in my voice.

"Why are you laughing?" – reprimanded me Peiu. "There is nothing laughable here. Pull this belly in! What is this relaxation, you look like a carcass."

I tried to follow his orders. I was now serious – not because I believed in it, but because I did not want to provoke further his anger.

"This is better!", he said and left me alone.

I could not hold my laugh. I turned a little on the side, silently laughed at the whole thing – I could not control myself – and after composing myself I continued the exercises. After he was satisfied by our performance, Peiu instructed us how to turn left, to turn right and turn about face. These exercises were repeated several days, until we had mastered every move. We no longer were ashamed, or found the whole thing a laughing matter. We were shouting like mad men. All officers were out watching us. Not everybody performed well, however. Some young men simply could not do it, and we continued laughing at the awkward performance of some. Some trudovaks never opened their mouths at all. I was watching a fellow from Goliamo Konare how he would open his mouth, say the command, but would not turn in any direction. Later, he was put in charge of the *butchonka* (the barrel by which he was bringing drinking water for the working men). That he managed to do, but when the thirsty trudovaks rushed to his charge, he would abandon it, run some twenty feet away, and would return only when the water was finished and the men had departed. He would take it back, to fill it, and drag it back to the project. That was Stoichko. Some of the men were making the motions with such a force, that they would pass the mark, almost lose their balance and take the appropriate position with difficulty. All this was done to fill our time – until all preparations were finished and we could start the work.

After days of such exercises we were ordered to line up in front of the barracks and split us in platoons. First they called those with university education, to step so many paces ahead. Then the High School Graduates – to form a second line. After them came those with elementary education, to 7th grade, and at the end – the illiterates. About nine men were with University degrees. Over one hundred had High School education, and over two hundred were virtually illiterate – gypsies and Turks. They formed three platoons of our company, of about one hundred or so men each. Those with university and high school education were evenly split among the three platoons. My platoon was designated as Transportation unit – in charge of loading, unloading and transporting all building materials arriving by train or trucks: cement, gravel, sand, wooden materials and anything else. When the division was taking place we attempted – my neighbor Todor, my old childhood friend, and Chakandrakov to get into the same platoon, but the commanders thought differently. Later T. and Ch. got in the same platoon, but I did not try to follow. I noticed that T. was trying to distance himself from me, apparently fearing that my political and educational baggage would rub on him. Besides, I sort of thought that the farther I was from people with whom I was one way or the other associated, it was better, because in difficult situations one ends up by losing his friends.

After we were split to three platoons, we were assigned to already appointed corporals, in charge of the units. These in turn proceeded to divide us to groups of ten and appoint one among them as the leader. I found myself leader of one of these groups of ten, of the first Platoon of Company 309. After this division was completed, our Platoon commander took us to the field to continue our exercises which I already discussed. These exercises were conducted in the established units of ten, and I found myself supervising my ten men, to correct those who were not so experienced, and demonstrate to them how to do it. I felt in a phoney situation, most inconvenient in relation to the ten men under my command, but there was no other way. The best I could do was to take my group as far away from the main body of men and get the boys around me and talk, pretending that I was instructing them. The repetitious exercise of marching, of turning left, right, and about face, was becoming a boring and annoying affair, but it lasted until April 17th, when all preparatory work on the field for building the runway was ready and we could start with forks, picks, shovels and whatever.

12. The Construction Project

Probably very few people ever stopped and thought of an airport runway. Until I was recruited to serve as a trudovak and was sent to Bezmer, I had not had the vaguest idea. In fact I had never been to an airport and the word runway meant nothing for me. Nor I had ever been interested in such subjects. But now it happened that I had to build one, along two thousands other trudovaks. This runway was to be of some unusual construction, as

I thought. This is what we were told. Its size was to be much longer than the ordinary runways and its intended use was for heavier and special planes, so its construction was a special assignment. First the trudovaks had to prepare the bed for the materials to be laid in. It was to be 2,500 meters long and eighty meters wide. This meant that a space of 200,000 square meters was to be dug 40 centimetres deep. This further meant that since the ground at the 1200 meter was low, and the section further up was higher, then in the first case the trudovaks were to dig the mound and transport the dirt down to fill the gap. It was an enormous task, it had to be done by hand, but then, so many trudovaks in one place, it was not an impossible task.

The bulk of the work, where mountains of dirt had to be dug out, was done by big machinery. It is for the first time that I saw this technology. Here an excavator was lifting four tons of dirt in one dig in five minutes, which a single trudovak could finish in a day and a half. There were all kinds of machines and the noise they all made was deafening, but it was advancing fast in doing the rough work to prepare the bed of the runway for pouring the concrete. So, in my calculations, some 200 000 square meters, about fifty acres, had to be covered with 0.20 meters concrete, which amounted to 40 000 cubic meters of concrete – to be prepared and poured in the pre-set sexagonal forms. That would require thousands and thousands tons of gravel, sand and cement, to be brought by trucks and freight cars. All the loading and unloading of all materials was assigned to our platoon of one hundred or so men. All along the runway were the units for mixing the concrete. Each unit operated nine concrete-mixing monsters and along them were the piles of gravel in three sizes – small, medium and large chunks of stone – mountains of sand and cement. All that stuff had to be loaded in the concrete-mixer and when the mix was ready, it was poured in metal tip-truck wagons, pushed by trudovaks over narrow rails to the pre-set forms. There it was emptied over a platform where other charge of trudovaks, with shovels, had to distribute the stuff in the hexagonal frames made up of lumber, each side of the form measuring 1.97 meters. I suppose the shape and the size of the block was designed in such a way, so that if the airfield was bombed, and the runway damaged, it would be easier to clear up the broken part and filled it up again with concrete as soon as possible.

The entire length of the runway was divided to several segments, and along each segment a place was prepared for a separate units for storing the materials. The concrete mixers were moved from the segments already completed to those further up. In addition to all that a drainage canal had to be dug out all along. At one time I watched how one single man, short, some five feet two, who had attempted to escape, was put to work there, digging a trench six feet and six inches deep and 36 inches wide, trying to throw the dirt in the form of mud over the side. One soldier with a rifle on the shoulders was guarding him. I felt sorry for the fellow, but to me he was a hero for having challenged the system.

From the very beginning of the work we were told that the runway was scheduled to be finished by November 10, but we were already one month behind schedule. For this reason in the months of July and August we had to double our efforts to make up for lost time. This only shows what a burden was being thrown on our shoulders. Our platoon was charged with the transportation job. We had to supply the men on the runway and on the concrete-mixing units with the building materials. There were units which were blasting stone from quarries somewhere around Bezmer, or as far as Sliven. Others were loading freight cars with sand. We were responsible for loading freight cars with freshly split rocks into gravel, piled two miles away from the projected runway, also to unload it near the runway. We also had to jump on the freight cars, a mile long trains, with our shovels to unload sand, so the cars could be freed in time and be taken back to be reloaded by units stationed at the river. When rocks were brought from the quarries, special units, with hammers, big and small, broke them to small pieces, and then we would go to load this gravel to empty RR cars for transportation to the units. The worst part of our obligations was to unload composition after composition of freight cars bringing in cement in bags of 100 pounds each. Before pouring the cement in, the men working on the runway would have spread several inches gravel, perhaps six inches of sand, have it stumped with vibrators and cover it with some plastic asphalt paper. Then the cement would be poured in, to harden in a day or so, the form would be pulled out the next day for another block. For several days the new blocks were held under water. When some aqueduct failed during the summer to supply enough water, five or six cisterns were brought in to carry water from the Tunja river.

13. The D – (Dog) Day Service

The D-day of my service as Trudovak was April 17th, 1950. After breakfast we were ordered to line up before the barrack. Near the camp we saw some trucks. The drivers looked with curiosity at us, smiled and said nothing. The officers were still in their office. At one point our Company commander came out with a notebook in his pocket and started reading the assignments. Our platoon was ordered to load railroad cars with gravel. Our Platoon commander, Shumerov, led us to the barrack where the instruments were held and we were given each one a fork. Then a truck took us to the railroad near Bezmer where a train of empty open cars in a sideline was waiting. We were assigned four men to a car, and ordered, by the end of the day, to have it loaded with gravel unloaded there sometimes before we had arrived. How do you shovel freshly broken stone with a fork? One should try it to see how it works. It was at first attempt an impossible task. You press the four-prong fork, it hits the rocks and does not go deeper. Then you dance around it, go from the bottom, one way or the other, you manage to pick a few rocks and throw them over in the car. You struggle with every new try, you listen to the platoon commander screaming at someone on the other end, you try again. You hear how your companions swear on mother or so, and then you forget all your education and your upbringing, your sense of dignity and your manners of decency and you too start swearing. If you do not try and if you fail to meet the assigned quota you will be punished.

So, you try, and try again and see how slow the car is filling up. In the beginning of the service we had no idea how hard the work was going to be, how much it was going to be. In no time blisters covered our hands, they broke and blood started running down our fingers. We tried hard, we were working like slaves. We feared the officers and, feared each other. Someone could report us. But in the course of time we got used to it all. We cared not that someone would report us, because we all felt the same way. In the beginning some appeared very anxious to please the officers and killed themselves. Others just took it easy. Very soon those who were so anxious to please, saw that it did not work and slowed down. Some, from the very beginning, did not put too much in it. I was in that category, but as little as I was doing, was still too much for me. If everyone was working as much and as efficiently as I was, the runway not only would not have been completed by November 10, 1950, but it still would be there to be finished next summer.

Something happened to me the very first time I leaned over the fork to scrape some gravel. In my shirt pocket was a little mirror, which was always with me. In those days, we all were concerned of some of these social ammenities, to appear always presentable. Now, leaning over, this little mirror, some two by three inches, without a frame, slid down, hit a rock and split in two. I did not think much of it as a loss, but I remembered the superstition that if someone broke a mirror inadvertently, that meant bad luck. I looked at the two pieces and thought: "Here ends one page of my life, and there begins the second." I could not figure what the second part would be, but as the days of my service progressed, and as I am writing these lines, I still remember this little incident. I have remembered it so many times in the last ten months, in the course of my worst experiences as trudovak.

The first day at work and our first assignment to load gravel was interrupted early in the afternoon. Someone brought a message to our commander Shumarov, and he called a group of us to go and load some traverses.. I was so grateful that I was called as one of those who were to do that job, but later realized that it was no blessing in disguise. Ten trucks were to be loaded with thirty pieces each of heavy oak logs. They were so heavy that I was making use of the last ounce of strength in me to hold the other end of the log, while my counterpart was holding his. My arms felt like they were going to be ripped off my shoulders, I hardly could stand on my feet, my hands were bleeding, but somehow I survived and all work was done. It was not easy for the other kids too. I was ready to cry. I began to hate the world, the service, and all that was around me. I gnawed at my teeth not to say a word and make things worse. On the top of it the truck driver, another trudovak, I suppose one of the privileged, was getting mad that the job was progressing slow. He was hanging over our heads and mercilessly urged us to hurry up. He was rushing us, so he could achieve his quota of work for the day. He threatened to report us to our officers. I was ready to curse him in the worst possible way, to swear at him, but I bit my lips.I thought that it was of no use to dignify this nothingness with my protestations. I wondered if he had any sense of feeling, of humanity, something to stop him from playing the role of a torturer! Couldn't he see our suffering? Was he so cold-hearted or derived some pleasure of our pain? May be he realized all that, I thought, but the greed for a few additional pennies which he might receive as bonus, the extra work above the quota, or, if he was a communist, our torturers gave him a pleasure. Was he an accomplice of the communist masters over our life, and was he being so cruelly merciless to add to our physical pain also his moral turpitude? Poor soul! In the one and in the other sense, he was a subhuman snake. Not only mine, but that of all my trudovak companions opinion of him, he was a despicable character. We stoically suffered him by ignoring his urgings.

Having finished this job, we were taken back to the gravel job. Shumarov simply nodded to us to pick up our forks and continue with our work. Now I did not find this job as bad as the loading the traverses. But I very soon, and the others, discovered the trick how not to overdo it. We followed with one eye the movements of Shumarov, and the minute we noticed that he would not be watching us, we just hung on the forks, talked among ourselves and cared not if the job would be finished. Thus, the very first day we discovered the way to dodge the work. Well, there were those who tried to impress Shumarov and hurried to finish their job, only to be sent to help on our assignment and all those who were falling behind. At about the end of the day, before finishing our work, I noticed a trudovak who the whole day was lying under the shade of a small tree. He was reading something. I decided to initiate a conversation with him.

"My friend", I began, "What are you, some kind of a big chief?"

"I am a trudovak like you", he replied, trying to be polite and friendly.

"But how comes, that you could stay under the shade the whole day and read, and we could not catch our breath? Come and join us, so you could understand what this service is all about." I continued to question him.

"Everyone does his job," he replied. "You do your job, I do mine."

"If this which you do is work, then, good for you!" someone of our group told him.

This trudovak, who spent the day under the shade, was some sort of a technician, in charge of measuring the amount of work done by us.

"Everyone knows himself," continued to defend himself our friend, focusing on his reading." At least you have your peace when you achieve your quota and nobody ask you any more for anything. I could go to jail any time."

"I am ready to go to jail instead of you," I retorted, "if you wish to change jobs."

Apparently this fellow had some big wig behind him, who had secured for him this privilege, not because of his technical qualifications.

As our exchange was taking place, I saw that Shumarov was coming in our direction, most probably noticing our diversion from the piles of gravel. We hurriedly stopped and picked up our forks to continue the work, which, according to our friend under the tree, carried no responsibilities. This evening we finished our assignment quite late. We lined up and marched back to the camp, while everyone had some complaint to share with his companions, about his pains and aches.

14. Sand! Sand! Sand!

This is how my service as trudovak began. The next day we were sent to sand detail. During the month before our arrival the trains had brought mountains of sand, which was unloaded along the railroad line. It had to be transported to the concrete units by trucks. We were assigned in small groups, three or four men to a truck, loading it on the truck, then unloading it at the unit. Most of our work during the summer was unloading railroad cars, brought to the runway, near the units where a narrow rail track was installed, then, when the composition was unloaded, moving it with wheelbarrows next to the units for use. In the morning we would line up at the tool barracks to be given a shovel (to be brought back and accounted for to the same barrack). We would jump on the railroad cars and start unloading the big piles. Loading and unloading sand was not as unpleasant as loading and unloading gravel with forks. The shovel smoothly would penetrate the pile, then lift it and throw the contents in or out of the car. But at end it proved much more exhausting than the gravel. The sand was heavy and the speed which was required from us, the amount of sand which was to be loaded or unloaded, broke our physical stamina. When we had to load trucks with sand, the drivers were after us, because they had to make a quota of truckloads; and since they did not have to do the heavy work, and their superiors promised them all kinds of privileges if they overfulfilled their quota; and further, since they did not have to do the heavy work and their own performance depended on us, they constantly threatened to report us to our officer for not working fast enough. They, among themselves, were in some sort of an emulation, competition stakhanovite style, so it was all done on our backs.

In the beginning some trudovaks paid attention to these threats, and even got themselves involved in this emulation movement to overfulfill their own quota. They worked like idiots, thus exposing the rest of us to the wrath of the officers, not to mention the resentment of the rest of us. Some of them gave up very soon their zealousness for more work, but some continued, and won the praise and the disposition of the officers to award them with better jobs. They came to be hated, despised and deeply resented by the rest of us. We secretly cherished a moment when we could deliver our vengeance on them. The officers and the truck drivers socialized under the small shade of the single pear tree, and gratefully rubbed their hands watching those who killed themselves to please them. The rest of us sought to cheat in any possible way, disputing the estimates of technicians, drivers and officers of our work.

I never tried to overexert myself, tried to maintain an average, moderate speed. And so did my group of ten trudovaks. If we would not be classed at the last place, we would be most of the time next before the last in performance as measured by the quotas set for the work. I was not worried and always ready to explain our lack of speed to any officer who questioned me. The *"oudarnitsi"*, those who overfulfilled the quotas, in the early days laughed at those of us who lagged behind them and acted as superior to us. It did not bother us. My indifference and composure on these matters gradually spread to more trudovaks, first in my group and then to the other groups. One trudovak, Vulcho Savov, reprimanded me once and urged me to work faster. I told him to mind his own business and not to interfere in my affairs, that he was free to do as much work as he wished. That was his business and I would not tell him anything about it. In due course of time, this same Vulcho, who had told me that he had to work more on my account, became such a lazy bum, that I felt it was my turn to return the compliment.

When we loaded trucks with sand, the *oudarnitsi* would end up with twelve to fifteen trips a day, my group never made more than nine, and most of the time seven to eight. In any event, we worked on these trucks, transporting the sand from the railroad station some eight or nine days. Whether because the work was so heavy, or because we were not used to it, during these first days I was scared that I could not hold it for long and that it would not end in anything good. I began thinking that this Diarbekir was to be my graveyard. The incident with my small mirror of the first day was assuming greater and greater significance.

Sand! Sand! Sand! There always was sand to be taken care of. It seemed to me that the whole summer we were loading and unloading, or pushing wheelbarrows with sand. Early in the morning the whistle of Diado Rusko would be piercing our ears, the barrack would be stormed by the company Day Duty corporal and the Trudovak on Day-Duty of the barrack, and with screams and pushing and pulling legs, nobody would be left to fall back in his bed for a second. One had to get dressed (most of the trudovaks slept in their daily dress), run to the water pump where a long pipe with ten or more spigots had to accommodate hundreds of trudovaks, who cared to wash and refresh their faces – most of them did not care – run to relieve themselves, and report to claim their shovels. Very soon everybody was on the sand mounds, shovelling in one way or the other. All this process, from getting up, to beginning work was an unbelievable chaos, pushing, pulling, swearing, cursing, yelling. First you wake up, and fight to stay in bed even a second more, which had no price, you want to steal from all other time, but there was no such luck.

You try to get dressed, but one pushes you from the left, another from the right, third screams at you. You look for your shoes, for your socks, if any, for your hat, for your shirt, while it is all a mess and everybody's things are there in the same pile. It was such a crowded place, you can't imagine. You struggle, you fight, you feel the humiliation, you think of the vast university halls, you think of the libraries which you have roamed in and the books you had enjoyed, and you see yourself in this human barbarity ... and you feel the pain rising in your throat, you think that death alone will save you from this hell, from this black fate, that this misery cannot continue for long... The first few days I had made the decision not to violate the principles in my civil life, poor as I was. I managed to pull myself out of the bedlam, run to wash myself and my teeth. I carried a towel with soap wrapped inside, but in a few days it was all forgotten – not that I was tired of it, not that I did not appreciate it, not because I was getting lazy, but because there was simply no time, the exhaustion was taking its toll. It occurred that for weeks I could not wash myself, covered with dirt and sand – on my face, in my ears, in my nose, over my entire body. I still looked for a little time and for some water to do even the slightest of these privileges which in civilian life are so precious. There always was sand to be shovelled out of the cars, or pushed in wheelbarrows.

One way or the other, the sand from the railroad station was transported by trucks. By that time a unit of trudovaks had built a rail line extension to the vicinity of the runway. But I had already come to the conclusion that something had to be done so that I could get out of this hell. I started bombarding friends with appeals for help. I could join the choir of the SLT conducted by Philip Kutev. Nothing came out of that. I was condemned to carry the heavy trudovak service as it was. Regrettably, the sand was not finished. During the whole summer, even to the last day of my service at Bezmer, it was still there and I had to be a slave to the shovel. After the installment of the rail extension, entire compositions of 15 to 20 cars would pull right near our barracks. There were two kinds of sand – from the sea and from a river, Tundja. I believe twice a day compositions of river sand of some eleven cars was arriving, from a station at the village of Tervel. Much more serious was the sea sand, arriving one composition a day, of some forty cars.

All these compositions, river or sea sand, had no strict schedule of time of arrival. Usually these compositions would arrive in our time for rest, evenings, nights, and early mornings. Apparently this is how it had been arranged. We accused the railroad officials, but the railroad men, also disgusted by this traffic plan, explained to us that the Commander of the project had requested from the Central Management of Bulgarian State Railways such an itinerary, so the compositions would arrive outside of our working hours. It often occurred that when we had to end the day's work at 6.00 p.m. the locomotive would whistle from the Bezmer Station – its signal that it was on its way to us. Exhausted and tired, ready to return to campus for supper, we were ordered to wait for the train to arrive. The piercing signal of the locomotive was penetrating deep in our souls and we silently cursed and swore at it. We observed the reaction of our officer in charge, and he pretended that nothing extraordinary had happened. It looked as if he was pleased by the event. Reconciled with our fate, we sat down and waited for the cars to settle at the stop. Then we would jump on the wagons and start throwing the stuff out. Still, this timing was the preferred irregularity for us. Most of the time, we would go to the camp, eat our supper, ready to go to bed, or already in bed, enjoying in full the few minutes which we had before falling asleep, and then, the whistle of the train at the Bezmer station announced the arrival of new sand.

"It is a composition!" someone would scream.

"No, it is not a composition. It must be some other train!," some would express his wishful thinking.

"No, it is not other train, this is our whistle," would continue the comments some pessimist.

"Eh, the devils take it", I would think and start getting ready.

Everybody was mad like hell – the optimist who tried to persuade us that this was not our train hoping to steal a few minutes of our free time, and the pessimist who saw only darkness in our destiny. Hatred and despise loomed all over us and in us. Then the tension set in. We strained our ears to see if the optimist was right, or the pessimist had got it straight. We were like wild animals fearful of a predator approaching us. We clearly heard the train approaching the camp, it was crawling on our line, it was not flying in another direction. We expected a miracle, but it did not happen. In a few minutes the train would pull in, and give a last whistle blow. Then we would start guessing, or rather trembling in expectation to hear the voice of the platoon commander to call us to proceed to the train. Sometime we were lucky. It was one or the other of the two platoons who had this honor, but more often than not, it was us who were the victims. We would engage in calculations as which turn it was, but the commanders had their own plans and our calculations most of the time were wrong. It was for us to go. Once the order was given, there was no way out. We had to get up from our beds, look for our shoes, pants etc. get lined up outside the barrack, and march in the darkness towards the cars. The march resembled more a funeral procession than a youth festival. Every step forward was uncertain, was boringly unpleasant, reluctantly made, as if we were going to face an execution squad. None had a word to say to the other. Everyone was deeply in thoughts which were not quite good. The faces of all revealed hatred, as if frozen on them, the fists were tightly squeezed and stuck in pockets, the souls of these young men were ice cold, and their hearts bleeding. The brain was refusing to work any more. Only the legs moved forward and forward.

"Miserable men", I thought to myself.

"Slaves!" was my following thought.

"Eh, would there be a salvation for us?"

"Would the sun shine on our street again?"

There was not one happy thought in my mind. The stars had already covered the dome of heaven, sprinkling the world with their brilliant colors, sparkling in the crystal clear skies, but for these miserable slaves, they shined with cold and mysterious light, lacking the beauty and the charm which under different circumstances would cause a delirium of excitement and inspiration. The moon, with its iced face transforms us from living creatures into dead apparitions.

We are marching... Marching silently. Some excitement occurred at the barrack where we would receive our shovels. Some eager beavers wanted to grab a better tool, but the majority did not care and took whatever was handed to them. With exasperating indifference we picked the tools and marched to the line, as if it did not matter with what whip they would have beaten us during this unlucky evening. In the grave silence, the disgust and despise in every step and movement, one could hear their, the trudovak, curses of their own fate. So, we climbed on the wagons. Depending on the size of the car and the amount of sand, we would be distributed in such a way that all cars are covered. Everyone preferred a smaller car, but there was no choice. And in a minute the action begins. Some men are better in handling the job, others are rather quick, or stronger, and rush to unload their car and return to the camp. Others were weaker, and in effect, unable to do the job. One wonders if they would be able to make a living in civilian life, but by a strange concurrence of events, had been thrown here to serve their military duty. Poor men! Their appearance alone made one feel sorrow for them. They are assigned on sand wagons too, and it is for those who happened to be with them, to carry their load. Then it is for every group to finish its assigned job and go to sleep. Everyone according to his abilities. Sometimes friends jump on the wagons of their buddies to help them finish their job. Most of the time, once a man has finished, he heads for camp. The help for friends is voluntary, not a duty. So, some would finish fast, and others would go well beyond midnight. But if necessity required that the unloading must be finished quickly, because the wagons had to be pulled out fast, then everybody would be forced to jump from wagon to wagon until it is all finished.

There were of course dodgers. Having arrived in the darkness of night, some would discretely disappear behind the wagons and after the distribution is made, they would jump on smaller wagons and help those in there, so they go free sooner. Some would pretend they had to go and relieve themselves and disappear in the darkness of night. When they would see a wagon empty, they would jump in there with their shovels, wait for the officer to show up and jump down as if they had done the job. Sure there were controls, but in the darkness of the night the officers were as annoyed at the whole thing as were the trudovaks. So, having made the assignments, they would go out there and take a snooze until the job is done. What do they care? Having unloaded our respective wagons we would proceed, in small groups, back to the camp, where the whole world was in slumber land.

Well, there were also morning surprises. It was an almost everyday affair. It could not be otherwise. The trains were to arrive before work time would begin, so that by the time the work day began, they would have been unloaded. The trains had their own itinerary, to move on the intercity one-track lines before the regular traffic begins. On the other hand, if they were unloaded outside of the working time, then the trudovaks would be used for their regular daily schedule. No free time was allowed for night work. The rule of 8-hour work day was not valid here. Those who planned the organization of time, counted the night work, or day work during hours for rest, as overtime for their reports. It was far more important for them to account working time above the designated quota, than the feelings of injustice in the minds of the trudovaks, their health or their humanity.

The trudovaks were cattle which was used for the benefit of the state. So, early in the morning, hours before we would hear the whistle of the Day Officer to wake us all up, we would in our sleep hear the signal of the train coming to the camp. Those who heard it would loudly curse, swear, grimace, and huff and puff. There followed the order "Get up!" and the tired, sleepy exhausted and poisoned by evil thoughts trudovaks reluctantly submitted to the order, one after another slipped into their raggedy clothes and ripped to pieces rubber shoes, hung around their necks the greasy bags and left the smelly barracks. The little sleep which they had had and the morning cool somehow helped the men to be a little more cheerful, in anticipating just another day. But again one had to go through the same bedlam in the barrack – pushing, pulling, screaming, swearing and cursing. There would be always some who would simulate stomach aches, claims of sickness, and all sorts of illness, only to be seen voraciously devouring their breakfast later.

These morning visits by the sand trains were quite frequent. Almost daily a train would arrive at three o'clock in the morning and the lucky platoons would march to unload it. It was a rare luck to go to bed at 9.00 o'clock. If no trains were coming, the officers would always think of some wild schemes to call a company meeting to discuss the work and the emulation movement. They would hold us until ten or eleven o'clock and then send us to the barracks where we would be given to the mercy of armies of bed bugs.

The officers had the sickening passion to order us to sing partisan songs, Russian hymns and party favourites. Lined up in front of the barracks , led by some musical genius, we had to shout with as much voice as had been left in us. Sometimes they would force us to sing in the morning on our way to the wagons. We did not feel like saying "good morning" to anybody, they asked us to sing. Our eyes were closing from lack of sleep, we were falling and staggering in the holes on the road, we were walking like blind men, our brains were still sleeping and some stupid corporal there asked us to sing. Nobody even tried it. With heads dropping down, with hands immovable, deeply in thought in ourselves, if we could think at all this hour, we were more like prisoners of war rather than men serving our military service, trudging the road to some foreign, far off, un-known land where we did not expect anything better than we were already suffering of. The corporal leading us to the train starts commanding:

"Left, Right, Raz, dva, thi!"

Nobody was listening to his shouts. We were dragging along like a tired herd of animals. We begin to wake up when we start working on top of the wagons. Usually these morning trains brought sea sand, loaded by a bulldozer, three piles on the long cars, each one of four cubic meters. You grab the shovel and start. But the work does not progress fast enough. One, two, three, five... one hundred shovels...the pile is still there. And you continue throwing and throwing. Your hands begin to hurt. Your muscles become weaker, your spine aches, you sit a little to rest, the corporal shouts in your direction to scold you, his eyes are the eyes of a predator. His look at you pierces your body, you sense his hatred, his evil and his utter resentment of you... You get up and you continue throwing the wet sand, but you feel that your head will explode. You think of life, you rethink your past, you venture to dream of your future, if there is any to think of at such dreadful circumstances, you focus on politics, you curse the Free World that does not do a thing, that drags its foot in fulfilling its numerous promises to free you from that slavery, that it does not give a hoot for your sufferings and you wish them, those beyond the ocean, to be visited by the same plague... Oh, no! If this happens, then who will come to save you?... Or you will be shovelling these piles all your life, until you drop dead, will sleep some three hours a night and return to the same misery, will suffer the same starvation, the same slave labor, the same katorga (the jail penalty labor camp). No! This plague should not take the whole world. The rest of the world will sooner or later realize that they have to do something, have to come in any possible way and liberate you...

Such thoughts, painful, intolerable, a million of them, upset you, are ready to explode your skull... and the pile of sand still stands there. You stop thinking, you feel your head is spinning, you grab the shovel, lean on it not to fall, and then you open your eyes and see the corporal still there, to return you to reality, that you are still a trudovak in Bezmer, that you are a helpless slave and you have to resume your work.

Fate!

Very often, after we had unloaded a train and time for breakfast had long gone, we still had to have it. We would prefer that it be brought to us, instead of marching back one kilometer at times. The Corporal never considered our preference, which, among other things, meant that we could rest until the food would be brought to us, and saved us the walking back for the cup of tea sweetened with marmalade, and then march back to move the sand to the complexes with wheelbarrows. So we marched back and forth.

Our noon rest hour, when we also ate our lunch, brought to the work place by the *bakkari*, was often interrupted by river sand, scheduled to arrive just after we would have finished our lunch. We would hear the signal of the train arrival at the Bezmer station. We hurried to finish our lunch, grabbed the shovels and soon all were on the wagons. The rest is a story which I have already told. What I have to add here is, that the unloading of these trains was taking place about noon time, in the hottest months of the year – July and August. The hot sun unmercifully poured its burning heat, penetrating to the heart of our bones, over our heads, our bare backs, our legs. Rarely we had water to wet our lips, let alone to cool off our bodies and very often in hours before our lunch or supper, totally exhausted, rushing to rid ourselves of this burden and go and look for food. Sometimes the food would be brought to us, open to all kinds of dust and bugs flying all around and drowning in the opened *bakki*. Winds sometimes would blow the sand in it. Too often we found in our cans things, which in civilian life would turn our stomachs and we would not eat it, but here? Here you eat anything that came around, because starvation was the second alternative which we could not afford.

Terrible!

Once I was looking at an egg brought for some reason, but covered with dirt.

"Will you eat it?" A corporal who was observing the act asked me. He apparently was thinking that I would not stop to this humiliation, in my capacity as a man with university education, that my pride would come between me and this indignity, even at the price of the unbearable starvation. He apparently enjoyed the pitiful state I was in, reduced to nothingness before him in this wilderness. Reason and compassion had no place where the two classes met – the class of the doomed, and the class of the privileged with party ticket.

Sometimes the train compositions arrived during working hours. We appreciated this good luck, because if it was our turn to unload, we would not be called in the evening, or during the night if other trains arrived. It would be the other platoons. And we did not have to make up the time with extra work, but would return to the barracks. If we finished before closing time, we would march to our camp. But be that as it may, we developed an organic hate for the trains, for the whistle of the locomotive, against the freight RR cars and our commanders who supervised us there. There were, of course trains, which we saw at the RR station when we happened to be working there. These were the passenger trains. Looking at the passengers while the trains would briefly stop, we recognized the still existing world of humanity and civilization and dreamed of the day when we would jump in and leave this hell behind us... And then we would again turn to our assignments. The word "composition" became the most hated expression in my vocabulary. It never led me to the world of music. For me it meant mountains of sand which was to be unloaded, unbearable pain and excruciating exertion. The late evening and the early morning hours, when in complete exhaustion, we had to grab the shovels and march to our circle of tortures, as if Dante's *Inferno* was gaping at us.

"Composition!"

"Composition!"

"Composition!"

This word for a long time was not forgotten. How much pain, how much horror and how many curses it would trigger in my mind in this desert of nowhere, near Bezmer! How many martyrs like me would recall in the future this soulless snake of wagons, this metal beast which huffs and puffs, intrudes in our hour of rest and devour a chunk of our flesh from this miserable life? I do not remember whether it was Stephenson, or God knows which Russian inventor, had created this guillotine, but I am certain, that if he had spent one day with us as a trudovak, he would have concealed his invention, he would not even had thought of such a monstrosity and such an instrument of evil designs.

Well, there was time, plenty of it, when the trains were not there, but we would not be let free to wait for them. In the regular working time, we had to move the piles of sand brought by the trains from along the rail line and transport it with wheelbarrows near the complex for use. And since the use was not keeping up with the supplies, we had to pile it into twenty or so meters high mountains of sand. Indeed most of our time during the summer was taken by pushing the creaky wheelbarrows. The wheelbarrow was a very liberal instrument*. It could carry one third of a cubic meter sand and could carry one tenth of its normal contents. The men operating them could, depending on circumstances, determine how much to put in. If the corporal was not watching, and the man assigned to keep record was negligent, easily distracted, or a good man, the amount of sand and the number of wheelbarrows fixed as a norm for the day would be proportionately reduced and overstated.

The corporals would be furiously mad when they saw a wheelbarrow half full, or when some of the men leisurely descended the mountain, taking their time to go back and refill. The man caught with half an empty wheelbarrow was ordered to go back and fill it up, the man going slow, one way or the other, was screamed at to fasten his walk. The officer would go to the record keeper and cancel ten, fifteen, twenty wheelbarrows from the record of the man caught cheating. The record-keeper would be instantly relieved of his privileged duties and sent to push a wheelbarrow, while the assignment would be given to someone else. There were of course those who went along with the orders and filled their wheelbarrows, ran up and down the mountain of sand, to finish their norm and be relieved from duty, which rarely occurred. Others were simply oxen and pushed, and pushed and pushed. That was their choice or their mental incomprehension of the job. The majority preferred to move half-empty wheelbarrows and take their time going up or coming down, except when a corporal appeared from somewhere. But it always happened that the officer thought we were not filling up enough our charge, the wheelbarrow. When I was stopped by sub-lieutenant Penev and asked if I considered the load in my wheelbarrow a full load to fulfil my norm, I assured him that instead of 120 courses, I would make 240, half-full, to meet my norm. He would look at me, incredulous, as if to say: "You, smart pants!" But he reported his conversation with me to the commander of the camp, who said nothing, except to look at Penev with a smile and leave him behind. This same Penev once had made a comment about me: "This guy, the Daskal (the teacher), is just a dzhezve (a pot to make one Turkish cup of coffee) of bones and skin." What then, he did expect from me? Well, that was the service.

---

* Our wheelbarrows were not running on rubber tires, but on plain metal rings – inch and a half wide.

The most difficult part of this operation was pushing the wheelbarrows to the fifteen to twenty meters high pile of sand. How do you push a wheelbarrow in the sand to any height? Then we had to set up wooden planks and push upward on them. It sounds neat, but sometimes the wheelbarrow's wheel would leave the track and you were stuck. You had to go to the front, lift it up on track and then continue. At the same time others were following and had no choice but to stop and wait. And if you could not lift it alone – this often was my case – they had to come and help you, or they would swear at you that you were delaying the operation at their expense of time. In such conditions the day's norm was eighty wheelbarrows. Who could fulfil that norm? Very few indeed, which maddened the officers. But what could be done? There always was plenty of sand to move, the work on the runway was not progressing fast enough, so they could not be short of sand supplies, and if we did not finish our norm a day, there was always the next day, the failure of the previous day was forgotten by the officers, and they were after us again the next day, and the next day, and the next day. There was no end to it. We did as much as we could.

The picture at work was horribly pathetic. We often counted some forty or fifty men at one sand detail. Blackened by the sun, our ribs counted like the fold of an accordion, raggedy, some of us looking as death shadows, others rough and wild, all exhausted, some loading their wheelbarrow, others pushing theirs up, falling, staggering, and then returning for another course, swearing and cursing, left and right, stopping to remove some small stone blocking the metal wheel, or falling off the track, and pushing, and pushing, and pushing... to build the runway for the Russian migs, to defend the empire of communism left behind the second world war which "liberated" us from fascism. Miserable creatures! We strain our muscles, our ribs shake up and down when trying to catch our breath, moving under our thin black skin, the tendons under our knees ready to snap, and huffing, puffing, cursing, swearing, crying and still making the last efforts to advance one more foot ahead to the peak of the sand mound. There, the record keeper was marking only the wheelbarrows, which reached him. Those spilled below the crest were not counted. I was so humiliated, and was so intimidated by the rest of those behind me on the line, because my wheelbarrow more often than anybody else's fell out of the plank and had to be lifted again and again. Indeed, anyone who like me failed was subjected to such humiliating adlibs from his friends who followed him. After the day would be over, it all would be forgotten and we again would be friends. What choice did we have? We were all here for the same reason: to be punished for our temerity not to walk in the right paths charted by the Party.

Sometimes I would stop at the bottom of the sand and look at the picture up the hill of sand. I would remember old films where penal work prisoners were shown in far away lands, on dry uninhabitable islands, and saw no difference between us, serving our military duty in conditions comparable to these penal institutions. I found one difference though. We did not have tattooed numbers on our backs. We were still seen as regular citizens who one day might return to society. But in fact, it was not so. We too, had numbers, but they were tattooed in our minds and in the platoon-company registers. Nobody was without a number. On the mound the record-keeper was noting our courses as per the number which we were assigned and we shouted at him the number assigned to us in the morning. Thinking again of the old films and the penal servitude quarries – they did not have to yell their numbers. Sometimes I had to repeat my number two or three times, while the record keeper was busy with somebody else: "Twelve! Number twelve!

Did you write it dear." Annoyed, the record keeper would give me a sour look and put another notch against my number. I would head down the mound for another course.

Meanwhile, the sun was pouring its burning lava from above, the sand was getting hot and unbearable, while we moved, slowly but surely, on the slopes up and down. So did the days, the weeks and the months of our service – May, June, July, August, September, October. The compositions never stopped arriving, the locomotive whistles never stopped irritating us, we unloaded them, we pushed the wheelbarrows up and down the sand mounds. Sometimes we had to begin a new mound and to open new pathways in the virgin fields. Then we had to go through holes and small embankments. We would strain again our muscles or suffer the sudden fall of the wheels in some small pits, shaken to our hearts, physically unglued, feeling as if our brain was being split from the skull, that our legs were going to break at the knees, our hands were to be pulled from our shoulders or our elbows, not to mention our twisting wrists, and at the end feeling our whole bodies falling apart. Some of the men were running like wild to beat the time and finish faster. Once I tried it, but did not last long and gave up the idea. It was just beyond my physical abilities, and that there was no benefit from it at all.

The number of wheelbarrows to be made every day was different. The terrain and the conditions were evaluated by the officers and determined accordingly. Some days we had to move eighty, some days one hundred, some days 120. It all depended on the arbitrary judgment of the officer in charge. We would ask in the morning how many were we expected to deliver. Some wanted it, so they could kill themselves to fulfil the norm and go and rest, others were simply curious and did not particularly care to reach the norms. Most of us managed usually half of the number announced in the morning. Then, at the end of the day, we would be left behind to finish and then return to camp for supper. The officers would leave, the record keepers would have had it enough, and in one hour, standing around, it was all recorded that we had fulfilled our norm and would go back to camp. Nobody ever asked what happened. The idea that we were left behind was enough to make the cowards to hurry up the next day, so they would not be left on evening duty. Usually the majority would meet the norms, but I was always at the tail and perpetually with the group left behind for evening duty. Sometimes the record keeper was a friend of mine and marked every course I made for two and so I would not be penalized.

It was quite often that we were penalized to work after hours. Some times the record-keeper was easy going and in a hurry, overstated five times the number of wheelbarrows we would bring, and very soon we were on our way. Once our corporal, Angel Khristozov, one evening, when we had fulfilled only half of our norm, stayed behind with us. At one point he had to leave for a while and handed over the record keeping to someone else. By the time he returned, we were all ready to go. The record-keeper fixed it all. That evening he stopped me outside the barrack and asked me:

"Spase, how many wheelbarrows you made today?"

"I do not know!" I answered him and walked away, leaving him to stare at me with hatred in his small eyes. He apparently was an activist in the Communist Youth Movement, who knows, may be some turncoat, heading for a better career, but now a blind tool in the hands of the officers. *

"Ivane," he turned to a student of Economics, expelled from the University some time ago, "did you finish your norm?"

---

* I recall now that he had come from a family of wealthy "kulaks". They had a harvester confiscated by the communists.

"The Record keeper helped me," told him Ivan Zhivkov, from some village in Pleven District.

"We will see!" ended the conversation Angel. We never saw anything as a consequence of this encounter.

The work of the record keeper was very delicate. Usually he was one of us, ordinary trudovak. I was picked quite often to do this job. This was during the time when our Platoon was under Sub-Lieutenant Sotirov. Sotirov was from the city of Sliven, not far from camp. He apparently was from some upper class and cultured family, with good manners, perhaps people from the old regime, with some good education, and, who knows, may be a former supporter or servant of the pre-Communists regime, but was somehow on the hysterical side. By calling on me he either wanted to free himself of my tardiness on the job, so he did not have to deal with me, a man of higher education, or secretly sympathyzing with me. So September and October did not turn too bad for me – on account of Sotirov. He might have been of Greek or half-Greek origins, because his name was of Greek roots. He saved me of much of the work beyond my physical abilities. It was much easier to stand on the top of the sand mounds, scratch notches for all wheelbarrows brought there and help your friends with the record-keeping. Many a time, I would mark twice the same course for those whom I considered my friends and who gave me little trouble. It was a different matter that the wheelbarrows had to be reasonably filled and be dropped at the crest of the mound. It was not quite pleasant to give explanations, sometimes to ordinary trudovaks, for their scores, sometimes to passing corporals who observed some cheating. Once a corporal, by the name of Gentcho Manev, looked at the wheelbarrow of a friend of mine as I was marking it and asked me:

"Daskale, do you count that for one course?"

"No!" I lied to him. "I count five such wheelbarrows for one course."

He did not believe it, but continued his walk. It turned out that when I was keeping record, almost all men achieved their norm, and for the sake of covering up the cheating, I would leave some of my friends behind, with very little left to be done. But there were times when there was little to be done. The cheating was so obvious, and the problem threatened to become a scandal and I would loose my privilege. Some of my friends literally stopped to work and the reports reached the officers. It could not be kept a secret. The mass of trudovaks who did their share conscientiously, would not tolerate that. The officers noticed the dragging of foot by my charges. Who could check it, anyway? But I soon realized that if I did that cheating early in the day, by midafternoon they would slow the work and it was noticed by the officers. So, I would hold until late afternoon when everybody would be close to fulfilling his norm with my help. I did not care for the work done or not done, but I cared to create the impression that the work was proceeding in a normal pace. Everybody benefited from my artful work. Those who had just finished their work, I would give them twenty or so more wheelbarrows, so they could claim the stupid title *udarnitsi*, and catch the attention of the officers. Those who were well under, with my over-writing avoided punishment, and those who were extremely low producing, I would leave for brief evening work, and everybody was happy. The technicians, who measured the work done, had little chance to apply any standards and to avoid problems with the officers went along with us, recording as much as was necessary, sometimes more, to meet the requirements for the day. Thus everybody was cheating everybody. I even suspect that the officers too looked the other way, for as long as the work on the runway was progressing. Still, at some time, my generosity to all was recognized and some of the officers demanded my demotion. A fellow by the name G. Russev, replaced me. He was the kind of a guy who thought that every wheelbarrow which was not full enough, or any "error" in recording more than in reality, was coming from his own pocket. So, I found myself again behind the wheelbarrow, to push, and push and push to some breaking point. Especially things were difficult when the officer in charge would stop at our place of work and watch our performance. Then we had to overdo it, so he should move to other groups. When it appeared, sometimes in August, that we were falling behind in moving the sand, the officers tried to change the system: instead of assigning us so many wheelbarrows per day, to assign us so many cubic meters of sand for a man, a day. The assigned piles along the rail tracks were arbitrary and it soon appeared that it was not going to work, because half of the sand which was expected to be moved in a day, was still there the next day. Soon this system of accord was abandoned and the wheelbarrow system restored, with all its imperfections and cheating going on.

In late September and October, when it started raining and the terrain for crossing with the wheelbarrows became a pool of mud, the management decided to install narrow-gauge rails, and the sand was to be moved by small wagonettes, each one about a cubic meter of sand or more. There was little chance to cheat here, because there were twenty or so wagonnettes on the same line, they all had to be loaded and pushed at the same time – two trudovaks to each one. The wet sand was heavy to load, the rail tracks were twisting, and breaking, and the wagonettes sometimes would jump off the tracks – two wheels on one side, outside the track, and two wheels inside the track. So what do you do? You try to lift it up on the tracks again. It was impossible for the two men in charge of it to do it alone, so all those from behind were to join in and lift first the front and then the back of it. You try with hands on the rims up, you turn around and put your back to it, one screams some order on one side, another screams other from the other side, the officer would come and give his orders, and after much effort, the cursed thing is on tracks and you push again. By the time the operation is finished, you feel that your tendons will snap, your back hurts, your muscles are reduced to shreds and your soul exhausted. I remember at one of these occasions a scene from the book of Mein Reed or Jules Vern, if I am spelling the names right, "Fire Land" I believe was the title, where a penal servitude prisoner was crushed under a rock he was trying to move with other prisoners, then the rock moved and rolled over this man. When I was giving a shoulder to this cursed wagonettes, I was careful not to find myself in a situation where it could fall over me.

This was a *katorga*, a penal servitude for prisoners in some far away island, described by the western novelists. I lasted there two weeks, and with a sense of utter humiliation, approached Sub-Lieutenant Sotirov, begging him to find some other assignment for me. I could not survive it. He gave me such a sour look, and grimaced in such a disgusting way, that I regretted for a moment asking this favor and humiliating myself. He looked at me and gnawing his teeth, said to me:

"Eh, Raikin, you are such a pest!"

But he moved me out of the wagonette detail. From a thorn to a hawthorn, as the Bulgarian proverb goes, or as the American word goes - from the frying pan into the fire. He assigned me to a detail which was in charge of spreading the sand unloaded by the wagonettes. While I was working on the wheelbarrow detail, sometimes I was assigned to serve as spreader of the sand on the top. It was not much of a problem. But, here, on the wagonettes, it proved an entirely different story. Sotirov assigned five trudovaks to every six wagonettes. The wagonettes were to make twenty five courses a day. This raised the amount to be spread when they were unloaded. It amounted to 20-30 cubic meters a day per man. It was a killingly heavy work. As the mound of sand was growing we had to throw the new loads above and beyond, to prepare the place for new loads. This misery had no end. After four days work in the new detail it appeared that our group could not keep up with the volume of sand which kept coming. The sand piles along the tracks were growing higher and higher, we had to throw the new loads over four meters or more over the hill and our shift was replaced by stronger men.

So, this is how my days of "military service" were passing along the sand dunes at Bezmer.

....

Here ended my notes in the gray notebook which I had received in December 1989 from Sweden. They were written in February 1951. I have stopped abruptly, without finishing my story for Bezmer, perhaps for lack of time when I received my order to report to Balchik. Or, maybe, anticipating nothing better at Balchik, I had lost my will to continue.

15. The Cursed Cement

Reading through these notes I remembered that I had not touched on the most frustrating experiences which I had at the time, the worst work that I had to do: the unloading of RR cars of cement bags. If my suffering at the gravel and sand mounds was bad, then the unloading cement was ten times worse. I have never cursed my fate more in those days than the days when a cement composition would arrive. It took precedence before anything else. I am mentioning this now, and adding it to my story after a forty six years long break, so it should not be forgotten. The cement compositions were to be unloaded at the Bezmer station. The rail tracks built to the camp could not sustain the weight of the cars. They would arrive at any time, day and night, every three or four days. And they consisted of tens of cars, covered wagons. They had to be unloaded immediately. How many tons were in every car, how many bags of these 50 kilos of stuff were inside of them, I have no recollection of that, but I have not forgotten that the unloading of each composition took an eternity.

We were distributed according to the number of cars, so that all cars should be covered. Two men would be inside, bring the bag, holding it at the door, and deposit it on the shoulders of the trudovak, standing under the opening of the car. You go there following the line before you, you position your legs spread for better support, you place your hands on your hips, to support better the bag, they place it on your shoulders around your neck, or sometimes drop it or toss it, and then you march to a shed or to the trucks parked some distance. The cement bags were covered with cement dust, you close your eyes and your mouth until the bag is placed on your shoulders, and when you feel that the operation is finished, march after the previous fellow, so somebody else to take your place. In day time we saw where we were stepping, but at night time we often dropped in some hole and if we did not fall under the weight of the bag, we continued with pain to get on the road.

How many times I crouched under the train door, how many cement bags I carried on my shoulders, and how much I suffered this hellish torture, I could not say. There were occasions when the bag was ripped open and the contents of the cement ran down your back, your chest, your face, your neck, your entire body, penetrating your private parts, clustering in your hair, plugging your nose, your ears, your eyes, your mouth and reaching your throat. We ran down the path to deliver the bag and relieve our pain... Only to end up with another bag. The officers were standing by and would not let anyone try to dodge the line. It was a murderous line. It was a murderous job. It was a living death. The unloading was to be done as soon as possible and no one could be left out of it. Swimming in sweat in these hot summer days or nights, with your bones aching, with your mind closing, amidst the suffocating smell and taste of this stuff, poisoning any feelings which you may have, sometimes overwhelmed by the sense of lack of sleep, you go on, and go on, and hope that the pile in the car is coming to an end... And in your ears sounds the call of the political officer that you are suffering all that in building socialism, to defend socialism with the Soviet Migs... Until these cursed cars were unloaded, our souls were going out of our bodies.

Here I was destined to live one of the most unforgettable, and most heart breaking moments of my life – the day my mother died, and later, learning of my father's death. It happened at the Bezmer RR station. One morning, about 10.00 o'clock, as we were bristling under the weight of the cement bags, a passenger train pulled into the station and stopped along our train. I looked up and opposite me, in the passenger train, in a pleasant compartment I recognize four of some of my best friends, students in the Theological School – three young priests and a layman. Comfortably sitting, engaged in a friendly chat, clean and well dressed, combed and radiating leisurely confidence, as the whole world was a happy place to live. One of them noticed me and for a second focused incredulously his eyes on me. In another second they all leaned at the window gesticulating with hands sending me their greetings. But I noticed how their faces froze.

I jumped out of the line, not thinking of officers, corporals and other men on the line. They had all stopped, looking at me. I jumped into the train and in a few seconds was with my friends in the compartment. I squeezed their hands, they squeezed my hand, I read in their eyes pain and sorrow for me. I do not know what I looked like, with cement all over me, over my head, my face, my back, my clothes. They were searching for words to express their shock and I was fighting the tears in my eyes. In a minute the RR station master's whistle gave signal to the passenger's train to move and I hurried out, only to look at them going away, and me, remaining at the platform. Was this the Raikin they knew as scholar, as President of the Student Theological Society who presided over public meetings, introduced prominent speakers, ravenously searching truth in the books for his research – honored, respected, friendly... or this pitiful creature covered from the top of his head to the split rubber shoes where his toes were protruding – blackened, painfully broken man, reduced to nothingness, on the brink of death?

I do not remember what we said to each other in the brief few seconds in the compartment. As I jumped from the already moving train, I climbed in the cement wagon, went to an empty corner, leaned to its wall, slowly sliding down to the floor, my knees trembling and refusing to hold me, and my hands covering my face where not only tears but uncontrollable cry was erupting from the bottom of my soul. I could not control myself. The men had lined up at the door, looking at me with sorrow on their faces, as if they had understood the revolt which had exploded in me against all that which surrounded us. One officer, witnessing the whole event, quietly, told them to continue their work and leave me alone. In this moment of supreme suffering, he did not reach to insult me and hurt me more than I was already hurt. In half an hour I had managed to compose myself and walked out. All cement was already unloaded. I marched back to camp with them. Nobody asked me a word. I did not say a word to anybody. But I read in their faces and in their looks their profound compassion. I heard only one comment which a fellow from Haskovo district, with whom I was not particularly close, to say: "Poor Spas! We here do not know anything else, but work. What is he doing here with us?"

As I moved around I asked myself: "Why did I cry so uncontrollably? Was I ashamed for my appearance, for exposing the nudity of my life before these friends of mine?" No! I was not ashamed of my appearance! I was crying on account of the cruel fate which had thrown me without mercy in this miasma of dehumanization, in this hell, on this mountain of the worst temptations to which the Evil genius of this world could have designed for me! Oh! I was crushed! I had lost every spark for life, every hope for resurrection and salvation. I was beaten by life and I was sinking deep into the den of the beast which was seeking my death. I lost all hope for survival!

The end of these Bezmer miseries came when the weather turn really bad, winter was coming and all work had to stop... We were surprized one day in November when we were told that orders had come from above that most of us should be laid off, go home and spend winter there. We would be called back to this or another place, to continue our service. When the order was implemented I was on the list of those who were to go home.

So, I arrived home unexpected. I had sent a letter to my father by Todor Vassilev, who was let out on a home leave sometimes in the fall. I had written that as things were going, I did not expect to come out of this service alive.

The announcement that we would be sent home on leave, until the spring, revived my hope for liberation. Arriving home my father was glad to see me. He told me, that while he was waiting in the village barber shop for his turn, Deniu Filchev, father of my late friend Yovcho, half-jokingly had observed to Mincho Kurtitsata, a prominent communist from the time when I was in the Seminary and he in the Business High School, sitting then on the chair for a haircut: "Why did you send Spas to be a trudovak? An educated man, a pride of our village, did he deserve that?" He had said these things discretely giving a look to my father. Mincho had answered: "Well, what do you expect, to send him with a gun to guard the border? Where are we going to find him to recover at least the gun?" So there was the explanation for all my miseries. Well, with gun or no gun, Spas found the border eventually, and vanished in the West. The reader will follow this story further.

X

TRUDOVAK II
THE PECULIAR INSTITUTION
BALCHIK

"… A peculiar institution, as they
called it… Negro slaves, unlike free
whites could be forced to toil… regardless of the effect upon their
health… A Maryland tobacco grower
forced a slave to eat the worms he failed
to pick off the tobacco leaves…"

Kenneth M. Stampp, The Peculiar
Institution, Vintage book, 1964,
pp. 3, 8, 16, 17, 172.

1. My Report to the American and the Greek Authorities

The second part of my military service as trudovak was spent in Balchik, some forty miles or so North of Varna, on the Black Sea coast. It lasted from sometimes in the beginning of March to May 6th, 1951, the day when we escaped, on our way to exile. I never wrote about my experiences at the Balchik camp, but two weeks after crossing the border, June 19th, at the Refugee Reception Center in Drama, Northern Greece, our hosts, Americans or Greeks – we could not tell the difference – asked me to write a report on my trudovak service and some other subjects of my choice. I was given a typewriter and in two weeks my reports were ready. I submitted them to the officials – police authorities, I believe, who worked together with the American intelligence services. Whatever they did with them I would not know. In my report on the trudovak service I took a different approach, more on the institutional set up than personal experiences. I treated both camps, Bezmer and Balchik. Whatever is in this report, will complement the Bezmer notes. My personal experiences at Balchik were not much different from those in Bezmer, and the institutional set up in both, Balchik and Bezmer, was essentially the same. I am including this Report here for a fuller understanding of the Trudovak service – as a personal story and as an institutional study. I have published this Report in my *Political Problems confronting the Bulgarian Public in Exile*, Vol. IV, pp. 334-357 (Sofia, 1993). It first appeared in the periodical *Free Agrarian Banner*, No.5, February 1980. Following is the Report:

TRUDOVAK

Objectives, Tasks, Organization, Cadres, Work Loads,
and Living Conditions of the Institution "Special
Construction Troups", formerly (Military) Labor
Service.

The Service "Special Construction Troups", the former "Trudova Povinnost", a sub-division of the military services in Bulgaria, which purpose was to supply with labor all construction projects in the country of confidential nature, more or less, which had to be completed in a hurry or a large labor force was required. One of the objectives of this service was to save money for the state, which, otherwise, would have to be taken from its already overburdened budget. The organization of work in "Special Construction" results in maximum production at minimal expense. The penal servitude in these "military" units would not have attracted free labor force in the numbers which would guarantee a speedy completion of the projects.

Along with this purely production objective, "Special Construction" was assigned special political goals. In the first place it was the discrete Communist intent to destroy physically the politically unreliable youth. The exhausting physical work, without the prescribed daily and weekly rest times, destroys the physical stamina of the draftees and they become victims of all sorts of diseases before completing their term. Considering that the service lasted thirty months spent in extremely heavy work and unbearable conditions of life, one could not expect that after completing his duty, the young man could have preserved even the minimum of his normal physical strength for a normal life.

There is a story circulating among the trudovaks, that some medical team has established that after twenty-four months of service, sixty per cent of them become sexually impotent. Even with this medical opinion, Tsola Dragoitcheva and Yanko Kaneti, in view of the international tensions, have forced on "Special Construction" to increase the service to thirty months. Recently the trudovaks in Balchik heard that a group of such servicemen in Dimitrovgrad have raised a slogan to extend the term to thirty six months. This rumor was never confirmed, but it fits the pattern of Communist tactics for promoting ideas which otherwise are unacceptable for those concerned, before introducing them as a policy.

Another goal of the Special Labor Troops (SLT) was to re-educate and politically redirect the young men having fallen in their hands, in most of the cases mentally inadequate and physically unfit minority peasants. They hoped that by incessant repetition of political clichés they could impress on their minds the Party line. All along with that this service worked hard to compromise politically the young men, inclined to be won by special privileges. The officers easily discover such young men and easily make them their tools. They elevate such trudovaks to less arduous work and turn them into stool pigeons, to report on their friends, or make them instruments for promotion of initiatives to raise the production. Such is the case with D.D. (I will use initials of the people named in my article) from the town of Maritsa, Company 309 of Bezmer, who in the very beginning of the service exceeded twice or three times the quota of the assigned work. He immediately won praise from the officers, but also resentment from his comrades. He had University education, and this made him even more hated by the rest of the trudovaks, so that they started mocking him. The same types I met in Balchik, 1951, where some individuals with university education, to win the attention of the officers, publicly argued that we should work ten hours a day and raise the production by 150 per cent. Privately they do not hide their anti-Communist leanings, but publicly, be it because of fear, or for selfish reasons, help the officers. At end they are compromised before the rest of the trudovaks, who come to consider them tools of the officers.

Still, along the same line, the officers of the camp conduct active political propaganda among the neutral or non-political youth, uncertain in its political orientation, to frustrate their expectations for liberation and, when possible, to win them to their side. However the communist officers are not very hopeful to achieve this goal. The political officer of Company 309 at Bezmer camp, after trying throughout the summer of 1950 to win such a group, including them in the propaganda circle of his office, eventually gave up. When they, for one reason or another, were late in coming to a conference of the Staff of the Company, furious at their impudence, vented his anger at them with the threat: "Despicable animals, oh, despicable animals" (*Gadove s gadove* – term reserved for the known enemies of Communism), "I will teach you a lesson." It was clearly demonstrated that the goodwill towards the educated trudovaks of this officer had been just a mask. Having failed to win the politically oriented educated young men for the cause of communism, the officers carefully monitored their activities, collecting information for their expressed opinions, so at appropriate time to be able to accuse them in subversion in the force.

The political education of the trudovaks is carried according to a confidential order which lists the themes the political officers personally should discuss at general meetings or delegate authority to the platoon corporals to take over. There is a program for these lectures to be followed every month. The sessions take place mornings, after breakfast, or during the short breaks after lunch. Often the trudovaks had to travel two to three kilometers for that purpose. After delivering the lecture, often interspersed with questions to the trudovaks, the latter are given opportunity to ask questions and offer brief discussions. When a man would ask a question, it is carefully recorded and his name taken down. The young trudovaks often fall for this trick and ask their questions, sometimes with not too well concealed desire to challenge the officer, to make him display his irritation and tell the truth as it is, or at least to make him show his incompetence and inability to defend the Communist economic and cultural policies.

Such incidents are not without consequences for the daring trudovak. He is discretely investigated for his political background and subsequently taken care of. All statements are carefully taken into account and the discussions end until the next time. All trudovaks are split in small groups of about ten men. The materials discussed by the leader are further elaborated on, before the smaller groups by someone with good political credentials. Every man had to record in a special notebook the key points made by the officers and go over them with his fellow members of the group. These small meetings are supervised by specially designated trudovaks, members of the DCNM (Dimitrovski Saiuz na narodnata Mladezh – Dimitrov's Union of the People's Youth). It should be noted, however, that the attention of the trudovaks and the individuals who were supposed to supervise these activities was negligible. Most of the time no such meetings were held.

During the general meetings which follow, the officers often ask questions to find out if the materials discussed before had been memorized by the trudovaks. On rare occasions, when held, they are turned into a farce. The officers soon realize that it was all waste of time. Those of the young men with education could easily repeat the Communist clichés, and those without education are still found to be a tabula rasa. At longer intervals such general meetings are held of all trudovaks stationed in the camp. There the Commander of the project, his political officer and the staff of the DCNM preside and ask questions. This is how they evaluate the political work among the trudovaks. The platoon corporals and company officers make sure that well prepared individuals are scattered in the crowd to whisper the answers to those who are called to talk. The subject for discussion at all political meetings gravitate around the theme of Socialism, to wit:

- BCP (Bulgarian Communist Party), led by Comrade Vulko Chervenkov, as leader of the nation for building of Socialism in Bulgaria.

- BCP – organizer and leader of the struggle of the Bulgarian people for liberation.

- TKZS (Trudovo Kooperativno Zemedelsko•Stopanstvo = Collective farm) – the way to build socialism in the Bulgarian village.

- The economic reforms of the Government of the Fatherland Front after September 9, 1944.

- Bourgeoise Nationalism, cosmopolitanism and proletarian internationalism.

- The contribution of the USSR for building of socialism in Bulgaria.

- The Labor Service Troops before and after September 9 and their importance for building of Socialism in Bulgaria.

The same clichés were being repeated ad nauseam in almost all lectures. The least literate individual could remember something. The mass of trudovaks is hostile to all this demagogary and sabotage it by abstaining to take any part in the discussions – with questions or expression of opinions. In so many instances the young man asked to answer a question simply stands there, not uttering a word, simulating ignorance, for which he will be punished one way or the other. Another part of the program of political education are the weekly sessions for political information, where the officers, and sometimes a trudovak, are assigned to discuss the internal and international situation as reflecting on Bulgaria. In these hours of information the political and economic state of the country is presented in glowing terms, while, when international events are touched, the Soviet Union is showered with compliments and the Anglo-American "imperialists" are castigated. *

Under this program, every trudovak is required every day to read the editorial of *Rabotnichesko delo* (The Party daily). This requirement is virtually ignored by all trudovaks. In addition to all that, the trudovaks are expected and encouraged to show their own initiative, inspired by the political officers. They were encouraged to form study groups, to prepare "Wall Newspapers" and study appropriate songs. The study groups would be discussing the biographies of Georgi Dimitrov and other party leaders, especially that of Lenin. Considering the nature of the work on the field, it could easily be surmised how little the trudovaks were interested and involved in such studies. They preferred to use every free minute of their time to rest and relax. At the same time the officers try to deprive the men of their free time, fearing that in such a free time the trudovaks would be contemplating subversive thoughts. The meetings of the study groups were reported on paper as having taken place, just to show that they are functioning. The political officer often scolds those responsible for such activities when they are not persuaded but they always have the excuse with lack of time. The Wall Newspapers are propaganda montages affixed to the wall, with portraits of Party leaders – Stalin, Chervenkov, Dimitrov and Party slogans.

The officers pay great attention to art self-expression and initiative of the trudovaks, which most often consist of singing Russian songs. It is difficult for a man, who has not gone through the trudovak routine, to imagine the comedy which the officers daily play with the poor men serving in this penal servitude, helpless victims of communism. Awaken from their sleep, they had to line up before the barracks and are asked to start singing. Physically exhausted, they have no voice left and no heart to join in singing. They are marching to the field to work and are asked to sing. Most of them do not open their mouths. The officer is furious and start reciting penalties. As they walk, the dust of the road rises in thick clouds, but they have to sing. They have to leave the impression that their service is a pleasure, not a penal servitude. At lunch, the few minutes left for rest have to be spent in learning new songs. Evenings, collapsing of exhaustion, dragging with difficulty their feet, they are asked to sing. Hardly the hypocrisy and the human demagoguery, the inquisitorial urge and inhuman cruelty could reach a higher pitch in their cunning outrage. The trudovaks sing! The penal servitude camp echoes their broken voices in which one could discern the cry of Bulgaria enslaved.

In 1950, the trudovaks of the platoon of lieutenant Sotirov defiantly refused to follow his order to sing. The punishment he imposed was that during the entire rest period after lunch they all had to stand against the sun and sing. The order was obeyed. For four hours the trudovaks involved stood under the burning summer sun and sang while he sat there under the shade watching them. In 1951 the trudovaks in Balchik, platoon No. 5463, returning from work, refused to sing. The officers called this sabotage, and punished the whole unit with denial of leave to free city afternoon in non-working days.

---

* I often asked questions to which I knew the answers, but phrased them in such a way, that I could not be accused of subversive attitudes. I loved to see them agonizing in trying to evade the trap which I was setting for them.

The repertory of songs sang in the entire SLT was the same. The most popular song was the Russian "Commander Chapaev" which Russian text is not understandable to any trudovak. Another such Russian song is "Krasnoarmeets bil geroi." A third such song is "Ves Soiuz Sovetskaya strana". Should mention here also the song "Mnogokrasnaya geroia". No Bulgarian song was heard in the camps of the SLT. On some occasions, as was the case in platoon No. 5463 in Balchik, when the trudovaks had to sing anyway, on their own initiative would take up "Zhiv e toi, zhiv e" (one of the best poems of Khristo Botev, the Bulgarian revolutionary poet before the liberation of 1878, for whom the Communists have the highest regard and consider him one of their predecessors. It is a song of the highest patriotic exaltation to which no political party can have claims.) They sing with full and shattering voices when they would reach the stanza:

> Harvest is now, sing you slave girls
> these sad songs! Shine also you, Sun -
> in this slave country will die murdered
> this young fighter...Keep silent, oh, heart!

Likewise, after this patriotic hymn, they would sing "Tikh Bial Dunav se valnuva" and with all their power would stress the stanza:

> For our dear Bulgaria
> we are flying to die
> and from unbearable tyranny
> we will liberate her.

The same stanzas in Bulgarian:

> Zhetva e sega, peite robini
> tez tuzhni pesni, grei i ti sluntse
> v taz robska zemia. Shte da zagine
> i toya iunak, no mlukni surtse...
>
> Nashta mila Bulgaria
> niy letim da mrem
> i ot tezhka tirania
> da ya otturvem.

The platoon corporal knew very well these songs and understood even better the exaltation of the singing trudovaks, his face would turn red, but on account of the patriotic character of the songs, could not say a thing. This particular corporal, Emil Alexandrov Popchev, from the village of Boshulia, Pazardzhik district, expressed his frustration by ordering a repeat of the Russian repertory.

First assistant of the political officer in every unit was the trudovak who was member of the DCNM. This organization has a well structured apparatus in the entire commanding hierarchy of the SLT – from the highest commanding staff, to the lowest working unit. In every platoon exists a DCNM group, which is a subdivision of the one in the company. The Company organization has its own leadership, consisting of the representatives of every platoon unit, and plans the activities of them all. From there, the Company leadership constitutes part of the entire camp body of youth leaders. From there, the higher leaders participate in the regional DCNM Committee of all camps, and from there it all is subordinated to the Central Leadership attached to the Supreme Command of the SLT in Sofia.

The main function of the DCNM, from top to bottom, is to win the political loyalty of the trudovaks to the Communist regime and lead in initiating and promoting stakhanovist emulation in the Labor Troops. The members of the DCNM are expected to report to the officers instances where they suspect some sort of an opposition, and also to serve as example for the rest of the trudovaks. But this does not last long, because, on account of the killing physical work, these young men are quickly appointed to better, sometimes confidential missions and jobs. Most of the times they are promoted to the positions of corporals and released from physical work. When this is the case, most of the time it is, they try to satisfy their protectors, the officers appointing them, by acting as real slavedrivers of the trudovaks assigned to their command. They are the most trusted agents of the political officers. They were always in immediate contact with the ordinary trudovaks, they ate with them and they slept in the same barrack. They had the chance to overhear conversations and keep track of who is associating with whom. Serving as spies of the political officers, they were very anxious to do their duty as stool pigeons, always hoping for promotions and small privileges. It often happened that on orders from higher officers, or even on their own initiative, sometimes they would call suspect trudovaks and subject them to interrogation. So it happened to this writer in Company 5463 in Balchik. He was called to a dark barrack and questioned about his past and his future plans in life. He was reprimanded for his opposition to the proposal of the officers that the trudovaks push alone a wagonnette, instead of two men. They tried to impress on him that if he continued the same road, he would not go very far, but, if he corrected himself he could be accepted as member of the DCNM. This instance is described elsewhere in this book.

The membership in DCNM is made up of those who bring notes to that effect from their local organizations. These are young men already committed, in many instances mentally or physically inept or most of the time, individually entitled to serve shorter terms and for convenience drafted in the SLT. There are also those who seek ingratiation with the authorities to cover their past. Some of them genuinely have accepted the new order of things, others hypocritically fake their support for the regime. A congress of the national organization this year apparently adopted a resolution for massovization of the organization, i.e. to recruit the entire youth population of Bulgaria. Sometimes, at the meetings where acceptance of new candidates is scheduled, one would witness pathetic scenes.

There was the case of this young man, Marin Donev of the village of Dermantsi, District of Lukovit, who from the very beginning of the service became a lackey of the officers, hoping to escape work in the field. He appeared at the general meeting – all trudovaks were invited to attend all the time that means, were ordered to attend. He stood there before all trudovaks and made extensive self-accusations, against himself, shed generously his tears as repentant sinner, made a fool of himself in the eyes of the trudovaks and the officers, and at the end, was not accepted for membership.

The DCNM was interested to recruit more members among the trudovaks, who, in their overwhelming majority, eighty per cent, were not committed. The objective of this interest was by engaging the politically alienated or opposition attuned youth in DCNM work, to offer them hope for gaining confidence with the ruling communists, which would open them the doors for better employment and careers in civilian life, including admission in higher education institutions. The officers expected to insure themselves against the opposition of the better educated trudovaks, even if the hope which they would create, prove subsequently to be false. But the trudovaks easily saw this phoney game and soon oriented themselves in the new set up of their service.

The political re-education of the trudovaks was carried also through activities in the so called complex GTO (Gotov za trud i otbrana – Ready for Work and Defence). This GTO in fact was a farce. To be awarded the badge of GTO, the trudovaks had to join in exercises of military nature, without ever touching a gun. They were expected to join in voluntarily. They were to practice running 100, 1000 and 3000 meters, to walk over what was called "partizanska pateka - Partisan pathway", (walking over a beam thrown over a twenty-thirty feet wide ravine, five-six feet; pretend of throwing a *granate* – a ball of rags – and using a shovel handle for bayonet, attached to an imaginary rifle; swimming, bicycle riding, callisthenics, climbing a rope or a pole, pretending of aiming and shooting. The officers held very much for such activities, which, if practiced in work time, did not relieve the trudovaks from doing their work to the level of their established norms. They had to finish their work in their free time, which infuriated the trudovaks when they were asked to do both – GTO exercises and work on the field. On numerous occasions a lieutenant in platoon 5463, Balchik, would say: "Nobody should assume that I will leave him to stay in the barracks while others are working to fulfill their norms. Everybody will run as a general rule." Nothing would force the trudovaks to follow such orders, and they – whenever possible – hide wherever they could. Somehow it worked out. When a trudovak would be called for GTO exercises, the score keeper would mark them as having fulfilled their norms.

Very important in the course of the camp events, were the so called meetings for discussion of the production. At these meetings the commanding officer would report on the results of the work of the trudovaks during the past week, pointing to achievements and failures. Trudovaks who consistently fall behind in meeting the required norms are publicity denounced by name. Those who had overfulfilled their norms would be publicly praised. These denunciations and praises are in line with the educational standing of the man in question. Thus, in platoon 309 at Bezmer, the reporting officer noted that an illiterate boy had made the comment: "They are going to skin us with work here." It all was the end of the story with him. He was mocked and just reprimanded publicly. Another trudovak, with high school education, having left empty cement bags in a RR freight car, after it had been unloaded, was called to a barrack, where the political officer had severely scolded him and had some unidentified corporals beat him up. The trudovak in question is Manu Manev, the political officer Dimitar Ninev Dimitrov from the town of Cherven Briag.

Discipline in the platoon or the company is another subject which is discussed at these meetings. Trudovaks cited for violations are publicly punished for their infractions of the established order. One of the punishments was placing the trudovak "Pod oruzhie" (under arms) for one, two or three hours. This meant that he would be ordered to stand up facing the summer sun, while the corporals would attach to his back a wheelbarrow and put over it shovels or forks. Someone would watch the man punished not to move until his punishment ends. In other cases, upon the shoulders of the man are placed five to ten shovels, picks or forks, which he had to hold for as long as his punishment last. Still another form of punishment is extra work from one to five hours in free time. At these meetings most often the officers address the question of stealing among the trudovaks. The thieves were threatened with severe punishment, which rarely was executed. Very often the thieves were turning into spies for the officers, reporting on their comrades. Sometimes these scenes were turned into parody.

There was this case in Platoon 309, in Bezmer. Lieutenant Sotirov once stood before the assembled trudovaks and, on occasion of one thievery, declared that the thief will not be penalized, he just had to have the courage to publicity admit his crime. There at once stood Dimitar Petrov of the village of Rodopi and with firm voice confessed what he had done. Sotirov congratulated him: "Bravo, Dimitar!" Dimitar responded: "We serve the People's Republic!" – the mandatory response to a congratulation by an officer. Someone ought to have added: "As thieves", so the comedy could be complete. The theft is something very common in the service. Most often it is theft of bread from the personal bags of the trudovaks, then work clothes, and when available – money. The theft of service items – hats, pants, clothes, tools – is not pursued. The officers would just advise those complaining to go and steal from somebody else what was missing in their possessions.

Such assemblies of the Company were held every week, and once a month all units on the camp would have a general meeting where the commander would account for all the work done and all political education activities. The Company which is reported to be a champion, would be honored by awarding it the Red Flag for the month. This red flag is carried to the work field every day by the champion platoon of the company. Every morning it is saluted by the entire company, and every evening at 9.00 o'clock review, the salute is repeated. To all platoons and individual trudovaks, who had especially distinguished themselves, the commander give a special letter of commendation – first class called "Pobeda", (Victory) and second class – "Trudov list" (Work paper) – both artistically designed letters of praise, to be displayed as diplomas, for propaganda purposes.

The SLT conduct also classes for illiterate trudovaks. They are more propaganda efforts rather than real teaching. Lieutenant Sotirov saw them as waste of effort. This writer has for a while been charged to conduct such a class in Bezmer and has come out with vivid impressions. The conditions under which these classes are conducted preclude any effect on the illiterates. They are held in the free time, when the other trudovaks would be resting. At 11.00 a.m., when the rest period begins with the lunch, the illiterate students had to march back to the barrack set aside for this purpose, sit around the table, and by the time the lesson would begin, they, exhausted and tired, either fall asleep or pay no attention whatsoever – physically and mentally unable to concentrate in the burning heat. The instructor, under the same conditions, is in no better shape either. He too has to give up his rest time and is subjected to the same inconvenience. With utter lack of interest and attention, having met the obligation, they all return to the field to continue their work. The illiterates are forced, threatened with severe punishment, to attend the classes. At the end of the month the political officer had to report how many classes have been held and how many trudovaks have attended. Most of the illiterates had been in school for two or three years, and had come out still illiterate. There was no hope that under these trying conditions they would have learned anything more. But they were attending for fear of punishment. The cause of their illiteracy – may be psychological, may be social – was still there. Most of them were minorities – Gypsies, Pomaks or Turks. The Communists view them as victims of the system which they had overthrown and are making Sisyphean efforts to do them justice. At end it all comes down to shallow propaganda effort.

The SLT is under the direct authority of the Council of Ministers and is under the command of Colonel Angel Tsanev – its general Director. In the summer of 1950 he paid a visit to the trudovaks at Bezmer and demagogically met with the men working for the future airport at a general meeting. When it was all over the trudovaks were left with the impression that he somehow laughed at their helplessness. In his speech, he said approximately the following:

I know what you are going to tell me. First, some will complain that the bread is not enough for him. Well, what we can do? It is not enough for me too. But take a look at the wheat in the fields how yellow it is turning? Now our entire nation is engaged in work to collect the golden grain. Let us gather it, and then we will think whether it is enough or not enough. Do not forget that our partisans for weeks would not have seen bread... Someone else will complain that there are no leaves to go home, to see his wife, and his kids. We should not divert now our attention to all sides. We should be thinking only of our work here, on this field. Still other may complain that the term of service is too long...

This is how the General Director was speaking, anticipating with an ironic smile, stupid chicanery and dangerous insinuations, warning and ready-made answers to eventual questions of the trudovaks. When he finished, the trudovaks were given opportunity to ask questions, but nobody dared speak, because all questions which possibly could be asked, and all complaints which could have been made, were already raised and answered with no difficulty by the Director himself. The trudovaks, with eyes pointing down to the ground, had witnessed the Communist hypocrisy displayed by the supreme commander of the SLT.

On November 5, 1950 the Bezmer camp was visited by the Political Officer of the SLT Yanko Kaneti. The occasion was the completion of the runway. He delivered a pompous speech, revealing his shallow intellect, his virtual illiteracy in the language which he used and the high opinion which he had of himself. Besides these two commanders of SLT, there was a third individual, colonel Zmiarov, exercizing the government control over the labor troops. His appearance on the camps is seen as deadly threat for the respective officers. They are usually tipped for his intended visits and then the burden falls on the back of the trudovaks, because they are awaken earlier than usual, asked to take their blankets out and dust them, to wash their pillow cases, to clean the entire camp of paper pieces and cigarette butts, and do a million other chores, so that if and when Zmiarov would come, the camp should be clean and shining like a glass. The trudovaks often mocked, privately, their officer with made-up stories for the upcoming visit of Zmiarov. Other chiefs of SLT, of higher echelons, visit the camp rarely, allow the trudovaks to ask questions, ask about the food and other impertinencies and leave without anything being done if there was any legitimate complaint. Such was the visit of Colonel Neikov and Major Zhelev of Shumen. Our camp commander were no better either. Major Stanchev, Camp Commander at Bezmer, was a cynical, brutal man without culture. Captain Stefanov, chief of Platoon 5461, in Balchik, a Shop (the shops are the peasants around Sofia) appeared as a more intelligent officer, but otherwise was merciless tyrant when it would come to work on the field. In the worst days, when a reasonable man would have ordered the trudovaks to go back to camp, in days of cold rain, mud, freezing temperatures and body piercing winds, he would not call off the work day, but would viciously explain: "You know", smiling he would say, "we, the trudovaks, do not stop at any impossible obstacles. We will fall, we will get up, and we will continue pushing..."

Deputy commander of the camp in Bezmer was Lieutenant Ivan Dinev Georgiev, from the village of Konyovo, District of Nova Zagora. He is not a party member, but for outstanding service on the front against the Germans has managed to win appointment in the SLT. In that service he has shown much dedication and has pleased his superiors. His father, thus the word goes around, had been a big landowner, a kulak. He has two brothers priests. With such reactionary background he has done well with the regime and the officers above him. He has a decent attitude towards the trudovaks, has expertise in his work and enjoys the respect of the young men. During his term in our camp there was no instance of making a big thing of political nature against any one of the working men. Still, he surrounds himself with stool pigeons and keeps away from trudovaks who could make trouble for him.

Commander of platoon 5463 in Balchik is Lieutenant Dimitar Ivanov Kotsev. Like Georgiev, he also is not a party member, but on account of his loyal service has gained the confidence of the communists. He is nicknamed "The Moving Madhouse" on account of his erratic behavior. In his line of service he screams like a mad man at the trudovaks for reason or no reason, for which he is mocked behind his back. If he had his way he would have the men at work twenty four hours a day and produce twice or three times more work. When a unit of forty trudovaks, as he noticed, never produce work to justify the capacity of the bulldozer, he came up with the idea of reducing their number by letting one single man push the wagonettes removing the dirt. This way the norm of the individuals working in the group would be doubled and he could report overproduction. The men of the unit tried to convince him that it was impossible, but he held on his idea and accused some of the trudovaks of sabotage. The reality is that they managed to work with great efforts as it was by two men pushing one wagonette. To have it his way, he ordered the platoon corporal to increase the working day to 10-11 hours. This way, calculated at 8-hour a day, the production was raised to 130 per cent. He used this to refute the "saboteurs" who considered his norm excessive.

This writer questioned his calculations at a company meeting, but on account of that was instantly moved out of his unit to another project "Where he could achieve his norm" as the commander said. (It just happened that this same writer and his two friends Zdravko and Stefan, for long time had been planning their escape, and right after the Company meeting, took off. The story is told elsewhere in this Memoir. I have wondered so many times abroad what the moral effect on the rest of the trudovaks in this company, and in fact on the whole camp was, when learning that after this penalty was imposed, he and his friends had escaped. The unit to which he was being moved was digging ditches, much more difficult work than on the wagonettes. Years later I heard that in a week's time, the trudovaks had been told that we were killed while trying to cross the border.)

Commander Kotsev has surrounded himself with corporals serving their regular military service and report to him every criticism of himself heard by them.

Deputy commander of Platoon 309 at Bezmer in 1950 was Lieutenant Dimitar Ninov Dimitrov, from Cherven Briag. Cunning, crafty, somehow educated and politically well versed, this Communist was doing his work well. When he would fail to bribe you with his smile and posing as gentlemen, he would skillfully remind you of his authority as a political officer. Later he was replaced by Lieutenant Atanassov from the town of Peshtera – candidate-member of the party. Ideologically and psychologically he was little prepared for his job as a political officer. It seemed that he was not in his heart a Communist, but just a turncoat. Deputy-Commander (political officer) of Company 5463 in 1951 at

Balchik (I am using Platoon for a unit of 40, more or less men, and Company – for three platoons) was a lieutenant Nedelchev from the town of Pleven. He was a Party member, was very strict, not quite cultured, inclined to impose the heaviest punishments for the smallest infractions of political orthodoxy. He is a sinister man, constantly threatening the enemies, the saboteurs and the instigators. He uses skilfully the DCNM in his Company, the corporals and the stool pigeons among the trudovaks.

One of the platoon commanders of 309 in the summer of 1950 at Bezmer was Lieutenant Shtilianov from Sliven. It seems to me, he had been an army officer during the war and his appointment was some sort of a demotion. He was a very decent man. He may have been able to get this appointment because his sister was married to one of the big guns in the service, Colonel Neikov. Once, unaware that I was overhearing his conversation with Lieutenant Ilchevsky, a counterpart of another unit of 309, and possibly purposely spoke loud, so I could hear him, he said: "The kids should salute me, because I am not a Party member." There was one occasion when he could have cited me and three other trudovaks for sabotage, but managed to cover up the infractions. For some reason we have been detailed to carry a truckload of can goods unloaded outside the kitchen to a shed. We thought that some of these cans may contain compotes and punctured probably fifty or so, only to discover that it was green beans. There were no labels outside. It was soon discovered and reported to him. He called us, looked at us, took us to the shed and ordered us to go over all cans, pick those which are punctured, ordered them cooked, and let us go unscathed. I never forgot this gesture on his behalf. Under other men I do not know what would have happened to us.

Shtilianov was replaced at some point by Lieutenant Sotirov, also from Sliven. He also had been an officer during the war, regretted his service but on many occasions admitted to the trudovaks that he was serving for one and only reason – for his salary. He had impeccable manners, but otherwise was very demanding, very cantankerous, suspicious, crafty, cunning and not stranger for bribery. Two trudovaks – Georgi Karamukov from Malevo, Haskovo district, and Khristo Zhechev from Konstantinovo, Harmanlyisko, were supplying him with butter and eggs, which he took home, so that they could be "taken care of" by him.

Sotirov was succeeded by a Lieutenant Penev, a rude, arrogant, bad-mannered hoodlum, swearing the trudovaks any place they met him in detestable pornographic ways. He represented the fullness of an ignoramus. In the lectures which he would sometimes deliver at meetings he often found himself in such embarrassing situation, that his face would turn red. On one occasion he was seeking to point to Korea in the Himalayas, and as he could not find it, told the trudovaks that it was not on the map. The better educated men had to hold their breath not to laugh at him. But when it would come to punishing somebody, he would use his fists and beat them like animals. It gave him some pleasure to torture his men. He publicly beat a trudovak by name Spas Penev. He did not conceal his desire to impose collective penalties by raising the hours of work. He would do anything to grab any assignment for his men, especially unloading cement wagons, which, done in minimum time as he would supervise it, offered him the highest percentages of exceeding the norms. He made his platoon a slave of the Company, always taking the most difficult projects. He was hated by his men, and the other officers laughed at him. But he was easily bribed. His protégés had their families sending him baskets of all sorts of goodies: Ivan Todorov Lozev from Uzundzhovo, Haskovo district, Georgi Karamukov, Khristo Mitkov Zhechev, Dimitar I. Daskalov et al. He would   threaten any trudovak who dared criticize him or his protégés. As to me, once he made the comment: "Daskala (the teacher) is a dzhezve (Turkish coffee pot) of bones and skin."

Lieutenant Ilchevsky of Plovdiv was an intelligent officer, though quite often he behaved as a cynic. He has a very positive and respectful attitude toward men with education, while Sotirov and Penev feared them, and sought to humiliate them in any possible way that they could. Ilchevsky protected his platoon in Bezmer, especially the educated men. With their help he always managed to fake the production statistics and his unit was always leading in percentages. He, like the others, accepts bribes, but more innocently. A group of trudovaks would purchase a lamb, wine and beer, prepare the party in his apartment in the nearby village of Boliarsko, he would take them out of camp and then they all would enjoy themselves overnight. In turn he does not put them to work in the field, but find supervisory positions for them, where they exercise strict control over the men under their assignment. To this group belong Khristo Chakandrakov of Dulgo pole, Plovdivsko and a young man from my village.

The sergeants play an important role in the SLT units. Sergeant-Major of platoon 309 at Bezmer was a sixty years old man, nicknamed Diado Rusko (Grand father Rusko). We had reliable information that he was not particularly fond of the communists. His sins, as keeper of the uniforms, gravitate more around the distribution of clothes. He would favor those who discretely offered him a bottle of wine, or brandy. Khristo Naidenov Yachevsky of Dermantsi, Lukovitsko, had distinguished himself in his briberies to Diado Rusko. Sergeant Peiu Peyev, ugly like sin, was known to protect his men, does not subject them to excessive work and is suspected of being an anti-communist. Sergeant Ganiu Manev was a known communist, but is moderate, defends the trudovaks when under some attack without justice, but often shows his illiteracy. He liked to support his statements with quotations, but would not remember if they come from Ivan Vazov, or Nastradin Hodja. He apologizes for not remembering his source. Most of the time he would quote Nastradin Hodja. The sanitary sergeant Encho (I do not remember his other names) had the mania that he excelled in knowledge. Once he lectured on Ivan Vazov and made him the greatest proletarian poet. Of the corporals, Angel Khristozov, of the village of Lozen, Svilengrad District, from a family which had been called kulaks, having their harvester tractor nationalized, had become member of DCNM, and from there had distinguished himself as a spy for the officers. He reports on all trudovaks who say something critical of the regime. When a young man, Georgi Markov of the village of Melnitsa, district of Elkhovo, instead of reporting to serve as a trudovak, had gone into hiding, caught and brought to our camp, Bezmer, Angel was ordered to beat him and had performed this task with distinction.

The sergeants in Platoon 5463 in Balchik are former trudovaks who had demonstrated high loyalty to the regime. Stoyan Ganev, of Draganovo, Gorna Orekhovitsa, raised in Czechoslovakia, does not hesitate to say that Bulgaria is far behind, culturally, of his birth country but acts as a lackey of the officers. He is very strict with the trudovaks under his authority, in discipline, and at work. Offended men threatened to beat him up. Once he told a corporal: "When you have a higher officer before you, you should shut up!" This evoked laughter among the trudovaks who listened to this threat. Sergeant Emil Popchev of Boshulia, Pazardzhik, Secretary of DCNM, despises any men with higher education. Once he scolded a graduate of the Law School of Sofia University, Angel Y. Georgiev of Pleven, who dared to challenge him, and as a punishment was transferred to another platoon. He was also deprived of leave to the town of Balchik. While all these penalties were spelled out, the Platoon commander felt embarrassed, Popchev turned red like a lobster, but the jurist boldly faced them, without giving a sign of regrets. It was a picture to be remembered: the officers looked defeated by the trudovak. Popchev often beats up some of the trudovaks. On one of these occasions the trudovak complained to the political officer, but nothing was done to the sergeant. The trudovak was Todor Zhivkov. He, Popchev, is not a stranger to bribery. On March 18, 1951, the trudovaks Bero Tsvetkov, Khristo Yachevsky and Krastiu M. Kalchev purchased a lamb, prepared it in the home of Sergeant-Senior Siderov, and Popchev was invited to join the company. The trudovaks who paid for the banquet subsequently were freed from heavy work. Khristo Yachevsky was caught stealing ten pounds of marmalade from the kitchen, but Popchev intervened and nothing was done to the thief. Last, but not least, reviewing the officers in SLT, I want to mention Corporal Rad Georgiev Lazarov, of Dragomir, district of Pazardzhik. He overworks his men, does not let them leave work in the hours set for return to the camp and forces them to march for him when going and returning from work. Corporal Tsoniu G. Stanev, using a trudovak, Peter Nedkov of Lukovit, served as the main agent for State Security. The same is true for Corporal Georgi Nikolov, a man with University education.

Finally, let me say a few words for the main body of the SLT – the trudovaks. They are selected for this service by the draft commissions, which make the decision who would serve in the regular army and who will be trudovak. The criteria they use are the political standing of the respective young man, as reported by the local authorities. Those who were reported to be anti-communists, as a rule would be set aside to serve as trudovaks. Trudovaks would also be those who have physical and mental defects, the illiterates, mostly of the minorities, and those who for a variety of reasons would have to serve a short term – nine months. Those judged as politically unreliable, sort of a lost cause, often try to play the game with the officers and emerge from the service as rehabilitated. Others simply do not care and just go and do their work as it comes. Two thirds of the politically unreliable, where the officers were concerned, are young men with high school or university education.

The overriding concern of the officers in SLT was to see that the production norms of our work were met and exceeded the ones set up by their superiors. This task was seen as a political and socio-economic obligation for all rank and file men in the service - from the commanders to the last trudovak. Every nuance in the evaluation of the production in terms of norms was a sign of political accomplishment or failure. Increase of production meant heightened political consciousness, expression of loyalty to the Communist regime. The other way, decrease of production, was a sign of low regard for the Communist regime which would amount to political sabotage by the men on the field. This attitude towards the norms of production has made the SLT an institution of penal servitude. If the norms are consistently under the set levels, the officers fear that they would be suspected in not caring for the political attitudes of the trudovaks and their own political evaluations, which affects their prospect for promotions. When they managed to exert enough pressure on the trudovaks and get the norms exceeded, they would be rewarded with bonuses to their salaries. Thus, politically and economically bribed, they ran their men to physical exhaustion in their own interest.

They do not choose the means to achieve their objective, to force the trudovaks to work more and more, beyond what laws and regulations stipulate. Threats, moral pressures, deceptive promises for privileged positions, insinuation for political rehabilitation of politically compromised trudovaks – all that is common practice for the achievement of the stated objectives. The officers who distinguish themselves as the highest producers with their units, are indeed promoted, or moved to camps of preference. They are publicly praised and rewarded. All this wins every careerist or greedy officer to overexert himself at the expense of the men placed under his charge. With very few exceptions, the majority of officers hold the trudovaks at work without mercy and human compassion, until some of them simply fall from exhaustion. The service to the state for military obligation is in fact a penal servitude. The work of the trudovaks is sometimes beyond their physical abilities. At the same time they are subjected to hypocritical propaganda that all that is a patriotic duty, which places the officers in comic poses. They never cease telling the trudovaks that the service before September 9th had been a concentration camp and that the most loyal to the people men had been sent there. It was a shameless distortion of facts, because all men knew that the term of service then was eight months, while at the present, under the communist regime, it had been increased to thirty months; that in the previous regime there were no norms of production to be forced on the trudovaks, that in those days only physically and mentally inadequate boys would be recruited for that service. Nobody could speak back and challenge the version presented to the captive audience, and things continued until the end of the service, with the understanding that the line of deception was ordered from above.

The economics of production processes are expressed in the striving for reducing, or as it was called "struggle for reducing the cost of production". So all that would contribute for such reduction of cost was permissible to the officers. It is economically justified to reduce the food rations for the trudovaks, it is economically justified that sick and inept trudovaks are sent to work, it is economics to exceed the norms of production, it is economics if the trudovaks are given military uniforms in appearance dating from the Balkan wars, to put rags on their bodies and go to sleep with blankets with large holes all over, it is economics to work with broken tools – shovels and forks with half handles.

Platoon 309 was in charge of transporting the building materials – loading and unloading freight trains with gravel, sand, cement, or wooden materials. For each kind of work they had set the norms – how many hours a trudovak had to work to complete such and such a work, how much, percentagewise, a trudovak had to produce in a normal, eight hour day's work. Norms are set up for all kinds of work by some special commission in the central office of SLT. They are mandatory for the entire system – no matter where a unit was working – on a canal, on an airport runway, in mines, in construction, anywhere where trudovaks were to do the job. Such norms are also set up for civilian jobs. It was said that there was some special Book of Norms. The officers were obliged to make every trudovak aware what was to be his norm for the day. When superior officers would visit a camp they are said to have asked individual trudovaks what was their daily norm. Formally, every trudovak is under obligation to achieve his norms. It is expected that if he did, he would be let go free or continue working for production above the norms. The reality, however, is quite different. If the trudovak achieved his norm in five hours, he is then told that he is under obligation to work eight hours, to fill up his working day, and produce three hours more work. If he fails to achieve his norm in eight hours, he is left to do evening work, for as long as it takes, to fill-up his norm. There is a tale of the wolf and the lamb. This practice was very much the same. You are damned if you do, you are damned if you don't.

But the norms set up by law, are not permanent. Every year they are increased. It is argued that the over-norm production shown the previous year was an indication that the previous norm had been underestimated, had been calculated wrong. It is not taken into account with what excessive pressures and with what abuse of the working men the over-the-norms production had been achieved – not to mention how much fakery takes place to report production above the norms for the benefit of the officers. So, the reported production for 1950, above the norms, gave reason to the Special Commission in SLT to raise the norms for 1951 by 15 per cent.

To the forcibly extorted overproduction from the Trudovaks should be added the artificially instigated extra work from the trudovaks through the infamous stakhanovist emulation movement. This is how this works: The company commander instructs the sergeant and corporals to inform the trudovaks that at the next general meeting he was going to propose that the unit under his command challenges a corresponding unit in some other camp to beat them in percentages of production. The promise has to be made and approved by the trudovaks to exceed their stated norms by thirty percent. If the challenged company, hundreds of miles away accepts, which it does, because these things are not initiated without pre-arrangement, the norms are automatically increased to 130 per cent. Our company in Bezmer went through this fakery in 1950.

Such a fakery was forced on us also in 1951. When the proposal is made at a general company meeting, the officers support it enthusiastically. Why not? They do not work! They are supported by previously designated lackeys and the proposal is put to a vote. Who would dare not to raise his hand? A special delegation is then sent to the challenged Company, there they would sign the agreement to compete with each other, and when it all is taken into account, the norms for 1951 jump fifteen per cent as the Book of Norms had stipulated, taking into account the achievements of 1950, and an additional 30 per cent as per the emulation challenges. It turns into a farce, but who cares? It only means that the trudovaks would be forced to work longer and harder to achieve as much as they can. In 1950 a man from another platoon refused to produce with the group under his charge the promised 30 per cent, making six blocks on the run-way instead of the nine as required under the norm, and in September ended in arrest. His fate remained unknown after he was taken away.

For working on trucks – loading gravel, sand and other materials, and unloading them – the time for completing the work was strictly determined. Otherwise, the respective officer would be charged for lost time, if it was not done according to the specifications. That makes him furious and he hangs over the heads of the trudovaks to hurry up. If the work was finished short of the time specified, he was given credit and both he and the driver shared in the bonuses. The same was the practice with unloading of RR freight cars, of sand, gravel and cement. If it was not done in the time allowed, the SLT was said to pay to the State, which operated and owned the RR vehicles, for the delays. If this was true or not true we did not know, but the officers nervously moved around to make sure that it all was finished ahead of time. Thus work which was to be finished in one day, was done in a matter of three or four hours. The rest of the day the trudovaks were shifted to other work, to fill-up the day. In such instances the production was calculated to be 200-300 above the norm. The trudovak would get nothing for that. One could not imagine a worse service than that of SLT. The service of the trudovak was a penal servitude.

Closely related to the production was the working time. Officially the work day was to last eight hours. In fact it was never shorter than nine hours, most often ten hours, and sometimes extended to eleven, twelve, fourteen and even twenty four hours, depending on circumstances. The officer in charge is supposed to periodically allow fifteen minutes break, but he usually forgets about it, and nobody dares remind him. He will not notice the time for ending the work and extend it at will. In Bezmer, Company 309, it often occurred that at about 3.00 a.m. a composition of sea sand would arrive and the unloading had to begin at once. By 6.00 a.m. the sand is unloaded, but the time for breakfast was over. After breakfast the men are sent to move the sand from the sides of the rail lines by wheelbarrows. By noon time, when the work is almost finished, there would arrive a composition of river sand. It too had to be unloaded by the same platoon. They would have lunch when they finish the work – about two or three o'clock, and then grab the wheelbarrows again to move it out of the way, because when the next composition would arrive there will be no room to unload it. The worst day for the trudovaks was the occasion, when, after they finish with this sand, a composition of RR cars carrying cement would arrive and the trudovaks are asked to show "work-heroism" and jump in the wagons. Sometimes this work heroism would last until 6.00 a.m. the next day. August 26, 1951, was such a day. We never forgot it. And nobody could protest this arbitrary abuse with the life and the health of the men, selected for this service to do their military draft. It was a sentence to death, without court and without defense.

There were many other ways to increase the work day. Thus, a month or so before May Day or September 9th, the trudovaks are asked to promise to work more than eight hours. On this occasion, on April 4, 1951, in Platoon No. 5463 the trudovaks witnessed an embarrassment for the officers which made them understand what were the true feelings of the camp. At a general meeting the political officer, giving himself appearance of exuberance on account of the announcement which he was to make, told the assembled men that at a conference the commanders had decided to increase the working time from eight to nine hours on occasion of the great day of festivities, May Day. He was now asking the trudovaks at this time if they would go for it. He asked anyone who desired to speak, to raise his hand. Nobody raised his hand. All men were dead silent. They all looked down at their feet. The buzzing of a fly would have been heard. The mood of the political officer began to change. He was nervously walking back and forth and watching the trudovaks.

At end a corporal stood up and praised the decision taken by the commanders. A sergeant followed him. But not one single trudovak, those who in actuality had to work, volunteered to speak. In vain did the political officer wait for someone to break the silence. Finally, a lackey, Marin Donev, of Dermantsi, Lukovit District, got up and solemnly declared that the trudovaks readily accept the proposal of the commanders. The rest of the men did not move. They watched Donev with despise for betraying them and cursed the officers in their minds. The political officer asked those who supported the proposal to raise their hands. All did. Did anyone dare to show opposition? Nobody!

Even worse is the case with the weekly rest. The law or the rule was that the trudovaks would work on Saturday half a day and be free on Sunday. But it turned out that the officers always found excuses to make the Saturday a regular working day, and most of the Sundays had to do the same. One Sunday a month, at most, would be left free, but it would be filled with meetings, lectures, cleaning the camp, checking the suitcases, etc. etc. It is all done under the watchful eyes of the corporals and the sergeants. With such overcharging with work, who cared for emulation and voluntary competition? Under the circumstances, the trudovaks tried to use every available minute for rest, for their personal needs and when it was noticed that no officer was watching, this was the time for all to stop working. All talks about voluntary emulation and exceeding the norm turn into empty words. The truth was that all work was done under threats of punishment. It indeed was a regime of a penal servitude.

The accounting of the production in percentages never corresponded to reality. All numbers are exaggerations by everybody involved – to receive a reward or to avoid penalty. The only available form of resistance was to work slow and not to produce what was required. Sometimes a way out of work was to go to the "bathroom" behind some ditch or some ravine and delay one's return for as long as possible. Lieutenant Penev often went out to herd those gone to relieve themselves and return them to work. In night time he would use matches to examine the evidence...

The conditions of everyday life of the trudovaks are diametrically different from those for the regular army soldiers, even though both are called to serve under the same law. This difference is due to the political discrimination which is deliberately practiced by the communist regime in considering the trudovak service as a penal institution. The uniforms which are given to the trudovaks are worn out rags condemned for destruction in the regular army units. Sometimes they are patched up in such a way that one could not tell which is the original material and which is the patch. Lately, because of the large number of young men being drafted as trudovaks, there appear some new uniforms, but they are kept in the stock, so when they have to send some trudovak for business somewhere, or on some emergency to go home, to give him a more decent attire. Sometimes, when someone bribes the sergeant, Diado Rusko, with a bottle of wine, he may be given some of the new items. The rest of the trudovaks, dressed in the rags given out, is a pathetic picture of a soldier. A Company commander could not restrain his laughter when he saw a trudovak walking through campus in pants which did not quite cover some private parts of his body. The Company Commander was Lieutenant Ganchev, the trudovak – Koliu Yanev Yovchev. In the hot summer this was bearable, but in cold spring or fall days – I could nor speak of winter – this put the trudovaks through severe test. It is not easy to stand to the cold weather. The trudovaks had to fight the cold with raggedy clothes, ripped shoes or rubber work shoes, which they easily lost in the mud or had them filled with water. Tragedy.

The living quarters of the trudovaks were another source of suffering. In summertime they live in barracks, but they packed the men so many in one barrack that the men had to stretch, in order to accommodate everybody. They could not curl to rest their knees and legs, they could not turn around without disturbing the next man; in the morning they felt as tired as they had been before going to bed. To that are added the herds of bedbugs which crawl all over the bodies and make the misery of the men so much worse. The burning sun during the day would warm the barrack so much that until late in the night it is unbearable. Some men would grab their blankets and stretch outside on the ground in dust and gravel, until some officer passes by and orders them inside. In Balchik the trudovaks for a while lived near a camp of regular soldiers from the airforce. Their premises were clean, aired, the single beds well arranged, the pathways wide, while they, the trudovaks, had to sleep on planked beds, man to man, on "mattresses" filled with straw from end to end of the barrack, in darkness, dust falling from the second "floor" to those sleeping under it, sometimes the pathway turning into a mud alley, or dust, and suffocating atmosphere... It was inevitable that any trudovak would make a comparison and would live through a painful sorrow for his fate. The trudovaks felt that they indeed were in a penal institution, ready at any time to explode in indignation and wipe out the whole structure of the Communist hypocrisy, which often compared the conditions in the camps of SLT before September 9th as concentration camps, while after September 9th they have been turned into "schools for education".

Food is the most serious concern for the trudovaks. Most of the conversations among them revolve around this subject and quite often most vivid protests were made. In the first place is the question of bread. Each trudovak is entitled to 800 grams of bread made of wheat and corn flouer. In most of the cases the weight does not measure up to the ration. Besides, the amount of bread so defined is far from enough to satisfy the men. Many of them receive packages from home to supplement their ration, and not starve. So, they work for the State but are sustained by their families. Most of the men consume the loaf as soon as they receive it and the next twenty four hours go without bread, eating only what is served as kitchen food. In reality, they simply starve, if they have no suppllement. It is a mass event. The officers know it, but there is nothing anyone locally could do. For breakfast in the morning the trudovaks get 30 grams feta cheese, which, by the time it reaches its destination is no bigger than a lump of sugar. It is improbable but it is a fact. The tea, the only breakfast menu for every day, is often served cold, tasteless, without sugar.

In March and April in Balchik instead of sugar they mixed it with plum marmalade, which more often would make it taste sour rather than sweet and turned the color of the tea to look like ink. Irony and grimaces were the only way by which the men could express their displeasure. For lunch the trudovaks were brought bean soup, or some sort of vegetable, rarely with some traces of meat. Quite often a real battle would be fought for one or another such piece, a war with serving spoons, with cans which every man had for receiving his soup and in any possible way. In our beautiful and rich with food country, our golden youth, condemned to penal servitude, was fighting for a miserable morsel of meat. All the times, at lunch or supper, if not in one group then in another, such fights were a regular scene. Poor country! The food situation became quite critical in Balchik. In quantity and quality it was under any minimum.

This writer raised this question at one of the company's meetings. He tried to use logic: energy was required for the work we were doing. Energy comes from food. We were conscious that we had to overfulfil our norms, but we needed more energy for that, ergo, more and better food. The company commander explained that the ration for the trudovaks was determined by a special commission of nutritionists, who have calculated that so much and so much beans were giving so many calories, which were considered sufficient. This writer had noted in his observations that the year before, in Bezmer, one could find more beans in the can, compared to the ration at Balchik. The Commander explained that as a result of the new studies and decisions at the Central Government of SLT, the beans have been reduced by 40 grams for a man. To this, this writer responded that he has not counted the calories in his can, but he had found that one hour after eating his lunch he was hungry. Even though the commander indicated that the order for rations was determined from above, soon we noticed that some change had been made. But it did not last for long and we went to the initial rations. Now the explanation was the lack of products. Whatever the reason, after 9.00 o'clock in the morning and after four in the afternoon, the trudovaks were literally starving and the pace of work slowed down.

Water was another problem which tormented the trudovaks, especially in the field. No provisions were made to have water at work, except what every man could carry in his small container, which was very soon emptied. There was a pump at the camp, but no one could go there, and it was at quite a distance sometimes. Drinking water was brought in a special barrel, mounted to two wheels, which was pulled by hands by an otherwise inept man. In the hot summer days, when our mouths were running literally dry, we waited for this man to show up on the road with his contraption. Hundreds of trudovaks would leave their work and run to intercept him and grab the precious liquid he was bringing. The officers were unable to stop the mob, or to supervise the anarchy around the "*buchonka*" where men acted as animals, screaming, yelling, pushing, pulling and fighting, much of the water being spilled. Those who would fail to get it, would have to wait for the next course of the "*buchonka*". This sorry scene is repeated every day. The man with the "*buchonka*" was "attacked" half a kilometer before he would reach the working place.

Hygiene, adding further to the everyday life of the trudovaks, was virtually non-existent. Lice were the worst plague that could not be eliminated. This writer was only three days in the camp at Bezmer, still wearing his nylon cream clean shirt when, checking, the officer discovered one on his chest, a little white thing crawling inside. I, like many others, was ordered to boil my clothes. I did. When I examined the "military" uniform which they had given me the previous day, I discovered clusters of eggs of this pest in the seams, left there by the previous owner of the rags. The facilities to protect yourself of such things were minimal. It is impossible to go further on the subject. We were turned into animals. There was a sanitary officer, not a doctor, who would not recognize illness of a trudovak. Many trudovaks are drafted by the respective commissions for reasons of health, but in the camp their physical condition was not recognized and they are sent to the field together with the rest, to squeeze from them even the last ounce of strength. A man, Maniu Manev from Possevina, Popovo district, certified with a heart condition by a medical commission, was sent to the field along the other trudovaks. The company commander did not recognize these certifications. At one point he refused to unload cement bags, tried to escape and seek medical help, but ended in being arrested, beaten and tortured by corporals, and when established that he was not going to recover, was sent to his father, back home. There was also the case of this young man, with stomach ulcers or something like that, may be gall bladder. Very early in the service at Bezmer he fell terribly sick, he could hardly stand on his feet, but was evicted in the morning from the barracks. I watched him for day or two sitting outside in the sun, with face yellow and drawn, supporting his chin with one hand. After two days he disappeared from camp. The word was that he had died. He was from one of the villages from my region, one of the catholic colonies around General Nikolaevo village (now city of Rakovsky).

So, this is what SLT is all about. Two and a half years thousands upon thousands of Bulgarian youths, were condemned to slow death. In the suffocating barrack and the empty bakka, in the pushing and pulling around the empty "*buchonka*", in the process of the penal servitude at the project, next to the good-natured, kind-hearted peasant boy, sent here on account of his honesty in criticizing the new order of things, or because his parents had been dispossessed from their land or had been involved in opposition activities, in so many instances having joined with the opposition parties of 1946-47, egged on by foreign governments to resist communism, were students expelled from the university. There were fading away ethnic minorities – Turks, Gypsies and Greeks. I will never forget a small group of three Greek boys, from Messemvria or Sozopol, golden characters, who had brought with them guitars and often played them for our supreme pleasure and sang some beautiful songs, which I had memorized – "Ta matia, ta dika sou ta matia", "As ta ta malakia sou, anakatimena" ("These eyes, these eyes of yours", "Leave your hair to flow with the wind"... Much later, in 1955, when I met a girl, of Greek parentage, in New York, I showed her my knowledge of these beautiful Greek songs. She flipped after me. After forty years plus, we are still together.)

Young men, with higher education, in the course of time having won each other's trust, gather in small groups, discuss their pitiful state, and share their hopes for eventual liberation. With expressions of utter disgust, they open the pages of *Rabotnichesko delo*, (the Party organ) and study the cartoon of some red propaganda artist, showing foreign individuals with signs hanging on their necks: "University educated man seeks any job", "Professor will work at any job", "Artist seeks a job". The caption below the cartoon reads: "The American Way of Life". The trudovaks glance at it with amazement, and try to understand if these things are written to deceive the Bulgarian people and world public opinion. Obviously this was a domestic interpretation of world affairs. They could not escape the ironic side of such comments. Perhaps the SLT was created, and they were brought here, so that men of higher education, university graduates and professors should not find themselves on the lines with those unfortunate American Intellectuals. Inferentially, these slave labor camps, because in reality that was what the SLT was: Penal Servitude Camps hiding behind the label of military service. Indeed, do these people, from the highest to the lowest level of commanding officers, even for a moment imagine that the forcibly enslaved young men would come to love this regime?

*

This writer, in the most difficult moments of physical exhaustion, pushing with his last ounce of strength the dirt-filled wagonette, or hardly able to stand on his feet under the burden of the 100 pounds bag of cement on his shoulders, covered in smelly, blue, cement dust, mercilessly urged by the officer to run faster, or sinking to his knees in mud in a deep canal – would remember what he was at the university, sitting in a library, surrounded by a dozen books of high scholarship, researching some complex subject; would see himself with Kant, or Hegel, or Spinoza, or Descartes, the giants of world civilization and culture; would relive the pleasures at the Opera where he saw seven times his favourite "Traviata", or the exhilarating excitement in the National theatre, though with the cheapest tickets of standing room only at the highest and farthest from the stage balcony; he would recall the scholarly discussions with learned and admired professors ... and would dream of the day when his liberation from this slavery would come! Nobody could frustrate his hope that the day of liberation would come and all those who had thrown him in this hell will be damned.

*

This is the world which I left behind when I took the first step in Balchik, seeking my freedom far away, beyond the oceans, if I could make it.

XI

ESCAPE FROM HELL

"Is life so dear, or peace so sweet, as to be purchased at the price of chains and slavery? Forbid it, Almighty God! I know not what course others may take; but as for me, give me liberty or give me death."

Patrick Henry

1. Excape from Hell – Balchik-Varna-Plovdiv

It is time to resume my story where I left it on page 31 to explore my reason for taking the dangerous road of escaping the hell which Bulgaria had become for me by 1951, beginning all the way to my childhood.

It was not by accident that we escaped from the Trudovak camp on the morning, immediately after the company commander reprimanded us for complaining that we cannot fulfil our quotas of work and ordered our transfer to another one of the company platoons, expressing the hope that we would be able to meet our obligations there. Our newly assigned platoons were on a digging detail and the work there was much heavier. He was being facetious. Very often, later, I would remember and wonder what the reaction of the trudovaks was when learning that we had defected right after our penalty was announced. We had taken our decision to run away the evening before, and we were waiting for the meeting to be over and take off.

We had planned our escape a whole month before the appointed day arrived. My own decision had been taken long before this day, ever since the experience which I had on July 21, 1944. I was only waiting for the convenient moment. When leaving for Balchik I told my father that I was going to run away. I had written him from Bezmer, sometimes in September or October 1950, that I did not expect to come out alive from the service. When I told him of my decision he looked in my eyes, turned around and started crying. I had taken with me all my diplomas, folded in a small packet and stuffed in my wallet. I did not have the vaguest idea how was I going to run from as far as Balchik on the Black sea, near the Romanian border. Perhaps I had the subconscious sense that this indeed was going to happen. In Balchik I kept my civilian cloths for a month, but loosing hope that I would need them, I shipped them to my father. Then we started talking for running away with Zdravko and Stefan. I had no more doubts in my mind. I have often thought, what would have happened if I did not follow up on my intended defection? My answer was that I could have kept quiet, I could have reconciled to my fate and let things go as they would, without exposing myself to the risks which were obvious. But my soul was filled with so much bitterness, so much humiliation and disgust, so much exasperation and sense of helplessness as to my future life, that I could not see how I could fit in the communist society. The most important, I could not act hypocritically forever, because every act of hypocrisy touched deeply my soul and I could not tolerate any more the pain. My soul was indeed poisoned against communism. It is true that everybody around me was engaging in hypocrisy, everybody was adapting to the situation, but I could no longer suffer it. The morning of the day when we were going to go, I did not have one ounce of doubt in my mind that I was going to go. All along, till we crossed the border, I never thought of turning back. The dye was cast. And I never regretted my decision and my taking the long road to freedom.

After the Company meeting was over, and all trudovaks were preparing and proceeding to the city's baths, as it was ordered, I wandered a little until Zdravko, and after him Stefan, had disappeared into the deep ravine, sloping to Balchik, some 150-200 feet away from the barracks. Then I entered the barrack, took the things I wanted to carry, and headed to the same slope where my friends were waiting for me. In the middle of the campus I met Ivancho – the half-blind kid from my village, mentally retarded, small for his age, pushing a wheelbarrow filled with wood. He stopped and started talking to me: "My fellow village man, they found it again." I asked what they had found. He explained that they had found lice on him and had ordered him to boil all his clothes. I friendly advised him to go and do it, and resumed my walk. Then I remembered the superstition, that if you are going on a long journey and in the very beginning someone meets you with full hand of something, you will be successful. I chose to interpret Ivancho's meeting me with the full wheelbarrow as a good omen. I remember the other occasion, which I may have mentioned, that when I was going last to the Seminary, I was met by Mailatova Pena with full buckets of water. She had told me of this superstition. Well, it worked for me in both instances. I continued to wander aimlessly around, not to cause anyone to suspect me. Then I met the platoon's corporal of my newly assigned place. He told me in which barrack I should move. I assured him that I will take care of it. After a few minutes I was behind the ridge, where Zdravko and Stefan were waiting for me.

Our plan was to cross through Balchik, to reach the Black sea beach, and continue on it for a while, until we are far from campus and there was no danger that someone would see and recognize us. We had to do that the previous Sunday, but for some reason, that Sunday was declared a working day – on account of the fact, that we had not worked on May 1st, so we had to make up for lost time. So, by necessity, we had to postpone the execution of our plan. Who knows, it might have been for the good. After crossing Balchik, we sprinted along the beach in the direction of Varna. Once in Varna, we were to take the train for Plovdiv. We were certain that our absence would not be discovered until 9.00 p.m. when the roll-call would be taken. By that time we would be far away from Varna on our way to Plovdiv. We further presumed, that having discovered our absence, they would post patrols to wait for us, thinking that we may have been drunk or in town. They would start looking for us in the morning when we would not have returned and would sound general alarm. By that time we would be close to Plovdiv, and with luck, out of the train on Philipovo Railroad Station. We would be met there by the Peltekov family and led to the Rodopi mountains, where the man secured by them would take us, and lead us to the Greek border.

After half an hour walk along the beach we met a group of five-six young kids, bearing huge bouquets of peonies. I again interpreted this to be a good omen. After another half hour we left the beach and crossed to the main road leading to Varna. But before venturing to that road, we found a ravine, left our military uniforms, and continued with our blue work pants and shirts. The idea was that on the main road we would be travelling as ordinary workers. It worked. We had to throw our good boots, which later would have been very good when we needed them in the mountains.

Once on the main road, we continued in normal speed, not to attract the attention of passing vehicles and occasional people whom we would have met. Stefan (I mentioned his book in the beginning) had invented a lot of stories, even suggesting that at time we have been running, but it was not the case. He had also made up all sorts of conversations among ourselves, but they very often trivialize what we were going through, and in some instances border to the absurd. We were gravely concerned about our chances and pushed further and further. In a few instances we were passed by military vehicles, especially somewhere midway between Varna and Balchik. A truck loaded with officers passed us. They looked at us with curiosity, and perhaps with suspicion, but continued without stopping. We had no choice but to go the risky road. There were occasions, someone would stop us to talk to, but I had to make up stories and avoid being involved. The weather was not all the time pleasant. We were pelted by rain quite a few times, then we were dried by the sun, there were winds, and on occasions quite strong and chilly, but we kept going. We had found out before that the train was leaving for Plovdiv at 6.30 p.m. and it seemed that we had time to make it. We were counting the kilometers as marked along the road. Tired, cold, wet, with blisters on our feet, we arrived in Varna. Must have been around five o'clock.

We saw thousands of people pouring to the main street. As we learned later, there was some big game of soccer and people were going home. Soon we were at the Railroad Station. We sent Stefan to purchase the tickets. Very soon he returned to tell us that we did not have enough money, short of some one thousand leva. According to his version, he got the information from the station master, which was improbable. The ticket, office windows were the place to go. The whole story of negotiations with the station master was either invention or he may have been confused and asked at the wrong place for the wrong things. Nevertheless, we were faced with a serious problem. We needed more money. I remembered that I had a friend in Varna, a priest, Dimitar Dimitrov. We were colleagues as students in Sofia. He was a man of happy-go-lucky disposition, proud with his conquests with the opposite sex, apparently adapted to the communist authorities, which I later found to be true, spending most of his time in Varna, while I was holding his student book and obtaining the professor's signatures for him. When I arrived in Balchik I had written him and he had answered me with promise to see each other some day. There was no other way but to look for him and try to borrow money from him under some false pretext. Asking here and there we found his house, a three-story apartment building. While my friends waited out in a distance, I went up there, rang the bell and a lady answered the door. I explained who I was, which she knew from her husband. She invited me in to wait for him. He was at the soccer match and was expected very soon. I had to wait for him. I do not know how I had looked in my wet and miserable working blue clothes. I do not remember what story I put to justify the 1000 leva I needed, but he was not coming as expected. My friends downstairs were worried and soon they were ringing the bell, so we could go.

With the lady was her sister, living in another apartment in the building. She offered to loan me the 1000 leva, and then her brother-in-law would reimburse her. I was extremely appreciative and took the money. The poor thing, if she is still living, which I do not believe, I owe her so much. But if my friend, the priest, had returned, he would have immediately guessed that something was terribly wrong and would not have given me the money, would have cornered me for a confession, and, who knows, how the whole thing would have ended. But now we had the money and flew to the Railroad station.

Stefan went to buy the tickets, only to discover that in fact we had had enough money to pay and the extra money was not needed. It was only a matter of minutes and we were on the train and rolling out of Varna. In the train we split, not to be in the same place, just in case they would start checking. Nobody was checking. We handed our tickets to the conductors, which from time to time changed, slept here and there during the night, and kept in touch from time to time. Our plan was working so far. At 9.00 p. m. we conferred and thought that at this time they were discovering our absence and were going to wait for our return. We were already many kilometers away. We discussed the possibility that they may have alarmed security forces all along in the country and probably it would be better if we disembarked at Belozem, or another station, and continued to Plovdiv walking. But at end decided to proceed to Plovdiv, to take our chances. When we arrived at Philipovo in Plovdiv, we looked for militia or agents – as we would have recognised them in the multitude of people, but in such circumstances one at least thinks that it is safe to look. We looked specially for someone of Stefan's family to meet us. There was nobody.

As it turned out they had expected us a week before, but we did not appear, and they had not been warned of our change of plans. Stefan's version places his uncle at Philipovo, where he had noticed suspicious people, perhaps militia agents, and had not seen us. We did not exit the main door, but had taken a side way out. What were we going to do? The plan had failed. So, as dangerous as it was, we had to look for Stefan's sister's home, not too far. While I and Zdravko waited aside, Stefan went to check it out. Nobody was there. Eventually his sister showed up. And then his uncle joined us. Now the question was where to go.

Relating these events Stefan tells some amusing stories. He tells of a conversation initiated by Colonel Zmiarov, Chief of Trudovak Service, telephoning to the Plovdiv District Office of our escape, giving him our names and ordering him to wait for us at Philipovo Station. The District chief supposedly immediately dispatched militia men to Philipovo and the Central Rail Road station. Another story told by Stefan is a conversation between the three of us:

"Here is Plovdiv, said Stefan to his two friends. We lived to see it again."

"It is possible that we may be seeing it for the last time.", whispered sadly Zaprian.

"Do not say such things, Zaprian," reprimanded him Slavi. "I am sure that we will succeed in crossing the border. One day we will return to our motherland."

"You may be, but not me", Zdravko had added.

Would the prediction of Zaprian come true? It did.

(Stefan has assigned to me the name of Slavi Radoikin, and to Zdravko – Zaprian)

I do not know when Stefan has written these lines, but eventually this is what happened. Zdravko never had a chance to return to Bulgaria. Last fall I was contacted by his niece, I sent her the first draft of my story and the other day, January 16th, sent her a better draft – the part of our escape – because her mother was with her in Chicago.

Stefan had a better recollection, if a little confused, how things went on. His uncle has waited for us at Philipovo, but as he did not see us, had left. He had seen agents waiting for us at the main exit, but we had taken a side door and so missed him...and, more important, we missed the agents. It happened that the chief of police in Plovdiv had been assisted when he had been partisan by Peltekov's family, suddenly realized who was one of the fugitives and hoped that we would not be apprehended. We were not. When Stefan appeared in his sister's house, his brother-in-law, Vassil, was surprized, having assumed that the plans were cancelled, since we did not arrive the week before. Now they had to improvise. He suggested that we go to a nearby small park, where he would dispatch their uncle to take us across town. We did go there, and after a while uncle Boris appeared. We had to cross Maritsa bridge and had quite a discussion how to do it – by walking or by bus. Decided on walking. On the other side we were met by Stefan's father. After a short conference, while we were eating bread and cheese which he brought to us, he suggested that we go near the High School nearby where Vassil, Maria and Boris would meet us. After a while we were there and waited for Vassil, Maria and Boris. Soon they appeared and we continued to find a safe place at the end of the Nebet Tepe. As we split in two, we lost Boris and Zdravko. Vassil then proceeded to tell us the plan. I will take this story from Stefan, which seems to be true:

"All were frightened from what had happened and feared the future... Vassil began explaining the situation."

"Listen now what you have to do. What I am going to tell you now, I will tell it later to Zdravko. After a few days you will be put in contact with the partisan "Goriani", if you have heard about it. It will be for a short time."

"Yes, we have," replied Slavi.

"The "Goriani" who broadcast by the underground Radio", added Stefan.

"Yes", continued Vassil. "We have people who will bring you to them. They will help you to cross the border. The first question is when and how to take you to them. So you listen well and memorize it. You have to get away from here, as soon as possible, to go in a field and then in the mountain."

Vassil instructed Stefan to take us to some locust forest near the Komatevsko shosse where that night we were going to be found by representatives of the resistance, to take us to another place. They were going to whistle a song as a pass word. We should come out only if they used the password. If someone is caught and the militia wanted to get to us, they would give them a wrong pass word. In the meantime we were to wait around the monument of Giuro Mikhailov, so when Zdravko is located, he will direct him to us. We spent the whole afternoon there, waiting. Stefan supposedly went to look for him in Philipovo, but returned without finding anybody. In the evening we decided to go to the Locust forest outside of Plovdiv, along the road to Markovo. During the night the promised contact was established, but the visitors were whistling a wrong tune. So we did not come out. Zdravko was nowhere to be found. The next morning, very early, Stefan and I decided to go back in the city to re-establish contact.

As we were approaching Plovdiv, I was taken aback. I came face to face with Geno Paralingov, Geneto, from my village, very close friend of mine, Brannik or Legionnaire in the past. He saw me in my dirty and crumpled blue working shirt, looking like I had just come from a sewer. He immediately understood that something was not in order, and before he asked any questions, I told him that I had just ran out of the trudovak services, that I was now illegal, and that with my friends I am on my way to the Greek border to leave the country.

"How could I help you?" he asked me.

I have no clothes", I said, "If you could find something I will be very grateful to you."

He lived nearby and led us back to his apartment. He explained to his wife briefly what was the situation and she in a hurry fixed us breakfast. He discovered some raggedy pants and a sweater and I put them on. At least now I looked like many people on the street. We had to hurry. He had been on his way to work, and this was his first day on the job. He was already late, but did not show any concern. With all his resolution, he wanted to help. What he did was heroic. He was risking his life. He had a baby in the crib. He could suppose that we might be apprehended and forced to confess our meeting with him. I never forgot this incident in all my forty years, and often prayed for him. I never heard what had happened to him thereafter. When I visited Bulgaria in 1991, he was still alive and came to see me at dinner in my nephew's house. Now we revealed the secret of our encounter so far back, in 1951. He never told anybody. In this most critical hour in my life, he was a godsend, a blessing. Later I was his guest and met his wife, his grown up sons who had good professional careers. Two years later he died. I never forgot his gesture and the determination with which he did it.

3. The Thrill of Holding a Gun

We left Geno's house and hurried to find Stefan's cousin who lived nearby. He was not home, but his wife was and he sent her to contact his sister. After one or two hours she was back, as we waited outside in the neighbourhood. After one hour or two, the woman came back with Vassil and Maria. Vassil led us to the Komatevo road as we split in small groups. He was looking for a hiding place. At one point he led us off the road, across a rye field, under a cherry tree. There was a small forest further up. He told us that the people had come for us the previous night, had whistled a different tune, having forgotten the correct password. We have done the right thing by not responding. We were to hide here. Boris and Zdravko are going to join us soon. They did. Vassil, Maria and Boris left and we had to spend the day in uncertain expectations. As things were going so far, it was not working very well, but it did not seem that our plan was failing.

In fact, that night, after midnight, the "goriani" came to our hiding place. They whistled the right password, and we responded. I believe they were three men – Peter Nikolov and Ivan Nikolov – brothers from Markovo, named by Stefan respectively as Ignat and Petko, and some young man whose name was, I think Angel. Vasil and Maria also joined us, and the whole group was led to the mountain. What was the most important in this meeting – for some reason Stefan does not mention it, he never mentioned it – is that they brought us guns – three carbines and ammunition.

One of the carbines, a short thing, apparently adapted by some local smith, with cut barrel, was assigned to me and the others to Zdravko and Stefan. They also brought us food. After a short conference, Peter and Ivan led us across fields and forests towards the mountains, where we were supposed to hide for the time being.

This night march lasted two or three hours. We could see nothing in the dark. Sometimes we were following in some paths. Other times we had to break through the brush and on occasions followed along embankments. I had no idea where we were and where we were going. Zdravko and Stefan were from the area and more or less were acquainted with the direction taken by Peter. In the beginning I was disturbed and uncertain, but followed the leaders on the assumption that they knew where they were going. At one point, while walking along an embankment, I was overwhelmed by a strange feeling of exaltation. Squeezing the small Turkish carbine in my hands, all of a sudden I felt free, I felt confident in myself, I had lost all fears in my soul, I felt proud, I felt secure, I was determined to conquer the world. In an instant I realized that I am no longer under the fist of my tormentors, that I had overcome them, and that if they showed up, even if armed with much superior guns, with this little thing I could take at least one of them with me. I became a different man, I had not felt this way for years. I was converted into something that I could not explain what it was. May be that was what St. Paul experienced on the road to Damascus. I felt that I could no longer be humiliated, I could no longer be intimidated. Even though I did not have the vaguest idea how to use this little thing in my hands - I had never touched any gun before, nor I have after I threw it in a ravine after I crossed the border - but for my spirit it performed a miracle.

This feeling has never left me. Time and again I would remember it. And again I would experience a supreme pleasure, would relive this supreme pleasure of this unrepeatable unforgettable moment. Many a time, when I would be lecturing to my students at the universities, goaded by them to tell them of my experiences, I would describe myself with a gun in my hands, something I could have never imagined myself before, and something nobody, knowing me what I was, would have imagined that I could have been that man. Well, life is a strange thing. Sometimes it confronts us with things which we have never thought about, have never been prepared for, and then we had to deal with it as best as we can. During my forty four days in the Rodopi Mountains, and many times later in exile, I would remember myself as the skinny bookworm in the School of Theology, with a dozen books opened and displayed in front of me, searching for the right explanations of the problems I was studying; I would remember myself furiously typing my manuscripts – the meek, the inoffensive young scholar, the Christ-loving, the God loving, the humanity loving altruist, the Tolstoist-like anarchist who would not step on an ant, the discard rag of society before and after September 9th, now crossing over fields and mountains with a gun in hand, ammunition in my pockets, looking for somebody else's or my own death... When I would recall this unforgettable moment in my life, and the feeling which overwhelmed me, my shoulders would shudder and goose pumps would spring on my arms. Then, and any time I recalled this moment, I experienced such an inner pleasure, such a thrill which I could never describe.

However, it puzzled me to no end when I read Stefan's book and found not a word about the guns given to us. How could have he forgotten such an enormously important event? Not a word! It took me three days, to remember, that about the end of his book he tells the story how he had joined the Theosophical society and how he had become a preacher and prominent speaker at their meetings. On page 276 he says:

> Stefan was listening with a thrill in his heart the philosophy of these people of the Theosophical society. New doors were opened before him, new thoughts, new ideas, new paths. Elena Blavitska, one of the founders of the Theosophical Society, had written a book *The Secret Doctrine*. Stefan found the book and read it. This book was giving him a true spiritual nourishment.

I did not know anything about the Theosophical society. I knew only that it was a religious cult. One of my dearest friends, Virginia Baslenkov, wife of my friend Stefan Baslenkov, was a Theosophe and I knew of her beliefs. Guns would be the last thing, and murder that is, that she would approve of. This was my explanation of the omission which Stefan had purposely made. With his new faith, he would not confess his past manipulation of guns. I have the same feelings about guns and murder, I had felt that thing before, and ever after, but human nature is something that dissolve all inhibitions.

It did not take long when the hour of truth came upon us. Vassil called us around him and told us that we were not going to go to Greece, but were to stay in the Rodopi and form a *goriani* group to fight the communists. More *Goriani* were to join us and we were going to start a partisan movement. We were the nucleus of this movement. This changed everything. We had left Balchik with the understanding that they had to insure our crossing the border. Now they were going to use us to form a Goriani group. This crushed all my hopes and sapped all my physical and mental strength. I saw the whole thing as foolishness. It had no chance for even the slightest success. With my little carbine, probably from the Crimean war, I was going to fight and defeat the army and the militia armed to their teeth? We were going to bring down a government which had murdered tens of thousands, who dissolved the old army and police without anybody firing a shot at them? They had initiated and still maintained a reign of terror over the country.

The communists had executed regents, ministers, generals,... officers, peasants; they hanged Nikola Petkov behind whom were the Americans, and now it came to Stefan, Zdravko and me to topple the regime? These people, I thought, did not have an ounce of brain in their heads. At once I realized that we have been taken in, that we have been misled, that we have been lied to, that we, especially me, had entrusted my life to a handful of fools, naive people, who either did not see the real situation, or themselves had fallen in the same trap. They apparently believed that if we started it, the people all over the country were to rise in a rebellion, that the Americans were going to march into

Bulgaria and the new liberation was to become a fact. One could fantasize, and fantasize, and fantasize much, but not to that extent. The million-strong political opposition in the country was crushed without a shot to be fired from anywhere.

The Americans and the British hid in mouse holes and no one saw them anywhere. They made speeches from far away, they attacked communism in blistering tirades, they screamed on the radio what an evil communism was, how the poor people of Bulgaria were enslaved and what terror and what horrible torments they were exposed to every day... As if they had to remind the people of his suffering from thousand miles away, to instigate this people to rise against its masters, without lifting a finger to help them. My feelings in this respect never changed.

Throughout my exile, after the Hungarian revolt of 1956, and the events in Czechoslovakia in 1968, I held the view, in opposition to other exile exaltations, and the official policies of the American propaganda agencies, that the Bulgarian people did not need to sacrifice themselves needlessly when they would not be helped. I recall that in 1956, I believe, after the Polish insurrections in Poznan, I was addressing a public meeting of Polish immigrants at the Polish National Home in Lower Manhattan. Then President Eisenhower had offered to send to the Poles bread, because in their slogans they had raised such demand. In my speech, I raised my voice and shouted: "Our brothers in Poland do not want bread! They want bombs to defend themselves." The rally exploded in ovations. The roof was going to collapse. My views were widely read in my publications. This is the reason why I was not very popular, in fact I was resented by the American agencies dealing with such matters.

So, this is what Vassil told us. We were now *Goriani*, as per his order. What could we say to him? We were in his hands. We were in their hands. We could lose them. They would easily abandon us. He, Vassil and Maria, were joining us permanently. They were to be with us to the end. We were going eventually to join another group of *Goriani*. For the time he was our commander. Until the next morning. One night stay in the mountains with his pregnant wife, was enough to change his mind. The next morning he called me on the side and gave me his instructions. He was appointing me commander of the group. I do not remember whether I wanted to laugh at the idea or to blast him out of his military uniform. I decided not to object to anything he said. It was for me and my two friends to reconsider our situation without anybody else present. I decided that I could not announce my appointment as commander of the group to Zdravko and Stefan. I was older than they were. I had university education, but otherwise I was the last one who would dream, or be aspiring to be commander of a partisan group. The whole thing was ridiculous. Under different circumstances I would laugh it out, but now I was a captive audience and had to swallow this stupidity. I was not made of that stuff. A commander of three men? He then called us all together and announced that he had to return to Plovdiv to make arrangements for us to join a bigger group of *Goriani*. He left us, and we never saw him again.

4. Haidut Dere. All Hopes Come Down Crashing

In his book Stefan has invented a number of stories which are pure fabrications. Zdravko supposedly went to visit his grand mother for bread. Peter Nikolov supposedly visited us with a man by the name Giuro, who in turn was going to come and take us to the *Goriani*, but, then, on his way to us, on May 16th, he was caught by the Militia when he was on the way to our hiding place, but did not reveal anything. Thus, the story which the Peltekov's put out to entice us to run away, was confirmed by this lie. Apparently there never was a Giuro. Then Stefan has concocted a story that Zdravko, in a moment of despair, for lack of cigarettes and all that was happening to us, charging that we have been abandoned, threatened to go back and surrender. Stefan's pleas to think of his surrender, how many people he would have to implicate in our escape, did not persuade Zdravko to give up. He just advised us, supposedly, to run wherever we want so they could not find us. When these pleas failed, I had taken the initiative to stop him. I had taken out a knife and had threatened to kill him if he persisted in his intention. I am being quoted as having said:

> Listen Zaprian, I have studied Theology. It means that I have been taught not to kill anybody! As until now, I have not killed anybody! But if you would make even a step forward to your village to surrender, I will kill you instantly. I will kill you for the sake of all those noble people, who helped us risking their life and whom you will, by your betrayal, destroy. Do you understand what I am saying? So, now, go ahead and decide what you are going to do! One step forward, and I will kill you.

I supposedly held Zdravko with my hand when I was delivering this threat. It is all a figment of Stefan's imagination. Zdravko was no less than 180 pounds, compared to my 135. He was a powerful man, and with one kick, or with one sway of his hand could have thrown me five feet away. That for the physical part. As for the degree of courage which I supposedly had displayed, it was far, far away from anything of the kind that I could have shown. Stefan continues with the story where Vassil reappeared again on the stage, and told us of the arrest of Giuro, who was going to take us to the big company of Goriani. Now, that Giuro was no longer in the picture, Vassil was going to Plovdiv, to find some other man, who knew the way to the border and was going to come and lead us there. In fact Vassil never came to us again. As Stefan tells it, he was lured back to his service by an officer and arrested there. Eventually he revealed all that there was to know about our affair. He ended in jail. When I visited my village a few years ago, a man, Maniu Poshtov I believe, told me that they had been together with Vassil in jail and that he knew our story. Recently I heard from Stefan that Vassil and Maria had divorced, and subsequently Vassil had died. Stefan never mentions in his book that Maria was pregnant, and whatever happened to her baby.

During the following days and weeks we had difficult experiences, which I do not recall now in details. At the time I kept a small diary, but during my peregrinations in America I lost it somewhere or inadvertently destroyed it. More often than not, we suffered from hunger. Stefan's father and uncle, and the Markovo people from time to time brought us bread and cheese, but most of the time it was a painful starvation. Stefan has filled this period with invented stories, some of them so grotesque that need not be repeated. One of these stories had Zdravko going to his village, then somehow attending a wedding celebration amusing himself for hours and then escaping back to us. What really happened in the early days was, when we were left alone, and knowing what we were told about the Goriani, we were bitterly disappointed.

We had taken the risk of running away from Balchik with the understanding that we were going to go for the border, not to join any Goriani group. We did indeed urge Vassil to seek the man who was going to take us across the border. More and more Zdravko and I were suspecting that we had been misled, but had no other way now to change things, except to follow orders. Many years later, may be in the 1980s or before, I read a story in "Anteni", a weekly published in Bulgaria by the communists. Someone from the secret services was reminiscing how in the early 1950s, (the story perfectly fits our case), they had used a device to trap fugitives and opponents, and send them to jails and concentration camps. They would recruite weak prisoners to cooperate with them, then let them out to freedom. In their turn these recruites would infiltrate or simply contact known anti-communists and persuade them to form illegal underground organisations, always leading them to believe that they would be taken to some Goriani groups, always somehow supplying them with guns, and when they would have thus formed such groups, they would prepare their transfer to the Goriani. However, before the transfer, they would arrange for them to leave their guns, to be used by other recruits. Then they would be gathered in some barn and wait there for the Goriani to come. Instead of Goriani, the Militia would come with their trucks, surround the building and march them all into the truck, and then strait to the Plovdiv militia.

When I read this story, I have the paper saved somewhere, and recalled all that happened to us, I came to the conclusion that we have been trapped in such a scheme, but for some reason it did not work. Maybe our failing to run away on May 1st from Balchik, may have confused their plans. If there was a man who was supposed to take us to the Goriani or to the border, he may have been one of those who had fallen in the trap and our people had never been able to get hold of him, and he, in his own turn, may have not revealed our inclusion in his plans. It crossed my mind that Stefan's father had been caught somewhere in this web, but somehow it did not work, and we ended alone in the Rodopi Mountains.

But the story in "Anteni" showed that this trap had caught many fishes like us. In this account Stefan speaks much of such a mysterious man, but his references are very vague and unconvincing. In any event I lost contact with Stefan. Sometimes in 1954, or about that time, I recall receiving a letter from him from France. He had changed his name to Etienne, the French for Stefan. I wrote him back with some reprimand that he had changed his beautiful name with some French distortion of it, which suggested to me that he was somehow running away from our Bulgarian heritage. He never wrote me again. When I read a couple of years ago in *Nova Borba*, published by Dr. Altunkov, where Stefan was mentioned as his assistant editor, I asked a friend, Iskren Azmanov, to see if he could obtain a copy for me. Iskren spoke to Stefan and he called me on the phone. We talked briefly and told me of his health problems. I started reading the book, was not impressed with his concocted stories, but made the effort to finish it. Soon after finishing the book I wrote him a letter with suggestions for a series of vitamins recommended by my friend Bai Dimitar. My letter may have hardly reached him when I heard from Iskren that he had passed away. I sent a sympathy card to his wife and this was all that I could do for him. Poor guy! Our paths had crossed briefly. We shared the most difficult and critical days in our lives, for good or for bad. My association with him turned my life in unforseen directions. God bless his memory! Otherwise, in his book he had referred to me in a very galant way and have nothing to make me keep any grudge against him. If I am making these remarks here it is only to put the record straight, and nothing more.

After we were told that we were to stay in the area and form a partisan group, and rejected it outrightly, we urged the father and the uncle of Stefan to make every effort to find the man who was supposed to take us to the border. They gave us all kinds of excuses, until, eventually, they admitted to us that the man had vanished and they could not contact him. Nothing was said about the mysterious Giuro, and that he had been arrested. But one way or the other we were now Goriani – with guns hidden nearby, hiding from the world not to be noticed by anybody, fully dependent on unknown to me people who apparently believed in their illusions. But we were Goriani not by our own choice, but by mistaken assumptions and actions of our helpers. We did not run away with such intentions. Neither we knew anything about the Goriani.

About that time, as I was to learn later, the American and Greek intelligence services were recruiting volunteers from among the Bulgarian refugees, trained them somewhere and then smuggled them back into Bulgaria for underground work. These groups of five to ten people would be instructed whom to contact and what missions were to perform. Some of these missions were to win new Goriani. One of these stories was that a group of them was dropped by an airplane with parachutes right in the center of a Bulgarian city, the local militia authorities had known well in advance of the plan, and shot or apprehended them at their landing. Another story was that someone, having gone this way to Bulgaria, subsequently had defied his superiors, had given up the service, had talked too much, had emigrated to Australia, and there the agents of the intelligence services had killed him – chasing him with a truck even on the sidewalk and had crushed him to death. I became very close with a group of young, educated activists of the IMRO, including the man who had gained the laurels of "The King of the Pirin Mountains", in the Lavrion Refugee Camp in Greece, 1951-52. He is still alive, in Canada, and all these decades we have preserved our friendship. They, the whole group, had been Goriani, but had dropped from the movement. The Americans had insisted that they reveal to them their contacts in Macedonia. On instructions from Ivan Mikhailov, they had refused to do that. The assumption was that the American services, the CIA, had been infiltrated by communists agents, who, once in possession of these names, would reveal them to the authorities in Bulgaria. They, in turn, would eliminate them. But in Bulgaria, the name Gorianin was big, was the synonym of freedom fighter, of underground organization for armed resistance to communism.

It seems to me, that the entire affair was staged at the time as a pressure group to be used in diplomatic exchanges with the Soviets. The Americans must have known that nothing would come out of that, that the young men they were recruiting for such activities were canon fodder for the communist Militia. Very often, I heard, these agents were returning from Bulgaria, bringing with them newspapers published in Sofia and the localities they were sent to as evidence that they had performed their mission. So many of them were shot at the borders when crossing into Bulgaria, and when crossing back. In many instances they were apprehended in the country. These were the true Goriani. We? We found ourselves in the web of this phoney guerrilla war, as uninvited guests – with no connection with the American Intelligence services, no organization to supply us and instruct us, all on our own.

We were not trained to use guns, we had no mission assigned to us, no personality predispositions to make the stuff that makes guerrilla fighters, not associated formally or informally with any Bulgarian political organization which could vouch for us, except on a very, very low, almost non-existent political level. We were outsiders of the gigantic confrontation between the Soviet and the American underground forces. It was by chance that we had fallen in the whirlwind of the current events, we were like fish on a dry sand, jumping in utter helplessness, without any idea of what we were doing and where we were going. If in the beginning we had this idea, only after a week or so, we realized the bitter truth that we were on our own. Our only contact with the world were the Peltekov family and the people of Markovo. If we were apprehended, or if they were apprehended, then the whole thing would collapse and we would rot somewhere in the communist jails or concentration camps, if we survived at all. But we never, never for a moment, entertained the idea of surrendering to the authorities. For as long as we were getting a bite of bread and cheese and kept an ounce of strength in our bodies, we were determined to go for the border.

5. In Limbo

From the very beginning things did not go well. Very soon we started loosing contacts with our helpers, and sometimes had to starve for days until they would reappear and revive us, and our hopes for change. The goods they brought us were often not quite tolerable for consumption but we devoured all that came our way. Most of the time it was bread and cheese. We could not judge them for their failures. They had to hide in their travels, they had to try to avoid being followed, they did not have the money that was needed to supply us.

Once Boris Peltekov brought us bread which was half-baked. I was so hungry that I ignored the imperfection and loaded my stomach. In a couple of hours, I was sick like a dog. I was in fever. By the evening when Boris had left, I was helplessly convulsing under the rock. I felt that I was going to lose conciousness. I felt as if I was dying. On top of it, it was cold, it was raining, it was damp and misty, and I did not have clothes to warm me up. I was pushing my knees to my chin to cover myself with a rag. At one point I asked my friends to make a fire so I could warm up. Zdravko gave me a sour look and observed: "Are you with your mind? They will see the fire from somewhere and come and get us alive like birds." I was begging them. Stefan was more sympathetic, but Zdravko was adamant. At end I told them, it does not matter for me now if they will capture me, or not. "I will be dead any way if you do not make a fire." They relented. Zdravko gathered the wood and soon the fire was under way. I relaxed somehow, turning back and forth to the fire. Besides, during the day Boris had brought a bottle of Plum brandy and they allowed me to take an extra gulp to what each one was entitled to. I do not know, was it the fire, or was it the brandy, but by midnight I was beginning to feel better. By the next morning – I had fallen asleep sometimes – I was as good as new. Nobody saw our fire, or the smoke of it. It was such a terrible night!

It was in connection with the supplies that Peter Peltekov once asked me if I would object if he tried to go to my village and contact my father. I was somehow concerned that he should be informed that I am safe in the Rodopi Mountains. I was sure that by this time he had been informed of my defection. But I feared to involve him into this. I told Peter my frustrations, and asked him that if he expected money from my father, he would not get any, because he did not have it. Peter assured me that he would then just contact him to tell him where we were and that they were taking good care of us. I suppose in this condition I could not even think straight and dropped my guard. I told him how to go to Zelenikovo, but instead of asking for my father, to look for Stefan Bambourov, husband of my cousin Tina. Through him he would contact my Aunt Plina and she would then find my father. But nothing was decided. I told Peter I still had to think about it. He did not wait for me to think of that, and without my authorization had gone to Zelenikovo and had followed my suggestions. What had happened after that I never learned – whether my father had given him money or anything else. Peter never reported to me in details. Later, in recent years, I heard about the meeting and apparently Peter got some money. *

I cannot help thinking that this was a grievous error on my part. I should not have fallen in this weakness to let Peter go to Zelenikovo. I do not know if anyone reported it to the authorities and what were the consequences. It is quite possible, that my father was murdered a year later, on the day of my anniversary of escaping from Balchik, because of this contact. But what can you do? I was never able to get deeper into that. Neither did I try to find more about it. The pain of the whole thing is so unbearable that I rather let old things be kept out of sight, rather than poke in old wounds. I only learned that in the first days, I do not know how long, the local militia or party functionaries had set secret observation posts around our house, just in case I appeared in the village and tried to get in touch with my family. I was not so stupid to fall in such a trap, but I still feel guilty for giving Peltekov information which he used to go to Zelenikovo.

The days in the Rodopi Mountains were following one after another – slowly, painfully, boringly, colored with total ignorance of our future. In vain we tried to extract some ideas from the Prltekovs. The people of Markovo, who appeared several times, were as little helpful as the Peltekovs, except for bringing us some food once in a while. We wanted to hear about some plan, but nothing was offered to us at least as a hope. There was nothing we could do, because we were at their mercy, which could at any time be cut off. When they could come, we would briefly talk about the conspiracy, some names were mentioned, but nothing definite was said as to their intentions. We were left to our own imagination, to wait for something to happen. During the days we hid in the brushes and the forest around our camp, always careful not to meet someone lost in the woods. We discussed our situation but could never come to any conclusion, in the absence of any information. It is amazing that all this time we never got involved in bitterness, and in disagreements among ourselves. Our friendship was cemented by our common unknown destiny.

It happened that this month of May was too rainy and too cold. Or it appeared to us this way, wandering in the high mountains. During the night we could sleep with back to back to keep warm. We dared not make a fire. We had more peace in the evenings, especially when it happened to be warm and comfortable.

---

* Just before the final reading of this text, before publication, I received a letter from my cousin, Todor , son of Stefan Bambourov. He had read this incomplete story in the Bulgarian version of this volume, published a few months ago. He had knowledge of what had happened Peter Peltkov had gone to Zelenikovo, had contacted his father, and my father, my aunt Plina and my cousin Georgi, brother of Stefan' wife, my cousin, had been called in. Peter had told them he personally had gone to Balchik to take us back to Plovdiv, had kept us in his house in Brani Pole, and was now making arrangements to send us to Greece. For this he had to pay money for a man to lead us across the border. They had given him all they had at hand – my aunt Plina 40,000 leva, my cousin Georgi 15,000 leva and Stefan 7,500. I never knew that. And I would not ask where this money had gone. But when I read this report, I felt as if I have been hit by a ton of bricks. The thought that my relatives had been put to this embezzlement in my name and that I had never been able to repay them, never knowing about it, is still killing me. I will see what I could do to those who are still surviving – my cousin Todor and Georgi's daughter Nedelia in Sofia.

6. Contemplations on the Rock

On many occasions I climbed above our rock-cave overhang, high to the top of the mountain, sat there on a rock and contemplated the entire Plovdiv valley, trying sometimes to recognize in the mist far far away, my Sredna Gora, some fifty miles north-East of Plovdiv. The Thracian valley, on both sides of Maritsa, was like a fairy land. The villages were basking in electric lights and could easily be located and identified, if one knew their geographical position. For as far as I could see, there were those islands of lighted nests.

But above all, there was Plovdiv. It was so near, and yet so far. The lights there were brilliant, the sprawling city was sparkling in the night's darkness and I could not take my eyes from it. Sitting on the rock, supporting my chin on my hand, I was taking it all in my soul and I was dreaming, or rather thinking of what was going on there. I was thinking of the Djoumaya, the Main Street, with its shops, with its sweet-cafes, I was imagining the people going in and out, I was following them in my mind in their strolling on the sidewalks, conversing with each other, sharing joys and difficulties. I was envying the young couples of boyfriends and girlfriends going to the City Park opposite the post office. And then I would turn to myself and see what a miserable situation I was in. I could be them.

But I was here! I was hiding in the brushes and under the rock-cave overhang, I was walking around, always listening for the slightest noise, and I saw myself as a wild animal, outside of the blessings of human civilization. Then I was thinking of my freedom. And I felt that I was not free. I was far from the reach of my tormentors, the communists, but I was not free. I was as free as a rabbit running down the slopes, as the birds were outdoing each other in their fantastic singing, I was free as the green lucious oak brushes, or as the mole that was crawling under the ground. I was not human. I was out of human society. I had escaped from Balchik seeking freedom. I had found myself a slave without hands, without legs, all of them tied together. I had my mouth shut, I was running like a deer hunted by a hunter in the woods. Was this the freedom which I was seeking? And a feeling of sadness would·overtake me. And I would end up with tears in my eyes, I would stay on this rock for as long as I would cry out my soul. Then I would descend to my cave and go to sleep, which my friends had for long been doing. One evening, while I was standing on the rock, in the moonlight, I decided to write a letter to my Father. I had nobody in this world for whom my life had any meaning and for whom I felt such a deep sorrow, than my father. I was not going to mail this letter, but I wanted to have it in my little diary. I have lost the text, with my diary, but somehow I have retained its substance in my memory.

My dear Dad:
I am sure the news of my escape has reached you already. My heart is breaking apart out of pain just thinking that I have brought this grief to you, but it is now too late to change things. I would not have come out alive from the service as trudovak and took this way out in the hope that by doing that, I would be able to save myself. Even if this hope proves false, at least I will speed up the end of my sufferings. My heart breaks for you and for Toncheto (my younger brother). For many years we all have lived a martyrs' life and it seems that there is no end to it. You take care of yourselves there, and do not be concerned about me. Whatever may come to me, let it be. Do not mourn for me! I alone decided to run away. You have no part in it. Let God's will be done! As for now, I just have to say: With God's speed!

Your beloved son: Spas.

Many times during these long nights I climbed to this rock on the top and ruffled the pages of my past. I could not see where I had done something wrong, or committed some crime, to have betrayed somebody to the authorities, to have caused pain to somebody, in order to deserve all that. The only sin which I believed I committed was that too often I spoke my mind, had expressed my views as my conscience dictated to me, but I did not see that to be a crime, even if those in power did not like it. I had never had a contact with the old police or the military in order to report on somebody. Why did they take after me with such vicious attacks, why did they make me enemy Number One? I could not find answer to these questions.

It often crossed my mind that I could have very easily relocated to some other place where they did not know me. I could have adapted to their ways and even gain membership in their party and with all my capabilities, might have advanced in its ranks. Sure, they would have asked for references from my village, my village enemies would have blackened me like a devil, but someone somewhere may have ignored them and would have trusted me on account of my hard work. But I could not follow the path of hypocrisy. What they had done in my country during the past six years was making me sick and disgusted. The senseless murders of innocent people, the inhuman tortures and beatings for which I heard from many people, so many times, the tortures they were said to have made a common practice, the beastly treatment of men and women in concentration camps which I heard so much, all that had poisoned my soul against them and I could not swallow it. It all was turning my stomach. They had started pressuring the peasants to form collective farms.

They used every trick in the book of injustice to suppress the resistance of these poor devils. Nothing escaped my attention. I remember one evening, in the fall, as we were eating supper in the open hallway, a young man from the neighborhood peeked over the outside fence and shouted to my father: "Diado Todore, how much land did you plow today and what did you plant there?" They were beginning to keep a record, then would total it and sent it to the central office in the Agriculture ministry. My father lost his temper and yelled back to him: "Get the hell out of here. If you want to know, go there and measure it." These were small things, but they were beginning to annoy the peasants. And as the pressure for organizing a cooperative was being pushed down the throats of the peasants, I heard a woman from the neighborhood, wife of Doncho Pundev, a heyward, commenting publicly: "Whoever has not eaten the bread of strangers, and whoever has not worked for somebody else, let him join the Collective." They were poor, they had been working on rented lands. Her husband was a flaming communist, but she was holding back.

Such thoughts were crossing my mind up there on the rock overlooking the entire Plovdiv valley. I remember a conversation in the village *Chitalishte*, where adult people, five to ten of them, had gathered around my old teacher in the first grade, Diado Georgi Bradata, (Grand Father George the beard). This was about November or December 1944. "This party " he said, meaning the Communist Party, "which claimed to solve all problems of the people, also failed." He was an old Social-Democrat, he was turning to the younger people and was beating his cane on the floor. All these memories did not make my misery lighter, but they were affirming my conviction that I had not erred in taking the road of resistance. I experienced an inner pleasure that I had defied them and they could not reach me here, in this wilderness. Deep in my soul I was cherishing the confidence that all this was going to end in a success, but I had to meet every difficulty with patience until eventually we would reach the border.

A lonely spirit on this mountain, listening to the fascinating concert freely performed by thousands of birds, I would again come to my contemplations of the meaning of freedom. Oh, I was not free! I was ready to run like a wild animal in the brush if I sensed a human presence around me, or to hide in the grass, in the shrubs, and hold my breath so that I would not be discovered. I sometimes thought, what would happen if I was confronted by some predator animal, a wolf. Such stories had been imprinted on my mind from the day of my infancy. Then the realization that the next day we may not receive bread and something to eat would dawn on me. So, there was to be starvation again. And then there were the frequent storms, and rains, and cold nights, which pushed us under the rock as a last refuge, shaking to the marrow of our bones. But I never lost my courage.

7. Fears of Abandonement

From early morning, we strained our attention to recognize a password from our helpers. When they did nor appear, we fell in exasperation, wondering what to do. One day we decided to seek provisions on our own. Zdravko proposed that he should visit a known anti-Communist, his friend. Perhaps he would help. So he did go to look for this friend. But he knocked on the door and heard the man from inside to ask who he was. Upon hearing his voice, this man had pushed the lock on the door and had yelled to his night visitor: "You should get out of here as soon as you can. I have paid my dues and I am not going back to the concentration camp. If you want food, there are my cherry trees, go there and eat cherries, but do not dare to come near me." So, that was it! The man had suffered and was not willing to go further. Who could blame him?

On another occasion, after losing contact with our people, we were alarmed that they may have been arrested and that we had to reconsider our plans. But we could not do anything until we found what had really happened. So we decided to go to Plovdiv and use Stefan's cousin again. We descended down the slopes of the mountain, dressed like poor working men. We found ourselves on a road and followed it to the East. Soon we found ourselves in a village which Stefan and Zdravko recognized to be Kuklen. For some reason we stopped in the middle of it and entered a tavern, I suppose for directions. Someone explained to us which road to take for Belashtitsa. From there I do not remember how we reached Plovdiv, but we found ourselves there and Stefan was able to re-establish contact. They were not arrested. It was just that they had to lie low, not to raise further suspicions. We were ordered to go back to our hiding place, but to take a different road. Some food supplies were found and we headed back. For some reason we found ourselves travelling on the road to Assenovgrad. We were walking leisurely, talking and making gestures, to give impression that we were an ordinary group of people on their way to somewhere. At one point we were passed by a truckload of soldiers, with their guns on their shoulders. We pretended that we did not notice them. But we realized that our luck may run out, and decided to leave the main road.

We could not make it to Haidut Dere that evening. When it became dark and we could not see where we were going, we decided to sleep in a wheat field. We just went in the middle of it, along a small inter-village road. And as we were settling down, from one end of the village on the right, a happy company of young men and women, led by a music of accordion, were going to the next village. We concluded that this was some sort of a wedding, kept quiet until they disappeared. It was frustrating to hear how the rest of the world was living, and where we, seeking freedom, were hiding. We decided that it was not so safe near the road and searched another place to sleep. Heading across the fields we passed through a wide and long field which had been plowed during the fall. At one end was a forest. We seemed to be in a safe place and, exhausted from the long day of travels, made our beds in deep furrows. We were thinly dressed, we had no blankets and pillows, but you adapt to the conditions.

The next morning we continued our journey and in a few hours were back in our familiar bivouac. That evening we were supposed to wait for our people to come. They did not show up, and we had to hope that the next day someone would be back with us. We were becoming weary that something was going wrong. We never thought of such prolonged delay in realization of our ultimate plans – crossing the border.

8. The Enemy on Our Tracks

Our situation was becoming more and more pregnant with dangers. Very soon, on May 22, we were told by Boris that Vassil was arrested. Vassil was our last hope. He was to track down the man who was going to lead us to the border. What was more ominous was the supposition that he was going to reveal to the Militia our hiding place. Sure enough. The next day or so, as we were resting under the rock, we heard hundreds of cow bells. Obviously a herd of cows was advancing in our direction. We left our camp in a hurry, running to the east of the mountains, some two-three hundred meters away. The man who was taking care of the animals was following them. A dog was running around, apparently picking our scent and barking wildly. As the herd passed up our camp, to the crest of the mountain, we assumed that they were going to go in other direction. We returned back to our rock, but had not yet sat down when all of a sudden, the cows were all over around us, the man had turned them back. We ran down a steep ravine and half way climbed the embankment, some ten feet high. All the times we heard the dog barking as if it was following us, Sure enough, it was on our tracks in the ravine, but as it reached the point where we turned right and climbed up the embankment, it skidded over the wet leaves some ten-fifteen feet in the slope. It lost our tracks and continued down with the cows. Was this man ordered to take his cows to our camping place to flush us out, or the dog to track us? This is what we suspected to be the case.

The next day a plane appeared over us and circled the place for one hour. Were they looking for us? Certainly they were. I have maintained it before that I was keeping notes, a diary, of our experiences, brief, but to the point. My little notebook could not last me for long time. I also mentioned that somehow, inadvertently, I had lost or destroyed this precious little record of my days in trouble. Recently, while looking over my correspondence with Zdravko, preparing for my visit to Bulgaria, May 2000, where I was going to meet his brother, I came across a letter I had written to him on May 23, 1955. In this letter I am quoting from my diary, after referring to our difficulties in exile. Here is part of this letter:

"Things indeed are difficult, but we cannot change anything. Is this what we were thinking when we were running from Balchik to Varna, or when we were in Haidut Dere?

I remembered these days and looked for my notes of that time. You will recall that at that time I was writing something. Once you asked me what was I writing. I did not tell you. I was keeping a diary. Under today's date I have written:

An airplane is constantly circling over us. I suspect that we are surrounded. But we are hiding in the brush. It is hot. Time goes killingly slow. I will explode. Every minute seems to be an eternity. In the evening came Boris. He led us somewhere to the East. We established our camp away from Haidut Dere. Now I was able to come to know bai Boris. Terribly naive man. I am beginning to regret that I have entrusted my life to these people. Now the only hope for us is the much talked about lawyer in Sofia Stefanov. I hope, he would come. We have lost all contacts with the people of Markovo. We learned that the night before, May 22/23, a big shoot-out between the Goriani and the Militia had taken place near the village of Ferdinandovo. Markovo, Ferdinandovo and Komatevo are surrounded. Eighteen militia men have been killed … if all this is true. Our information sources were Peter and Boris Peltekovs, and they lie as gypsies. Stefan tells us that he knew Stefanov personally, that he has contacts with English and American diplomats, his house is something like an embassy… To believe, or not to believe, I do not know.

This is, Zdravko, what we were faced with four years ago… Well it is a finished story. What is left to us is to look at the present, at today, and think of the liberation of our fatherland. I see no meaning in anything else, and I do not hope for anything else amidst the troubles which I am going through. You think over all that I wrote you above and if there is anything that I could do to help you, I will not hesitate to do it…"

Let me continue the story.

That evening Boris arrived. He had traveled long paths to make sure that he was not followed. He had been told by Markovo people that we had to change place, to go to another area, miles away, known as *Yurukalan*. As the darkness of night fell we gathered our things, uncovered our guns and started the journey where Boris was going to take us. We struggled in the darkness, in our search for safer camp. Sometimes after midnight, we found ourselves in a cemetery. Cemetery in the middle of the Mountains? I was trying to read the names of the gravestones in the nightly darkness, but could not make anything of it. Grass was growing over all of them and one could see only the tops here and there. There must have been over one hundred of them. I was told that this was the cemetery where the battalion of Anton Ivanov, a leader of the Communist partisan group, was fought and defeated by the Bulgarian army in 1944 or before. They were all killed. This field of the dead made me think of Bulgaria and the Bulgarian people. The men who were buried there were Bulgarians killed by Bulgarians. I was thinking of us too. We could be killed here by other Bulgarians, the heirs of Anton Ivanov and his company. They were victims of a truth in which they believed and had sacrificed themselves for. So were we, believing in another truth to which we also could become victims. Does this make any sense? Where were our bodies going to rot in this virgin forest if we were detected by the Militia?

Silently I was reciting Pushkin's *Stancy*: "And where the fate will sent me death…" I saw the whole thing a total nonsense and crime against humanity. I could not hold my indignation and whispered loudly: "Bulgaria, Bulgaria, why are you devouring your own children?" Zdravko alone responded: "Fate!" he said. And we continued our march to somewhere. I did not know where. Early in the morning we settled in a small forest block, which was utterly indistinguishable, where no one would supposed that some terrible people like us were hiding*. Boris had to leave us and inform the Markovo people who were the next to come and bring us food.

It turned out that our new hiding place was not as safe as we had supposed. Sometimes in the afternoon, we heard a group of people approaching our forest. They were travelling along a road which was going to take them down in the valley. As we learned later, we were in the vicinity where Party functionaries had come to the Resort to organize celebration for the heroes in the cemetery we had crossed during the night. The next day was the day of Christo Botev, June 2nd, and such celebrations in honor of partisans lost in battle had become a tradition. The group was somehow boisterous, from time to time shooting in the air their revolvers. We were stunned. They were going to pass not farther than one hundred feet from us. There was a dog which was barking and running loose in the brush. We decided that the best way to weather this danger was to lay low. All of a sudden I looked ahead of me, following some noise. There was the dog. It stopped, it looked me in the eyes, I looked it in his eyes, we watched each other for a few seconds, then it jumped back and ran after its masters. It did not bark. It did not jump on me. What a close call this was!

This evening a man from Markovo came to us. We told him what had happened. He explained that these people were probably the Committee preparing for the celebrations. He then took us and led us to another place, some four-five kilometers away, in a *sechinak* – a young forest, growing out of trees which had been cut for fuel. No animals were allowed there until it would grow. The spring leaves were so fresh, so lucious and the whole surrounding was so pleasant that we liked the place. Here were we going to stay until our next appointment with our helpers. The password was going to be by hitting rocks in an established way. At this point I am not sure if anyone came to us by the fifth or sixth of June. We were so exasperated, and believed that we had lost contact with all of them. It did not occur to us, that they may have been arrested and that the search for us might be entering the last and critical stage. In our confusion, made worse by the erratic events ever since we had left Balchik, we still believed that we could find them. So, we decided, on June seventh or eighth, to go to Markovo to look for Nikolov brothers.

9. In the Jaws of the Shark

Somehow we arrived in Markovo in the evening twilight. Entering the village, I noticed on the wall, high on a ridge, perhaps one hundred feet up, a man who was leaning and looking at the road. As he saw us he pulled back and disapeared. It crossed my mind that he might have been planted there to watch for us. But we had gone too far now and there was no way of turning back. We had made a risky move, and we had to carry it to its end. Further down in the village Stefan asked where the Nikolov's house was and we headed there, in a side street, branching from the main street. We knocked at the door. Ivan opened. As he saw us, he turned white. He pulled us inside and in a few words

---

* In my dossier we are consistently called "Bandits".

explained the situation while his mother was preparing some supper for us. He had not seen Peter since the day before. This evening, returning from Plovdiv, he had met a truck. Peter was there with many people, but managed to yell at him: "Wait for me at home tomorrow."

Peter had been home, had waited for him long time, but he, Ivan, was late. Peter told his mother the story, so she could relay it to Ivan. Peter had been arrested the day before. He had been told that they knew everything. Vassil had confessed and implicated everyone involved. Realizing that denying the truth would take him nowhere, he had decided to admit it all and express regrets for what he had done. He had claimed that he had been misled, that he had been deceived, that he had been taken in, that he was young, that he had made a mistake, but that from now on he would do anything to win the forgiveness of the government and the authorities. He was so convincing that his interrogators asked him if he would help them to apprehend us. He readily agreed to do that, but told them that he did not know where we were. Only the Peltekov's knew of our whereabouts, and he had to go and ask them. He would send his brother to find out.

In his mind Peter was thinking if he could only be left free, then it all would have changed. They insisted that we should be persuaded to leave our guns up in the mountain and bring us down under pretext of turning us over to a group of *Goriani*. Peter assured them that if he could see us, we had complete trust in him and we would do whatever he suggested. This was, indeed, true. Then they asked him to sign a declaration that he was going to collaborate with them. Under such conditions they let him go free, to seek his brother and send him to Brani Pole to Peltekovs. He was told that if he did not returned by 6.00 p.m. or so, he will be considered a fugitive and if caught, would be executed without trial. He accepted the conditions, went home in Markovo, waited for Ivan, but Ivan was late in coming. Since his hour to return to the militia was fast approaching, he caught a truck in the center of the village and when he saw his brother walking with friends, going to Markovo, he could only yell to him to wait for him at home the next day.

When we showed up at the door, and Ivan saw us, his mother had just finished telling him of the happenings. We had walked in the wolf's mouth. But Ivan was determined. He was going to alert the Peltekovs, he was going to collect additional guns from friends, ammunition, carbines and grenades, and the next day, if and when Peter came home, they would head to our camp, and then, we all would head for the border. Now the question was how to get us out of Markovo. Meanwhile his mother had prepared the supper – it was the last supper I had in Bulgaria – we ate in a hurry and then … out of the door. Ivan did not take us to the main street. It just happened that their house was near a small brook, all surrounded by high trees and brush. As we were walking up the brook, I was listening to the songs, it sounded like a million, nightingales. Never before in my life had I noticed such a festive concert of a myriad of birds, as if it was a heavenly music sending us to our long dangerous journey. Were they telling us that all was for the good and that it was to be successful? We made our way through it all led by Ivan, until we were two miles away. Then he instructed us how to reach our camp, and returned home to wait for Peter.

That night he went to Peltekovs, told them the story and the plans, and asked them to join us and them in the camp in the mountain. Then we all would head for the border. From there he went to his friends and gathered all the guns and ammunition. The next day Peter returned home. Having returned to the Militia he had convinced them that he was serious, that he would bring us down from the mountain, without our guns and surrender us to them. They should have known! Then both brothers sneaked out, through the same brook, straight to the mountains and by noon they were with us. Soon after them came Peltekov's – Peter and Boris. We had to wait until dark, to continue our journey.

10. At Last… off on Our Way

That evening, I believe it was June 11th, we were on our way. It was dark, it was difficult but one of the Nikolov brothers knew the way to Chepelare, deep in the Rodopi Mountains, and we were going to head first there. Peter knew someone there, was to look for him and get instructions how to proceed further. I suggested that we reach Chepelare, then we just had to go South and eventually would hit the border.

Yet, we were in a state of shock, and in a state of hope and expectations. I never lost confidence that we will succeed. Sometimes after midnight we had to stop, to catch our breath and try to take a little sleep. Then in the early morning, continued for two hours and for the rest of the day we had to hide, to avoid meeting people. It just happened that in the afternoon a rain began falling, it was cloudy and drizzly. We decided that we could continue safely in that weather. Besides, it was about six o'clock in the evening. But no sooner we began, and we found ourselves crossing a wide field, no brush or tress, just many pathways leading from one end to the other. We were already one third of the length, when from the nearby curve, from behind a cluster of trees and bushes, a herd of cows emerged heading across the way to the village somewhere down there in the East. We could not run back to hide, we could not proceed further, and we had to stop.

In a moment, there appeared a young man, with blanket on his shoulders, walking behind the cows. He saw us, but immediately froze his neck, looked ahead of him and never turned in our direction. How did we look to him, with our guns on our shoulders, in our raggedy appearance, in our awkward looks and situation, we will never know. But this was a dangerous moment, How did we know that he would not immediately go to the authorities and report what he had seen? How fast the authorities would react? What choice did we have? As I am writing these lines, it crosses my mind that we could have captured him, we could have tied him to a tree until we would have been a long distance away, when someone would have discovered him. But we did not think of that. We were no enemies to any one. We were not out to do harm to anyone. We were just fugitives and our only enemies were the authorities.

So we continued on our journey. The drizzle continued, but we would not stop. In a couple of hours, already dark, we reached a river which had to be crossed, but was big enough to make it impossible. This was Chaia. We had been following a path and this path led to a bridge – perhaps twenty feet long, and about ten feet wide. It was a perfect place for staging an ambush. We feared that it is just that what the Militia would have done. But, we had no choice, and had to take the risk. Peter started towards it, we were supposed to wait for him to go to the other side, but we instantly forgot the instructions and ran after him almost immediately. There was some pull in us which was irresistible, the hope and the urge to reach the border as soon as possible. Peter was an agile man, sharpshooter, bold, fearless and athletic like a goat. He was a man to be trusted. Once he shot a rabbit with one single bullet in the head, so we could satisfy our already advancing starvation.

Our first goal was to reach Chepelare. We did. We settled in a small pine forest on a slope, above which was a wide and long meadow. Peter and Ivan went to the town to seek the man they knew from the past. I do not exactly remember, but somehow several attempts to contact him did not work. We stayed in the vicinity of Chepelare for two and a half days. Finally, they found the man, were well received and supplied with food for us. The man also directed Peter to contact some man, way out near the border, who had a sheep pen and a hut, who could further help us. So, that evening we continued our trip beyond Chepelare. Sometimes in the morning we came close to some village which later turned out to be Shiroka Lucka. We had to stop and hide during the day. It was decided that Ivan and Zdravko were to go there and try to buy something to eat. We had some money left. Back in the *Haidut Dere* days I had given to Peter Nikolov my savings account book to withdraw what was left there. He withdrew 1,500 levs, leaving some 150 levs in the balance, just not to raise any suspicion. He was successful in taking the money. I never asked for it. What did I need it for in the mountain? Besides, these boys were doing a lot for us.

So, Ivan and Zdravko left. We heard their story later. They entered the village and immediately realized that it was packed with soldiers. They were walking the streets, they were crowding the stores and the bars. When asked here and there who they were, they simply said that they were working on the nearby forest tree-cutting mill, or something like that. They leisurely entered a grocery store and saw some biscuits and few other edibles. They played curious and asked what is this, what is that, let me have one of those, oh, let me have two more. They bought quite a few things, and then as leisurely as they had come, managed to leave the village by way of side streets, bordering with the forest. Ivan told us that he had two grenades in his pockets and if things had turned out hot, he would have pulled the strings and would kill himself and those who would have tried to apprehend him. What a courage! It worked.

When they came we divided the biscuits and whatever was there in equal shares among ourselves. It was to turn out that several days, when we were about to cross the border, having had not a bite of solid food for days, I was chipping small grains of the last biscuit to keep me going. But the report of the army and the number of soldiers they had seen there was a scary thing. We concluded that the whole region was occupied by the army, that wherever we went, in whatever direction we turned, we would be met by soldiers and either killed or apprehended. So that evening we held a conference to decide what to do.

11. Exasperation and Resignation

Peter Peltekov, in exasperation and resignation, advised that we should return back to our villages and continue the struggle there, with the help of anti-communists and friends. He was supported by his brother Boris and Stefan. Peter and Ivan Nikolovs and I were adamantly determined to go ahead for the border. Zdravko was staying on the side, listening to the discussion and saying nothing. We were divided three to three. We all turned to him. He turned his head and firmly spelled it out: "I think we should head for the border, whatever may happen, let it be. Let God be with us!" Perhaps this strengthening of the border troops was the result of the report of the young man with the cows. After Peter had not returned, and his brother and the Peltekov's had disappeared, the authorities may have concluded that we were heading for the border. Much later, a week or so, when we were already in Greece, fugitive soldiers, who heard our stories, simply commented: "So, it was you which caused the military to declare general alert and did not let us sleep for two weeks and brought large army contingents."

So, having decided to continue for the border – Peltekov's did not put any opposition, they themselves were scared to return back - we had to determine which way to go. From this point on, nobody knew where we were and what direction to follow. I suggested that we just go South and eventually we would reach the border. But where was south? Someone went to the trees and started touching the trunks for moss. The moss would usually grow on the north side. Nothing was discovered. I said that we should follow the stars. Nobody in the group knew anything about the stars. But I knew which was the polar star - straight line from the back paws of the Big Bear constellation. I had no difficulty pointing it out and we headed in the opposite direction.

How long we have travelled? I do not remember. At one point we were crossing a small river, some ten-fifteen feet wide and perhaps a foot-foot and a half deep. Stones were sticking out here and there and my friends started jumping from stone to stone. So did I. I was carrying a stick to help me walking. At one point while I was in the middle of the river, my stick broke, I had been leaning too heavily on it. I was with my two feet in the water, up to my knees. Well, I came out, my feet freezing, I took off my pants and my socks, twisted the water out and continued. Much later, when I was teaching my classes in the universities, and came to discuss Emperor Barbarosa, drowning in a small river in Asia Minor while going with the Third crusade, I never missed the occasion to tell the story of my own adventure. Happily, I was crossing only a small river.

After many hours of walk without any knowledge where we were going, but continuing to the South, we stopped, to rest and sleep a little. It was a warm night. We were awakened by the rays of the morning sun. Ivan got up and took a walk some forty feet ahead of us, where, it seems, the meadow was ending. He reached the end, peeked down, and abruptly turned back and ran to us. "There, down" he explained, "are the Smolian Military Barracks". We had slept in the vicinity where the enemy, armed to their teeth, was located. In a hurry, we jumped and ran as fast as we could from the place. Soon we found ourselves on the crest of some mountain. It did not seem that there was much life around here

After some hours we noticed in the distance a sheep pen and a hut. We approached it with great caution. It did not appear that someone was there. It was decided that it should be inspected for food. And while we were hiding, Petar and Zdravko went to check it out. May be this was the hut, which the man from Chepelare had mentioned. No, it was not. They came back and said that apparently the shepherd was bringing his flock there during the night, but there was nothing for food. With some hesitation, Peter mentioned that he had seen a wooden bowl, with stuff in, corn flour mixed for the dogs probably. Hungry as we were, we urged him to go and bring it. He did. It was a mush, with a brown crust on top, but when it was mixed up, the whole thing, perhaps half a gallon, appeared very attractive dish. We put salt in it, we pulled our spoons, and then ... there was such a hearty gobbling up that dogs would have lost a competition for speed of consuming food. It was so delicious, and it was so nourishing that we were eternally grateful for it.

During the same day, while we had to rest a little at the bottom of a ridge, with brush and trees growths above us, we heard a man following his cows and calfs. He began whistling: "Quiet white Danube is mildly waving, joyously it is making noise…" (Tikh bial Dunav se vulnuva, Vesselo shumi…) a beloved hymn in Bulgaria, written in commemoration of Khristo Botev's (the greatest Bulgarian Poet, killed while leading an insurrectionary band, after crossing the Danube, in the Vratsa Balkans) crossing of the Danube. We kept silent until he had disappeared. Many a time I have remembered this incident and thought if this man had seen us from somewhere, and by whistling this particular song, was giving us a sign that he knew what we were, and was trying to give us courage. I still think that this was the case.

We continued our journey. Sometimes in the evening we found ourselves in a brushy mountain, criss-crossed with well traveled paths of animals. Apparently we were near some village. We could not go further and decided to stop here and sleep. But before that we had noticed another sheep-pen and a hut, sort of a small room. Somehow Peter Nikolov, concluded that this must be the hut and the shepherd recommended to him in Chepelare. He decided to go and check it. He did go. He knocked at the door and heard a man asking from inside: "Who is there?" Peter introduced himself and told him that his friend in Chepelare had directed him to his hut and see if the man would help him with bread. All of a sudden the door had opened and Peter saw a barrel of a gun pointed in his face. "Get the hell out of here or else I will blow your brains out." Peter did not need a second invitation. He did not have his gun. He turned around and disappeared in the darkness of the night. We found ourselves again in a quandary. Not because of the incident, but because we did not know in what direction we should go. We moved out of the tracks where animals had traveled the pathways, retreated some fifty feet to the left and decided to wait until the morning.

12. Crossing the Iron Curtain. "This Day, Created by God, Let Us Rejoice and Cheerfully Celebrate It!" (Resurrection Hymn)

That morning, June the 19th, we were awakened by cow bells. A big herd of cows apparently was approaching us, having left the village. They sounded very close. A thick fog, like pea soup, was all around us and we could not see a thing. We grabbed all that we had and started running East. All of a sudden the fog started lifting. It was lifting so fast that the cow herd and the man or men after them could see us any minute now. We ran with all our strength and speed which we could muster. In no time we saw ahead of us the end of the field, a forest in the slopes after the ridge and in seconds we were jumping over the ridge, down the slopes. We were skidding on the leaves, but we were out of sight. By that time the fog had vanished, the cows were marching where we have been resting, and we waited until they had gone far enough. We recovered and pondered the direction which we should take. I suggested that we follow the slope. We walked a long, long tree covered slope and by three P.M. reached its end.

Opposite of it was a long mountain, very long, but not very high. It was perpendicular to the slope which we had followed. Down the slope where we were standing, was a valley, not very wide, but stretching from East to West, for as long as we could see, crossed lengthwise by a good well traveled stone road with curves here and there. The most important sign which we saw for the first time, was the presence of military forces in the area. Five minutes had not passed when we saw a military cart, big and long, emerging from the East drawn by big horses. Three soldiers were in it. We hid in the bushes while it disappeared behind a curve. We looked at the mountain ahead of us. Far to the East we noticed a structure on the top of it, but even if we could not make what it was, I was certain that this was a military border post. A similar structure was seen to the west, perhaps five or less kilometers, also on the crest of the mountain. This further confirmed my contention that this was the border and that we had reached the most critical point in our journey.

Ivan was still skeptical, but we all accepted this assumption. What was left to us was to cross the valley, that means the military road, and proceed to the crest of the mountain. There was no time to waste at this point and we risked running across the road. In two minutes which felt like an eternity, we were over the road, out of the valley and in the bushes again. We walked some fifteen minutes more and found ourselves at the foot of the mountain. We were extremely tired and stopped for a brief rest. While all of them stretched on the grass, I got up, took my can and told them that I was going for water nearby, some twenty feet or so.

When I was out of sight, I fell on my knees, raised my hands up, looked over the tips of the high pine trees, to the blue skies, and prayed as hard as I had never in my life done it. "God, my dear God," I was silently saying, whispering. "extend Your right hand over us and take us out of here without being harmed." And I recited my favorite prayer from the beginning of the 18th Psalm, used also in the Liturgy in the beginning of the dogmatic part, of St. John Chrysostom's Liturgy. I prayed to the Holy Theotokos. I prayed that she intercedes for me before God, the Lord Jesus Christ. I pulled from my shirt pocket a small icon, which I had discovered in a small field shrine in proximity to our camp, in a wide meadow, a pile of stones, with room large enough to place this icon in – three by three inches. Then I repeated my favorite prayer to Her: "Under thy mercy ..." Tears were running down my face. I knew that this was the ultimate moment of decision of my life. I composed myself and returned back to the group. In a few minutes we were up and heading for the crest of the mountain, climbing up in a ravine. It was difficult, but we were advancing. In twenty minutes we had reached the crest.

We could not see from there what I had taken to be military posts. Ahead of us was a meadow, some two hundred feet wide, and long, from end to end, as far as we could see. Apparently it was purposely deforested. I was certain this was the border. We looked left, we looked right, we did not see anything, and we charged across through the waist high grass. I never stopped looking left and right. I remember crossing in the middle a well traveled path, two feet wide, not a speck of grass on it. I figured that this was the path the border guards were constantly walking on to protect Bulgaria – rather to prevent people like us from escaping from Bulgaria. At the other side of the meadow began another brushy area, and then a forest, where we rushed to disappear as fast and as far as we could. Unexpectedly, about fifty feet from where we crossed, we came out on another meadow, with no high grass. This meadow was covered with wild flowers, whose root, a miniature bulb, big like an almond, was edible. Ivan had taught us that but we came by very few of them in Bulgaria. Now this meadow was covered with clusters of them. "Good, said Ivan, let us stop and dig out as much as we can." "Ivan," I told him," even if I know that I will fall dead a hundred feet from here out of hunger, I would not stop here." "You are a coward," he said, to which I responded: "Call me any name you like, but don't you realize that we are at the very border and that we should run away from it?" "What border?" He objected, "I saw no soldiers to guard it." This exchange ended when I told him that we should consider ourselves lucky that we did not see any, and none had seen us.

We continued speeding. It was about 6.00 P.M. * After two hours, we reached the end of the mountain extension to the South, and reached a place where we could not go further. Ahead of us was a precipice, probably two miles downward, and at the other side wide beautiful and long meadows. We decided to stop here and see if we could find the miraculous flower. We found very few of them. Later in the night we heard back where we had crossed the border the bark of dogs. These were the frontier dogs. We were quite a distance from there, but we still feared that the dogs would follow our scent and the soldiers after them. Nothing of the kind happened. We theorized that they were as scared to go in the dark as we were. They in all probability knew that we were armed, and that the group which had crossed the border was, in all probability, our group.

So we fell asleep, ready for the next day to continue our trip. What still puzzled us was that we did not see any Greek soldiers on the other side. But we were confident that the next day we would find them. Until I fell asleep, I prayed silently and thanked God for our successful crossing the border. I have never forgotten His mercy for me, and sometimes had the daring to think that He had some mission for me, and this is why He was protecting me. I felt as some heavy burden had fallen from my shoulders. I was certain that we had crossed the border, though I still was hearing doubts from one or another of my friends. Some of them had not even noticed the deep path in the middle of the meadow. As far as I was concerned, it was all over.

13. The Final Days

The next day we had to descend the precipice before us. It looked dangerously steep, bare, here and there some small growth, dry roots and rocks. There was no other way. To the East was the military border post. The same to the West. How did we end up here? No man with all his mind, would have ventured to cross the border here. Only someone who had lost his way in the mountains would have ended here. Well, we had lost our way. And ended over this precipice. Thank God! It was not a curse! It was a blessing. Sure, there was some fear in taking the first step, but we made it and proceeded, step by step, sometimes sitting and sliding ourselves down, holding to rocks, roots and small growths, everybody his own way. It went slow, but we were advancing. When we were almost in the middle of the steep slope, we were faced with a difficult problem.

Before us was a steep smooth rock, as it looked from end to end, water running on its surface, not much but enough to make it slippery. The slope was some ten feet downward

---

* I may be guessing the hours wrong.

incline. After these ten feet there was a sort of a terrace, three feet wide, and then further down a vertical water fall, some one hundred or more feet deep, with sharp rocks sticking out at the bottom. What were we going to do? Peter Nikolov went first, crouching step by step, he managed to reach the terrace. I followed him, but midway I lost my balance, fell on my stomach and slid down the rest of it. For my luck I ended up in a puddle of water in the middle of the terrace, in a hole probably eroded there for centuries by running water. There I was with my feet hitting the side behind which was the precipice. I stopped there. If I was two feet to the left, or two feet to the right, I would have gone over the edge and down the pit, a hundred feet, over the rocks. There probably my bones would have rotted for the next decade, after vultures would have feasted on my body. Apparently my candle was still going to burn in this life. I was followed by Ivan and Zdravko. They made it safely to the terrace, but the Peltekovs refused to take this road. I should have stayed with them. Nevermind! They walked to the East and we lost them for the time being. We continued our descent down on the slope. It became easier. If we had started et 7.00 o'clock in the morning, we arrived et the bottom of the mountain about 2.00 p.m. It was the most treacherous segment of our long journey. But we made it.

Before us was a long and wide meadow, rising slightly to a crest some one kilometer up where we noticed a tree. It seemed to have been touched by human hands. We believed that we were near the Greek border guards. And we hurried to reach the crest of the hill. It was a disappointment. We reached the crest, a road seemed to be leading East, but no sign of life or human presence. We were tired and hungry. We sat there on the top to rest a little. After some 20 minutes we saw the Peltekovs come out the same way as we had travelled. We waited for them. When they arrived, Bai Peter excitedly started telling us that they had found an easy way some hundred or so feet East of our waterfall. Well, we were glad for them. There was no word ever uttered that they deserted us or we deserted them. They showed to have been wiser.

After a while we resumed our march. We followed the old road.

After some half an hour or more we suddenly saw red tile roofs and exclaimed of jubilation that we had reached a village. Not so soon. It was an abandoned village of no more than ten or so houses. There was not one living soul there, and apparently there had not been any living soul there for long, long time. Some of these villages had been abandoned during the German occupation, or during the Greek civil war. We found plenty of plum trees, with the new crop plums no bigger than large corn grains. We attacked them with ferocious appetite, but they proved so sour that we could not eat more than a few of them. We continued down the same road. No sooner had we left the abandoned village, that we stumbled upon a turtle on the very road we were following. She was big - some 10 inches by eight, one of those field turtles. It was Godsend gift. In a few minutes we discovered seven more of the same in the clearing along the road. But how one could eat this thing? It would pull its legs inside and becomes invulnerable. But Peter Nikolov was the man to look for. In no time he had killed them, we had made a big fire and threw them to be baked. The fire burned their shells enough that it was not difficult to crack them up.

Very soon all the meat was out of the shells, many of them with good size eggs inside. They filled the entire pot which we had with us. Peter salted them and then we were invited for our first dinner in the "Free World." We were grabbing the juicy morsels of meat, we were devouring the precious stuff with the speed and the voracious energy of hungry dogs, we were cutting the pieces with our teeth, we were behaving like vicious wolves swallowing the half chewed chunks like animals. Well, there was enough and more than enough for everybody, but we could not control the animal instincts which were taking over manners and deference for the others. Well, we had not eaten a thing for four days. The last crumb of the biscuit I had I licked a few minutes before crossing the border. As we had descended the precipice, relieved of all fear of the previous days and exhausted of the last ounce of my strength, I felt that I could not longer keep with my friends, and followed them with difficulty some forty feet behind. But I was not panicking that at some place I could collapse and would not be able to go further. It was all over. After this banquet I was reinvigorated and could march.

In later times, when I was lecturing to my students in America, I never lost the chance to talk about my experience with the turtles. I would tell them how the prehistoric man would crush with a rock the shell of the turtle and proceeded to describe the progress of technology (I was trying to illustrate my concept of culture), and then ad lib and tell them my turtle dinner. I would also mention that when I was working on the New York piers as a social worker, having free dinners on the Queen Mary or Queen Elizabeth, reading on their two-foot long menu about what they were listing as Forest Chicken, I would ask the waiter what was this forest chicken. He would explain to me that it was a turtle. I would pass to the next item. I had never before eaten turtle. And have never touch it after, but in that critical day of my life it was an experience never to forget. I would end my story by saying that when man is faced with death by starvation, he would eat anything, no matter how disgusting.

We slept that night at our first restaurant where we had our turtle dinner. The next morning I was in a better shape and could keep up with my friends. We followed the same road and passed two more abandoned villages. It was near noon when we crossed over a small river. Under the bridge, and a little downward, we saw a deep pool of water. Large fishes, some up to fifteen inches long, were playing and swimming in it. We had not eaten breakfast. So here was our chance for lunch. Peter Nikolov suggested and we agreed that he would throw one of the grenades which he was still carrying, in the pool. We would line up below the pool and when the current takes and carries down the stunned fish, we would catch them and throw them out on the grass. We did that. The grenade exploded in the water, the current carried a lot of fish and, we caught as much as we could. We then made a fire, cooked the fish in the pot, and had our second dinner in Greece.

When we finished it, I suddenly realized that I did not need my little Turkish carbine any more. But I decided to shoot it out at least once. Aiming at a tree, perhaps four inches thick tree, I pulled the trigger, and, to my surprise, it gave the shot. I had hit the tree. But the shell of the bullet was not ejected as it should have. It had to be pushed out from the front of the barrel with a stick. So, what good was it for? It is good that I did not have to use it in Bulgaria. I asked permission and my friends agreed that I should throw it away. I said "good bye friend" and leaned it on the tree which I had hit.

Not far from this place, we reached a big river, perhaps one hundred feet wide, but only about two feet deep. So we crossed it, and at the other side found a much better road going south along the river. We followed this road for some time. All the time I wondered, why we had not come across Greek border guards. Where were they? After some hours of walk, we suddenly came out of the forest, still on the same road, and saw before us lands planted with agricultural vegetation. A little further we noticed a woman bent over the ground, hoeing some plants. This was the first human being we met in Greece. We stopped. Peter mounted a white shirt on the top of his bayonet. It was decided that I should go and try to talk to her. When I approached her she first appeared frightened. I smiled to her and said: "Vulgari" (Bulgarians), and pointed to my friends.

My Greek was very limited. She first looked incredulously, and after a few seconds of hesitation, apparently having realized who we were, began jesticulating, pointing to the nearby bridge, some five hundred feet. Behind the bridge was a cluster of trees and we could not see anything. She hurriedly took off her apron, led me to the road, and made a motion with her hand for us to follow her. In a few minutes we were standing at the end of the bridge. She made a sign for us to stop and ran across the bridge. On the other side several Greek soldiers were already coming out of the military post. Peter was waiving the white shirt on his bayonet. The Soldiers made a motion to come to them. They were all young, well shaved, in attractive uniforms, curious at our sorry appearance, but understanding that we were fugitives from Bulgaria. We surrendered all our guns and ammunition and they gave us bread and cheese. We had not eaten bread for quite a while. It was delicious. They made us to understand that they had telephoned to their commanders for a truck to come and pick us, to take us to headquarters.

I felt infinitely happy and relieved. Our Rodopy Odyssey was over. We were safe. We were far and away from the Bulgarian communists. They could not touch us any more. We were at the doors of a new life. A new page was opening before us, full of unknowns, but overflowing with exuberant hope for the future. This new page I will continue in the next volume.

XII

APPENDIXES

"And you shall know the truth, and
the truth shall make you free."

John 8:32

I

BULGARIA I LEFT BEHIND
1951 *

---

* This Report was written in the middle of July 1951, in the Drama Refugee Reception Center, Greece – three weeks after crossing the border, while my memories were still fresh in my mind.

BULGARIAN CULTURE
IN THE GRIPS OF BOLSHEVISM

1. Freedom and Culture

Every culture is expression of the inner essence of its individual subject. In as much as this creative agent - be as a social-historical, or personal psychological potential force – expresses itself in concrete intellectual, aesthetical and ethical generalizations, it is a creator and carrier of its own, typical for it, culture. In this sense every nation, in the course of its historical existence, creates its own, materially identifiable culture – the result of the creative efforts of many generations of talented leaders in the realm of science, of arts and social and political life. The national culture is a historical achievement, and result of historical processes which last centuries and accompany its creator and carrier. On this account this process is not determined by some temporary conditions of political and economic nature. It represents a natural free expression in the development of the national spirit. If the culture, then, in a historical perspective, represents a free creative flight of human thought, emotions and actions, even more, as a manifestation of the inner sense of its individual subject, the human personality, as a natural process, is subordinated to the laws regulating the evolution of the spirit. So it can develop in conditions of full freedom. This makes clear the connection and the exclusive significance of freedom in the evolution of culture. A broad political democracy, which insures the freedom of its citizens, is the first condition for the progress of culture. Democracy and culture are correlated, inseparable one from the other. Every democracy is distinguished with remarkable achievements in the field of culture. And, to the contrary, every despotism is characterized with the death of culture. (Now I would add: Athens and Sparta, the Italian renaissance and medieval Spain, are typical examples in this case – STR). History is replete with proofs of this truth. This truth, however, is disputed by the champions of the most fearsome dictatorship in the life of mankind - the Bolsheviks. They are obsessed with the mania that they are creating a new, Bolshevik, Communist, so called Socialist culture. *

2. Bolshevism and Culture

The Bolsheviks know perfectly well the force and the significance of the cultural values achieved by the past generations and the benefit from their implementation in the

---

* In later years, when teaching courses in Western Civilization, and for a short time Introduction to Sociology, I was to define culture as the sum total of the ways and means devised by man for a better survival, consisting of four components – technology, social relations, economics and ideas.

life of man. Even more, they loudly proclaim that they alone are in a position to offer a true and worthy understanding of culture, that all other political, scholarly and philosophical interpretations or attitudes towards culture are in lesser or greater extent mercenary self-interest oriented views, and by that, they are an obstacle for the cultural advances of mankind. Posing as the only defenders of culture, they mascarade everywhere and for everybody as men of scholarship. Especially their leaders – Lenin, Stalin, Georgi Dimitrov, Vulko Chervenkov et al. – like to be photographed, or to have their appearances painted in big libraries, before shelves loaded with books, or sitting at table with opened thick books, over which they adjust with deep concentration their glasses, or mingle with writers, artists et al. Their portraits and pictures are painted in the millions and are disseminated among the people, so that they are seen, being the leaders as they are, as men of scholarship. This way they seek to win the Bulgarian people, who, indeed, look to science and its representatives with deep respect. The Communist leaders never realize how silly they appear to the people, who know them well, attributing to themselves what they do not have.

The Communist leaders, in addition to that, are anxious to pose as macenates of the arts, of sciences, and their leading representatives. Soon after September 9, 1944, they realized that they cannot fill the empty seats of the humiliated, of the expelled, or simply criminally murdered Bulgarian scientist, writers, artists, et al. with their own Party cadres, and reconciled with the thought that they had to take back men compromised with their collaboration with the regime prior to September 9, 1944. These men, thrown on the street, had to trample with their feet their own honor and dignity for a morsel of bread, and shake hands with the red bloody butchers. With their reputation in the past, they sought to conceal from the people their criminal excesses. The Communists indeed needed the reputation of an Elin Pelin (a classic in modern Bulgarian literature), for whom they knew that he was personal friend and buddy to King Boris, but they had no other way. In the eyes of the Bulgarian people Elin Pelin was worth more than the entire Politbureau of the Party, more than all Communists writers of the category of Lamar, Burin, Polianov and Bozhilov. They had to arrange for a celebration in honor of Elin Pelin in order to throw dust in the eyes of the people. But, did they believe that the people trusted their sincerity to Elin Pelin? Likewise, was he sincere with them? We have our doubts! With the name of Elin Pelin, as well as with the names of all other well known Bulgarian scientists, writers, artists and actors they simply orchestrated a comedy. In this comedy the Communists, who for decades had vilified Vazov, Elin Pelin, Yovkov and the entire Bulgarian intelligentsia which now follows them, so not to starve, they seek, and they successfully play as maecenas. They created the so called "Dimitrov's awards" and proclaim as laureate every incompetent man they meet on the street, promising them to write a lousy poem for Dimitrov, Chervenkov, the state achievements, for the village kolhoses, for the USSR, and everything that they want glorified. Every artist who paint something on these subjects is given the title of Laureate of the Dimitrov's award. For the same reasons they reward composers, pig breeders, kolkhose leaders, scholars etc. The themes, the paths of Bulgarian culture, are strictly defined. Everybody who stays in the framework of this decreed culture, sooner or later will receive the Dimitrov's award. Anyone who ventures to come out of this framework and stands on the side of the people, will join Stilian Chilingirov in the concentration camp, or will be taken to Court like Trifon Kunev.

There is no other way. Everything in the Communist society is planned in advance. The directions of culture are also planned. It has to move within the frameworks defined by Stalin, Dimitrov and Chervenkov.

It is a strange logic when the Communists have to justify their attitude towards culture. Their starting point is their philosophy – Marxism-Leninism – which is proclaimed to be the only scientific ideological system, while everything else is a deception. Naturally, society has to be saved from deceptions, which implies a license for the communists to force their philosophy and their attitudes upon all individuals who have not, as yet, grown up to accept this truth without questioning it. In view of that, there could be no word for freedom of thought in the sphere of culture. Nobody would need, on the contrary, any cultural achievements in science, arts and politics, not in accord with Marxism-Leninism, or, God forbid, contradicting it. Such culture would be harmful to society. It is inadmissible. But then, even in the interpretation of Marxism-Leninism, not everybody is allowed to have his say. The best authorities there, in science et al. are Stalin, Dimitrov and Chervenkov. Nobody else is allowed to discuss it. There is nothing else left to philosophers, scientists, writers, artists, composers, actors and conductors, except to regurgitate thousands of stupid quotations from the works of the Communist evangelists, to engage in fights over these quotations among themselves what exactly Stalin intended to say in this or that word. They will go to him and the question will be closed. If, however, God forbid, the scholar, the writer or the artist ventures to offer his own opinion on the subject, he is finished.

To all this one need to add the espionage, having grown deep roots among the Bulgarian cultural activists. They are afraid of each other, do not dare to confide their intimate thoughts and views on the questions under consideration. They have to refer to Lenin and Stalin on the most trivial matters. They are biologists, physicists, composers, literary critics, philosophers, linguists, etc. If it happens that someone develops a thesis at a meeting, before his colleagues, and expresses his own opinion, he could not trust that his views would not be reported to KNIK (Committee for Science, arts and culture). He immediately would be called to give explanations.

This is the environment where the Bulgarian scientists, writers, artists and composers create their works. Bulgarian culture is placed in such limitations. The Communists cannot understand that with these limitations they are providing bear's service to the Bulgarian cultural development. With their cares for science, arts, and the progress of our nation, they are doing nothing more than to suffocate every progress and destine the Bulgarian culture to an inevitable decay. This decay is openly visible and could not but be noticed by any careful observer of our cultural life.

3. The State of Bulgarian Education

It is not possible to discuss Bulgarian science, without touching on the subject of educational institutions. Immediately after September 9th, the communists seized the educational system. The first thing they did was to throw out the prayer before and at the end of the school days and expelling the teaching of religion in the elementary, middle and high schools. Next, they moved on the subject matter and purged the material contradicting their philosophy. The principal task of the educational process was set to supply the Communist Party with loyal and educated cadres. The elementary and middle schools were transformed into Communist educational institution. Anything that was not in accord with the Communist pedagogical concepts, philosophical doctrines and political system was thrown out and replaced by the pedagogic, the philosophy and the political views of Bolshevism. It was decided that the classical languages are a burden for the students and were removed from the programs of the schools, ignoring the fact that all civilized nations value very highly these languages. It was further decided that there was to be a Communist youth organization which purpose was to spy and terrorize teachers and students. For this purpose they created the EMOS – Edinen Mladezhki Obshto-Ouchenicheski Sayuz (United all-student Youth Union), then the DCNM (Dimitrov's Union of the People's Youth), et al. It was decided that the cadre of teachers was to be purged of secret, concealed "fascists". This way they threw out of the schools the highly trained teaching personnel. The pioneers of Bulgarian education, who for decades had served with dedication the younger generations were thrown out on the street. Thus, the director of village of Zelenikovo schools, Plovdiv district, Peter Videv Zidev, who had made out of his school a model for the whole district, having been expelled first from the class rooms, was called back to take his office. However, humiliated by the shameful treatment he had received, and offended by their lack of recognition of his decades long service, in spite of the apologies and the pleading, he refused to return and serve communism in the school.

The most dedicated and honest workers in the field of education were chased out of the school. Inept and incompetent party members took charge, regardless of their lack of pedagogical preparation, which stood under the line of any minimum. Thus, in the newly opened high school in Brezovo, Plovdiv district, after September 9th 1944, they appointed Bogomil Bossev, former teacher in the Plovdiv Theological Seminary of this writer, who had been defrocked for his moral escapades. One year after his appointment, he was dismissed on account of his involvement with female students, many of them pregnant. In addition to that, the authorities discovered embezzlement of funds on large scale in the student cafeteria. He was also accused of receiving bribes, borrowing and not paying back large sums of money from private individuals. He was dismissed, but soon emerged as director of the School in the village of Svoge, near Sofia.

The same happened to his successor, Ivan Pukhtev. He was brought to court for having violated six female students. This is what the Communist School directors do. Maybe this is the way they see their pedagogical responsibilities to the youth, which, of course, would take them as roll models. Who dares oppose them and criticize them? This is how they train the cadres of the Communist Party and claim to raise the culture of our people. Most of the teachers are not trained in the profession of education. They are High-School graduates, never studied in pedagogical institutes, and are appointed to positions in the Middle Schools, or, with strong political support – also in High Schools. To all that, the low pay which is short of insuring a decent life, and the excessive work required to be done outside of the schools – in the Chitalishte (Public Library), the local cooperatives, the Kolkhos, the Soviet-Bulgarian Friendship Society, the Fatherland Front Organization, variety of Circles etc., – force the dedicated workers in the field of national education out, and their places are occupied by dilletants.

4. Purge of the University of "Reactionary" Students and Professors

Quite delicate was the situation with the communists in the University. There they had the least influence among the students and the faculty. The first problem was easily solved. In 1945 they opened the doors of the university widely and registered as students all their protégés, without securing for that all conditions necessary for a qualified student body, on account of which a great many of those unprepared, soon left the school. Subsequently they introduced the system of the so called "F.F. Notes" – references from the local authorities guaranteeing the loyalty of the prospective students to the new regime, without which nobody could cross the threshold of the university in Sofia. Gradually it was being turned into a Communist University, transformed into a Party School. Next they proceeded with revising the syllabi of the individual subjects and dismissing professors who did not tow the ideological line. The academic independence of the University, strictly protected in the civilized nations, was trampled over quite unceremoniously by the communists in quite a brutal manner. Rector, Academic Councils, Deans and professors, were instantly transformed into State clerks, fully dependent on some party functionary, sitting in the chair of the KNIK, who, ordinarily, had never stepped in the University.

In 1946-47 the faculty of the Sofia University appointed election for a new rector. The entire academic staff, with exception of the communists, voted for Ivan Sarailiev. The Government vetoed the election because Sarailiev held idealistic philosophical views. The election was repeated, but the results did not change. The professors did not bend under government pressure. Neither did the Government retreat. It trampled over all traditions and legal rights of the highest institution of national education and extended the mandate of the old Rector – Dr. Orakhovats. Then it purged the academic faculty. Professors who demonstrated irreconcilable attitude towards Marxism and the Communist political programs were removed and in their place were appointed men of their own, without the necessary qualifications. At once the academicians Prof. Mikhalchev (The most prominent philosopher in the country) "aged" and were retired. Thus, the men who in effect created the Bulgarian university – Mikhalchev, Romansky and Arnaudov – had to accept their fate. After this the Sofia University was tamed.

But in the student body there were "reactionaries" who had "sneaked in". The Communists could not sleep with the thought that there were young men in the university auditoriums who were not with them, and schemed to expel them. Thus in early February 1949, five thousand names were plastered in long reams of paper on the outside walls of the university, the names of those students who refused to join them, had to collect their text books and ordered to leave Sofia in three days. Is it possible for the civilized world to imagine such barbarian policy in regard to the education of a nation? Is it possible, say France, to watch how five thousand young men are thrown out on the street because of their political convictions? Hundreds or more of these young men were about to graduate, had taken their final exams, or to take one or two more, in order to receive their diplomas were expelled from the University. This did not bother the communists. The most important test for them is the political attestation. It is received elsewhere – not in the university class rooms where the student can demonstrate his intellectual abilities; not in the "volunteer labor brigades", where the entire Bulgarian youth population passed through, hoping to gain some degree of political rehabilitation. Such attestation was received from the village council, somewhere, in some hill-billy village in the mountains, where on a dilapidated chair sits a man who can hardly write his name, but on which conscience weigh heavily two or three assassinations. The political attestation weighs much more than all the A-s, B-s and C-s achieved with great efforts and noble impetus to the crests of science. But the man with the bloody hands who writes the attestation of the young worshipper and victims before the altar of Science, and the parvenu "comrade" who will read this attestation-verdict of "guilty", so to cut the wings of the rising on the way to progress creative personality, commit the worst crime against the Bulgarian people: they destroy Bulgarian culture. They expected to see before the gates of the university rivers of tears. In contrast, these young men separated from their Alma Mater with gnawing teeth, looking to the future when they would return to judge their judges and defend Bulgarian science and culture.

But with the expelling of the "reactionary" students, things did not straighten out. The Division of "Cadres" carefully continued to investigate the rest of the students and nobody is sure that after taking the exams and take all required courses, will be awarded a diploma. Such was the case of Angel Yordanov Georgiev, of Pleven, a law student. After taking his last exam in January 1950, he goes to Division "Cadres" to obtain the permit for diploma. There, a clerk pulled the file, stared for long time over a green piece of paper containing two or three sentences, and, scratching his neck, tells him: "You will have to wait a little." This meant that the "Cadres" attestation nullifies all his exams and deprives him of the rights of a man with higher education. He is a jurist. He tries to remember somewhere in his studies of jurisprudence such a legal enactment, and failing that, comes to the conclusion that such a vile act could be committed only by the communists, that communism is nothing less than a new multi-headed and much worst Herostrat in world culture and civilization.

5. Dialectical Materialism – The Procrustean Bed of University Studies

Having freed themselves of professors and students opposing communism, the Bolshevik satraps of culture, turned their attention to the contents of the academic studies. In all subjects, including Mathematics, they implemented and implement, rather forced and continue forcing, the doctrines of dialectical materialism. The University Science was crucified, or rather layed on the procrustean bed, and all that contradicted their views, was cut out, and what was outside of that framework and could not be tied to dialectical materialism they tried to color with political and philosophical nuances, in a way which would make it irrefutable proof of their thesis.

Having suppressed all resistance in the Academic Council, the communists had no difficulty in laying hand on the educational processes of the University. Immediately they introduced the subject of "Dialectical Materialism" as required course, with final exams, for all students in all School of the University. All students, willy-nilly, managed to conform to this order, but it put the students of the Theological School in a difficult dilemma. The majority of them were priests, and to satisfy the professors they had to argue and refute Christianity, and compromise themselves as church servants. Thus, a priest, Chavdarov, from the village of Tanturi, district Gorna Orekhovitsa, appeared for final exam before Prof. Zhivko Oshavkov, and after having presented in a brilliant manner the substance of the questions posed to him (the examinations are oral), is subjected to a cunning look by the professor and asked: "Are you convinced in the truthfulness of all that which you said?" The priest responded: "This is what dialectical materialism teaches" The professor then interrupts him: "In such a case we shall see each other once more, in the fall" (Meaning to repeat the examination.) and failed the priest. Presently the students in the university take Marxist polit-Economics, History of the BCP et al.

Having abolished the administrative independence of the Sofia State University, having fired the independent professors and having expelled the "reactionary" students; having further forced the philosophical doctrines of Marxism on all branches of scholarship, the Communists reached to put an end to any free interpretation of all subjects, in all branches of science. All of a sudden it appeared that until that time, there have been no pure science, that it all have been deceptions. Therefore all that had been achieved prior to their time had to be corrected. But where the true science could be found? Sure enough! It was in the U.S.S.R. And together with the harvesters, in Bulgaria arrived also Soviet Science.

In the first place attention was paid to biological sciences. Naturally, this question was raised not because there was no biological science in Bulgaria until now, but because it was discovered that it was full of "deceptions". At about that time in the Soviet Union they were purging the professors who were working in the field of Biology. So, Bulgaria had to turn with the wind in the same direction. Our Biology scholars, led by the world renown Prof. Metody Popov, member of the Academy of Sciences, had to bow down their heads, to engage in self-criticism, to renounce their views defended for decades, and to accept the theories of Michurin and Lisenko, supposedly the greatest geniuses in that science. Metody Popov, who, from the height of his chair in Biology in the School of Medicine had previously laughed at Michurin as a simple gardener, in no way a scholar with recognized credentials, now had to publicly recant before a general conference of the biologists in Bulgaria his former views and humiliate himself, like an eighth grade student, who had been told that he had made a mistake, and had to correct himself standing before the scholarly audience, he publicly tore two chapters of his recently published *Introduction to Biology*. All of a sudden the numerous volumes of his publications, of hundreds of pages of scholarly discussions, were judged to be useless paper. He had to write, and he wrote, a new text for biology students, following the line of Michurin-Lisenko theories. At this same conference, many of the green scientists showed their horns. Assen Kisselinchev, at that time assistant professor of philosophy, claimed that he had predicted the Michurin-Lisenko theory two years before, for which Todor Pavlov, the patriarch of Bulgarian marxists, mocked him. Serious scholars, like Doncho Kostov, chose to absent themselves from this conference, claiming "illness", but letters purportedly written by them supporting the Michurin-Lisenko thesis, were read before the Conference. They renounced their former views and adhered to the new science of Biology inaugurated by Michurin-Lisenko. At the same time, another pillar of biological science in Bulgaria, Dr. Konsulov, was rotting in jail.

The same type of comedy was played in the other fields of scholarship. The communists appeared much interested in the subject of History. Immediately after September 9 they started re-examination of the various periods of Bulgarian history and offered their interpretations. They criticized mercilessly and rejected long held firm views on the values of the Bulgarian past. A Bolshevik agent climbed at the top of the University facade of the

Sofia State University and with great zeal chiselled the letters "Sv." (St.), deeply rooted in the historical conscience of the Bulgarian people, as part of the name of St. Clement Okhridsky, patron of the University. These same letters were erased from all textbooks and all writings about the creators of the Slavo-Bulgarian alphabet, Sts. Cyril and Methody, the saints venerated as "Equal to the Apostles". Finally, obsessed with this re-evaluation of the values of Bulgarian national culture, the Communists tried to cast a shadow over the shining images of the saintly pioneers of the Bulgarian renaissance, Paissily Khilendarsky and Sofroniy Vrachansky. It became necessary Prof. Derzhavin from Moscow to intervene and restrain the illiterate appetites of our new home-grown Bulgarian historians. The climax of this obsession was reached at a conference in the auditorium of the Sofia University in 1948, where the state and the tasks of Bulgarian History Science were subjected to a thorough examination. Vulko Chervenkov, the third highest functionary of the Communist hierarchy, after Dimitrov and Kolarov, sat there, near the lectern, with a notebook in his hands, carefully keeping notes of the comments of every professor. All leading professors vied with each other to glorify Soviet historical science and castigate our old professors of history, indeed founders of the studies of History in Bulgaria, like Professors Vassil Zlatarsky and Peter Mutafchiev. The new stars of History in the Communist era were emerging to be Alexander Burmov, Dimitar Angelov et al., which scholarly work is exhausted with a few articles in periodical publications, and some light weight studies. In their writings they are falsifying irresponsibly established truths in Bulgarian historiography. Students mock the scholarship of Burmov, but the Party has placed high hopes on him, because he is a loyal agent of the new regime. In the course of this conference it was revealed that they have appointed a special commission in the Academy of Sciences which was charged with the preparation of a "full" and "purified" of falsehoods History of the Bulgarian people. We do not know if it has been published by now. However, it seems that the question is very difficult, because it is very hard for a professor to compromise his conscience and subscribe to a falsified record, which would expose him before the academic community and the rest of the world as a traitor of his professional qualifications.

Some amusing things happened at this conference. Thus the professor of Church History in the Theological School, Ivan Snegarov, respected member of the Bulgarian Academy of Science for many decades, had to explain at length to the well heeled with dialectical materialism juniors in the profession, how this method fits in the historical science. He was subjected to sarcastic remarks made by a young professor of History in the Military Academy, Yono Mitev. Snegarov did not take offense, because with his discussion of the dialectical materialism, even though inadequately, and sometimes ridiculously treated, never implemented by him in his voluminous historical writings, he tactfully fulfilled his obligation to speak, without touching on the dangerous topics where the pulse of a professor was checked out.

6. Attack on the Bulgarian Alphabet

In the field of linguistics the Bolsheviks did not abstain of their wish to impose their authority. Soon after September 9, the Bulgarian public was surprized by the intent of the communists to reform the national alphabet. They apparently discovered enemies in some of the Bulgarian letters. It was indicated that the letter Ѣ (double ER, pronounced in the Eastern Bulgarian dialect as YA, like CAT, and in the Western region of the country, from Ikhtiman to Okhrid, as E, like PET), had to be abolished. The letter ъ (pronounced in the middle of the words as U in MUCH), never written at the end of words ending with vowel, but always in words ending with consonant, where it was not pronounced, was to be omitted in the last instance. So, also was the case of ь (Little ER, as distinguished from the first called Big ER, used after soft consonants at the end of the words, and pronounced as YA as in CAT when an article for nominative, or adjective was used. In the first case it was to be omitted at the end of words, and in the second instance was replaced by Я (YA). The fourth letter eliminated completely was Ѫ, remnant of old Bulgarian language, used where once the letter 8 was used, pronounced as OO in LOOK, having suvived in new Bulgarian as Ѫ, in the middle and the end of the words. It was replaced by A in the plural of the verb съм (to be) – sa (a pronounced as in TAR which was a gross violation of the principle of the phonetic pronounciation in spoken language. Prof. Stoayn Romansky, Dean of Bulgarian slovistics, dismised from the University School of Philology because of old age, when asked by this author, his student at the time, what was the reason of eliminating the Ѣ, with unconcealed pleasure replied: "Because it stood above the other letters and was poking them in the eyes." The whole class exploded in laughter. Romansky was a member of the respective commission and had fought the change. Perhaps this is the reason why he was soon judged to be old, and was retired, in fact, expelled from the university. The fact that he was and remains one of the greatest Bulgarian Slavist, is unimportant.

As of now the greatest authority for Bulgarian Slavists is the most renown genius ... Stalin. His pronouncement on linguistics is mandatory manual for every professor, for every student, for every scholar. The studies of scholars like Prof. Mladenov pale before this remarkable work of a genius, the thesis of Stalin in world linguistics. Perhaps Stalin was too late in coming out with his work of a genius. May be this was due to the complexity of the subject, requiring long research, and promenades in history, delicate analysis of the many languages in the world ... Who knows? ... If he had come out with it earlier, our letter ™ could have been pardoned, because, if the commission honored the phonetic principle, the third person, plural, of the verb BE would still be written as C™, and not as CA.

7. Law, Philosophy and Pedagogy

No better is the case with Law either. We will mention only two names here. The place of Prof. Venelin Ganev was taken by a High School teacher from Yambol, Vassil Zlatarev. The illeteracy of this "Professor" is imprinted on his face. At first glance, the observer discovers his intellectual poverty, and if one listens to his lectures and his speeches, which he sometimes happened to make, he would in puzzlement ask himself if Bulgaria has reached to such ends, so that out of six million people, not to be able to find at least one decently educated man to take the chair of Ganev?

Another personality in the School of Law is Prof. Yanko Athanassov. In one of his lectures in 1946 this hero of Temida declared to his shocked audience of students and professors: "The German people have given nothing to world culture." Many of the listeners were shocked. Questioned on the subject, he repeated that he meant what he had said. When the names of Goethe, Kant, Hegel, Schiller, Beethoven and Mozart were mentioned to him, he turned red in his face.

In the very beginning, after September 9, 1944, the communists paid great attention to the teaching of Philosophy in the Sofia University. They had their universally recognized theoretician of Marxism – Todor Pavlov, but opposite him stood Prof. Dimitar Mikhalchev, who was dispatched to Moscow as ambassador of the Fatherland Front Government. Mikhalchev did not stay for long in Moscow. Well informed circles spread the rumor that his reports from Moscow had described the lamentable conditions in the Soviet Union, the Soviet economy in the first place, and that Bulgaria could not count of them for her own recovery, even more, the Bulgarian economy might be harnessed for the restoration of that of the Soviet Union. On account of this report, Mikhalchev was recalled to Bulgaria and re-occupied his Chair in the Department of Philosophy. Before his students, dazzled by his courage, he develops theses and concepts which entirely reject the official philosophy imposed after September 9th. He subjected the system of dialectical materialism to devastating criticism as a system of knowledge. He criticized Todor Pavlov in such a merciless attack, that his students admired him. Even more, at that time appeared his book "Philosophy as Fundamental Science", where he, step by step, demolished the positions of the Communist philosophy. The return of Mikhalchev in the University confused the communists. They had no man of his stature to oppose him. Todor Pavlov appeared as a simple sophist, not a philosopher of the class of Mikhalchev. At the time, when the entire press – dailies, monthly periodicals and all printed literature were singing hymns of the new "revelation", the Marxist philosophy, when all philosophers and scholars were frozen in apocalyptic fear, Mikhalchev, from his Chair in the Philosophy Department was thundering and demolishing to the grounds Marxist metaphysics, epistemology and logic as philosophical system. This all led to his "removal" and dismissal from the University. Later on, there appeared a rumor in informed circles, that out of the university he had written a review of Todor Pavlov's book "The Theory of Reflection", but is held in secret, because the communists would not allow public criticism of their positions.

Undisputable authority in the field of Psychology in the Sofia University was Prof. Spiro Kazandjiev. However, he found it more convenient to bend before the powerful forces in the country and in the new editions of his "General Psychology" some pages appeared in quite different form from earlier editions. Whether his research discovered new factology, or his scholarly conscience was sacrificed – it could not be established. Still, very soon, he had to yield his Chair to another high School teacher from Shumen district – Assen Kisselinchev.

In the philosophical circles in Bulgaria, before September 9, 1944, the most popular and highly regarded by the Bulgarian public was the periodical "Philosophical Review", which editor was Prof. D. Mikhalchev. With its serious, scholarly philosophical articles it had gained a wide circle of readers and selected highly competent contributors, even though sometimes it suffered the weaknesses characteristic for every publication with firmly established line of discussion. After September 9 1944, this periodical was discontinued, and its place was taken by "Philosophska Misul" (Philosophical Thought) with Todor Pavlov as its editor. It is hard to recall any other case like this. The themes, the contents and the editorial policies of the publishing collective of philosophers, revealed on its pages their narrow-mindedness, their philosophical poverty and ignorance, that hardly ten percent of the copies published are read. The authors of the articles never came out of the stereotypes and cliches, the party slogans, the quotations from Lenin, Stalin, Dimitrov and Chervenkov, as if sucking out of their fingers profound discussions about the philosophical views of the leaders of the communists mentioned here.

The intolerance of the communists to any philosophy which opposed their ideological system, is proverbial. It was assumed that only the orthodox marxists were entitled to exist. However, things have reached the point where it is difficult to decide who upholds the orthodoxy, and who is a sectarian, becauce Todor Pavlov seriously mocks and criticizes Kisselinchev, who, in his turn, does not let him have his way and engages in strong polemics with him, in defense of his views.

In the system of national education, the science of pedagogics passed through serious changes. It fell completely under the influence of the Soviet system. The objective of the process of education was no longer the well prepared citizen, but the fanatic communist, which mentality does not go beyond the party slogans, and whose political and civil views begin and end with political intolerance, psychological and physical terror, tortures, gallows and bullets. This type of pedagogics cultivates in the younger generations the ethics of the Communist morality, not permitting freedom, or brotherhood and equality, not knowing any human compassion, civil courage or gentleman's honor.

Such is the pedagogical and moral foundation of Bolshevik culture. Sure enough, the civilized world fears such a culture.

As to the material conditions for the educational work in Bulgaria it is not necessary even to mention it. It is enough to say that there is no paper in the country to print textbooks for the elementary schools. Every pupil at the end of the year has to return his books, which he/she used, to be entitled to textbooks for the next grade, as those which he/she returns are passed to the incoming new class. Has Bulgaria ever been in such a state? No one remembers that. As to living quarters, food, lightening, and heating for the students in the high schools and in the university, one could not even venture to speak.

Bulgarian science agonizes in the hands of the communists. It is on its way to perish. It is a duty of the civilized world to save it.

8. Arts, Music, Graphic Arts, Performing Arts

Bulgarian arts are in no better shape. Bolshevism cast its deadly shadow over the works of Bulgarian music composers and performers, painters, sculptors, actors, writers, men of sports, movie producers and all. The communists are well aware of the enormous influence of the music, and particularly of the Bulgarian passion to sing, so they did everything in their power to kill this art. They accomplished that easily. The slave conditions of life cause more tears to be shed from the eyes rather than stimulate desire to sing. But the communists fear the Bulgarian man when he is silent, and for this reason they attempted to make him sing again. But they would not allow him to sing his folks songs idolizing his land where he spent his day from early morning to late evening in gratifying labor; they would not allow him to sing folks songs where he idealized his home and his possessions earned with heavy labor. They would not allow him to do that, because they felt guilty in their high noon robbery before the eyes of this nation. They sought to write new kind of songs, still mascarading as folks songs and played them on the national radio, so that they could teach the people to sing them. Recently we heard one of these songs, about some "Biala Rada" (White Rada) who had won two red flags (awarded for excelling at work) and is conversing with "Georgi the Brigadirs". They know how the people look at such "White Radas" with black souls, all kinds of girls, members of the Party, and how the people look at such "Georgi Brigadir". They know that the people despise them, but they are guided by Goebbels's formula: One lie repeated ninety nine times will be accepted once as a truth. The communists know that the folks songs are created by the people, not by composers and poets on government pay or for awards with titles. They should have learned that from the fate of the "Brigadirsko Horo" (Brigadir's dance – Brigadir is a group supervisor of laborers) of Asen Karastoyanov, which is not danced anywhere by anybody, and only complement the repertoires of the concerts of a variety of choirs.

The communists could not allow the Bulgarians to sing their marches, by which they once glorified and expressed their love for the motherland. This is why they, after running down their partisan marches ad nauseam for the people and for themselves too, turned their attention to Soviet music. All of a sudden the country was flooded by Soviet marches. You try every school, military barracks or Trudovak camp, you are going to hear only Soviet marches. Everywhere you will hear "Komandir Chapaev", "Mnogo Krasnaya Geroya", and many such songs which neither as musical compositions, nor with their contents represent some beauty. Better say, the contents is not understood, because it is distorted to such an extent that a Russian would have hard time to guess the original version of it. As far as the soldiers and the trudovaks are concerned, they sing these marches with disgust.

After September 9, 1944 the communists threw out on the streets our best choir director – Sasha Popov, but when they realized that they had nobody to replace him, and that they could use him for their propaganda, they called him back. We have our doubts that Prof. Sasha Popov is sincere with the communists, and, on the other hand, we doubt if the communists are sincere with him. However, since this is good for the purpose of their propaganda, and even more so if they made him a laureat, everything else could be put aside, including the 300,000 levs salary skinned from the backs of the Bulgarian taxpayers.

In 1948 Prof. Pancho Vladigerov visited the Soviet Union. In conversation with Prof. Boris Marinov of the School of Theology, the brother of the leading Bulgarian composer, Prof. Todor Vladigerov, has revealed the profound sorrow of the first for the miserable fate of the two hundred millions of the people of Russia. From this, one could guess what is the attitude of the prominent Bulgarian music specialists towards communism.

The communists make great efforts to place Bulgarian Opera on solid foundations. It must be admitted that in this respect they have remarkable success. However, it should be added at once that they are appropriating someone else's credit. The only credit they deserve is with the Russian classical operas – "Ivan Suzanin", "Queen of Spades", "Prince Igor", "Boris Godunov" and "Eugine Onegin", as well as with "La Traviata", "Trouba-dour" and "La Boheme". They castigated the arts of the tsarist regimes, but, having nothing else to present themselves with, they offered that art to the Bulgarian public. In 1949 Todor Mazarov, during his guest appearances in Sofia, stated that if some of our performers, like Pavel Elmazov, Mikhail Popov, Raina Mikhailova et al. appear on the European stage, the Bulgarian public would not see them again. This is true! The communist know it and would not permit to any of our leading actors in the Opera, to cross the "Iron Curtain".

The management of the Opera adopted a policy, known nowhere, as to how to disrtibute the tickets for the performances. Large blocks are given to Labor Unions, who in turn distribute them to loyal to them workers - communist agents in the working class. And because those who get the tickets are not really interested, not understanding music, they would go at the doors of the theater before the show would begin, and sell their tickets to people unable to obtain them on twice or three times the original price. Some made a good money of this scalping the tickets. It is a shameful trade! This writer once had to pay three hundred leva for a ticket which original cost was eighty leva.

After coming to power the Communists decided that the Musical Comedy was a bourgeois art and abolished it, but soon realized their mistake and restored it. Soon after, they decided to make it a State musical theater, which buried it forever and ever, because its Russian repertory did not permit the actors to develop their full musical talents potential and the public to find in it what they were looking for – light enterainment.

This is what the state of music in Bulgaria is.

Side by side with the art of music the communists paid attention to the graphic arts. It is not necessary to point out to the failure of their effort. This was already done by one of their orthodox communists – Alexander Zhendov. The devastations which were inflicted on the graphic arts in Bulgaria were so forcefully expressed and openly revealed that the whole pack of lackeys and spineless critics was frightened. Nobody dared respond to Zhendov, because his charges of the failure of Bulgarian graphic arts were irrefutable. Of their fall in a blind alley, of stereotypes, of a catastrophic decadence and irretrievably lost direction, were an ugly, undisputable reality. What could be done? Nothing, but to admit failure of the official court policies of the paid artists. It could not be done. There was only one exit: To kill Alexander Zhendov, to destroy him politically, ideologically and professionally. There was only one who could do that: Vulko Chervenkov. And he performed his assignment with distinction. He gave the tone and the chorus of lackeys picked the tune, using his.

Nobody, who has not had a chance to visit the art galleries may have an idea of the Bulgarian painting and sculpture of that era. (I will venture to interject here my impressions in 1991, when I visited Sofia for the first time after forty years. I was appalled, looking at numerous art works of those decades, removed from public display in parks and squares, piled as garbage behind the former royal palace. I speculated publicly, in print, that this art should be preserved in special museums as an illustration of the arts of the times.) But if someone has a chance to visit an art exhibit, could not but be left with the impression that Communist art has not progressed beyond two or three themes, developed in thousands variants. One of the themes is "Georgi Dimitrov". It is apparent how much effort has been wasted by the artists in trying to invent a new pose of the "wisest one". Other theme is "Vulko Chervenkov ..." – the same efforts of the artists. The third theme is "Stalin". The same! Nothing new. Side by side with these figures one would see tractors, buildings in progress, factories, kolkhoses, public political rallies, library shelves with books, young kids and dams. These are the themes of the graphic arts

- in portraiture, in landscape, in engraving, in sculpture, etc. The Bulgarian artists are getting convinced more and more that they are not men of the arts, but ordinary hired hands. Their works are no longer bought, because no-one, with an interest in the arts, would like to bring home something that is despised. These works of art are purchased by the KNIKs, by municipalities, but never by private individuals. The trivialization of the arts expands wider and wider, because the artists more and more see themselves painting placards for meetings and slogans for public demonstrations promoting communism for which they are well paid and see no need to engage in serious art of aesthetics.

Next we have to turn to Theater as a performing art. Here too, the Communist policies inflicted heavy casualties. The Actors' guild was purged. Actors of the class of Panteley Khranov, of a Petko Atanassov et al. were physically destroyed. Actors of the class of Nikola Balabanov were doomed to oblivion and men like Stefan Savov et al. were to be subjected for long time to re-education, to be rehabilitated. Others, of the class of Krustiu Sarafov, in their old age had to join the Communist Party, to add some glamour to it, so that it could show its face that in its ranks are some of the most talented masters on the Bulgarian stage. They arranged a jubilee to him with a Russian play, Griboedov's Ninteen century classic *The Misfortune of Being Clever*, produced and edited in such a manner that the great Bulgarian actor appeared like a bird in a cage. It may satisfy the taste of the communists, but it could not contain the talents of a Saraffov, or to satisfy the Bulgarian public. This showed that the management of the theater either did not appreciate the fullness of Saraffov's talent, or did not know the taste of the Bulgarian public. Several attempts were made to stage Bulgarian plays. So, Andrey Guliashki was represented with his piece "Blato" (Swamp) by which our dramatic arts, and the purse of the national theater, fell in a swamp, and with difficulties pulled themselves out after several showings. It was a painful experience to watch how a collective of our best actors tortured themselves on the stage before a public of no more than twenty spectators. If the National Theater experienced such difficulties in preparing its repertories, one may imagine how much more difficult was the task of the provincial theaters in seeking something decent for their stages. The public was saturated with plays with sub-machine guns, shooting inept police officers, executions of helpless men, arrests and tortures. In the village of Zelenikovo, a collective of young communists, in their helpless efforts to find something satisfactory for the ideological line of thinking and the taste of the public, had the ill advised venture to stage a play of Yordan Kovachev, a most respectable writer in the country, but a leading figure in the political opposition led by Nikola Petkov. At that time they had no idea who the author was and how his ideas fit in the new Communist society. It was with chagrin that they had to find later in what spot they had put themselves.

The only production on the Sofia stage of the National Theater, which would not have been closed for lack of public, was Ivan Vazov's "Khushove" (Bulgarian exiles in Romania), but the communists hurriedly stopped it and took it away from the admiration of the spectators. They saw something dangerous in it, the public was ecstatic in applauding the actors, thus expressing their suppresssed feelings of nationalism and patriotism triggered by the national art. It may be that their chiefs from Moscow saw the rebirth of nationalism which they feared for their own sake. Our theater has undoubtedly talented actors, which are a pride for our national theatrical art. It is to be regretted that with their repertorial policy of ingratiation, the management of the theater, and the whole Bolshevik pack in the KNIK, are killing these talents. If the latter find satisfaction in the titles of laureates of "Dimitrov's awards", "People's Actor" and "Deserving Actor"; in other words, if they think that their worship in the Temple of Art could be bought and sold for money, then their chiefs do not have to worry, but Bulgarian art is doomed to a slow death, the Bulgarian National Theater will be empty, sooner or later, in spite of its artificial support by the Party organizations in the factories, who keep a record who attends and who does not the Russian plays. Of course, no actor is pleased by such honor – to show his art before audiences brought in from the villages by force, who fear more their political officer, rather than to admire theatrical performance. Kissimov would not accept such honors, and so is the case with Ivan Popov, Nikola Popov and Balabanov. But if the soul of the Bulgarian actor is burning in a fire of indignation, if he, under the yoke imposed on him by the Bolsheviks, keeps his spirit for the tomorrow's day of freedom, then his present chieftains should worry from now, because their evil doings and crimes against the Bulgarian people will be revealed best by these same actors, on these same stages, where they are now forced to serve the bloody butchers of the Bulgarian people.

Our theatrical arts will be resurrected again!

Together with the theater, we must mention also the cinema as a form of the performing arts. The Bulgarian public is shown only Russian films. It is interesting to know whether the leaders of KNIK, also their Russian chiefs who force upon them the Russian productions, are aware of the catastrophic defeats of Russian cinematography on the Bulgarian screen. One wonders if they are aware that the vast theater of "Dimitar Blagoev" cinema, the former "Evropa Palace", justifies its existence where between ten and fifteen visitors go to see the projected films. This may seem improbable but is an everyday reality. The other smaller cinemas need not even be mentioned, because the picture there is even worse. It is interesting to know why scores of cinemas have been closed. Well, the answer is easy: because of the lack of films and public. One Russian film is projected for years on the Bulgarian screens. It is nauseating to the public, but the cinema operators have nothing to replace it. They have the choice to either close the cinema houses or to keep repeating what they have.

9. Sports

Sports hold the exclusive attention of the communists. But here, too, they do a bear's service of the sports life in the country. This is particularly true for the Capital, and especially for soccer, where they make every effort to suppress the political preferences of the spectators, favoring one or another club. Some interesting scenes happened in recent times. Even before September 9th, the club "Levsky" was in existence, all nationalists were "Levsky's" fans and thunderously applauded its victories. It was the best soccer team anyway. This is how they expressed their anti-communism, at the sports stadium. Gradually it became clear that all applauses for "Levsky" were in effect expressions of opposition to communism. The victories of "Levsky" made the communists mad. The anti-communists were jubilant. The climactic point in these games was reached in May 1948, the competition for the Award of the Red Army, when "Levsky" had to play against the favorite of the Reds, the team of the Bulgarian Army. Over 30,000 spectators filled the stadium to capacity. The game was fought not only on the playground, but also at the benches of the public – on the field the soccer players, in the public – communists vs. anti-communists, who would outshout their opponents. From the sport stadium the excitement passed to the streets, to the restaurants, to the barbershops, to the offices, to the student class rooms. The struggle was a political confrontation concealed under sports fanaticism, it was blue against reds. The blues, delighted of the victories of "Levsky", were branded as "fascists", while the "fascists" could not allow themselves to attack the other side of opponents as "reds". It was too risky an affair. The public was going to these games not so much to watch the round ball, but to vent their political frustrations. Which meant that not all those who went to the soccer field understood the rules of the game. Many went to the stadium just to shout their opposition to communism. It was reported of a man asking someone: "There was one man there, dressed in black. He never kicked the ball. Why?" He was speaking of the Referee. Usually the "blues" far outnumbered the "reds" and out-booed them. This spirit spread to the provincial towns. The overwhelming "blue" public at the games and the victories of "Levsky" were perceived as victory of the anti-communists vs. the communists. To put an end to all that the authorities initiated changes aimed at the destruction of "Levsky". First they changed its name to "Dinamo". Subsequently, down the invisible paths of Communist persuasions, numerous players of "Levsky" were transferred to other clubs, but still "Levsky" continued to be favorite of the anti-communist public and the communists continued their efforts to put it down.

Here, in the sports, the communists continued to force their poisonous sting, seeking to suppress every independent free activity.

10. Literature

We have to say a few words of Bulgarian literature. Here too, the communists suffered humiliating defeats. The Writers' Union was transformed into an obedient army of propaganda agents. These propaganda agents travel across the country, read their works before peasant and urban workers' audiences, and earn their living. This is the only preoccupation which they are engaged in. It is no wonder that during the last seven years nothing of substance has been created in the fields of novel, drama or poetry, least of all as literary criticism. It is pitiful reality. The themes treated by the writers in the published works do not break out of the decreed clichés and sloganeering in favor and glorification of the Party. Nothing can be said for free creative art. The case of the Communist writer Pavel Vezhinov, with his novel "The Blue Sunset" published in 1947, is a typical example. "Rabotnishesko delo", the Party organ, came out with an article which denied its value as a Bolshevik literary work. Two days later, the novel disappeared from all bookstores. Apparently it was last read in some factory for recycling of paper. It was destroyed and the public never saw it. These are the conditions under which the Communists write. These conditions, necessary for the free expression of a literary artist, could not even be discussed. Who is going to be the publisher? The State, of course! The State publishing houses would not accept a work for publication, without receiving a permit from the KNIK, which is the official censor. This is why our literature since September 9, 1944 is in a state of lethargy. In vain

Vulko Cherbenkov calls upon the writers to go to the factories, to the kolhoses, to the motor-tractors stations to study the new life, the new relations among the people, the new conditions of living and then sit and write novels, plays, short stories and poetry. It is easy for him to give such advice, because he himself would not go and be confronted with the ugly reality in factories, villages, and work places. The writer who would venture to follow his advise, should have a heart made of stone, to be able to look at reality with indifference, to suffer the tears, the curses and desperation of the country. If someone, after all, goes there, then he, if nothing else, will be plagued with doubts about the Party system and its social an economic policies. If he does not express his doubts, he would never have any desire to create works of literary art where he would have to "reflect" the life of the peasant or the worker. Of course, there are such writers who give a distorted picture of life in the country, but who reads them? The writers themselves, at public meetings where workers and peasants are herded together by force. They are highly paid to act as propaganda agents for the government. But does anyone listen to these works? It is doubtful. Propaganda literature is not an art, especially when it is constructed of banal phrases and political clichés The current Bulgarian literature, as much as it exists, is nothing more than a propaganda publicity of the lowest quality.

Much worse is the case with poetry. Having taken the line of total rejection of everything not conforming to socialist realism, the communists brought poetry, like all other arts, to a catastrophic failure. There is nothing of significance created during the past seven years, artistically and in ideas, even in tune with Communist standards. One needs only to open the pages of the Communist publications, to read which ever poem he finds there, in order to convince himself, that indeed, our poetical heritage of the past has not been surpassed, has not kept on the same level, but, in fact, has fallen below any tolerable standard. The most recognized poet at this time is Lamar – Laliu Marinov. In one of his poems he uses, purportedly to make some comparison, the expression "delvata na zhivota" (the crock of life). For this gem in our poetry, and for many of the same kind, the humorist "Sturshel" tried to give some figurative explanations, so that if someone of the readers does not know what is "delva" (crock), to educate them.

No better is the case with literary criticism in Bulgaria. Not that there are no authorities to handle this not very pleasant art. On the contrary, in the present state of our literature and our poetry, an honest reviewer, who values highly the prestige of our national art and the progress of our culture, will find a lot to say. But who will dare? The great literary figures are all important Communist leaders. The entire cultural and administrative power is in their hands. To that, everyone seeks membership and stability in the Party. How this same Party would allow that its weakness be shown to the public, when all its leading functionaries support each other and would not undermine anyone among themselves? They all are partners in the plot against Bulgarian science, Bulgarian art and Bulgarian culture. It is better to keep silent. Those who are outside of the Party, are no threat to anybody. It is much more to their advantage to praise every ineptitude, to bow down to any Party celebrity and get appointment to some lucrative job, rather than tell the truth. This is what happened with Elisaveta Bagriana, Dora Gabe, Pavel Spassov et al. When they were writing their Indulgentsias, Fani Mutafova was passing away in jail. Stilian Chilingirov painfully suffered and still suffers his old age in the concentration camp ... Prof. Mikhail Arnaudov ... we need not mention them all.

Literary criticism in Bulgaria does not exist today. Ivan Bogdanov attempted to defy circumstances by publishing his "Lost" (Lever), but was soon stopped, because it was muddying the water of the only literary weekly, "Literaturen Front". There is a periodical now, "Izkustvo" (Art), but we could not say anything different from what we said about "Philosophska Misul" of Todor Pavlov.

The question of our drama we mentioned when we discussed the theater and it is not necessary to return back to it.

So, this is the state of Bulgarian aesthetics. Every free initiative in any art which one considers, if the orders of the CC of the BCP are not followed, is doomed. It does not reach its destination – the reader, who alone is the fair judge. If an art does not find a worthy judgement in this Court of last resort, it will die. If it is accepted by this Court, contrary to the verdict of the BCP, it will also die because it will be suppressed. In the final analysis, this means depriving art of its most important aspect – its free character. In such a case it could be anything else – profession, publicity, propaganda – but never true art. At its best it is a bad model, and at its worst, it is a parody of art.

11. Humanitarian Movements

The question of humanitarian activities in Bulgaria is in a worse shape. The civilized nations are proud of their humanitarian cultural achievements and do everything to facilitate every philantrophic initiative. Well, Science and Art are in a way attempts to express the perennial search for ideal intellectual and emotional perfectness of man and society. But side by side with the development of his reason, with the perfection of his aesthetical feelings, man has always searched for better relations between the members of society, to build such a social order where Hobbse's dictum *Homo Homini Lupus* be replaced with other principle: "A man for another man is a brother". If one has to identify the supreme ideal of culture, it would be the ideal for humanizing man, for liberation of man from the chains of his animal nature. Neither science, nor aesthetics are possible and have any justification outside of this ideal. With its iron logic, and with its beauty they may satisfy the biological egoism. But they could not direct man towards the supreme ideal of culture. This egoism logically leads to an inevitable conflict between man and man, between nation and nation, between one class and another class, one race and another race, between individual and society, law and moral precepts, civil and international wars, which in the final analysis destroy culture and civilization, and interrupts progress. The supreme ideal of culture, therefore, is a total rejection of any – personal, national, class, race, etc. egoism – the creation of a society where science and arts serve man to survive, not to destroy him. We are far too far from this ideal, but the civilized world is marching towards it, led by its most prominent representatives. Quite different is the case with the countries conquered by Bolshevism. In those countries they cultivate emotions, moods and habits, saturated by beastly hate towards individuals and societies. Science is harnessed to serve such hate. Aesthetics is developing under the shadow of death, destruction, conflagration and ending the life of civilization. In view of that every humanitarian initiative, which raises the flag of humanitarianism of "justice" and brotherhood, is rejected as a counterpart to the class struggle, upon which the Communists build their social philosophy.

The Bulgarian people, this should be underlined, have always had close to their heart any humanitarian acts and have never refused to support the cause of humanizing of man. However, it is not possible to speak out for humaneness in Bulgaria today. In order to suppress and uproot all such views, the Bulgarian Government prohibited, or appropriated all humanitarian organizations. In this respect a mortal blow was dealt to religious education, the League for Defense of the Rights of Man and Citizen, the Abstinence movement, the Vegetarian movement, the Red Cross, advocating pacifist ideas, the Public Libraries as free institutions, et al.

The communists always have feared Christianity and its impact on people. Where the civilized world is proud of the Christian spirit of their culture, the Bolsheviks do all that is in their power to discredit it and persecute the Christian preachers. They cannot listen to church sermons against murder without becoming nervous. At the same time, when they are preaching murder. They cannot listen to sermons against theft without fretting, or sermons for freedom, for brotherhood, for equality, when they themselves are thieves, enslavers, engaged in fratricide and themselves represent a new class of aristocrats. So, they have set for themselves as a goal that in the course of the next 50-60 years, to put an end to all "religious deceptions". They hope that after they prohibit preaching of Christianity among the young, it will disappear. They prohibited the organization of Christian Youth groups. Prior to September 9, 1944, there was a well organized Christian youth movement under the auspices of the Bulgarian Orthodox Church. Thanks to the efforts of its leader in those days, Mr. Stoyan Petkov (presently Metropolitan Andrey, head of the Bulgarian Orthodox Church abroad) it scored remarkable progress, but it was outlawed by the communists. An attempt at restoring it in 1947 ended with the internment of its new leader, Arkhimandrite Myron.

Having prohibited the Youth Christian movement, the communists outlawed any propagation of Christianity in any form among the youth. The Bulgarian Orthodox Church had already attempted to hold Sunday Schools for young kids and appropriately intructed the priests to proceed with this undertaking. The result of that was that the priests who had followed the orders of the Synod, ended in concentration camps. Many of them, realizing that they could not follow the orders of the Metropolitans for the stated reasons, advised the parents from the church pulpits to educate their children in the Christian precepts. Having done that, a priest, Peter Yotovsky, of the village of Tsalapitsa, Plovdiv District, was called by the mayor of the village and strongly warned not to repeat such advises. At the same time the authorities ordered the publishing of a periodical, "Zov", intended for education of the High School youth and the universities, stopped. Other periodicals, "Christianche", intended for elementary school students, "Bratstvo" – for the family, were also ordered stopped. "Naroden Strazh" – a high level publication for the intelligentsia, was also done away with. Not different is the case of the treatment of the religious denominations in Bulgaria. World public opinion is well acquainted with the staged court trial of the evangelical pastors. The newspaper of the United Protestant churches in the country, "Zornitsa", was stopped in 1948, during the editorship of Trifon Dimitrov who until now (if my information is correct) is in concentration camp, if not executed. The state of the Catholic Church is no better. The organ of that church in Bulgaria, "Istina" (Truth), was also ordered stopped in 1948.

In 1948 representatives of the Orthodox Student Society in the School of Theology established contact with Protestant and Catholic groups with the intent of sending a representative of the Bulgarian Christian Youth at the Congress of the World Christian Movement. The project had to be abandoned, following the delicate warning from the police authorities.

The bolsheviks are conducting a strenuous underground struggle with Christianity. The Constitution strictly limits freedom of religion, but widely publicizes the activities of Patriarch Alexei I of Russia and the Englishman Hullet Johnson, who actively promote Communist causes. At the same time they do all they can to suppress every Christian sermon in the Captive Nations. Christian love for man and Communist hatred are incompatible concepts. It is an open question who serves culture – hate or love. And if they persecute preaching of love between men, while the cult of hate is strongly defended, encouraged and planted in the hearts of the growing generation, it is not difficult to understand the direction of our culture.

The Bulgarian government unhesitatingly outlawed the League for Defense of the Rights of Man and citizen. There were people in Bulgaria who were ready to defend the rights of man and citizen, proclaimed by the "Declaration of Rights ... by the French Revolution, highly valued for building a moral conscience in a civilized nation, a document highly respected among the achievements of the civilized world. But this had no meaning for the barbarian communist world. Its apostles trampled over this document, dissolved the League and sent to concentration camps its leaders. Venelin Ganev, who was one of the leaders of the League, in spite of his errors in politics, in 1948 gave a classical discussion of the principles of democracy in his work: "Democracy. Basic Principles" (which this author read with great interest, and received his first acquaintance with these matters), but later had to experience personally the charms of the Bolshevik "People's Democratic" order in Bulgaria (which he helped to bring to power by becoming Regent in the first Fatherland Front Government on September 9, 1944). In jail he met the other advocate of the objectives of the League – the well known to the Bulgarian people humanitarian activist Yordan Kovachev.

The Bulgarian communists decided that the existence of a vegetarian movement was rather dangerous, especially of the Abstinence movement, which they used before September 9th as cover for their activities, until they came to power. They thought that there was no need of these movements, because they cultivated a spirit of independence of judgment and many of their members were anarchists. They proceeded cautiously. First they closed the existing clubs in the provincial towns, and when this process was accomplished, they simply ordered their centers in Sofia dissolved. This ended the whole affair. They ordered that nobody should discuss these actions of the government any longer. Thus, without much fanfare the humanitarian ideas in the country, having gained thousands adherents and enjoy great prestige in society, were suppressed.

Further on the Communists laid their hands on the Bulgarian Red Cross, turning it into a state organization, in service to the Communist party. Its supra-party, non-class and international character were rejected and it was transformed into an instrument in their hands to serve their purposes[*].

Before September 9 the Communists were the most active in the *Chitalishte* movement (Public libraries, privately organized by educational societies). They used them as a cover for their activities. After the coup they abandoned this field. But they did it in

---

[*] Years later, in fact on May 15, 2000, this author, perusing his dossier in the archives of the Bulgarian State Security in Sofia, discovered the original of a letter he had written to his father through the International Red Cross. No explanations are necessary as to how this letter had found its way to the offices of the Secret Police. From the Bulgarian Red Cross, it had automatically been sent to State Security.

such a way that neither they worked for it, nor they allowed others to get involved in this useful for the people educational institution. President, secretaries and librarians of the *Chitalishte* became the most prominent Communists, preoccupied with many other responsibilities along Party lines. In time the "Reading rooms" were turned into stock rooms for the materials requisitioned from the people for the army. The Libraries were carefully cleaned up of literature condenment by the communists and stuffed with Bolshevik publications, which, outside of the system they called "self-education", no one was interested in.

It is appropriate to say something about the attitude of bolshevik propaganda towards pacifism. To say that they prohibit campaigns for peace among nations, is definitely untrue. On the contrary, there is nowhere such an obsession in propagating the ideas of peace than that of the bolsheviks. They have a large net of agents in the whole world, organizers of the World Peace Movement, collect millions of signatures for peace under various proclamations, have won otherwise honest intellectuals, educators and men of the cloth around the world – naive as they may be in their hopes, reliance and believes in the pacific intentions of the USSR. If they do not do that for reasons of monetary awards, bribed agents of Moscow, they, most certainly are victims of deceit. The signing of the various appeals in the countries behind the "Iron Curtain" takes place under force. God help those who do not appear in time at the designated place and sign the prepared text. To all people of the "People's Democracies" is clear that the word "PEACE" in the vocabulary of the Communists means "WAR". It is not difficult to understand that. One could only read a speech, whichever and by whomever, where the stereotyped clichés of the Communists are reduced to Party slogans, dedicated to peace; that under that word the bolsheviks in fact mean war. The widest propagated slogan in Bulgaria is: "Stalin, in all languages means Peace". If the word "Peace" in the Latin wisdom "Si vis pacem, parra Bellum", is replaced with the word "Stalin", it will be translated as "If you want Stalin, be prepared for war". No dialectics are needed here. The truth of the Bolshevik love for peace shined with all its disgusting nature. When a Bolshevik speaks of peace, he understands this word to mean war. One could speak in Bulgaria only for peace which means war. To speak of peace in its true sense, is an act, which will send the pacifist to a concentration camp.

Such is the state of the Bulgarian humanitarian cause. Preaching the ideal of Christianity for universal brotherhood, for love among individuals, propagating the ideas of vegetarianism or the abstinence movements, of the ideas of the Great French Revolution, of pacifism – in general the ideas of any humanism, and opposing them to hate, evil, misanthropia, to the morality of the beasts, of wars and destruction, is viewed as a crime under the laws of the country.

CONCLUDING REMARKS

We began this essay with a discussion on the subject of culture and its essence. Having established that it is manifestation of the inner soul of its individual, social-historic and personal-psychological creator and carrier, we concluded that there is an absolute connection between culture, civilization, democracy and freedom. This conclusion, that dictatorship destroys culture, is proven by the Bulgarian realities. Bulgarian science, Bulgarian arts and Bulgarian humanitarianism are doomed to a slow death. Would the civilized world allow them to die? In the name of world culture, in the name of world civilization and in the name of world progress, they should be saved. The Bulgarian scientists, terrorized by barbarian bolshevism, the Bulgarian composers, conductors, artists, writers, philosophers, the Bulgarian professors, students, the entire Bulgarian society expect that day and hope that it will not be delayed for too long. The Bulgarian cultural community is not bolshevised, it is only enslaved. Will the world continue tolerating in the XX century the tears and the sobbing of the unfortunate slaves? Will the civilized world fulfill its duty? We pose these questions and we are justified to expect an answer to them, because we like our motherland and our people, because we dearly love our national culture, as Americans, or Englishmen, or Frenchmen dearly love theirs.

*

We crossed the border on June 19, 1951. This article was completed in the middle of July, in the city of Drama, Greece. Now, forty seven years later, we would change very little, mainly stylistically, but we let it stay as it was originally written. We may have changed, as our views have changed, about our faith and reliance for our liberation on the Western nations in the free world. We may have been naive placing too much hope on them, but were sincere in our hopes and expectations. To this author, the essay, as it is, speaks a lot about his growth since he left Zelenikovo in 1939 and then in 1945. As to some of the ideas expressed here, the reader must keep in mind that this essay was written in 1951, a month after escaping from what, in effect, was a slave labor camp, and a forty four days odyssey in the mountains.

II

BULGARIA I SAW AGAIN –
FORTY YEARS LATER
1991

IMPRESSIONS OF BULGARIA
MAY – JUNE 1991

E P I L O G

Forty years ago, on May 6, 1951, this writer, author of the Report which follows, defected from a military labor camp, near the town of Balchik in north-eastern Bulgaria. He and his six companions formed an armed underground resistance group. After forty four days of wandering in the Rodopi mountains, near Plovdiv, in the vicinity of the village of Kuklen, in a gully known as Haidut Dere (Rebel's Gully), on June 19th, sometimes about 6.00 p.m., they crossed the border into Greece. For forty years he played an active role in the resistance movement abroad, in and out of political organizations and as editor of numerous publications. This is now history which will be read and commented upon in the future, if anyone is interested in it.

On May 21st this year he returned to Bulgaria and spent three weeks there. His "Impressions of Bulgaria", as the title of this Report reads, are not flattering for the country. All through these forty years, in hundreds of articles and essays, he has never flattered anybody, always seeking the truth, always guided by Henry Clay's maxim: "I had rather be right, than be President." He sees no reason why in this Epilog of his political life he should be untruthful to himself. However, he entreats the readers of these *Impressions* to be more lenient to Bulgaria and less generous to him. What he found in Bulgaria – legacy of forty five years of communist rule there – he saw as vindication for his decades long struggle for the liberation of his land and his people.

Bulgaria today is a half dead corpse. To those who have undertaken to revive her he most sincerely wishes success. Nothing, nothing will please him more than to see his motherland resurrected for a new life, like the mythical Phoenix emerging from its ashes. In the autumn days of his life, spent abroad where he has grown deep roots, he continues to be concerned of his native land and his heart bleeds at the site of her agony in these post-communist days. He hopes that if and when he has another chance to visit there, he will not be as bittery disappointed as he was the first time, and that his pessimism of today, as to the future of Bulgaria, will yield to a more optimistic outlook for brighter and happier days,

Spas T. Raikin

Stroudsburg, PA.
September 1, 1991.

1. To Return to Your Home Land ... and to Find it Devastated

To return to your homeland, to your home town, to your father's house where you had been born and grown up ... after forty years; and to leave it after three weeks, with feelings of depression and bitter disappointment, is an experience that one does not wish to repeat.

It is with such feelings that we stepped on Bulgarian soil on May 21 of this year, in Sofia, and left it on June 6th for Greece. After crossing the border at Kulata, a fellow passenger turned to me with a question: "How did you find Bulgaria now, after four decades abroad?" I did not spare my disappointment and my pity. "I found her badly bruised, badly beaten, worn out, robbed, exhausted, dispirited, dirty, improverished, numb, overgrown in weeds, embittered, arrogant, ingratiating, and utterly confused. I found it on a death bed and see no prospects for her revival." My interlocutor never uttered another word. My wife suspected that he was a Greek communist.

2. Places and People – Revisited

I spent these three weeks in my native village Zelenikovo, Plovdivsko, in Plovdiv and Sofia. I visited the Bachkovo and Rila Monasteries and the city of Gorna Djumaya. I met and conversed with relatives, with friends, with communists and anti-comunists, with independents, officials of the BANU and The BANU "Nikola Petkov", the Deans of the Plovdiv and Sofia Theological Seminaries, with the Dean of the Theological Academy of Sofia, the Abbot of the Bachkovo Monastery, with Metropolitan Pimen, the oldest Member of the Holy Synod, with my good friend Gancho Velev, professor in the Sofia Theological Seminary and Prof. Radko Poptodorov, emeritus professor, of the Theological Academy, a former classmate of mine four decades ago, and very dear friend. I met numerous lower clergy, I met the leadership of the BANU "D. Gichev", and participated in consecration of the flag of the local chapter of the BANU "Nikola Petkov", to which I am a member. I met the Director of the "Tsanko Tserkovsky" publishing house of the BANU and the spokesman of the newly formed group of parliamentary members of the BANU "Nikola Petkov", Lazar Dulgerski. I met a former friend of mine, a professor in the Faculty of Agronomy in Plovdiv and leader of the young Communist league in my village in old days. I visited the local TKZS in Zelenikovo and met many of the workers there. I met hundreds of peasants, former friends and classmates of old times. I met a Dr. Kiril Popov, and his charming wife who drove us to the Rila Monastery and Blagoevgrad. I visited and took part in the religious services in the church in my village and visited the cemetery where my parents, who lived and died as martyrs. I met former prisoners of the regime of Todor Zhivkov, inmates of the notorias Belene Camp. My cousin Slavi Maximov and my brother have experienced the bitter-sweet menues of this institution. I visited my favorite hangout in my village, the *Chitalishte* (reading room) "Rakovsky," where I have spent countless nights in old days, reading voraciously every book and periodical I could put my hands on, and the new house of splendor built by the communists to replace it. I was interviewed by a reporter of the BANU organ "Zemedelsko Zname". In view of my comments on Agrarian politics, I doubt that this interview will ever see the light of day. I visited numerous churches, museums and exhibits. Among them the Ethnographic Museum in Plovdiv, the National Art Gallery in Sofia in the former Royal Palace, the National Historical Museum in Sofia (the former Palace of Justice), the Museum of the Theological Academy where the curator gave us an extended tour and fascinating explanations of practically every item, the Museums in Rila and Bachkovo Monasteries and other archeological sites.

There was much more that I wanted to see, and many more people I wanted to talk to, but time limitations cut my desires to a minimum. Even the people with whom I had made arrangements to meet, I could not keep my commitments. But, God willing, there will be a second time. Everywhere, and everyone I spoke with, I discussed, and heard about ... Bulgaria in times past.

So, I had many impressions of Bulgaria and the Bulgarians. I summed up these impressions in the few words, in my response to the fellow traveler in the bus taking us to Salonica. Understandably, these few words may offend, may delight, may shock or may intrigue the reader of this report, and it is only fair that I should elaborate a little.

3. The New Accent of the Spoken Bulgarian Language

Probably the most innocent, but impressive change that a returnee to Bulgaria, where people are concerned, is the change of language – not so much of vocabulary, but of accent. The present day Bulgarians speak with a distinct accent which is different from what one would hear some forty years ago. The present Bulgarians have acquired a staccato type of pronounciation, coupled with an extension of the western dialect over the eastern part of the country.

4. Optical Illusions. Reduced Dimension

There are also other changes of appearance. First, Bulgaria seems to have shrunken in so many ways. The distances from place to place seemed shortened. Travelling from Sofia to Plovdiv by car and then off to Zelenikovo takes an hour and a half, when in times past it would take a whole day. The streets in the village seem to be shorter, perhaps five times shorter than they were four decades ago. The buildings of old times seemed also to have shrunk. What once were the biggest houses in the neighborhood, which were considered to be houses of the wealthy, now seemed to have shrunk so much, to be so small, that they looked like relics of the past. What

425

once was a large village reading room, the Chitalishte, now appears to be a small, a very small tiny enclosure. The public squares where once the entire village used to gather to dance *horo* on holidays, now seemed very small where people no longer gather to dance and celebrate. The streets in the village have changed. They are paved, they have sidewalks – narrow, but sidewalks. The main street is generally empty – here and there some humans would appear, and from time to time some bus would buzz from nowhere, or a donkey-driven four wheeler with grinning faces of one or two shabby figures.

5. The New Bulgarian Village

All houses, one floor mostly, are well hidden behind solid mud walls, well plastered. Here and there one would see iron fences. They all look like small fortresses, as if to protect the inhabitants from barbarian invaders. Through the doors, or through the fences, one inevitably observes the front yards turned into small farms – well planned, neatly maintained, growing all kinds of green vegetables – potatoes, green beans, tomatoes, peppers, squash, onions, garlic, lettuce... Somewhere in the corners of the yard are the pig pen, the chicken coup and the sheep pen, if any. The rooms of the houses are of moderate size, Spartan furnishings, but neat and with large windows. Surprisingly, in my village there is indoor plumbing – running water, which was the privilege for one to five homes in old days. The shower which was unthinkable in old times, was installed in a separate room, attached to the house. There was a furnace, cylindrical metal contraption, probably five feet high, 12 to 16 inches wide. At the bottom one would light a fire. In five minutes the water was hot enough. It was a blessing. Another attached room was the kitchen, a small neat little room with a stove alternately fired by wood or electric – more wood than electricity – but radiating pleasant heat in the cold days that happen to be there at this time. There was, of course, the outhouse, but it was neat, clean, and a toilet American style, with water tank above it. The only problem was that one had to go out during the night. The water was excellent for drinking and I was surprised. I could not have enough of it. I was told that it came from the place where we once had an orchard and garden down the river, the place where I had spent my summer days for a decade or so. The only problem was that there were not enough electric bulbs. I suspect during my stay there my brother had borrowed some from relatives, so the place may be well lit for as long as we stayed there.

6. And the Old Bulgarian City

Indeed, the change in the village was for the good and was very impressive as compared to what it was some forty or fifty years ago. I suppose, this change would

426

have come any way, no matter the regime, the system and the people who governed the country, and perhaps even more might have been accomplished. But I did not see much change in the cities. Except for center city in Sofia, where they have built the Party headquarters, the Presidency, the Council of Ministers, the Sheraton Hotel and the Tsum, (Central Universal Store), and the conversion of the center into a mall. The rest of the city is old, dusty, aging, with its old landmarks – the National Assembly, Alexander Nevsky, the Tsar Liberator Monument, the National Theater, the University. And there is the Palace of Culture with the impressive display of its massive torso, reminding the remains of an ancient ziggurat. But the rest of the city did not have much to show besides its never painted, never cleaned, graying buildings, never washed windows, empty shelves, poor appearances and endless streams of exhausted humanity.

7. A Nation Without Smiling Faces

And so, there is this humanity... the people, the peasants, the townspeople, the workers, the intellectuals, the clergy, men and women, the politicians. When you have seen them all, and talked to them all, your impression is that the entire nation is in a deep mental depression. What is totally absent is the smiling faces. There is no display of energy, there is an all pervading look of pensiveness, occasional irritability, sometimes outright rudeness, arrogance and universal pessimism. Some taxi drivers spewed so much venom, expressing themselves in lewd language that sometimes I was quite uncomfortable with the thought that my wife would catch some of it. She did!

8. Peasants ... Human Ruins

Walking through the streets of the village I was greeted by hundreds of peasants, mostly women, dressed in black, and occasionally men, moving around as ghosts – worn out, bent, exhausted, shrunken, prematurely aged. They all asked me if I recognized them and much of the time I had to pretend that I did. They recalled old long forgotten events which I did remember and then still could not remember them. But it did not matter much any way for the brief encounter which I had with them. They all seemed to remember me as a very slim boy, always with books in my hands or in my bags while tending sheep or cows. But I could not recognize them. They all looked like human ruins and the town looked like a dodge city. They were shadows from the past and ghosts in an abandoned city. They came from all over to greet me in the center of the village square. And they told me names which I had long forgotten. And everybody seemed to have some nickname – Taniu Katranya[1], Payaka[2] Ivan, Gugenya[3] Slav, Riapata[4] Groziu, Patkata[5] Marin, Kuchenya[6] Ivan, Byalata Shapka[7], Mincho Kisselekya[8], Ivan Patura[9], Ludiyat[10] Denu, Dyavola[11] Encho, Matiu Palamuda[12], etc.

But they all looked like ruins. After talking to a man sitting on a bench for five minutes, he was trying to explain to me who he was, and I had difficulty recalling – it all suddenly dawned on me. He used to be my best friend in old times. We were buddies. As a young man he was bursting with energy, strong, athletic, handsome. Now I was in the presence of an old man, he was five or six years younger than I was. He was toothless, he was shrunken, he was stooping, he was grey, he was not shaven. A physical ruin. A physical wreck. After talking with him for five minutes something clicked and I recognized him. I recognized his voice, the peculiar staccato which I well remembered was still there. So much change! I imagined for a minute the life this man must have had all these forty years. There were so many old men and women. I stared at their unshaven faces, in their sad impressionless eyes, in their twisted grimaces ... they moved one after another ... In my sleep during the first night their faces haunted me as ghosts. I felt that I was in a land of apparitions. All along I saw men who did not come near me but I noticed them staring at me from under their looks. I read resentment in their expressions. I was told they were the Communists. When I was surrounded by well-wishers, a uniformed man drove in very close on a motorcycle, pretending to be doing casually his work but stealing looks at me and apparently trying to catch a word. I was later told he was the local militia man, son of a friend of mine with whom I used to sing with in the church.

9. Bulgarian Peasant Women ... or the Women in Black

I saw more women than men. They were all old, all in black, all widows, all toothless. Emerging from behind doors they ran to me. I was accompanied by my entire family, holding hands with my wife. They all wondered how well I looked, how young I looked – twenty years younger than I really was. They were all my age or older than I was, but they all looked twenty years older than me. I did not see much of younger men and women in the village. I could not help seeing them every morning from the window of my bedroom hurrying in their blue work clothes down the street towards the farm. One

---

[1] Taniu the Tar.
[2] Ivan the Spyder.
[3] Slav the Baby Rabbit.
[4] Groziu the Turnip.
[5] Marin the Goose.
[6] Ivan the Potato.
[7] The White Hat.
[8] Mincho the Sour Weed.
[9] Ivan, the Male Goose.
[10] Crazy Deniu
[11] Encho the Devil.
[12] Matthew the Big Fish.

evening I stopped two elderly women passing by the house, obviously neighbors, with heavy hoes thrown over their shoulders. I was curious how old they were. One of them said she was 75. I wondered why she was still working. She tried to explain that she was bored staying home and so she was working in the collective vineyards, cultivating the soil to kill the weeds. She was carrying a big heavy and hoe. Both of them were clutching bundles of branches under their arms. "For the goats", they explained.

I was told that all women worked. In mid-winter and early spring they trudged through snow and mud to the vineyards and trimmed the vines. All women! I saw them in Plovdiv wielding big brooms sweeping boulevard Yordanka Nikolova. I tried to catch one with my video from the eighth floor of my apartment... I thought of the feminist movement in the United States. I recall how many of my students, middle age women, passionately defended the woman's cause and how many thought that women in the Socialist system had progressed so much better than in the United States. I could not help contrasting their claims with the pathetic pictures of the women I saw in the village and in the city. And I saw plenty. I saw woman who were driving a donkey-pulled four-wheeler collecting or carting milk barrels – loading and unloading them, I saw them working in the collective farm trudging in the mud with big buckets or bales of straw on the dairy farm. I shook their hands, I heard their stories... I took their pictures... in their blue robes, in their muddy boots, in their masculine walk... and never forgot the seventy-five year old woman with the hoe on her shoulders. What a marvelous condition socialism had created for women! Everywhere I heard the complaint that the men were all supervisors and officials in the big castle in the center of the village at their desk jobs...

10. Bulgaria: a Country Without Children

What I did not see, or saw very little in Bulgaria, are children and young people. Where had the children gone? The Bulgarians had stopped reproducing? I walked the streets of my village and I hardly saw any children or young people. The school where once I was, with two to three hundred kids, age seven to fourteen, filling the rooms and splitting the heavens with our screems, was closed. I saw a few kids coming out of an evening bus – attending school in the next village. I looked in my neighborhood, specifically the little corner where way back we played ball, and danced horo, and horsed around wildly in the evenings, twenty to thirty kids... now there was nothing to be seen. No young girls, no young boys, no young people anywhere to be seen. You don't see many of them in the cities either, in Sofia and in Plovdiv. After Bulgaria I was in Greece. It was another world – kids and young people everywhere. Not in Bulgaria. Is the nation dying? I heard a political leader – I believe the president of the Democratic Party – delivering a speech at the steps of what once was the Royal Palace, on September 6, saying "The country is dying. If nothing changes, the country will be dead in two years." I was not sure what was he referring to, but in the absence of kids, the country certainly is moving to its death.

I recall that one of the theories of the fall of the Roman Empire was the depopulation factor. I am afraid, Bulgaria is deeply in trouble if my impressions are correct and if things do not change. I read somewhere that the country is in the first place or somewhere near the first place in the number of abortions. I suppose, if I was in authority in Sofia, and if I was really concerned for the survival of the nation, the first thing I would do is to outlaw this wholesale massacre of Bulgarian babies. The nation is killing itself with open eyes. What a picture of decline of a nation indeed! Where once swarms of kids bubbled everywhere, now it is a desolate land of old and aging people. Occasionally I saw a few kids. What a joy! There was this kid, Tsanko, in the old section of Plovdiv, and there was this boy on a bicycle, Peter I believe, at the Palace of Culture, whom I chatted with. To see a child in Bulgaria is a rarity. Indeed Bulgaria is threatened by physical extinction from inside. Socialism has done a good job in digging the grave of the Bulgarian nation.

11. The Best Memories from Bulgaria

But there is something good to be said about the Bulgarian people in general. My wife wanted to visit the country to see the people, to find out what the real Bulgarians are like. She has met many Bulgarians abroad. I had to explain that they were not the typical Bulgarian. The typical Bulgarian in the past stayed home. Only rarely they ventured to seek life in foreign lands. What is abroad is a more adventurous element and the privileged classes of the past. What we found in Bulgaria in abundance, in impressive abundance, is the essential goodness of the average Bulgarian. We met them all. We met the peasants, and we met the people of cities, we met intellectuals, and politicians, we met clergy, upper and lower, we met men and women, and we were overwhelmingly impressed by the goodness of soul and character. The hospitality, the courtesy, the attention, the friendliness, the charm, the humanity we witnessed are the best memory which we took with us out of the country.

12. And the Worst Memories from Bulgaria

Of course there were disappointments. There was much rudeness and discourtesy especially from people who held some official position. Whether it was a telephone operator, or a hotel clerk, or a tourist guide, or a sales clerk, they all seemed to be full of themselves, outrightly rude when approached for information. If you did not quite understand the telephone procedures and asked for more explanations, the telephone operator would half-scream, half reprimand you and angrily answers you. If you asked a sales clerk for an item, it is thrown at you and God help you if you want to see another model. You don't ask a tourist guide for something if you do not want to hear a wisecrack behind you. But stop any one on the street and ask for information or talk to anyone on anything, you will see so much courtesy, so much compassion, so much readiness and humanity that you are delighted to be among such kind people. Ordinary people will go beyond themselves to accommodate you. We met so many of them, everywhere, from the upper to the lower and the lowest class. And it was not all because we were Americans.

13. Bulgaria in the Stone Age

ECONOMICS was the key question which came up in all discussions wherever we got involved with people – taxi drivers, peasants, professors, chance meetings. After forty years absence from the country a returnee is faced with signs of economic depression at every turn. And if one happened to be travelling from Bulgaria into Greece, as was our case, one is appalled by the contrast between Bulgarian Socialism and Greek Capitalism. Economics is the most disappointing aspect of the socialistic reality in Bulgaria. Looking back, I cannot free myself of the contrast: the abundance of goods in Salonica and Athens, and in Seres – a stone throw from the Bulgarian border – and the extreme poverty of goods on the Bulgarian side. One has the painful realization that economically Socialist Bulgaria – it is still a Socialist Bulgaria where economics is concerned – has returned to prehistotic times. One does nor need the statistics of those who defend socialism. The Socialists (former Communists) still defend it in their paper as the greatest achievement in the history of modern Bulgaria and make every argument in defense of it.

The Bulgarians call Bulgarian statistics – stakmistika (something that is concocted to deceive you). One need not study statistical charts of critics of the system. One need throw one look only over his immediate surrounding, no matter where – on a street of Plovdiv, or Sofia; in the fields where his road takes him through, in the yard of a TKZS (The Bulgarian Kolhose), in the windows of a shopping street, or in a farmers market. Everywhere one sees the economic collapse of the country. It is a devastated land, far worse from what it was prior to the war, or even in war times. When I was sharing my impressions with a lady in America, born and raised in Seres, she was astounded to hear the story, recalling that in old times Bulgaria was considered to have been a rich country, while Greece was the poor land. All good things in old times used to come down from Bulgaria, she said. Now it was the other way around. A fellow traveler on the bus to Greece, a Karakachanin, confided to me that as an agricultural worker in Greece he makes in a day what he would have made in Bulgaria in a month. The wealth visible in Greece all around suggests that Greece may buy off Bulgaria a hundred, a thousand times. And if one need to compare Socialism to Capitalism and needs to illustrate his conclusion with examples side by side, one just has to look at Bulgaria and then look to Greece and let the facts speak for themselves. Socialism has indeed devastated Bulgaria economically. This Karakachanin, an illiterate middle age man, told me the whole story.

14. Bulgaria and America. Price Equivalents

The question of equivalency of pay in Bulgaria and in the United States came up too many times. They all wanted to know how much a working man makes in America. I just pointed to what my plumber charges me – $34.00 an hour. It did not take them long to compute it in Leva – 612 lv., a two month salary at the time of my visit, or more. I treated on every occaison friends in top restaurants with dinners. They fought to pay the bill. I would not allow them. When they saw the bills, they were appalled. A three hundred leva for a lunch or dinner was an entire month's salary for them. For me it was a mere $15.00. There were so many embarrassing moments for them when they felt that their hospitality required them to pick the bill and I had to reassure them that I was in a good condition to afford it. Many times I had to take taxis. Where in the United States for a similar ride in New York I would have to pay $10-$15, in Sofia I was paying the equivalent of one dollar, and that was doubling the charge on the meter. Imagine, what a deal it is to make your fortune in America and spend it in Bulgaria. A man with a Social Security check of $250.00, the minimum pension in America, could live like a king in Sofia... if he could take it. My friend, Bai Dimitar, is in that state. I provide him with a free lodging, and he helps me here and there, I suspect he would not be able to stay there even with such a state of the equivalency of the Bulgarian money of the country for whatever purpose.

15. Production of Goods. The Achilles' Heel of Socialist Economics

So, this is where Socialism has taken the country economically. I was well aware where Socialism had failed most. It was in the production of goods, not in the system of distribution, defective as it may be. It all depends on the availability of goods, and goods become available either by being produced in abundance in the country or being imported. But to be able to import them, one should be able to export goods to obtain foreign exchange. Failing to produce goods, the system turns into a total failure. And you see the lack of goods in Bulgaria at every turn. You walk the streets of Sofia and Plovdiv and you see signs like "Fruits and Vegetables", but you see no fruits and vegetables. You see signs for dairy products and you see no dairy products. You hardly see any windows with men's and women's wear. Hardware stores are nowhere to be seen. Bakeries with steaming goodies are nowhere to be seen. All you see is rows and rows of dusty windows never washed, with nothing to show, but garbage filled niches. If you see anywhere anything exhibited, it is of such quality which outside of socialist Bulgaria will be thrown on the garbage heap.

16. Free Market ... or Flea Market

Walking up the "6 September" Blvd. In Plovdiv we by chance ended up in something like a farmers market, or a free market. There was much *flea* market stuff here and there, mostly empty tables. What we saw there on sale – fruits and vegetables – was below any standard sold on the western market and would not be touched even by the least discriminating shopper in America. What appeared, however, in abundance there, piles of it, was sunflower seeds. So much sunflower seeds! Many of the merchants had placed small portions of it in small funnels. Why so much sunflower? Was this the substitute for food? But even this was of a substandard quality. Probably 60% of the seeds appeared to be empty. Piles and piles of it. In Sofia we visited the TSUM – the central department store, the Sofia equivalence of Macy's. It was mobbed. People were running up and down the escalators mostly empty handed, browsing all over, looking for things. There were some things, but again, the quality was substandard. What I saw in the clothing sections – this was actually in all sections – looked like a store going out of business. Everything for sale looked like left-overs – from clothing to electrical supplies. I needed an extension cord to recharge the batteries of my video. The clerk told me that they have not had such extensions for three months. Wow!

17. The Lines in Bulgaria

Much has been said for shopping lines in Bulgaria … forty-five years after the war. In the rest of Europe such lines are a forgotten thing. But in Bulgaria they are pretty much visible everywhere. People were patiently waiting for one thing or another. For bread … of all things! After so much "progress" in socialist agriculture, people had to present ration cards for bread. I saw these cards in the village. The bread is trucked in from a distant town. Every evening about 5:00 p.m. the truck arrives and the whole village waits for their daily bread. Sometimes the truck may not arrive or be late. People wait patiently in the square. When someone waits at home and asks early birds about the length of the line, he inquires if the line is a "rabbit's tail" or "fox's tail", meaning short or long. When the turn of a man on the line comes, he goes to the small window at the distribution shed, produces his ration cards, pays, and gets his loaves of bread. A lady from inside pushes the loaves to him and there he goes to dinner with waiting family. I videotaped the scene.

Watching the event I asked why, for God's sake, now, that they are liberalizing things and allowing free enterprise, someone does not open a bakery for bread and bread products? I was told that a local Secretary of the Party has declared to those who had inquired: "For as long as I am in this place, there will be no private bakery." I was told, however, that lately the quality of bread had improved. There was a time when the bread was either half baked, or of such poor quality that it was inedible. I did not see the situation in Plovdiv or Sofia any better. No products for breakfast – thinking of hard rolls or croissants… Forget it. I ate at Hotel Leningrad in Plovdiv and the Sheraton in Sofia. The pleasant luxurious restaurant in the latter was not up to par where food was concerned. One could dream of breakfast in Vienna, or Geneva, or Rome, or Paris and London, not to forget Salonica and Athens. This is a dream of fantasies if applied to Bulgaria.

I mentioned the farmer's market in Plovdiv. There was a long line at the corner of the street. At the head of the line a truck had opened its back doors. It was loaded with eggs. It was probably a private enterprise. Hundreds of people were lining up there. Near the Orlov Most I saw another truck, on Dragan Tsankov Street, parked and unloading enormous carcasses of beef. The line was forming there too, waiting for the distribution to begin. We saw the bread lines everywhere in the city.

18. A Promenade in TKZS (Bulgarian Kolkhos – Trudovo Kooperativno Zemedelsko Stopanstvo – Agrarian Cooperative Labor Farm)

The backbone of Bulgarian economy is agriculture. And the backbone of Bulgarian agriculture is the TKZS. When I fled the country the Bulgarian village was in the midst of collectivization. It was an outright imposition, forceful collectivization. The present "socialists" try to claim that it was not all done by force. Those who remember these days know otherwise. I did not have a chance to visit many TKZS, though I saw them from the road outside of every village. But I visited the one in my village – not to gather information, but to see where my brother works. What I saw was nothing to talk about. All over I saw abandoned broken machinery, rusting and fading in weeds. No less broken and seemingly forgotten in the world were the few shadows of men and women tending the farm. I met them all, I shook hands with them, I heard their lamentations, I pitied them. They were dirty and filthy, unshaven and muddy, worn out and burned out. They were trudging in mud and broken grounds in something that was supposed to be internal road. And there were weeds, and weeds, everywhere, overgrowing and giving the place the drabbiest look. The cows did not look in a very good shape either. And so were the young calves. Though not appearing as starving, there was much to be desired.

In times past the village was packed with cattle and sheep. And the environs of the village had plenty of room to graze the animals. Way back I tended sheep, over a hundred of them, of my neighbors. One year I tended the cows of the neighborhood, from early spring to late fall. Now they had destroyed the natural pastures and keep the cattle inside on a constant feeding. How could they succeed? The surrounding fields of the village were pastures for sheep. Once they were clean and pleasant places. As kids we played here in freedom and enjoyment. There we took the lambs for grazing. Now all this was gone. It is all broken and useless ground, overgrown with weeds. And so overgrown with weeds are the winter crops. For as far as one could see, the vast fields were ruminating in red wild poppies. One wonders whether they were going to reap poppies or wheat and barley. And so also looked the vineyards of which they have planted so much, overgrown with weeds, drabby and uncultivated. What one could expect if the work is left to 75 year old women? You travel through Italy, France, Greece and England. This is what I did a year before and my eye was caught by the cleanliness, the orderliness and the prosperity of the fields in these countries. When looking at the pathetic state of Bulgarian agriculture one wonders how the country will get out of the present crisis. All you see in Bulgarian fields is weeds, and weeds, and weeds… and broken and abandoned machinery… and vast fields of poppies. Once Bulgarian agriculture was the pride of the nation. Once Bulgaria was exporting grains, and fruits, and vegetables. Now they do not have enough for themselves. I was told that the fruits fall from the trees on the ground and rot because there are no people to pick and transport them for processing. What socialism in agriculture had brought to Bulgaria! Bulgarian agriculture presently is in ruins. It is sick, and sick, and sick.

19. Animal Husbandry in Bulgaria. The Lost Paradise

In old days my village was full of livestock – sheep, goats, cows, oxen, donkeys... There were so many shepherds – tending fifty or a hundred or more sheep, taking them to graze in the morning and bringing them back in the evening, sleeping in the fields in the summer. One summer I was the shepherd of the neighborhood sheep. Another summer I tended the neighborhood cows and calves – some fifteen to twenty of them. There were also the swineherds, the goatherds, the donkeys and their attendants. Every morning the village of some two to three thousand people, was a pandemonium with all kinds of animals crowding the streets. Socialism put an end to all that by establishing the collective farms, gathering all privately owned animals into one village joint enterprise. This has led to a virtual disappearance of animal husbandry in Bulgaria. The number of animals of every breed has fallen to the level of extinction, and has crippled the national economy. And they wonder why they are in such a crisis. But this all has changed the profile of the village environment.

They have destroyed the pasture fields around the village and have uprooted the forests to plant new vineyards. The fields surrounding the village, which in old times, from early spring to late fall, served as pastures, clean and pleasant, are now turned into a desert, overgrown with weeds and scatterred rusting metal skeletons of abandoned machinery, a million gaping holes and ugly thistle. I climbed my favorite Thracian mound where in old times as kids we have played and run around to our heart's content. The view from the top, for miles, around the village, was a view of a deserted place. Four decades ago all this was a pleasant and beautiful plain, leveled, clean, covered with close cropped green grass and full of life. In July and August those lawns were turned into threshing fields. In spring time we grazed the lambs there. In the fall we grazed the sheep in these endless fields. Now it all looked like deserted lands. Next to these fields now is the collective farm. I made my acquaintance with the cows and the little calves. I felt sorry for the little babes stuffed in small cells. Where was the old freedom, and beauty, and joy? Only Sakar Tepe was soaring above the village as a silent witness of the village pain and misery.

20. Sakar Tepe

Sakar Tepe! It has become a part of me. I could not have enough of it. For two decades when I was growing, from the days I do not remember, until my early twenties, I have stared at its peak, from early morning to late evening, standing at the window of the family living-room-bed-room of our house falling apart from age and poverty. Sitting at this window, learning to read, and to write, studying and dreaming my youthful dreams I stared at it in early morning hours waiting for the sun to rise above it. How many times I had climbed to its top, proud of my achievement and delighted to embrace in my gaze the vast Plovdiv expanses unfolding in front of me. How many times I have sought it in the distant horizons from the top floor window of the Seminary and from the hills of the city? How many times I have remembered it in my nostalgic days in foreign lands? God willed that I could see it again! So magnificent, so proud and so much the same! What a sense of victory and what inner delight I experienced in old days, running up its slopes to reach the summit! Now, standing in the middle of this desert, after forty years of wandering in the wide world, after attaining many peaks in my life, I was looking at Sakar Tape again, as if I was meeting my dearest friend from the past. Instantly a thought flashed in my mind: Maybe it was this peak, maybe it was the same excitement and pleasure which I experienced in old times trying to reach it gave me the strength and the energy which supported me all these years of exile torment! Who knows!

21. Bulgaria – a Wild Field of Weeds

In Zelenikovo, in fact throughout the country which we travelled by car, I saw fields of winter planting, but in too many places I could not decide whether they had sowed wheat or wild poppies. Endless fields were ruminating with wild poppies, as if they have been deliberately planted there. What crops are they going to reap out of this wild field? One can only imagine. Where the wild poppies had not taken over, the wheat fields were lying on the ground, apparently from excess of water. Vast fields in my village, once grannary of the province, have been turned into vineyards, but it was difficult to see the vines among the overgrown weeds of an abandoned field. Some forty years ago someone told me in Greece, that the Bulgarian vineyards look like soldiers in a parade – lined up, clean and cultivated. Today this visitor would say that our vineyards look like a raggedy gypsy mob. Indeed, the entire country looks like a raggedy mass of people. Everything seems to be vanishing into weeds and abandonment. What else could be expected when all the work is left to seventy-five years old grandmothers? You travel across Italy, France, England, Germany, Switzerland and Belgium – not by plane – you cannot see anything from the air, but by train or bus (This is how we traveled in Europe, for this reason, last year), you observe and then you compare what you have seen there with what you see in Bulgaria ... and your soul start suffocating, you feel pains for Bulgaria. To what a brink of disaster have they led our once beautiful country? You think of the orderliness and the wealth of other countries, you stare at the pathetic picture of the land in Bulgaria, and you begin to wonder whether she would ever come out of her present pathetic state. There was the time when Bulgarian agriculture was our national pride. Now it seems to be entirely ignored more than any other aspect of national life. There was the time when Bulgaria was exporting grains, and vegetables, and fruits. These days the country hardly produces enough for its own needs. A decade ago I was reading the speeches of Todor Zhivkov where he was boasting that in a few years Bulgaria would produce so much food that she could feed three Bulgarias. As of now, she does not seem to be able to feed herself. I was told that much of the crop cannot be gathered in the fall and that fruit and grapes are left to rot in the fields – for lack of workers and for inadequate transportation.

436

22. Ideological Foundations of the Economic Catastrophe of Bulgaria

This is what socialism has done to Bulgarian agriculture. Some forty-five years ago the Communist Party introduced Socialism, defending it as public ownership of the means of production. Every available means of force and violence was used to dispossess the peasants and push them all into the collective farms. This was accomplished. And it is this kind of Socialism which killed Bulgarian agriculture. Interestingly enough, in their self-flagellation now, they never mention this fundamental error of their socialist economic philosophy – public ownership of the means of production – as the root cause for the economic catastrophe of Bulgaria. Prominent philosophers of the Party today continue defending this principle of their economic program. They see the merger of the Party and State authority as the root cause of the economic ruin of the country. This is what they call deformation of Socialism and charge it all on specific individuals, led by Todor Zhivkov – not the system of public ownership of the means of production. In their endless befogged speculations on market economy, which they now see as the magic trick which would lead them out of the crisis, it becomes clearer and clearer what they consider to be the right way to come out of the crisis in the country – not by liquidating public ownership of the means of production, but letting the free market determine the prices. This way, the collective farms controlled by the nomenclature and all state enterprises will preserve their socialist character – in one form or another – but will sell their goods on the free market. At the same time anybody willing and able may start his own business. This way they will continue holding under their control in a feudal manner the weakest unorganized elements of society, using them as slaves, to vote for them in the elections and to maintain Socialism in Bulgaria.

On all other matters they – present day socialists, yesterday's communists – are in agreement, in words and deeds, with the opposition parties on all measures proposed in Parliament, and create the impression that they have changed. They are now for free elections, they restored all liberties in Bulgaria, they voted the law of the land, they even voted for a referendum on republic or monarchy. All along, they were convinced that as long as their doctrine of public ownership of the means of production remain intact – in theory and practice – all concessions which they have made will amount to nothing; that at the ends they still hold the economic levers of power in the country, and as long as they hold these levers, they will control political power. Their play with democracy until now is only a farce for as long as they do not undertake a drastic, immediate and total dismantling of the economic system of Socialism – in theory and in practice – liquidating public ownership of the means of production.

23. Confusion in the Agrarian Policy of the Bulgarian Socialist Party

I met an old friend, a communist. He is an agronomy professor in Plovdiv. I asked him to give me his assessment on the question why Communism failed. He delivered a long windy tyrade, and without answering my question tried to tell me that Stalin had come to the conclusion that socialism was a failure and was on the way of scrapping it… but then unexpectedly he died. They know that the system failed and are looking for way to get out of it, at least not to be blamed for its survival. Yet, they are convinced that it is still a good system, that it is going to work, that it is better than private agriculture, but have no argument to defend it. They see no alternative to it and while giving the impression that they are for its dismantling, they still cling to it. This is why they voted for the law to return the land. They know that this is not going to work. How is this going to work when the basic structures of private agriculture are destroyed and cannot be revived? First, there are no people in the villages who will go *en mass* to claim their land. Who is going to work on that land, who is going to plow, to cultivate, to sow, to collect the fruits, to reap, to thrash the crops? The old men and women who are now retired? Or all those who have fled the village and one way or the other have settled in the cities and made their life there? They are going to go back? And how many of those who joined in TKZS have abandoned agriculture forever and ever? And how many are going to opt to sell the land to the cooperatives? And where anybody willing to buy land will find money to pay for it? Where the agricultural machinery is going to come from and how is it going to be paid for? A return to the old private agriculture is a mammoth job for which the most important element is no longer there – the human resources. Sure, there will be those who are going to claim and obtain land, but over all, it will be a small, very small minority, and it will not change the profile of Bulgarian agriculture. Which leads nowhere but to the consolidation of the old Communist nomenclature in agriculture, where the BSP draws its strength to win elections. The law to r e t u r n the land carefully provided for the preservation of the present organization of agriculture in the form of cooperatives as one form of agricultural enterprise. Whether these cooperative farms will be organized as private enterprises, and when, and by whom, and how – is still an open question. As of now they are firmly entrenched and serve as a base for operation of the Communist-appointed nomenclature. It is not surprising then why the SocComs so readily agreed to vote this law. Given the present state of the Bulgarian village – depopulated, childless and aged – where people, because of their low intellectual and social experience are least able to question the nomenclature, accustomed to cling to the very little the SocComs could give them, and therefore easy to manipulate, the communists always will be able to insure their dominance on the political scene. Under this law the majority of people will opt for selling their land, or abandoning it to the TKZS Communist-controlled cooperatives, or simply go to waste.

24. The Old, Small Village Farm has no Future in Bulgaria

The land reform in Bulgaria will not work. Two overriding factors militate against it: 1. The need for small agricultural machinery and fertilizers. They could be supplied only from the West, and then only free of charge, because there is no foreign exchange and the peasants could not pay for it. Only some sort of a Marshall Plan could meet this need; and 2: The need of working hand, of peasants who would engage fully in agriculture. There is no Marshall Plan to meet this need. The young have left the village and have settled for city jobs where they are paid salaries and hold nine-to-five jobs. The village could never compete with it. They will not return to cultivate the land – in collective farms or in small private farms. The small private farm is no longer a viable alternative. It is a thing of the past. It is gone forever. The future farm in Bulgaria must be income producing enterprise, not self-sufficient unit, as it had been for millions in the past. Of course, there will be thousands of farmers, who, in small ways, will try to ekk out a living, and some will be successful, but this will not solve the problem of national agriculture as a competing economic sector on the international market. Reviving of agriculture in Bulgaria will come when the city virtually is reduced to starvation and a wave of migration back to the village materializes. Otherwise, things are going to linger for long time in the twilight of ebb and flow marginal existence.

It is worth noting that the problem has become very accute in the last years of Communism when they had to suspend classes in high schools and universities to send swarms of students to collect crops in the fields. Their attempts to retain youth in the villages had dismally failed. I heard some of the methods used by the authorities to keep the youth in the village, especially young men and women of unreliable political orientation. A young man whose story I heard, discretely took courses to qualify for a city bus driver and obtained a job in Plovdiv. When the authorities in the village learned that, they used their political clout to have him fired. His mother went out fighting for him. She went as high as the Minister of Transportation, who intervened and the young man was reinstated in his job. Similar stories I heard from others. There was this case where the parents would not let their daughter go and work on the pig farm. When told to send her child there, the mother, who herself had worked for years on that farm, told them that she will support her daughter for life, but would not let her work there. And so, the depletion of the village farm of work hands is an ongoing process and – with the extinction of the present working generations – the Bulgarian collective farming will be even in worse condition. It will depend on the least educated, the least mentally, socially, and physically fit, not able to find other work. Until, and unless farming in Bulgaria, private or collective, becomes a lucrative occupation and attracts people from the cities, it has no future in Bulgaria. And agriculture being the backbone of Bulgarian economy, will drag on for some time, and it will drag the country down the drain.

25. Economics and Politics

It is politics, not economics, where national attention is focused today. The problems of economics and their solutions are seen as political problems. In this respect for as long as the SocCom party controls all levers of power, the economy will go in one direction or another, by one speed or another, as the SocCom Party allows it. Therefore it is the political solution which is the key to the overall recovery of Bulgaria. But politics, too, like economics, is going nowhere. Most certainly, things have changed since November 10, 1989. There is an unquestionable explosion of political freedom – freedom of the press, freedom of speech, freedom of organization, even though the social controls are in the hands of the SocCom Party are effectively limiting the access to this freedom for the broad masses of the population.

25. The Terror of Fear

During my stay in Bulgaria the country was seized by a great fear that the SocCom Party was preparing a general massacre of its opponents. I heard this from many people of different social status, intellectual level and political orientation. I had to reassure them that such a course of events is out of the question, that the Bulgarian Communists are so cornered, now, that their leaders would under no circumstances allow a new red terror to take place. They are trying desperately to make a good impression in the West in order to qualify for help to pull the country out of the catastrophe. They would do nothing to spoil their chances. I believe that to be the case. However, I was under the impression that they themselves, the Communists, deliberately fan such fears. Time and again, I read in their paper *Duma* references to danger of bloody confrontations in Bulgaria – as if such rumors circulate in the country – with warnings that they, the SocCom, would make sure that if such a thing is initiated by the opposition, they would meet it with firmness and determination. Thus, by maintaining such fears in the people, they let everyone make – up his own mind – to come out against the Communists or keep quiet in his corner while they battle the opposition on the political front.

27. Freedom of Press in Bulgaria

There is freedom of the press in Bulgaria. If the former Bulgarian Legionairs, could publish their paper, *Prelom*, and have their organization, "Democratic Forum", then freedom of the press in Bulgaria, and freedom of organization and parties is beyond question. Such freedom did not exist neither before, nor after September 9, 1944. However, it is a different question how far the free press reaches the people. The opposition papers in Sofia are sold very early and late comers cannot find them. It is even worse in Plovdiv. There are no newsstands in the village and the only way to get hold of a paper is through the mail. Who would dare to subscribe to *Demokratsia* when the *nomenklatura* would know it immediately, and it controls the bread? The only paper I saw in the village reading room, in the *Chitalishte* of Zelenikovo, was the SocCom organ <u>Duma</u>. When I inquired why the other papers were not available I was told that for lack of funds they could not subscribe. I donated $50.00 and indicated which paper they should subscribe to – *Demokratsia*, <u>Ze</u>m. *Zname*, *Narodno Zem. Zname*, *Svoboden Narod* and a few others. The next day the entire village knew that I had been so generous to the *Chitalishte*. But if this holds true for the rest of the country, then free political education is still far away from the people. Their only source of information then is the Television, which I had very little chance to watch and appreciate. Another way of limiting the access to opposition papers is the price which the government charges for paper. As soon as a free market for paper was declared, after SocCom party had secured its papers for years to come, the price jumped astronomically, and in a short while it will obliterate whatever papers do not march to the tune of the party.

28. The Bulgarian Socialist Party: "Le plus ça change, le plus ça le même chose"

The political stage in Bulgaria is dominated by the Socialist party - the old Communist party under a new name. It has the same membership as before, though half their numbers. But it is the same organization, with the same leadership, though somehow rejuvenated. This party condemned its own past and past policies as having been saturated with deformations, but it still claims credit for the past, under its old name, and still defends the regime overthrown on November 10, 1989. The errors and the deformations are attributed to half a dozen people, while the Party is exhonorated and is placed in the first place among all other parties. Especially they emphasize the struggle of the Party against fascism. They never miss the opportunity to take pride that they have been shoulder to shoulder with the United States, France and England. In this respect, September 9th is glorified as a victory for the democratic revolution. They also claim credit for pulling Bulgaria out of the position of a backward country and lining her up with the advanced states. All evils in Bulgaria are ascribed to a few individuals. These individuals having died, or the statute of limitations having expired, then it is time to forget it all. It is the future they are interested in, not the past. They are making every effort to claim the heritage of the founders of the Socialist party in Bulgaria, pre-Leninist socialism, which rightfully belongs to the Social Democratic Party. The violent nature of Communism they attribute to Stalin and Stalinism, but they ignore the fact that they practiced the Stalinist violence in implementing the basic doctrines of Socialism. Had they ventured to confront history and ideology, they would have been forced to account that the core of their errors and deformations is in the ideological and philosophical underpinnings of the system of Socialism, not this or that individual. But, even without admitting this, they no longer openly advocate this principle – "public ownership of the means of production". But they have never renounced it.

29. Free Market: The Bulgarian Model

Publicly the new Socialist party is committed to free market economy, but nowhere has it spelled out the mechanics of this free market economy: are the goods to be marketed going to be produced by private enterprises, or by publicly owned enterprises? The most that the Party has done is to express its preference for a mixed economy – publicly owned enterprises in industry and agriculture, and privately owned business, on equal footing. But, the Party stands for publicly owned enterprises, while allowing for those who want to start private industry and agriculture to go ahead. As a consequence of this policy the Party is not particularly interested in pushing the private sectors, and is still wedded to the socialist sector. Since they hold all the levers of power to speed up or delay the dismantling of socialism in Bulgarian economy, they are not budging to advance the private sector in the country. The result is that Bulgarian economy continues in irresolution. Since the power of the Party is dependent on its power over the economy – through the economy they control the state and the nation – they will not move a finger to dismantle socialism and the reform in Bulgaria will remain a dead letter. Their professions for reform are only lip service, and their commitments to market economy are commitments to a free market of goods produced in socialist enterprises. The competition in the free market they see as competition between publicly owned, state controlled production collectives. They do not see that a free market economy and socialist production of goods are as irrelevant to each other as light is to darkness. Either the one or the other will be defeated. In the final analysis, it all ends up as a farce. The biggest loosers in such a system will be the workers, because, being the weakest social class – factory and agricultural workers – they will find themselves enslaved again, in fact, never liberated from the feudal power of the socialist production collectives. It is exactly because of this confusion of terms and ideas, Bulgarian economy presently is in shambles, the reform in Bulgaria is deadlocked, the country is fast approaching an economic collapse, and no prospects of change are on the horizon. This is the policy of the Party. Party economics is Party politics and Party politics is socialist economics.

30. Nothing New in the New Economic Philosophy and Practices of BSP

Two years since November 10, 1989, the reform has not made a dent on the system of economics in Bulgaria. Peripheral, trifling private initiatives, are undoubtedly a fact, but they have amounted to nothing of substance. The communists have it in their power to initiate a drastic change, a rapid transformation of Bulgarian economy from socialist to capitalist lines. It is in their power to do so. But they stay put. If this attitude they take now is dictated by the consideration that the worse things turn out to be, the greater the nation's disappointment of democracy would be, they are badly mistaken. Because the worse things turn out to be, the sooner the upheaval of the nation will throw them out of power.* But, at the same time, if they opt for a drastic transition to capitalism they will lose their power base in the economy and with that all gone, they will be out of power in no time. It is a catch 22 situation – damned if you do, and damned if you don't. If they lose their control of the economy in a capitalist economic organization, the vast nomenklatura which they now control by economic means, will vanish from their support lines. This nomenklatura is in its overwhelming majority opportunistic and will serve anyone who holds their life sustenance and privileged position. Now they support the communists, fearing that otherwise they would lose their jobs and privileges. Without that threat, this opportunistic crowd will be glad to free themselves from the oppressive undignified slavery to the Party which it has suffered for decades. The Socialist leaders knew that and they know that once they let things slip out of their hands, they will be gone forever. So, the establishment, the socialist establishment, will not move to disestablish themselves.

31. The Political Power of BSP Continues Intact

As to the political power, the November 10, 1989 palace revolution left it intact in the hands of the Socialist-Communists. Prior to that revolution they argued for the "guiding role of the party." Now they argue that in the context of democratic reforms it would be a partisanship to seek personnel changes; to dismiss the army of party appointees and replace them with anybody else is violation of the principle of democracy. Ergo! The nomenklatura continues in power. The entire Communist staff of the administration, the economic institutions, all public and cultural agencies, receive the status of civil service. The opposition cannot respond to this argument, and is otherwise powerless to effect any change. Very soon they were satisfied by a few bones thrown to them by the Socialists – the presidency of the republic was one of these bones - and they accommodated to the power of the ruling Communist party. They divided the spoils and simmered down. But, even if this was not the case, the opposition parties, if given the chance to restaff the governing apparatus, do not have the human resources to replace the vast administrative machinery with non communists.

At end, the bureaucracy will be left in the hands of the same people who by hook or by crook have the advantage that they now stand as protectors of the positions held by every clerk and every official of the governing class. In fact, they do not conceal this advantage and consider it as their obligation to hold power in order to protect the beneficiaries of the regime. These beneficiaries have faithfully served the regime of Todor Zhivkov, they have been raised in the Komsomol and the Party, or have been recruited from the transmission-belt organizations. It does not matter whether they believe or do not believe in communism, or in the mission of the Party as it styles itself now as Socialist. They have been comfortable under the regime of Zhivkov. They are comfortable now, and they

---

* As it really happened on January 10, 1997.

would not buck the Party which guarantees their jobs. They do not want the situation disturbed for something that is called democracy.

32. Bulgaria and the Story of the Sinking Titanic

All this means that we are going to have more of the same. The country may go down the drain. The nation may be driven to disaster and the abyss. But they are not expected to move to save it if saving it by toppling the Socialist party means threatening their position. The ship is sinking, they are sinking with the ship, but at least they have the advantage of being one notch above the rest of the people. The rest of the people will sink one instant earlier than they themselves. Bulgaria now is like the Titanic. It is doomed, and while the doom is coming, the captains will go to the bottom with it. They do not see that they too may have a chance if they release the levers of power, that Bulgaria too may have a chance, but they are determined to cling to the end, hoping that something may happen and they will be saved. What they need now, what Bulgaria needs now, is a second revolution, a second push to throw them out.

They, the communists, read the omens the same way as everybody else. The omens are that the opposition is loosing momentum, that it is turning against itself, and that the coming elections seem more and more winnable for them. There is no chance for a second revolutionary push. The opposition with its wishy-washy leadership of 1989 days missed the opportunity to topple communism. Subsequently, it was either bought with the presidency and a few glamorous but meaningless positions, or manipulated to serve inadvertently the cause of the Party. The Socialist Party is firmly entrenched now, and there are no chances that it will be toppled from power soon … while the ship of Bulgaria is sinking to its doom, and then, God knows who will rule in Sofia – Turgut Ozal or some regent from Belgrade.

33. UDF (Union of Democratic Forces): an Example of Political Amateurism

Side by side with the Communist Party is the UDF (the Union of Democratic Forces), seeking to replace it as a governing party. This Union was formed in the wake of the November events. A few intellectuals, fallen out of the then Communist Party, and a few remnants of the old Socialist and Agrarian Parties, joined together as an opposition group. To protect their leadership and exclude the rest of the Bulgarian political world, they reserved for themselves the right to veto admission of new members. Then they were engaged in protracted round-table negotiations where the communists outmaneuvered them in many ways. They went to elections controlled by the Communist Party and lost them. They could not have won them. Then they started bickering among themselves. The core political parties, Social Democrats and Agrarians, seeking to perpetuate themselves in power, formed, what they called, "U.D.F. Center". The multiple groups of intellectuals, seeking to build themselves on the back of the former, tried to establish themselves as a separate party – "UDF – Movement". The Communists bribed the intellectuals by giving them a slice of the bacon, including positions on local level, thus isolating the politicals. Then the communist socialists derailed the Grand National Assembly into bottomless legislative interests and obscured the real purpose of its mandate, to prepare a constitution. In the process the internal crisis in the UDF became even more acute. It was politically fragmented and ideologically pulverized. At the present there is a see-saw struggle between the two tendencies; the political and the intellectual wings of UDF. They are desperately trying to stay together at least until the next elections, but it is doubtful that they will be able to hold together. Irreconcilable political tendencies are shaping up in the background, which, if allowed to take the upper hand, will divide the opposition and will benefit the Communist Party. The politicals are definitely republican, historically tied to the old fatherland front. The intellectuals and the second and third rank UDF forces tend to be royalists.

34. UDF: Monarchists vs. Republicans

The political and ideological evolution of UDF slowly but surely leads to differentiation of the republican and the monarchist forces inside it. On the one hand are the Social Democrats and the Petkovist Agrarians, representing the republican wing, while the Democrats and a number of other small groups stubbornly promote the monarchist cause. Behind the attempt to transform the UDF into a movement were the monarchist forces, who sought to take the leadership of the opposition. Dr. Trenchev, President of the Labor Union *Podkrepa*, and Rev. Christophor Subev, parading as representative of the Bulgarian Orthodox Church, which in fact, has nothing to do with him, lead the monarchist forces in the country. If this process continues, there will be three factions on the political stage. The Socialist Party, the Communists, will occupy the left. The center will be occupied by the Republicans, the BANU and the Social Democrats (United), while half a dozen to a dozen of small parties will take their place on the right for the monarchy. Some of these parties have old names, but new leadership. Others have proclaimed themselves as parties only recently, without any following in the masses. To these groups belong the Democratic Party, the Radical Party, the Liberal Party, the Conservative Party and half a dozen of societies and clubs, including the Democratic Forum of the Legionairs and a small group of Gichevists, all or some of them gathered in the "Svoboda" group, outside the UDF.

35. Who Destroyed the Bulgarian Political Parties

In this respect there is something to be said about the political parties in Bulgaria. One of the prime specific political objectives of the Communist Party in the 1940s and throughout the regime of Todor Zhivkov was to squelch all political parties. The parties outside of the Fatherland Front were outrightly outlawed and castigated as fascist by a special agreement, signed, among others, by Dr. G. M. Dimitrov and Traicho Kostov. Next the FF parties and groups were dissolved and absorbed by the Communist party. Only the BANU survived under its name but was transformed into a peasant section of the Communist Party. The memory of the old parties was obliterated from the mind of the growing generations of young Bulgarians or was cast in the darkest of colors. So, when the November events occurred, there was no sign of the traditional political parties to be found. The brutal physical and intellectual suppression of the political parties was one of the achievements of the Communists.

In parallel with this Communist persecution and suppression of political thinking in Bulgaria, the propaganda agencies in the West, Radio "Free Europe", the "Voice of America" and all other second rank informers of the Bulgarian people, adopted a misguided policy of rejection of any party political organizations or associations. They mistakenly proclaimed any reference to political parties and work with political parties to be partisanship and they would have nothing to do with it. They promoted non-party people, non-party activities, non-party policies. Thus, instead of preparing a team of political leaders, with defined political ideologies, programs and staffs where Bulgarian political thinking would have crystalized, ready for a moment when it could take over in Bulgaria, they prepared political eunuchs unfit for any role in a situation with restored political freedoms. When that time came in November, 1989, Bulgaria was found in a political vacuum – with nobody prepared to take over or to guide the political evolution along established political lines. Since there was no political leadership steeped in political ideology and political traditions, the vacuum was to be filled by intellectuals hatched in the Komsomol – utterly ignorant of political traditions in Bulgaria, ignorant of political events of the past fifty, or rather seventy years, and burdened with the political baggage of the Kommunist era, not knowing if they are still in the era of socialism, or have entered the age of democracy.

A number of old parties were resurrected, and a number of new parties appeared as offshoots of old movements and parties, but so far they have shown themselves to be only amateurish initiatives, weak and pale images of old traditions, with little, if any, resemblance to old veteran parties. Sometimes, most of the times, they all are so much alike that the observer gets the impression that they are twins, their programs and charters written by the same hand somewhere in the background. They all parade under the banner of UDF or of "Svoboda", and they all easily could be merged in one single party, all-national democratic party. Their best chance of success – it would be for the good of the Bulgarian people – would be if they all merged into one single Bulgarian Democratic Party. But, as of now, their fragmentary character will only help the Socialist Party to retain power. The major obstacle for their unification is the ambition of the founders of these splinter groups to act as national leaders. The political groups they have created are, in their hands, only jump boards into national leadership.

Obviously, the way out of this political fragmentation and meaningless diversity in unity is in a system of elections where the political ambitions could be displayed on local level, in some sort of primary elections, where the party banner of a united opposition is won in primary elections according to the majority system. If and when they devise such a system, then a two-party system may emerge and it would serve Bulgarian democracy and the Bulgarian people better than in the present state of political fragmentation and confusion.

The American system of primary elections is a ready model which they could follow. If, and when they choose to follow such a model, Bulgaria may develop a two-party system, controlled from below. Such a system could take Bulgaria out of the permanent crisis which the present system will perpetuate. In the framework of such a system the present Socialist Party undoubtedly will occupy the left position, but, to win votes, it would have to move to the center. Likewise, the new national Democratic Party will have to stay closer to the center, in order to pull votes out of the left center. But this scenario still leaves the question of the BANU open.

36. Wings, Winglets and Feathers in BANU

The BANU, both factions – the Petkovists and the Vulkovists (those who stood with the communists all through 1944 to 1989) – taken together and considered separately, are the largest political party and parties in Bulgaria today, after the Communist Socialist Party. Presently the BANU is split three ways, the former Traikovists-Tanchevists, the Petkovists and the Gichevists. The first two factions have their common origins in the Pladne group, once led by Kosta Todorov and G.M. Dimitrov. This group split in 1945-1946 when the Communist Party moved to take control of the Union. In November 1945 the Petkovists and the Gichevists proclaimed their unification. At that time the Petkovists had already passed to Gichevist positions but still keeping the leadership in Pladne's hands, and still defending the policies of Pladne, as espoused prior to 1944. The Gichevist line differed from that of Pladne in two most important aspects: In foreign policy Gichev held to the end the traditional Agrarian view that Bulgaria should stay neutral in the struggles between the great powers. At the same time the Pladne Agrarians departed from this principle, formulated by Alexander Stamboliysky, and oriented the Union towards the Allies – England, USA and USSR – always under the leadership of Yugoslavia's royal Court. Where domestic policies were concerned, Gichev categorically refused to cooperate with the Communist party, while the Pladne leaders for many long years collaborated with the Communists and at the end joined with them in the Fatherland Front, to topple the Government headed by the Gichevists on September 9, 1944.

After November 10, 1989, the Petkovists quickly revived their Union and became the core party in the UDF. Exercising a veto power over admission of new members of the UDF, they blocked the admission of Gichevists, thus becoming the major opposition force in Bulgaria. The Gichevists, too, attempted to revive their Union, but, for lack of experience, incompetence, ineptness and wrong political moves, were isolated and reduced to a small sectarian faction in the Agrarian movement. It did not take long time for the Petkovists and the ex-Tanchevists to occupy the key strategic positions in the Agrarian movement. The former Tanchevists had already split with the Communists.

Thus no place was left for other Agrarian groups. The Gichevists joined with the Legionairs in "Svoboda" and a sundry of monarchist groups. But the Petkovists soon realized that they are a minority in the UDF where a number of other small groups had been admitted. They saw themselves as being used by phantom groups, building national reputation on their back. They tried to form a political center in the UDF with the Social

Democrats, but the latter soon fell on bad times among themselves and the position of the Petkovists further weakened. To make things even worse for them, the former Tanchevists seized the initiative for unification of all Agrarians by orchestrating the return of a few leading exiles, former lieutenants of Dr. G.M. Dimitrov, led by Tsenko Barev. If this initiative succeeds the United BANU will re-emerge as a major third force in Bulgarian politics. Thus, the Petkovists were isolated in the UDF, and isolated in the BANU. They had no other choice but to stick it out with the UDF. It would have been an ideal situation if the two Agrarian factions – the Barev-Vulkov group and Drenchev Group – joined together. But the personal animosities among the leading personalities will not allow that.

In the final analysis, on the road to Agrarian unity one will meet the two personal enemies – Tsenko Barev and Iskur Shumanov (Not Milan Drenchev and Victor Vulkov), the two pretenders for the mantle of Dr. G.M. Dimitrov. There was much bad blood between these two and the conflict will be resolved only when the one or the other is removed from the political stage in Bulgaria. Arithmetics of biological nature suggest that this day is not too far away for the one and for the other, but until this day arrives much mischief may be done on the Agrarian field. Such a day will also remove the major obstacle for the unification of the two Pladne faction – the Traikovist-Tanchevist-Vulkovist group, and the Petkovist-Drenchevist group. It should be noted that the Petkovists would not forgive their opponents decades-long cooperation with the communists, but do not mind to collaborate with such Communist fellow travelers in the UDF. All this puts the question of reunion in a different light: it seems that the question is: who will lead the Agrarian Union, not who has cooperated with whom. Otherwise, the desire of the rank and file Agrarians is the restoration of unity in the BANU. The unification of the two parliamentary groups – that of the Vulkovists and some of the Petkovists – paved the way for the unification conference scheduled for July 27th. For sure, such a Union will be proclaimed, but it is an open question how many Petkovists will join in.

37. Gichevism

We have stood politically and ideologically behind the positions held by D. Gichev. Between 1978 and 1990 we have published a quarterly, "Free Agrarian Banner", espousing these positions. It is our conviction, and we have defended this conviction with arguments from history, that among all political parties in Bulgaria, including every shade of Agrarian opinion, the BANU "Vrabcha 1"[*] alone has never been compromised in foreign policy and in domestic policies, before September 9; that it has always stood closest to the positions of Alexander Stamboliysky and in every respect may be characterized as *Sine Macula et Ruga*, without a spot or a wrinkle – to use the words of Pope Nicholas of the 9th century, addressed to King Boris I. History has confirmed to the last detail the policies of "Vrabcha 1", and has refuted to the last point the policies of Pladne, Dr. G.M. Dimitrov's group.

[*] Name used by the Gichevists, "Vrabcha 1", after the address of Agrarian headquarters at the corner of Rakovsky and Vrabcha Streets.

At the unification conference in November 1945, turning to the Pladne agrarians, Gichev said: "We saw 9th of September prior to that date what you saw it to be after this event." By these words he summed up his entire political philosophy prior to September 9th. After that conference the Petkovists, the Pladne leadership of the Union, having passed all the way on Gichevist positions, continued and continue defending the Pladne line prior to the split with the Communists in July 1945. Today they do not forgive the Vulkovists for having cooperated with the Communists for 45 years, but they would not even mention that they had become an instrument in the hands of the Communists, that they have been a Trojan Horse which brought Communism to Bulgaria. Since 1945, however, the policy of BANU "N. Petkov" has been Gichevism at its best. The controversy between the BANU "N. Petkov" and BANU "D. Gichev" is a question of political consistency and a question of responsibility in the face of history – not a question of politics and ideology in the post 1945 era. It was a matter of policy before 1945. When Gichev (prior to 1944) declined to cooperate with the Communists and bring Bulgaria into the war between the great powers, Dr. G.M. Dimitrov and N. Petkov did the one and the other: they departed from the traditional policy of BANU for neutrality, and they cooperated with the communists.

But Gichevism dismally failed in the aftermath of the November events. In fact as a force for political action it has been moribund since 1944. The outlawing of the parties by the personal regimes of King Boris in the 1930s destroyed the infrastructures of all moderate democratic parties. What was left of them by September 9th was a skeleton leadership. That leadership was further decimated by the brutal treatment at the hands of the Communists, and by November 1989 was non-existent. Unlike the moderates, the extremists of Pladne, supported by foreign intelligence services for their own purposes, survived. They joined in the Fatherland Front, but subsequently walked out of it and became the catalyst for all opposition forces. Lacking possibilities for political expression, these forces joined *en mass* the Agrarian party.

This is how the Petkovists built a following, which, as much as it has survived the Zhivkov regime, revived the BANU "N. Petkov" after November 10, under the leadership of Milan Drenchev. Claiming also the legacy of Gichev, but parading their Pladne ideological and historical credentials, they conquered the terrain for political action on the Agrarian front. Thus the Petkovists and the Vulkovists conquered the strategic positions in the BANU and left no room for Gichevism. The Vulkovists gladly welcomed the remnants of Gichevism, but it amounted to nothing. The Drenchevists pointedly excluded the Gichevists from their ranks and blocked their admission into the UDF. They argued that Gichev and Petkov had united in 1945 and represented this union as a complete political and ideological abdication by Gichev, while, in fact, it was a political and ideological abdication of "Pladne".

It was Strahil Gichev, son of Dimitar Gichev, who attemped to form a Gichevist Union, but his inept leadership, faced with the formidable array of Petkovists and Vulkovists, failed to create anything credible. His untimely death in October 1990 virtually obliterated the movement. It just happened that until my arrival in Sofia, I did not know what had happened to the Gichevists. One day I noticed on the newsstands a paper – *Zemedelsko Edinstvo*. That is where I discovered that there was a Gichevist faction, and went looking for them. I found them… lost in a dingy building on "D. Blagoev" Street. There was one secretary there. Soon, there appeared the leaders – Georgi Andreev, Marin Duparinov – uknown to me if he has something to do with the old Duparinov of Stamboliysky times *,

---

* Much later I learned that he had nothing in common, where blood relations go, with Spas Duparinov.

and the editor of the paper... Nikola Kurtokliev. They claimed that they are the United Agrarian Union: Andreev, representing the Tanchevists, Duparinov – the Gichevists, and Kurtokliev – the Petkovists. They have heard of us in the United States. Andreev reluctantly admitted that he had seen the *Free Agrarian Banner*. No more comments. We chatted for an hour and a half. They seemed well oriented on Gichevism, but overall they appeared to me a pathetic trio – without organization, without following, without money, without headquarters to speak of, without a decent paper. *Zemedelsko Edinstvo*, except for a few articles on political subjects, seems to be filling its pages with ... fables. Besides, they have joined a group of rightist organizations, called "Svoboda" – monarchists and Legionairs. Question was raised if I could give them straight answer as to my intention to come and live permanently in Bulgaria. I did not encourage them to expect me. I spelled out my reasons. Question was also raised, and they asked for my opinion, if the unification of the BANU comes up, which side they should take – the Shumanov or the Barev side. I recommended the Barevs. I do not know how long they will be in existence, and if they have not by now ceased to exist for reasons of lack of money. They promised to send me their next issue of "Zemedelsko Edinstvo". I never received it.

38. For a United BANU

It is a pity that the BANU has come to be in such a state of disarray. Otherwise, as an united party, it would have played, and still could play, a major role in Bulgarian politics. As of now, it is only a beast of burden to all kinds of small parties and individuals who are building careers on its back. When a reporter of the Vulkovist paper asked me of my opinion as to how BANU could achieve unity (I have every doubt that this interview will ever see the light of the day), I told her that the only effective way is to go back to Stanboliysky and Gichev, a return to the ideology and the philosophy of the founding fathers of the Union. History and practical politics would suggest such a track toward Agrarian unity. Otherwise the BANU is in danger of becoming a coterie of hungry politicians.

It appears that all present Bulgarian parties are moving in that direction. More and more the political parties are only a means to coming to power. The trend in the leading factions of the BANU is not much different from that of the other parties. The Vulkovists have already adapted to the status of a small party which will always be sought for a parliamentary majority in a combination of parties in a coalition government, where they would place their leaders in comfortable prestigious positions. The Petkovists, unsure of their ability to elect members in the Parliament, will bow down on their knees, humiliating themselves with small groups and clubs in order to stay on the surface of Bulgarian politics. The Gichevists, showing their face here and there, will trudge on the periphery of Bulgarian politics without direction and without principles.

The leaders of all Agrarian factions fail to comprehend that the BANU may become a leading force in Bulgarian politics only if they all join together, that to achieve such unity, many of the old hands should step down and let the younger cadres take over the Union. When writing these lines, July 20, I have the impression that they are on the way of uniting: the Vulkovists and the Barevists – on July 27th. This initiative was taken and is led by an old veteran in the Agrarian movement, a universally respected poet, Iosif Petrov. To him we sent our congratulations and wishes for a successful end of the attempt to recreate the unity in the BANU. In him we have placed our last hope for a United BANU. *

39. BANU and the New Social and Economic Profile of the Bulgarian Village

Finally, the Bulgarian Agrarian movement need to be seen from one more perspective. The social and economic conditions in present day Bulgaria are radically different from those of the 1920s and 1930s when the BANU grew up to be a dominant political force. Communism changed profoundly the social and economic landscape of the country. The demographic map of Bulgaria has changed beyond comparison. The migration from the country to the city has depopulated the villages and changed their social composition. There was the time when the village was a fortress for the BANU. But the last elections of June, 1990, showed that it has become a fortress for the remnants of the Communist Socialist Party. It has become a haven for pensioners, not a hub of agricultural producers. It is in the hands of the ruling nomenklatura. If there still is a social base for the BANU in the village, that base is very thin indeed. All this moved the BANU from the position of champion of peasant interests, to become a champion of democracy in a wider national context.

This is not necessarily an abdication of the organization from its primary mission to serve peasant interests. The BANU has accumulated enough credits as defender of Bulgarian democracy. But this confronts the Agrarian movement with the dilemma of becoming a third force in Bulgarian politics, focusing on peasant interests, in the middle of the political spectrum, between the Socialist Party representing the left, and the UDF, representing the Right, or at joining the UDF and becoming the backbone of a mass democratic party in a two-party system. Only the future, and the new forces emerging now in the BANU will solve this dilemma.

As things stand now, the Vulkovists seems to have come to the conclusion that a national party on the scale of BANU of the times of Stamboliysky and Gichev they will not become, that they will just wander on the periphery of Bulgarian politics. In such a case there is no better policy for them than to remain a small independent party and serve as arbiter between the big blocks. In that capacity they would be sought by all other parties to form political coalitions and to be included in the government. In old times Stamboliysky attacked such parties with derogatory names. The Petkovists on the other hand, realizing their inability to rise as an independent national party, are seeking to cozy with elements in the UDF. The necessity to achieve a unity among all non-communist forces to topple the Communist Party, forces them to stay with the UDF. There is a very little chance that Vulkovists and Petkovists will unite, but there is a greater chance that the inclusion of Tsenko Barev in the Vulkovist BANU, as their President, will temporarily strengthen this faction. From there on, only time will tell which way the BANU will go. If the new united BANU under Tsenko Barev does not find some way to enter the coming national elections

---

* Nothing of this expectations came true.

with the UDF, or whatever combination may be formed, and thus prefer to go it alone, it will again play the role of a spoiler and thus insure a victory for the Communist Party, as it did last year. Hopefully Tsenko Barev would prevent such a gaffe.

40. The Present Bulgarian Political Leaders Run Away from the Judgment of History

There is one problem which weighs heavily over the entire length and width of the Bulgarian political panorama. It is the problem of historical responsibility. Otherwise defined, this problem is reduced to the lack of a sense of history in the political circles in Bulgaria. Either they know nothing, or they know very little, and care even less, or perhaps deliberately run away from history, to avoid confessing their sins. Having followed wrong policies and having brought catastrophes for Bulgaria and the Bulgarian people, the present leaders in Sofia, both communists and non-communists, at least some of them, cannot stand the verdict of history and will hide their face from it. History is the judge of the past. They tremble before its judgment the same way as criminals before a tribunal of justice. It may be that the length of time will make some crimes to loose their weight. May be they prefer ignorance of history where the Bulgarian people have suffered catastrophic defeats. It is most probable that the window to history in many of the political party headquarters in present day Bulgaria is positioned in such a way that the errors of the past are excluded from their view.

An observer of Bulgarian political life today will have no difficulty noticing that in as much as the new generations in Bulgarian politics are concerned, the past has a very little meaning, it is not too much of a concern as to which party has done what in the past, or who is responsible for the miserable conditions of the Bulgarian people today - in international and in domestic affairs. Otherwise, many of the idols and the heroes of the leading political groups in Bulgaria would be cast to the heaps of history. In the first place the heirs of Dimitar Blagoev, in the second place all those who took part in the September 9th affair, and third the entire monarchical click which today glorifies the grave digger of the Third Bulgarian State, Ferdinand. Knowledge of history will frustrate anyone who witness the political progress in today's Bulgaria. The political leaders of present day Bulgaria are little interested in accounting for the lessons of the past and are laying the foundations of a new Bulgaria over the same rotten beams which ruined her in the twentieth century. Opportunism, apathy and ignorance, all three together, stand as an impenetrable curtain between the past and the present in Bulgarian political life and augur nothing good for the future.

41. The Honeymoon of the Monarchist Propaganda in Bulgaria

A great number of monarchist groups – old and new parties and clubs, in and out of the UDF – have emerged on the political stage. Arriving in Bulgaria I was impressed by the heightened enthusiasm for the monarchy. The Bulgarian television, especially Kevork Kevorkian, a top star in leading public discussions, were having enormous impact in popularizing the royalist cause, even making available the national screen to the King who had expectations to recover his throne. Radio "Free Europe", long before Kevorkian, had become a mouthpiece of the royalist propaganda in Bulgaria. Numerous pamphlets and books, subsidized from abroad, were being published and sold in the market after the November change. Kevorkian has published a collection of his *Interviews with Simeon II*. The *Memoires* of Queen Yoanna are displayed in every bookstore in Sofia. The *Instructions of Ferdinand to his Son*, which are drafted after the model of Machiavelli's *Prince*, have been published by the prestigious University Publishing House, though, judging from the contents they seem to be a fakery. Such is also the book of a Mikhail Topalov *Conversations with Prince Kiril*. There is an outpouring of publications about King Boris. The most important work, however, is Stefan Groueff's book *Crown of Thorns* which publication by the University's Publishing House is imminent. It is a much praised, cake-like biography of the late King Boris, which, as work of history, is full of fallacies and misrepresentations.

The royalist propaganda in Bulgaria had reached its climax during my visit there at the end of May and the beginning of June 1991. On May 28th the national Assembly, acting on a motion by Petar Dertliev, leader of the Social Democratic Party, passed a resolution appointing a national referendum where the citizens of the country were going to express their preferences for Republic or Monarchy, so the popular vote be reflected in the new constitution being prepared. The Referendum was to be held on July 7th. This sudden turns of events shocked the nation. The President of the Republic, Prof. Zheliu Zhelev, publicly declared that if Simeon had applied for a Bulgarian passport, he would endorse it. The people were confused and disoriented. The most astonishing was the fact that even the Socialist (Communist) Party voted for the resolution. But this euphoria of the royalists was cooled off when a week after, the same National Assembly rescinded the resolution, realizing that this was too much, too soon. Simeon himself, in apparent embarrassment, and a sense of reality that he would loose, approved of the cancellation of the referendum. The hopes of the royalists have never been better, and for sure, will never be better, than in those hillarious May and June days. *

All this fever of royalism was sparked by the visit of Princess Maria Louisa to the country. The national adulation had no limits. It just happened that I visited the Bachkovo monastery days after she had been there. The Abbot showed me with pride her signature in the Guest Book. She had not written one word of a message after or under her signature. I wrote a whole page in praise of the Church. Everywhere, during my entire stay in the country, I was asked questions and had to listen to comments about the monarchy. Some of them I dodged, to some of them, out of consideration for the people who sought my opinion, I made non-committal responses, and sometimes could not avoid expressing my negative views of the founder of the dynasty, Ferdinand. During this fever of royalist excitement the organizers of the campaign had arranged an exhibit of pictures on a street corner near the former palace, a gallery of portraits of King Simeon from his childhood, of the royal family, of King Boris, King Ferdinand and numerous memorabilia from the past. The attendants were surrounded by a great number of interested persons, some of

---

* Recently, December 1999, reviewing all these events, I came to the conclusion that the UDF used the King to legitimate them as anti-communists. They were all former communists and who, but the King, with his ex officio position, could legitimate better their credentials? After accomplishing all that, they turned their backs on him.

them irate of the royalism of the exhibitors, others supporting it, and all arguing and shouting at each other with utmost seriousness. I enjoyed standing on the side and watching the show. I felt that I was at Hyde Park Corner. Among the other things I noticed that several of the enlarged photographs of the wedding of Simeon were doctored. Metropolitan Andrey, who performed the wedding ceremony and was prominently present in these same pictures which I was familiar with, was carefully excized by the exhibitors.

42. The Monarchy and its Sins before the Bulgarian People

My question, all along, was why so much attention for a cause which I have always considered dead, once and for all. Surely, King Simeon was on his exemplary behavior abroad.

In fact he had been making meek eyes to the Communists for decades, and I have written many times on the subject. He clearly sought to impress all – foreign and Bulgarian political observers – with his moderation, with his maturity, with his intelligence, with his tolerance. He had indeed succeeded to fool so many people. He has never fooled this writer who had a good understanding of Coburg policies in Bulgaria, policies which have brought nothing but disaster for the Bulgarian nation and State. We have for a long time recognized in King Simeon not only his physical likeness to his grandfather Ferdinand, but also his manner of treating the political forces around him.

We have been always afraid that once more, in manipulating the political processes to his advantage, like his father, he would erect a personal regime clothed in democratic vestments, nothing different from the regimes preceding September 9th. History, if it says anything about the Bulgarian monarchy, it proves that the Bulgarian national cause has been ruined by the founder of the Coburg Dynasty in Bulgaria, King Ferdinand. He came to Sofia with the blessing of the Habsburgs, to serve the interests of the Hapsburgs, which interests included the destruction of Bulgaria, the prevention of her rise as a power in the Balkans. He served these interests well. Even if the Central Powers had won WWI, the Austrian-Hungarian interests would have prevailed and Bulgaria would have been sacked. Queen Ioanna, in her Memoirs, à la Marie Antoanette, tries to tell a different story, but Ferdinand's hate for the Bulgarians, reflected in his "Advises to the Son", recently published by the University Publishing house in Sofia, tell a different story.

43. The Failure of the Opposition to Topple Communism and the Hopes that the Monarchy will Complete the Job

But why so much adulation for the fallen monarchs and their descendent in Madrid? Certainly there is this thing about kings and royalty. But it does not explain it all. In every question and comment which we heard from common people all over the country, there

454

was something unsaid, something not spelled out, something not articulated in so many words, but something deep and powerful which was very important to them. I hope I have guessed it right. The country and the people were deeply disappointed from the current opposition political leadership. The Opposition has failed to dislodge the Communists. The nation has lost confidence in the Opposition. With their political wheeling and dealings – from the Round Table, to the elections, to the Grand National Assembly – the politicians have lost the confidence of the people. The King, as a symbol of everything that is not communist, as the one leader who has never compromised with the communists, whose status of a king stands on the opposite pole in the spectrum of political forces in Bulgaria when matched with communism, is perceived as the knight on a white horse who would save Bulgaria.

This could be read all over the faces of all those who talked to me about Simeon. I observed this, all to my regrets. Zheliu Zhelev flunked the test of history and politics. The mind of the ordinary Bulgarian, yearning for greater security and insurance against the return of communist times, in his quest for a final, complete, unquestionable and definitive demise of Communism, is turning to the monarchy as the one effective means of accomplishing these objectives. How well will Simeon satisfy this yearning, with his wishy-washy politics in exile for the last forty years, will probably never be known, because the climax of the monarchist cause in Bulgaria has passed, and it is doubtful that the monarchist fever will ever rise to the level of May 1991. Stung by the resolution of the Grand National Assembly of May 28, the anti-monarchist forces lined up their battalions to oppose it. But, when everything is said against the monarchy - and nobody has spoken in stronger terms against Bulgarian monarchy in exile than this writer, and nobody will more forcefully criticize the Coburg Monarchy in the future than this writer – the failure of the opposition forces in Bulgaria to dislodge communism, their failure especially to elevate a symbol, which even only for appearance purposes would persuade the nation that communism is gone, once and forever, lead this writer to conclude that if the opposition fails again in the next elections, there will be nothing left to oppose to communism but monarchy.

44. King Simeon: The Symbol who Hid Under the Crown During the Days of Slavery

Back in the 1950s and the beginning of the 1960s I most energetically stood for restoration of the monarchy in Bulgaria. Having accepted the Turnovo Constitution as a program for uniting all political forces against the communist regime, we viewed the monarchy as a constitutionally mandated legal basis for such an objective. For us it was the only symbol around which we all could join to topple the communists. But we were disappointed by the attitude of Simeon. He did not come forward to take over the leadership of the political exiles. He did not come out to stand at the front lines for the liberation of Bulgaria. He isolated himself in Madrid behind the walls of his palace, away from the masses of Bulgarian exiles, surrounded by a small group of lackeys. With a few inept moves which he made in trying to constitute a Bulgarian Political Representation Abroad he proved that either he was little interested, or did not understand the principles of democracy, preferring to follow the practices of his father of functioning in Bulgarian politics in the circle of a group of people outside of the political parties. He did not come openly and categorically against communism in Bulgaria. At the same time he made some muddled statements which suggested that he was seeking an understanding with Sofia. All that forced us to review our attitude towards him and the monarchy in general. In addition to that, the Turnovo Constitution had lost its relevance, and together with it the King as a symbol, got lost in our exile controversies.

It is very doubtful now that the monarchist fever of May and June could again excite the Bulgarian public. Bulgaria is again faced with elections next September. These elections represent a new chance for the opposition to defeat the communists. But if it fails, the question of the monarchy as the only trump card and the only symbol for a national unity to topple the red regime may again surface in Bulgarian politics should such a contingency develop. I would not hesitate to give it my support. There is the Bulgarian wisdom of: "Come along evil, because without you, it is a greater evil."

THE BULGARIAN ORTHODOX CHURCH

45. My Alma Mater

The affairs of the Church have always been of special interest to me. A graduate of the Plovdiv Theological Seminary and the School of Theology of Sofia University, I have made close and lasting friendships with many of the present leading figures in the Church administration, the Holy Synod, the Seminaries and the Theological Academy. In exile I have written extensively on Church matters and revealed the plight of religion under the communist regime. I have vigorously attacked the accomodationism of the Church hierarchy in Sofia and abroad. Arriving in Bulgaria I did not know what to expect, since I had cut all my ties with friends and foes, not knowing who had done what all these years. I knew next to nothing as to how they faired in the conditions of the post-November era, how were they treated and what part, if any, they have played in the processes of revival of freedom and democracy. Only dribs and drabs have reached me by way of occasional mentions in the press. I had written to some friends that I was coming to Bulgaria. Some responded to my letters. Most did not. As it turned out some of my letters were not received. I tracked down the people I wanted to see through friends and acquaintances. I had a number of extensive discussions with a few of them.

46. A Village Church

My first impressions of Church life in Bulgaria came from my village. I met the local priest, a gentle, pious, down to earth, intelligent, and practical servant of God. He lives in what once was the Church hall, now transformed into living quarters for him. In the church yard, some 20 feet from the Church, was a pen where he keeps his goats. During the service one could hear the bleating of the kids. The Church yard is overgrown with weeds. They have no lawn mowers there, and they do not cultivate the yard for any other purposes. The Church is the same as I left it, only 40 years older ... presently over two hundred years old.

The bell tower seemed to have been painted. The cross, which was destroyed by gun fire in 1944, is now restored. Except for a new wooden ceiling, all is the same – the icons, the frescoes, the furniture. In my lectures in America I often mentioned a painting of St. Elija. It is still there. I have also mentioned a fresco where the eyes of a saint were gouged out by the Turks. It is still there. Father Nikolai explained some of the icons. He conducted the services with impressive mystical concentration. He was assisted by a small female choir of old women of the village. I have always wondered how do they conduct services in the absence of trained psalt. They are doing very well indeed, under the circumstances. It pleased my heart. I noticed that the Gospel was rather raggedy. In

Sofia I purchased a new Gospel. Will have it specially bound here and will send it to them. Father Nikolai is a district priest. I was told that he serves seven or eight villages who have no priests and some of the churches are used as warehouses. The worshippers were old women, all dressed in black. I saw no young people entering the churches, lighting candles and praying. My impression was that the Church is virtually dead and it will take a lot to revive it. As of now it looks like a lost cause. If the situation is the same in all villages, and I suspect that it is even worse, there is a little hope. There are no priests. To be an optimist on that score is naïve, and to be a pessimist feels like a betrayal. But this is what the situation is and there is no point of misrepresenting it.

47. Bulgarian Church Art – an Unsurpassed National Treasure

A great deal of my time in Plovdiv and Sofia was spent visiting churches, chapels and cathedrals. They are national treasures of art collections. The iconography, the wood-cuts, the frescoes, which make no impression to the natives, are masterpieces for appreciation by foreigners. In some instances one begins to wonder how neglected these monuments of Bulgarian culture are. The wall painting outside of the Igoumen's headquaters in the Bachkovo Monastery, and the paintings in the entrance of the same Church, as well as those surrounding the central Church of the Rila Monastery, are masterpieces of art which, if not comparable to Michelangelo, considering the difference of culture and versions of Christianity, deserve better attention and special studies as to their meaning and the moral concepts of artists and societies of the times. The iconostasion of the Bachkovo Church, and in a number of other churches in Bulgaria with their wood work, are masterpieces of art indeed. The Church museum in the Theological Academy, especially if one is lucky to see it with the help of its curator, is a treasury which one rarely has a chance to enjoy. The performance of the church choir of St. Alexander Nevsky in Sofia is out of this world.

48. The Compromised Leadership of the Bulgarian Orthodox Church

Yet, the state of the Church, as we were able to assess things from our conversations, was nothing to be excited about. It appears that the leadership of the Church is deeply compromised with its cooperation with the Communist regime. The Dean of the Theological Academy admitted: "Compromises were made. Compromises had to be made in order to save the Church." Our charges made in exile for years that the leadership of the Church had become a lackey of Zhivkov's regime, were confirmed. We heard these charges from many quarters – clergy and laymen. High and low clergy are being accused in dishonest compromises; theologians of the highest ranks were said to have been police informers. When the Dean of the Theological Academy was making the above admissions,

458

he appeared to have been quite embarrassed. The previous evening when we had met him briefly, he received us with great enthusiasm. This morning he was somehow restrained. Apparently he had discussed with someone our prospective meeting and had been warned of our positions. So there was the admission: compromises had been made in order to save the Church. We were told by other circles, close to the Church, that those who had compromised had done it for their own personal benefit – for career advantages and for all sorts of personal privileges. Protecting the Church was only a cover-up for their ingratious betrayal. From all the stories which are circulated it will appear that these same people who were saving the Church were deeply involved in persecution of other members of the Church hierarcly and the theological world who courageously had stood for the Church. The case for protection of the Church remains to be substantiated, because not much of the Church has survived. Except for the full complement of Bishops, and a skeleton of teaching staff in the Academy and the Seminary, the clergy had been decimated, reduced to an absolute minimum of pensioners and, due to circumstances, unqualified men ordained to the priesthood just to perform religious rituals and services. No reliable statistics exist as to the number of believers and church goers. The art treasures of the church, pilfered by thieves, are being peddled in antique shops in the West.

49. Along the Dark Alleys of the Compromise

We listened with pain to some of the charges made by the critics of the accommodationists. All criticism was directed against Patriarch Maxim, the Metropolitans of Stara Zagora Pankrati, of Vratsa – Kalinik, and their proteges. We heard of a staged shameful incident involving the former Rector of the Seminary Bishop Gerasim and his subsequent trial. He has been accused that in his capacity as Rector of the Seminary, had been recruiting candidates for the priesthood and for monks among the students, plus other unsubstantiated charges of moral deviations. Some leading clergymen had been involved in this staged trial. He had been acquitted. The trial judge had reprimanded an witness testifying to the first charges with the words: "And you think this is a crime? But isn't it part of his responsibilities?" Gerasim was an old friend of mine, but for some reason for long time I thought that he had been dead. He is still alive and is assistant to the Metropolitan of Sliven. We also heard the story of the gruesome death of Arkhimandrite Gorazd. I also heard some disconcerting comments about my old, very close friend, Theology Professor Todor Subev, General Secretary and for sometime one of the Presidents of the World Council of Churches, protégé of the same people serving the regime. He was said to have been an agent of the KGB, reporting and receiving his instructions straight from the KGB in Moscow. I heard similar stories about the Dean of the Theological Academy. He had closely cooperated with the communist authorities at the expense of other theologians. We wondered how the Theological Academy had survived for so many years. But he himself revealed the secret. The Academy had been a free ticket for provincials to live in Sofia for a few years.

50. Behind the Wall of Silence After November 10, 1989

Indeed the immersion of the Church leadership in the Communist swamp has been so total, without parallel, so self-effacing, so deeply penetrated by a spirit of betrayal that when the regime collapsed on November 10, 1989, the Church was nowhere to be seen joining with the jubilant masses to celebrate the liberation. One would have expected that the Church would have been in the front line and leading the nation with crosses and holy banners, with tears and exaltations of joy. Instead, it was nowhere to be seen. The Church hid, and is still hiding from the faithful. It was a refractory monk, styling himself as a Hieromonach, Christophor Subev, who posed as spokeman for the Church. All along his activities on the opposition front somehow did not inspire confidence. Sometimes he commited such gaffes, which made a mockery of the Church and demeaned the clergy. He contributes nothing to help the church in these transitional times. In spite of all that, if it is thought of the Church in Bulgaria today, it is judged by the antics of Christophor Subev. It is a tragedy that in these decisive days not one single prelate of high rank appeared to speak on behalf of the hundreds of priests and clergymen who have died as martyrs for the faith during the difficult days of slavery under communism. Presently all over Bulgaria, when the voice of the church should be heard, it is the voice of Subev which is heard, and he assures all that for everything he consults with God directly. Not one Metropolitan, no Bishop, no Patriarch Maxim himself, appeared on the stage of history to castigate the tyrants of yesterday and catalog their crimes against the church. As the Bulgarian saying goes: "Neither they have eaten onions, nor had they smelled it." We raised such questions but never got a satisfactory answer. As if Church discipline and subordination had placed a zipper on their mouths – from the metropolitans to the lowest clergymen. The Holy Synod is stubbornly silent. How could they open their mouths when all along they have been faithfully serving Todor Zhivkov? The President of the Priest Union, Radko p. Todorov, a close friend of this author, Professor of Canon Law in the Theological Academy, forced to retire before November 10, appeared to be making great efforts to break through this wall of silence and isolation of the supreme government of the church. His Union, dissolved by Vulko Chervenkov (Comuunist Prime Minister in the early 1950s) on demands from Patriarch Kiril and Bishop Jona, restored by PopTodorov, has not been, as yet, recognized by the Holy Synod.

51. "The Secret History" about Patriarch Kiril

The leadirship of the Church is viciously attacked in the daily press for its accommodationist policies. We were given a book written by a priest, Dr. Yanko Dimov, on Patriarch Kiril. The author has documented an exposè which, if not repudiated by a credible authority, reduces the former patriarch to a faker, an Albanian, an atheist and the lowest moral character under the sun. If the book is let to stand, with its documentary evidence overwhelmingly damnable, the Bulgarian Orthodox Church and the reign of Kiril will appear to have been the most shameful period in the history of the Bulgarian

Church for 1,150 years.* Patriarch Maxim ought to appoint a special commission with a mandate to verify the quoted documents, to publish them in photostats and let the general public judge him for themselves. Such a commission should consist of trained historians. If Maxim and the Synod fail to do that, the book, reminding us of the *Secret History* of Procopius, will be left unchallenged and the honor and dignity of the Bulgarian church tarnished forever. If it turns out that the documented facts are authentic, Patriarch Kiril has no place being buried in the church of the Bachkovo Monastery. We do not know Dr. Dimov, but the information which he gives of himself seems to confirm his claim to speak on this subject with authority. He has occupied a prominent position in the clergy of the capital.

52. Patriarch Maxim. The Liturgy Which he Never Celebrated

There is no such "secret history" about patriarch Maxim, and we do not believe that there will ever be written** for the simple reason that deeds ascribed to Kiril are of such a magnitude, that only a charlatan could commit them. Maxim, whom we do not know, is not of the class of Kiril. The articles against him appearing in leading newspapers are left unanswered. He is said to have retreated to his patriarchal quarters in a mood of depression and inaction, apparently in confusion and dispassion for all that is happening around him. He may be regretting the good old days under Zhivkov. I suggested, wherever I discussed the Church, that he should have taken the lead right after the days of change by holding a Patriarchal liturgy in Alexander Nevsky, together with all Metropolitans and Bishops, and should have anathematized communism and the communist regime from Kimon Georgiev to Todor Zhivkov. Having done that he should have confessed and explained publicly the sins of the Church and the clergy during the past forty five years, should have layed his scepter of the patriarchal service and retire to some monastery in prayer and remorseful contrition for all that had happened during his term. Not only he did not do that, but he allowed the biggest demonstration of hypocrisy, having Prime Minister Lukanov, President Peter Mladenov and Defense Minister Dobri Djurov***, the three arch-Communists, to kneel down for prayer a few feet from him in the patriarchal cathedral. The case of St. Ambrose and Emperor Theodosius I in the Milan cathedral after a massacre of 15.000 men in the circus of Salonica, ordered by the Emperor, is appropriate to mention here. Ambrose stopped Theodosius at the doors of the cathedral. Maxim did not rise to the occasion and the Church fell even lower than it was in Zhivkov's time. I suggested to Metropolitan Pimen that probably the only dignified way for Maxim now, one by which he would best serve the church, is to retire and let the Church find its way on the road to renewal. I did not tell Pimen what rumors had it, that he was being considered as a Pro Tem President of the Holy Synod until a new Patriarch is elected.

---

* In subsequent years we discovered that Dimov was in the habit to doctor the documents which he used. We discussed that with him on the telephone. He denied doing it but we had a proof in our hands, and we could not be sure of the veracity of his studies.

** I was wrong. A year later Dimov published such a scurrilous biography of Maxim.

*** Recently we found out that Maxim and Djourov had been classmates in the Seminary.

53. About the Servants of Communism Under Bishops' Crowns

After the Patriarch, the criticism centers mostly on the Bishops of Stara Zagora and Vratsa. Pankrati of Stara Zagora was a former classmate of mine, whose ordination in February 1949 in the Bachkovo Monastery, for some strange coincidence of events, probably (most probably) saved my life. If I had not gone to his ordination with a company of students and professors from the School of Theology, my bones, for long time by now, may have rotted in the cemetery of some concentration camp ... or I would have been served for breakfast to the pigs of some swine farms operated by the then Angel-Protectors of the Bulgarian people.

But, enough for our old friendship with Dyado Pankrati. He rose as the most passionate, the most dedicated, the most fanatic supporter of the Communist regime in Bulgaria. It is claimed that documentary evidence seen in Sofia confirms that he and his closest friend, Todor Subev, had been members of the young Communist League. I am not inclined to believe it, but who knows? After the change of November 10, Pankrati had continued serving the Communist Party. In fact, he had been elected to the Grand National Assembly by the Communist Party under a concocted Fatherland's Party, in opposition to the UDF candidate. In the Grand National Assembly he continues serving the Communists, which is an outrage in itself. The Bishop of Vratsa, Kalinik, is said to have been the most fanatic and loyal servant of the Communist Party and had used the State Security to manipulate diocesan elections, where, with the help of the government, had prevented the election of Bishop Arseniy for the See of Lovetch. Arseniy subsequently had been elected as Metropolitan of Plovdiv. Some suggestions were made about the new Metropolitan of Varna, his name escapes me now, as having close ties with the State Security, but I could not very well understand what the case was. The Metropolitan of Sliven, Ioanikiy, who some years ago delivered a shameful speech to ingratiate himself with the authorities (I published it), was excused to us as being young, inexperienced and lacking of good advisors.

54. On the Scandal with the Visit of the Ecumenical Patriarch, and the Need for Retiring Patriarch Maxim

Public indignation against the present leadership of the Church had been demonstrated recently during the visit of the Ecumental Patriarch Dimitrios to Sofia in the beginning of May. A public protest meeting had been held in front of the ancient church of St. Sofia. After the service, when the two patriarchs had appeared in their limo, the angry crowds had pounced over the car, demanding the resignation of Maxim, and asking the Ecumental Patriarch to free the Bulgarians from this "Communist agent". It appeared to me that this had gone too far – not on account of Maxim, but on account of the Ecumenical Patriarch, who, after all, should have been spared such uncivil treatment. This incident, at end, reflects on the civility of the Bulgarians. Church circles viewed this incident with dismay, but this further indicates how untenable the position of Maxim in the Bulgarian Church is. Indeed, the only dignified way for Patriarch Maxim would have been if he layed his office down and let the Church select a new leadership, in the spirit of the new times. The longer he stays the longer this thing will fester and will poison the Church presence in Bulgarian society. I suggested this way out to Metropolitan Pimen, but I am afraid it would not go very far. [*] He would not venture telling the Patriarch to go. It is for the Patriarch himself to make this decision. The best way left to him to serve the Church is to retire. Should he still persist on holding his position, then the Bulgarian Orthodox Church is a lost cause. A big man and a great leader recognizes the time when he should step down. For Patriarch Maxim time has long passed.

55. The Fear of Protestant Propaganda in Bulgaria

All those with whom I discussed the Church were keenly aware and alarmed that the new conditions in Bulgaria open the possibilities for foreign propaganda – coming from Protestant and Catholic missions. Already the *Church of God*, or the *Assembly of God*, had held a mammoth meeting in Sofia where thousands had attended. The Church leaders are stunned by this event. They are afraid that these Protestant missions, generously subsidized from America, may find a fertile ground in Bulgaria. I did not try to alleviate their fears because they are well founded. The Protestants will come, and will come big. Who will stop them? How could they be stopped in a system of democracy? Besides, the Bulgarian people at the present are so poor in religious knowledge and religious orientation, and the Bulgarian Church follows such obsolete methods that any propagation of any faith by any Church is better than the religious apathy, the atheism and the indifference to religion.

Maybe this is what the Church and the nation need – a challenge. Such a challenge may stimulate Church and nation for a new life. It would be preposterous to claim the Bulgarian people for Orthodoxy, when there are very few practicing Orthodox Christians. That the Orthodox Church is far superior than any other Christian confession with the richness of its spirituality, its ritual, its worship, its history, and in the case of Bulgaria – with its enormous contribution to Bulgarian survival in its 1300 year history – is above any question, in view of its past glories. The truth, the bitter truth is that this Church has lost touch with the soul of the nation, and, what is worse, is not interested in seeking the way to it.

56. The Fear of Catholic Propaganda in Bulgaria

The Church circles do not hide their fear of Catholicism. Bishops and theologians look with alarm to the ingratiating catholic jestures to Bulgaria. The possibilities of restoration of the Bulgarian monarchy – a traditional stronghold of Catholicism in the country – further stimulates these fears. On one hand the Church community feels great sympathy for the monarchy as a symbol against communism. But, on the other hand, when the question of Catholicism is raised, the monarchy is seen as an agent of a foreign religion. The Bulgar-

---

* It turned out later, that the time of my conversation with him, he and Radko Pop Todorov had been plotting with the government to depose Maxim and appoint him, Pimen, as *pro tem* President of the Holy Synod. This plot was executed on May 25, 1992, leading to the still continuing schism in the Bulgarian Orthodox Church. I most vehemently opposed this uncanonical procedure and to this day condemn the schismatics.

ian dynasty, the Coburgs, is of Catholic origins. The Bulgarian bishops have been on the front lines of the opposition to Tsar Ferdinand a century ago. Ferdinand forced a change of the constitution in order to baptize his son, heir to the throne, Prince Boris, as a Catholic. Later, to obtain Russia's recognition, he had him chrismated into the Orthodoxy.

Ever since, the Bulgarian Court pays lip service to Orthodoxy and the Church. King Simeon, in his long years of exile, never appeared to be interested in the Orthodox Church. He has never appeared in any of our Churches abroad to worship with the Bulgarians. All these years he was living in a Catholic environment. His gracious wife, the Queen, has never joined in to form a Bulgarian Ladies Society and engage in charitable work for helping needy, sick and handicapped Bulgarians. We deafened the world with the very little voice that we have. But we never saw our royal family to join us. Where were they? They were in the Catholic Church. Probably they have never believed that the day may come when they might claim the throne of Bulgaria. This is why they never cared about the Bulgarian Orthodox Church. For appearance purposes the royal family should have shown some interest towards the Church, to establish at least a Bulgarian chapel in Madrid and show up there once in a while. The King could have established a Bulgarian Bishop in Madrid. His princes and princesses remained aloof from the Bulgarian Church. We do not know if Marie Louise has ever stepped in a Bulgarian church. All royalists in exile have been on the payroll of the Vatican. All this is a good reason that the Church in Sofia should fear a forceful Catholic propaganda in Bulgaria, especially when the democratic forces are flirting with the Catholic Church. And, on the top of it all, taking into account the sorry record of close cooperation of the Orthodox clergy with the Communist authorities, and the fact that the Catholics gave as a sacrifice their Bishop in Bulgaria, Rt. Rev. Bosilkov, one will understand why the orthodox leaders would fear the Catholic propaganda.

57. The Old Holy Men are Dragging the Church to the Grave

I brought these points out to all those with whom I had contact on the highest and middle level of church leadership, but I saw no signs of any policies as to how to confront the challenges. I could not stress enough the need for a new, younger, dedicated leadership, free from the compromised past of collaboration with the Communists. I could not see how these men, in their 80s and 90s, were to respond to these challenges. Nothing seems to move them, to penetrate their aging brains. They cling to their thrones and will hold on to them to death, always expecting sycophantic praises and adoration for their rank, their age, and their crowns. They live in another world. They are dragging the Church into the grave and refuse to see it.

58. Bishop Kiril of America and his Synodal Judges Tied to State Security

The question which I was to raise with the church leadership in Bulgaria, particularly with Metropolitan Pimen, was that of the status of Metropolitan Kiril in the United States.

Before leaving for America I spoke with him on the telephone on the subject. He told me that he was in contact with Pimen. He had asked that they, the Synod, lift the condemnation imposed on him in 1963 when he, in opposition to the subordination of the American Diocese to the church authorities in Sofia, in effect to the State Security service there, had left the Synodal jurisdiction with many parishes, in fact splinter groups, and joined the Russian Synodal Church, outside of Russia, subsequently going to the Orthodox Church in America. All parishes in America were torn apart, some of the parishonors following Kiril, forming new parishes, while others remained loyal to Metropolitan Andrey, accepting the jurisdiction of the Holy Synod in Sofia. Those following Kiril founded their own new diocese. Those with Andrey, having passed under the direct jurisdiction of the Synod, which until 1963 was formally recognized by Andrey, without any contact with Sofia, in effect found themselves under the shadow power of the Bulgarian secret services which were hiding behind the Holy Synod. Kiril was condemned to argos – prohibition to serve in the church – by Metrpolitan Andrey. The Holy Synod confirmed it. The Synodal churches of Andrey became nests for communist propaganda, making the altar pulpits available to communist officials to speak to the exiles and the immigrants, and to all sorts of emissaries coming from Sofia. At the same time the members of the Bulgarian community, opposed to communism, followed Kiril. But all this, after the change of 1989, became past history.

Kiril was now interested to have the condemnation lifted, without him returning to the clergy of the Bulgarian Church. Such an act by the Holy Synod would restore his canonical position within the Orthodox Church. Even more important for him was that he would be able to visit Bulgaria in honor and dignity and concelebrate with the Bulgarian clergy. Now the tyranny in Bulgaria has been ended, the Church was again its own master, was free, and no longer under the wardship of the Secret Services. But formally the division in the Church in America is continuing. This is one of the legacies of the Zhivkov's regime. Kiril, and the Bulgarians who had followed him, and are still following him, continue to be under the condemnation imposed on them in 1963. For whatever reasons, the Synod continues to follow the policies of communist times. Two of the Synodal Bishops – Simeon and Joseph (Dikov) – who fought bitterly with Kiril and his churches after 1963 – are still in charge of these policies, and it may be that they are opposed to lifting the condemnations. So, my mission to Sofia, in agreement with Bishop Kiril, was to see what could be done on this subject.

I raised these questions everywhere I had a chance – with the church hierarchy. But it appeared to me that there was not enough interest and understanding of the problem. Metropolitans and Bishops completely ignore the role played by the Communist State Security in opening this wound in the body of the Bulgarian Orthodox Church. They see in it nothing more and nothing less than a deviation from the canons of the church, simply a matter of insubordination. As if this crisis did not exist! I raised this question with Metropolitan Pimen. He assured me categorically that it all will be resolved during the current session of the Synod – "by the end of June," he said. Now it is the end of July. The question has not been settled. While I was in Bulgaria, someone quoted to me a statement made by Prof. Todor Subev about Pimen, which, out of respect for the old man, I could not repeat, something of the nature that he never tells the truth. I stressed it before His Eminence what a great importance a move to that effect on their behalf will have for the Bulgarian community abroad. Before meeting him at dinner in Sheraton, I had prepared a draft resolution, a proposal how the Synod could justify lifting the argos of Bishop Kiril. This draft was in my pocket, but after his solemn assurances that it all will be taken care of, I felt disarmed. How could I be so impertinent to question the assurances of my prominent guest? This question, regarding the status of Bishop Kiril, is very important for us, the emigrants.

59. Bishop Kiril was with Us, Against the Communists. The Synod still Holds for the Anathema Dictated to Them by State Security

After all, when we fought the Communists in defense of the Bulgarian Church, Bishop Kiril was with us. He was our leader. He encouraged us, he supported our faith in the liberation. Today, when the penalty imposed on him on orders from the Communist authorities continues to hang over his head, apparently there is no amnesty for him, and the executioner of the verdict against him is none else but the Holy Synod. Then, it could justifiably be argued that November 10th has never arrived for the Church. Irony of all ironies! My impression, after many discussions on this question, is that the rank and file of the Church hierarchy are deadly afraid to challenge the Metropolitans on this subject. Hoping for promotion in the hierarchy, eventually to become Metropolitans, the lower echelons are reticent to take a stand, out of fear of retaliation from above. I was left with the unmistakable impression that the most fanatical opponents to the restoration of Bishop Kiril with dignity in the Church are the Bishops sent by the Synod to administer the diocese abroad*. They, the emissaries of State Security, dispatched to America to rein the exiles, even now, after November 10, continue to be in charge of Church affairs abroad. The hand of Todor Zhivkov is still hanging over the head of our Church in exile. Indeed, the solution of Bishop Kiril's question is a test for the Bulgarian Orthodox Church where the Synod stands to prove to our community abroad whether it has freed itself from the guardianship of the Communist Party, or it still is under its power. If this question is not settled, then the Bulgarian Orthodox Church is to be pitied.**

60. The Harvest is Big. The Harvesters Are Old

After spending some time in Bulgaria, I could not escape the conclusion that the Church functions, even after the big change, in a spirit of business as usual, as if Bulgaria had

---

* As recently as December, 2000, I discovered in my Dossier in D.S. that the late Bishop Joseph (Dikov) had submitted a report about me in 1978, used by that agency, the State Security office, to declare me "hostile adversary", and deprive me of sending money, packages and letters to Bulgaria, and of permits to visit my family there. This confirmed my numerous public statements that all clergy sent from Bulgaria to the United States were agents of "D.S."
** The condemnation of Bishop Kiril was lifted a month after Pimen's uncanonical, government ordered a coup, by the Holy Synod of Patriarch Maxim. I often wonder if this move of the Synod was in any way connected to my comments in this Report, which were by that time published in the Bulgarian press, duplicated and widely circulated by the schismatics.

466

never been under Communism, or as if Communism is still there. Much of the criticism these days is directed against the Holy Synod as a company of old, very old men, with diminished physical and mental capacities. One of the Metropolitans, Sofrony of Russe, a 93 year old man, a good friend and a classmate of university times, when told that he should retire because he cannot walk any more, he has retorted that he administers the diocese with his head, not with his legs. Another bishop, Stefan of Turnovo, my teacher in the Plovdiv Seminary, also an old, decrepit and senile man, rules the diocese with the help of some old lady, aunt or something, who acts as a vladichitsa (Bishop's wife). No one seems to have a say as to how to handle this situation. I made these points clearly and emphatically to Metropolitan Pimen. But he, too, is in his eighties. Once he was my teacher of literature. When I insisted that there comes a moment in one's life when he no longer has the physical and mental abilities, that when such a moment arrives it is better for the Church, and for any institution, that such people should retire, that the age of 70 is a reasonable time for aging officers to go, he responded that this had been tried in Romania, but it had not worked. I got the impression that there, in the Synod, they do not think of retirement. I spoke of the isolation of the Church from society, of the need to have a younger leadership, with contemporary ideas, to respond to the challenges of the new times, especially after the big change. In the face of all criticism of the Church, the Church has to respond to it by a publication of its own, addressed to the general public. I saw the latest issue of *Tsurkoven vestnik*. It is still published the way it was published during the long years of the Communist regime – stale, hagiographic, and empty. Pimen told me that I may have listened to their discussion in the Synod today. But he admitted they have no people. And I am sure they do not have people. I wish I had been free to take over a paper like that. Well, the present leadership of the Church would not accept me, and I would not move to Bulgaria after all that I saw there. I would have to give up my family and my career, and my comforts in America. It is too late for me to even think of that. I told Pimen that at this station of my life I could not move to Bulgaria for whatever purposes. Besides, they have all wallowed in accommodationism, so much that they would be the first to feel the blows of my criticism.

61. Elegant Smiles and Intellectual Vacuum

In general, my meeting with Metropolitan Pimen, over dinner, at the Sheraton, went well. I said everything that I wanted to say, but I said it in the spirit of respect and courtesy. He was very receptive, elegant, attentive, a gentleman, captiviting with his goodness, but nothing suggested that up there, in the Holy Synod, they entertain any ideas as to how to pull the Church out of its present debacle.

Frankly speaking, after my meeting with Pimen, I see no hope for the Bulgarian Orthodox Church for rising out of the present ashes in foreseeable future. The signs are just not there. I did not sense any energy, any inspiration, any dynamism in their present attitudes. This may go on for a whole generation, until a new generation takes over, or until some new regenerative movement explodes in some unsuspected corner. Hopefully, such a movement should start in the theological circles, among the students whom I did not see and could say nothing about. If it does not start there, then one has to look far ahead indeed, something to come out of the younger generation. I base my pessimism on my observations on the old leadership of the Church and the second rank prelates who did not impress me as very dynamic personalities. As of now, the old ones have the church by the scruff of the neck and would not bend a bit. It is to the theological schools that one should look for a new leadership, not to the clergy and the establishment, the nomenklatura.

62. Meeting an Old Friend Rreturning from "Exile": The Plovdiv Theological Seminary

The Theological schools have scored the most significant gain for the Church after the November revolution. During decades of exile in the Tcherepish Monastery, reduced to a priests' school on reduced curriculum and number of students, the two former seminaries had ceased to exit. It was not clear to me how they had recovered their buildings after the change. I understood that soon after November the Seminary staff had simply taken over the building of the Sofia Seminary and told the tenants, some youth organizations, to move out. No one had stopped them, and it was all over. I saw that part of the Plovdiv Seminary was still occupied by some musical school, but the wing of the Administration was already in the hands of the Rector, where he held his office. The front yard was stacked with building materials. They were in a process of restoring the Dining Hall. The south-western corner, where the medical center once stood, was now a gaping hole towards the street. I could not see the class rooms and the dorms. They were still occupied by former tenants. The Rector showed me the chapel. It had been restored, or had been left intact. I was moved deeply by sorrow and sadness. In an instant I felt the tragedy of the Church for forty-five years, now in the past. I felt as if I was in the presence of a long lost friend, having returned home from exile. I stood on the spot where for many years, in times long past, I had stood in prayer, morning and evening, in our routine seminary life. I could not control my emotions. Outside the chapel I broke down. I could not help it, I could not hide my feelings. I have always felt attached to that place. I have always felt that whatever I have achieved in life, I owed it to the Seminary. So, after so many years, I was again in the presence of this old friend returning from captivity in a lamentable state, needing my help. I reached in my pocket and gave what I could to the Rector, firmly determined when I return to America to do what I can to further help them. These people, here in Plovdiv and Sofia, are faced with gargantuan problems in their efforts to restore the schools.

Discussing the seminaries with the rectors, I suggested, quite spontaneously, on the basis of my experiences abroad, where and how the Theological schools could improve their curriculums of old times – 45-50 years ago. I see the mission of the Seminary in contemporary society not as a school of general education, but as a school preparing missionaries for a reconversion of the Bulgarian people.

63. An Island of Silence … and Exasperating Poverty: The Sofia Theological Seminary

The Sofia Seminary is an island of peace. Its parks are a natural preserve. The blooming shrubbery of bridal wreath, the luscious greenery and the concert of nightingales which I captured on my Video make a visit there a dream not to be forgotten easily. But the state of the buildings had much to be desired. Certain parts of the structures are of dire need of repair. They are uninhabitable and unusable before they see the hand of a plumber. But the most precious treasure which I saw there were the sparkling eyes of some one hundred young seminarians. I have been with young people for over thirty years in my professional capacity. I am meeting them every semester, every week, every day in my classes, but for the first time in my life I had the delightful unique experience of reading in the eyes of so many young people so much innocence, curiosity, hope, bewilderment, irresistible quest for fulfillment in a meaningful calling. I was grateful to Father Vladimir for the invitation extended to me and my wife to visit the school.

We met Father Vladimir across the street from the university. I saw him from a distance crossing to reach the trolley stop. I stopped him and initiated the conversation. In a few minutes I learned all about the Seminary. He soon understood that I have had extensive contacts and great interest in church affairs. He invited us to join him in Vesper service in the Seminary. He was on his way there. He presented himself as Chaplain of the Seminary Chapel. Later on I was informed that he was Deputy Rector and a Medical Doctor by profession. After a stroll in the park, meeting in our path many shy young seminarians, we entered the chapel for service. The Vesper service was indeed a religious feast. Nowhere do they sing the "O Gladsom Light" as they do in Bulgaria. I was touched. I was introduced to the kids by Father Vladimir as an American Professor who was a once seminarian, graduate of Sofia Theological School and instructor in Sofia Theological Seminary …

I had to respond with a few words. I told them how pleased I was to be among them. I spoke to them for the value of the education received in the seminary, for the magnificence of Orthodoxy, for the need of missionaries of the church among the Bulgarian people, for the nobility of the priestly service … all in this spirit. I was looking at these raggedy young boys and compared them to my students in America. We too, were raggedy and sloppy as students some forty-five years ago. Nothing has changed. So are also our American students. But in America it is out of negligence and slovenly attitude, and as a defiance of social conventions, rather than out of need, as is the case with these youngsters standing in front of me. Here I met face to face the misery, the poverty, the dire need. This misery showed its ugly face in the dining hall where Father Vladimir invited us to join them for supper. At the door he reprimanded several students who had rushed from the chapel to grab a bite before the rest had gotten there. "Because we were hungry" one of them explained. The dining hall had a Spartan look. The supper consisted of … lentil soup, rice pudding and bread. The bread was stale, like the bread in the whole country. But it all tasted good. Only I do not know how long it would take for these young kids to get hungry again.

We were shown the class rooms. They exhibited the ultimate of poverty. The chairs and the tables looked like they had been picked from the Salvation Army. The text books and the notebooks were some pitiful remnants of loose pieces of paper. Time and again, in Plovdiv and in Sofia, I heard desperate pleas for maps and the usual school supplies. I felt then, and feel now, that something should be done to help these schools. One may have one or another opinion of the bishops and the metropolitans, of policies and politics in the Church, but one could not argue against the proposition that the future of the Church is in the hands of the few who would come out of these seminaries and theological schools, in the hands of these kids who, against all odds, for one or another reason, have come to study theology. If the Church is worth anything for anyone, he should come now and help with whatever he can. Such help could come only from America. And as I saw the value of the dollar at this time in Bulgaria, a few dollars would go very far, $100.00 much further and $1,000.00 is a fortune. I hope I could do something about this right after I conclude this report.

64. In the Theological Academy

I did not form any opinions of the Theological Academy. Currently, it is being restored to the University of Sofia and being renamed School of Theology. The few glances which I got from the theology professors while talking to the Dean did not leave me with much comfort. A clergyman, a Prof. Shivarov, apparently knew of my activities abroad and seemed embarrassed. The other man, Prof. Hubanchev, was very polite but restrained. All along I felt some sort of a tension in the office. The spirit of accommodationism had penetrated deeply into the Academy. I heard that when the question of reinstating the Academy as a School of Theology in the Sofia University had been raised, there has been some sort of resistance from the academic body. An argument had been produced that they would be better off if they stayed under the Synod. Firmly established in their freedom, they apparently feared the possibility of interference in their work, that somebody might reveal their ties to the regime of Zhivkov.

I did not meet the students. The professors, the Dean particularly, (should explain that he was the Deputy Dean, in fact) who was moved very much when we met the previous evening, now seemed cool and reserved. He may have discussed my visit with someone prior to my coming today, had been informed of my attitude on accommodationism and was reserved. He could have arranged for me to meet the students, in one form or another. Nothing of that sort happened. God knows what might have transpired and who would have been embarrassed at such a meeting. Let them have their peace. And it is possible that it may have never occurred to him. I was told that during the dark years of the Communist era they did not suffer from lack of students. In fact they had more than they could handle. The number of 400 was mentioned to me. I might have misunderstood. Provincial young men used their status as students in the Theological Academy, when not admitted in any of the university Schools, to spend a few years in Sofia. Now they again seemed to have a flood of students. With pride they speak of some 35 or so men holding bachelors or higher degrees, but studying theology as a matter of personal interest.

The latest development which was reported to me by the Dean of the Theological Academy was an initiative taken by faculty and students at the Turnovo University for opening a Theological School at that university. The project is now under discussion and the Theological Academy is called to assist with its organization and initial staffing with teaching personnel. I was told that they might need some extra help.

A TOURIST NOTEBOOK

65. About the Small Things

I went to Bulgaria as a tourist. I did not go there to gather information to criticize the regime, or whatever. I went to Bulgaria to see my native country, to see relatives and friends, to visit the places where I had grown up. That I paid so much attention to their politics, their economics, their church affairs and social problems is only a reflection of my special interests which have accompanied me all these forty years in exile. I could not suppress them when I arrived in the country. So, let me take a look at Bulgaria through the eyes of a tourist.

When we were leaving Sofia, for Greece, sent away by Dr. Kiril Popov, a research fellow in the Bulgarian Academy of Sciences and his charming wife Mariana (they were our guardian angels, sacrificing so much time and means to make our visit most comfortable and worthwhile) – I was asked by Dr. Popov what I considered to be the most striking failure in Bulgarian life today. I thought a little bit, seeking to find the best words to sum up my worst impressions. My answer was: "The greatest failure of Bulgaria, the cause for the deepest disappointment for a foreigner, is the obvious lack of attention to the small things." Thinking my answer over, later, I found it to be, indeed, all – encompassing. They may build Palaces of Culture, imposing Party Headquarters, they may saturate the country with monuments of their heroes, kolkhoses, industrial complexes, vast condominium complexes; they may show mind – boggling statistics of the progress of Bulgaria, but when they neglect the small things needed by the individual in his daily life, hour by hour, then everything else, the big things, become lost in the travails of everyday life. When these small things are noticed and experienced by the tourist, how much more they may be torturing the ordinary Bulgarian citizen? It is an easy way out to say that the Bulgarians are used to the small nuisances. Poor Bulgarians! Poorer those who believe this nonsense! These small things leave lasting impressions on the memory of the tourist.

I would not be surprised if I heard that Bulgaria is on the black list of the world tourist agencies as a place where they would advise their patrons not to go. Having heard about tourism so much abroad, my impressions of Bulgaria were not too good. It appears to me that the Zhivkov regime had seen tourism and tourists as vacationing animals interested only of beaches and sands like cows for a midday resting place. They have built expensive hotels and resorts along the Black Sea, now empty, but they have done next to nothing to make their cities attractive places to visit. Or where they have attempted, they have looked at the big things only, and have omitted taking care of the small things. Sure there are wealthy tourists who could afford the Sheratons and the Otanis, but the average tourist is not a wealthy man, and is not looking for beaches and sands. He is interested in less expensive but comfortable and decent accommodations. We were impressed by the Sheraton in Sofia, by the Leningrad in Plovdiv, but we, frankly, did not feel like paying their prices. In Plovdiv we settled for a private apartment offered by a relative, and in Sofia we stayed at the Park Moskva, recommended to us by a friend.

66. Electricity, Elevators and Candles ... "Bulgarian Job."

So, let us speak of these small things. You are visiting the Bachkovo Monastery and on your way out stop at the local restaurant down the road. It is a charming place, but right in the middle of your dinner, the electric power is cut and the manager brings the candles. If you ask to visit the rest-rooms, what do you think the hosts are going to do? Well, they light candles all over and settle your problem. But what do you do in Plovdiv when you happen to be in the elevator on your way to the apartment, or going down? The power is cut by some invisible hand, without notice, and then you get stuck there for two hours, until the same invisible hand turns the power on and you proceed. So, what do you do? You slouch on the floor and wait for two hours for the power to come. It did not happen to us but we were made aware that such incidences occur. The rest of your stay in Plovdiv you dread such an experience sometimes, climbing to the seventh floor, with all the years on your shoulders and the tiredness of a day's walking. If they had any schedules, or any signals alerting you that the power will be turned off, then you would follow and avoid all the unpleasantry. Otherwise, if you chance it, you just stand in the elevator and count the floors. If you are a Bulgarian, visiting home, you may dismiss it with the traditional "bulgarska rabota," but if you are a foreigner, it is another matter.

67. Of the Jammed Lock and the Elevators which Never Come

Or you may be in your hotel room, ready to leave for town in the morning when you discover that the door is jammed and would not open. The cleaning lady outside hears your struggle with the key and tries to open it from outside. She fails and calls the mechanic. After twenty minutes he materializes from somewhere and lets you out. While waiting you look down from the 13th floor and think of a fire... then you go to the elevator and press the call button. You wait fifteen minutes for the elevator to come. In exasperation you call the office. They go out there, press the 13th floor up, and finally you make it. The elevators go up only when passengers start from downstairs. Most of the time you take your chances and when nothing happens you call the office downstairs. Sometimes you think of the Plovdiv elevators. But your are assured that they do not do that in Sofia. Otherwise, the hotel is good, European style, modern, vast halls, lobbies, elegantly furnished. You feel that you are in Western Europe or the United States. It is when you move a few floors up that the problems start. If you happen to drop your glasses between the night table and the bed and start looking for them, you will be disgusted by the things which you may find there. And if you want your peace, you simply do not tell your wife of your discoveries.

68. The Shower and the Bath Tub with the Yellow-Greenish Water like ...

The bathroom where they put you is another catastrophe. It all appears elegant, but after you leave the mirrors and the sinks, all else suggests difficulties. You do not see the

472

traditional shower. Instead, you are treated with an European hand-operated douche and if you are not used to it by the time you learn how to handle it, you will see the entire bathroom flooded. You take a look at the towels and you wish you had brought your own. They look rumpled, as if they had been used, and look dirty- and dirty-gray. You look for a hook to hang your pajamas or robe, and there is nowhere a hook to be seen. You ignore the hand operated douche and decide to take a bath instead. But as you see the water accumulating, you loose your desire to refresh. The water turns into a yellow-greenish color, looking like … let us forget it. Later on someone explained to me that the color of the water changed when it rains in the mountains. What filters do they have … or no filters at all. Imagine (you don't imagine, because it happens) that you have to have a drink of water from the faucet. Do you drink it? No! You don't! You will drink anything else, but not water. In Blagoevgrad a waiter brought us a glass of water, took a look at it and took it back. We settled for a glass of carbonated water. Mostly you drink *Zagorka* (Bulgarian Beer) or some other beer imported from Yugoslavia.

69. Of the Telephones, of Courtesy, Rudeness and … the Golden Heart of the Ordinary Bulgarian

A word should be said about the telephones. Bulgaria is said to have the greatest number of phones, per person, than any other country in Eastern Europe. But there are no telephone directories in the hotel rooms. In fact we never saw a telephone directory in the country. I was told that the last telephone directory was published in 1979. So you try to call information. But how do you dial information? There are no instructions in the hotel room how to dial information, how to dial the operator, how to dial long distance … Eventually you find out this information. You call the operator in the hotel, you ask for instructions, she tells you something, you don't quite understand it, and ask again. Then she takes over and with all the rudeness and discourtesy which she could muster, she gives it to you. You start regretting that you had asked. Well, you apologize and try not to tangle with her again.

Courtesy is something that the Bulgarians seems to have erased from their vocabulary and to have forgotten to practice it. You will not find anywhere in the world so much rudeness as in Bulgaria. This is not the case with people you meet on the street when you stop them and ask for directions or for any other reason. It is there that you meet the true Bulgarian and you will fall in love with him. So much courtesy, so much humanity, so much friendship, so much attention that you never forget the pleasant experiences. We met so many such Bulgarians. I will never forget the manager of a tavern between the village of Rila and Rila Monastery where we stopped for breakfast. Mrs. Popov had prepared home made food, including coffee. The tavern was not yet opened but the manager, a charming young man, opened his heart with his charge. He refused payment and gave us full attention. You do not forget such generosity. But try to speak to someone in a position of authority – be he or she only a telephone operator… you will feel that you have found all your troubles in the world. But let us forget the telephones.

70. The Delicacies of the Bulgarian Breakfast

Let us go for breakfast at the Hotel restaurant. Here you meet with another Golgotha of your visit to Bulgaria. First you look for the coffee to wake up. At a table near the entrance you see a table with a pot which they bring from the kitchen. A pile of espresso cups (four of these will fill a good cup of American coffee) are ready to satisfy your thirst. Further down you notice a big samovar of ready made tea, perhaps five gallons or so, and plenty of coffee mugs, a decent size for your habits. So, you pick one of these big tea cups and go to the coffee pot and pour yourself a good cup of coffee. You then discover that you have practically emptied the pot. Your wife, who follows you, does not get her cup filled and is disappointed. Then they take the pot to the kitchen to refill it, but it takes them forever to come back. And when you have to have a second cup, and the pot is empty, you practically have to beg them to fill it up. Then you pour yourself a big cup and feel their resentful looks nailed on you. You feel like a thief. Sometimes, as a last resort, you reach for the tea. You will have difficulty deciding if it is tea or some sort of a soup made out of herbs, but you drown it in sugar and it goes down. Then you approach the delicacies for the breakfast fare. The eggs are edible – if you decide to abandon your diet and take up a large dosage of cholesterol, under the disappointed eye of your wife. You pick some of their sausages and soon try to decide whether they are made out of meat or out of salt. It is good that they had domestic feta cheese, which saves the day. Otherwise, you have to depend on all that stuff they call ham and kielbasy. It is living death. Regularly you will find their domestic cheese danishes – banichki – but they were always dry, and stale, and salty, and if you pass them up, you do not feel cheated. But if you are used to starting your breakfast with orange juice, you will be better off if you drop the habit. On the table you will find small bottles of orange soda – if you care for soda on an empty stomach. If you do not go for the soda you may try some of their juices, served in large pitchers. First you try to decide if it is prune juice or pear juice. You will have difficult time discovering its nature and at the end will decide that it is made out of boiled dry pears. There is no toast, but the ever present stale bread. You chew on it and dream of your favorite donuts or danishes, bagels, croissants, or just a toast. One gets the impression that the Bulgarians have either never learned, or have simply forgotten the art of baking.

71. Broiled Veal ... Shopska Salata ... and the Famous Bulgarian Napkins

Somehow you get through breakfast. Lunch you eat if you manage to find time. In a few instances we landed in some decent restaurants. "Bulgaria", opposite the city park in Plovdiv, proved to be a nice corner for respite. Dinner was a more serious problem. It seems there was a conspiracy that all waiters in all restaurants will give you a menu and immediately proceed to tell you what they have ... which excludes most of the items listed for the day, and from all the items that they mention to you their *teleshko na skara* – broiled veal – sounds best. So, you order it. But it comes dry, and tough and tasteless. You eat it once, you eat it twice and the third time you look for another restaurant for better luck. You hit upon the Hungarian restaurant on Rakovska, hoping at least for Hungarian goulash… Again you end up with *teleshko na skara*. Not a bit different. And everywhere they serve you *shopska salata*. By the time you have to leave Bulgaria the *shopska salata* will come out of your ears. Either they have no cooks in Bulgaria or we just had bad luck to pick the worst. Everywhere the bread is the same … stale. But the best part in the restaurants are the prices. If you could treat six people for a dinner and it cost you three hundred leva, it is a bargain. Not so for the Bulgarians though. At that time this was a month's salary, and not the lowest to that. In the United States you probably would have to pay the minimum of $150.00 in similar circumstances. But some small things surrounded the dinner which need to be mentioned. They, of course, serve the silverware with napkins. You wonder if this is a napkin, or some sort of a joke. In first class restaurants the napkins seem to have been made out of recycled cartons. Most of the time the napkin is 4" by 4" and made out of a hard pink glazed paper which some 50 years ago was used to cover our school books. In every restaurant and hotel, on our way to the bathroom, you have to stop at a table where the attendant lady will hand you over a little napkin for the toilet – after you pay your 20 stotinki (Bulgarian pennies).

72. Bulgaria at the Head of the Black List of the World Tourist Agencies

The ordinary tourist is interested in tasting a little bit of the culture of the country: to visit its museums, its libraries, art galleries, theaters, opera, archeological sights and bookstores – as much as his time permits. It is not possible to see it all, but one hopes to satisfy his interests to some degree. After one sees what one can in Bulgaria, he will reluctantly conclude that the country will probably be placed at the end of the line for the amount of care which it's government takes to present the foreigners with the values of Bulgarian culture. You hardly see any visitors in the Bulgarian museums and art galleries. Only here and there someone may loose his way and end up in there and enjoy the accomplishments of the Bulgarian masters. In Athens, Paris, Rome and London, one cannot break through the mobs. In Bulgaria one needs a candle to find them. Of course, Bulgaria is not Athens, Rome, Paris and London, but still, even the little of what she has to show, remains unknown. Often it happens that after meeting some foreign tourists in Plovdiv, you bump into them in Sofia – in museums or in galleries – and everywhere you listen to the same complaints they have: the poverty of the Bulgarian exhibits. The Ethnographic Museum in Plovdiv appears to be run by amateurs. Everything there is overaged, worn out, dusty: the floors, the walls, the doors. The objects exhibited often impress you with their primitivism and lack of taste in the arrangement. It all looks like it has been picked out of piles of discarded items. The service (this was the case everywhere) has no class and leaves you with the impression of indifferent heywards. The National Gallery of Arts in Sofia, the former Royal Palace, impresses you with its appearance outside. For some reason, however, they have painted it some greenish color and thus have killed its beauty as an architectural monument. Inside, it is not God knows what a miracle. The elegance of the old palace is not lost, but everything seems to have faded out and shabby. Together with the tasteless arrangements it could hardly attract visitors. I had the impression that besides me and my wife there was nobody in the gallery for at least half an hour. Excepting the works of some great old masters of Bulgarian art, we did not see anything else to leave a lasting impression. Among the other things I admired two paintings of Boris Denev – the Blessing of the Soldiers and the Golgotha. The later painting I had appreciated some fifty years ago in the Plovdiv Museum. It is set against the background of Turnovo, all in a redish hue, which I had then interpreted as a symbol of the tragedy of Bulgaria after the wars of 1912-1918.

73. The Dark Halls of the Bulgarian National Museum of History

The National Historical Museum overwhelms you with the grandiosity of its structure in the heart of Sofia. It is now housed in the former Palace of Justice. Whoever conceived the idea of converting this building into a National Museum of History must have had a very high regard for the value of Bulgarian history. The credit for it is given to Ludmila Zhivkova. Whatever she was as a Communist, she has left an impressive legacy behind her and deserves all the credit for it. I heard that the legal profession in Bulgaria today feel cheated, having been forced to disperse their departments all over Sofia. But may be they deserved to be expelled from the center of the capital. The record of Bulgarian justice during the Communist era is probably the most inglorious, in reality the lowest manifestation of Bulgarian conscience. For this they do not deserve the attention which possession of this building would bring to them. They have to atone for their sins. Perhaps the Party Headquarters, in the Center of Sofia, should be given to them. It is there that they should repent for their sins and erase their sorry record of the past. So, the old Palace of Justice should stay as a Historical museum. *

Let me say a few words about this Museum, and the Bulgarian museums in general. Bulgarian history and Bulgarian archeology have some things worthy of showing to the outside world and the new generations in Bulgaria. The Museums are temples of Bulgarian culture and they should be looked after as shrines of our past. Some years ago, walking down the street in my little town of Stroudsburg, I overheard a conversation of people unknown to me, standing on the sidewalk. An elderly man was praising in glowing terms and admiration the ancient "culture of Bulgaria". He was telling his friend about an exhibit he had seen in New York. At this time a New York Museum was showing the Thracian treasures, on loan from Bulgaria. He was urging his friend to go and see the exhibit. How much more of that could be seen in Sofia?

But there are a few things which do not stand too well and should be mentioned. First, the lighting system is quite inadequate for a Museum. The exhibit halls are quite dark and the objects of interest are under spot lights. A well lighted hall would leave much better impression. We did not see this Bulgarian practice repeated in any of the European Museums. The curious visitor will find it difficult to acquaint himself with the nature of the objects exhibited in the halls. The brass plates, inscribed with small letters, with the

---

* This Spring, in May 2000, I saw the big crates sliding down a ramp on the steps of the old Museum, loaded on trucks and taken to Boyana palaces built by Todor Zhivkov, where very few, if any, visitors would venture to go and see the glory that Bulgaria has been in the past.

dimmed lighting, are invisible and unreadable. Probably this is why the patrons are so scarce and the guards are so visible, that the visitor feels as if he is followed by police dogs.

Before leaving the Museum the visitors may desire to purchase literature on some of the museum objects, cards and souvenirs. But what is offered at a little table at the exit, served by a few girls, is so poor, so tasteless and so inadequate that the imposing structure of the Museum and the overwhelming impression which it makes, is melted away and gets lost in the memory of the foreign tourist. The same is the case with the Rila and the Bachkovo monasteries. Rila Monastery was still under Government control. The day of our visit there, it was being transferred to the Church and the Abbot, Bishop Ioan, a classmate from old times and a close friend, had been called to Sofia for the transfer. Our scheduled meeting with him, which both of us looked forward to with excitement, could not take place. The Monastery kiosk does not have much to offer, and the gift shop outside, apparently government run, is a flea market place, stripped by Greek tourists, hungry for small things to buy. All that is left there deserves to be thrown in the waste basket. We found the same situation in the Ethnographic museum in Plovdiv and the National Gallery of arts in Sofia. What an opportunity to disseminate information on the Thracians or the First Bulgarian state, or the Monasteries in Bulgaria, and how little it is taken advantage of! What a national treasure these monasteries are and how little attention they are paid when there is a chance to acquaint the foreigners with them? It was an exception that we found a book by Slavcho Kissiov which my wife purchased at the Bachkovo Monastery.

The Ossuary of the monastery, with its famous medieval frescoes we found padlocked. And so was also the Assenova Krepost near Assenovgrad. The access to this medieval monument is another question. Probably when it was being built, by our ancestors of 800 years ago, there must have been a better access to the place than today. Now you have to climb and crawl over goat paths, over gaping precipices of rocks. And you climb these cliffs, paralyzed by fear of loosing your grip. At the door you will be greeted by an enormous padlock. Later on, it will leave you with a pleasant memory, but at the time you struggle to reach the fortress, it is a real challenge.

After going through all this you begin to wonder whether the Bulgarians care for tourists or not. How could they place an officer of the State, a lady sitting on a table at the entrance of every rest room in the restaurants to collect your 20 stotinki, but cannot station a volunteer guide to introduce you to the cultural treasures of Bulgaria? If they paid more attention to tourism, they could have more tourists. They could at least print a tourist guide for Bulgaria and display it at places where foreign visitors stop. They do not seem to have even domestic tourism organized. All over Europe one could see swarms of school kids led by their teachers from one site to another. But you do not see such things in Bulgaria. Maybe it is too much to expect from the Bulgarian tourist industry officials. When my wife made an observation to a journalist that Bulgaria has much to do to attract tourists from abroad, the lady did not conceal her annoyance: "Bulgaria now needs much more important things, than tourists." Probably she will be complaining, like many others, that the country does not have foreign exchange, or valuta, as they call it. They built luxurious hotels in Varna, with the narrow vision that foreign tourists are more interested in beaches and sands ... and staffed them with prostitutes for such purposes, to attract tourists. Some years ago we saw a postcard, sort of promotion literature, where the center piece of a hotel terrace is reserved for a prostitute waiting for patrons. Now we read in the papers that after the fall of the Iron Curtain, these hotels are empty and under lock – for lack of visitors. Foreign tourists are interested in more diverse things, not only in sea waters and sands. Somewhere the Bulgarians have mixed their priorities. Hopefully, they will straighten them out in the future.

74. Classic Antiquity on the Back of Nebet Tepe

We could not visit some of the archeological sites in Bulgaria, except where by chance we hit upon them. Climbing the slopes of the Nebet Tepe in Plovdiv, on the side of St. Marina Cathedral church, we ran into a magnificent Hellenistic or Roman theater, with the stage fully restored, making Plovdiv look like a true classical antiquity. I could not remember such a spectacular monument in this city just behind the hill where my school was. I had spent six years there. How did it happen that I had never seen it? Minutes later I learned from a lady passing by, that the entire complex had been discovered in late 1960s or so.

The old city of Plovdiv is a beautiful attraction as a tourist site. But I was somehow disappointed when I reached the top of Nebet Tepe. It was my favorite place during my student years. It is now inaccessible from the Seminary side where once there were steps leading to the top. How many times I had climbed them! We reached it now from the Old city. From the top you have a marvelous view of the Thracian plain, North, towards Sredna Gora, across the Maritsa River. Somewhere to the north-east I could contemplate my village in the distant mist. In old times the top of Nebet Tepe was a small beautiful park. Now, it is all dug out, turned into gaping big hole. You have to struggle through these craters to reach a place from where you could see the Seminary and the rest of the Thracian plain. Apparently the excavations have been done by archeologists who abandoned the site midway. It is a pity.

75. Of the Raggedy Bookstores in Sofia

I will conclude this report on Bulgaria with a few words about the bookstores. This is another wasteland. There was the time, and that time was some 40 years ago, when the bookstores in Sofia were loaded with paper: Books, and books, and books. Well, they were all Communist literature. These were the times of Stalin. One could go in the vast bookstore at the corner of City Park and Rakovska (or was it at the Russki Blvd. and Pl. Narodno Subranie?) and have a wide choice of publications. One could browse to his heart's content and pick and choose. The bookstores in Plovdiv and in Sofia now are poor relatives of what once was the book trade. I discovered that all bookstores have numbers, that the literature exhibited there was scanty, not much to choose from. Sometimes one could get a better choice of titles at the tables of the street vendors than in the state stores. In the underpass in front of the University in Sofia a young men had spread hundreds of volumes of books. I noticed a two-volume large English-Bulgarian dictionary. The young man asked $20.00. The price was tempting but I would not carry it all over Sofia a whole day. He offered to bring it to me to the hotel that night. Soon I forgot the incident. That night, at 10:00 o'clock, I was stopped at the elevator by the same young man. He was waiting for me for hours and hours. What a patience? I suppose it was worth it to him. I got my dictionaries and he got his $20.00, the equivalent of 400 leva – a one month's salary.

Looking over the titles displayed by these private entrepreneurs, one may indeed find items not found in the bookstores. That is where I found a book from my old professor and friend, Dr. Ivan Panchovsky. I came across a book entitled "King Ferdinand: Advises to his Son". If it did not sport as publisher The University of Sofia Press it would have been taken as a pure forgery, even though it correctly portrays Ferdinand as a super-Machiavelian. It claims to be a copy from the archives of the Coburg, written by a prominent historian, Prof. D. Yotsov. I still think that it is a spurious work, written by some Communist to discredit the former Bulgarian monarch. It is Machiavelian in style, and far surpasses the Italian renaissance political philosopher with its arrogance. It may be retitled *The Prince II*. It is a parody of an ancient Egyptian text, *The Maxims of Ptah Hotep*. Another spurious work, though identified as written by a legitimate author, purporting to be authentic, claims to have been *Conversations with Prince Kiril*. Its purpose apparently is to promote the cause of the monarchy in Bulgaria. The same kind of literature is a biography of the late Patriarch Kiril, but it is another matter, and we have already commented on it.

76. The Most Luxurious K.... in Europe

Since we are on the subject of small things which impress the tourist, let me add a few words about some trivia which characterize the political culture of the Bulgarians in these transition times. The foreign visitor in Sofia will not miss the most outstanding graffiti in the center of Sofia, spread over the facade of the Mausoleum where once the body of Georgi Dimitrov rested – Lenin style. Just under the platform where for four decades Communist dignitaries received spectacular parades, the hand of some zevzek (muckraker) has spread from one to the other end of the facade a sign reading "THE MOST LUXURIOUS TOILET IN EUROPE". Perhaps, if the place was not so prominently located, the City Council of Sofia could be well advised to convert this architectural monument into city rest-rooms, facing the former royal palace and the present Communist Party headquarters. But since the square is a focal point in Sofia and important national events are often celebrated there, rest rooms at this location might be quite inappropriate. It is a better idea, that this place would be turned into a national pantheon for the victims of political intolerance in Bulgaria since the liberation of 1878. They could erect memorials dedicated to assassinated Bulgarian leaders of the rank of Stefan Stambolov, Alexander Stamboliysky, Nikola Petkov et al. as a lesson to future generations – not to allow a repetition of the bloody tradition in our national history of politics. *

77. Brother Bulgarians, Bulgaria is Dying!

Sitting at an open air restaurant facing the Salonica Gulf, Aristotelous Square, My wife and I were reminiscing over our impressions of Bulgaria. We felt sad that my country was in such a sorry state. I was wondering if Bulgaria will ever revive, will ever resurrect to a new life. I must admit that my wife was more optimistic than I was. I thought, and still think, that the country is dying and that nothing could be done to regenerate it. When I was in Sofia, the last day of our trip, sitting at a curb to the far left of the Dimitrov Mausoleum, near Vasil Levski Street, facing the palace, some political organization – I believe it was the Democratic Party – was holding a political rally, to be followed by a Memorial Service for Prince Alexander Battenberg. The key speaker, I think Stefan Savov, leader of the Democratic Party, was shouting in the loudspeakers: "Brother Bulgarians, Bulgaria is dying. If something is not done in the next two years, Bulgaria will be dead." Nobody stopped to attend the rally. People were just passing, listening and going away. So did we in a few minutes. I could not help hearing these words again, now, in this cool pleasant evening at the Salonica Bay, amidst splitting voices of boisterously enjoying themselves Greek kids playing in the large square. These words still reverberate in my ears.

It is with these words of sorrow and regret that I am going to finish this report, the words of Stefan Savov: "Bulgaria is dying. If something is not done in the next two years, Bulgaria will be dead." I did not see anything, and I did not hear anything in Bulgaria to dissuade me that these words do not express this sad, but inevitable fate of my dear country, if things do not take another course.

*

The English text of this document was completed on October 15, 1991. We beg the reader's indulgence for the technical defects of it. From September 1, when the Bulgarian text was published, to this day, much has happened in Bulgaria. To a certain extent it has only confirmed our impressions. We saw no reason to change anything that we had originally included in our observations.

---

* Early this year, seized by a zeal of partisan passion, the government moved to erase the Mausoleum from the center of Sofia, as if to wipe out any memory of Dimitrov. Several attempts were made to dynamite it, but the cursed thing stubbornly refused to fall – for the delight of the communists and the embarrassment of the officials who hoped for a triumphant victory.

XIII
PHOTOS

My father (1890–1952)

Seminarian - 1939-1945

Theology student - 1945-1949

**Shortly after crossing the border –
June 19, 1951.**

483

Parish "Brotherhood St. Petka" My mother is on the extreme right of the second row from the top. Sitting in the middle is the village priest, Stefan M. Tafrov, executed by Communists on October 7, 1944.

My brothers Petar and Stoiu - with their families, 1962.

Student, c. 1946

My seminary classmates (Top)
With friends in the village - (Middle)
Seminary Graduation pictures.
(Bottom)

Seminary Diploma

Memorial service for my parents. With Father Kolentsov, 1993. (Top); Same service. From left to right: my wife Ruby, my niece Minka, my brother Stoiu, his wife Vessa, with their grand-daughter, Vesselina and their daughter Penka. (Bottom).

An old pear tree - planted and grafted by me, c. 1939.

A willow tree, planted by me, c. 1938.

Sakar Tepe, 1992.

Remnants of the old oak tree. 1993.

The old "Chitalishte" (Public Library), 1992.

War victims memorial in Zelenikovo.

With Patriarch Maxim, 1992.

Celebrating our 35th wedding anniversary with my wife - 1991.

Zdravko with his family in Brazil, c. 1970.

The Village church in Zelenikovo.

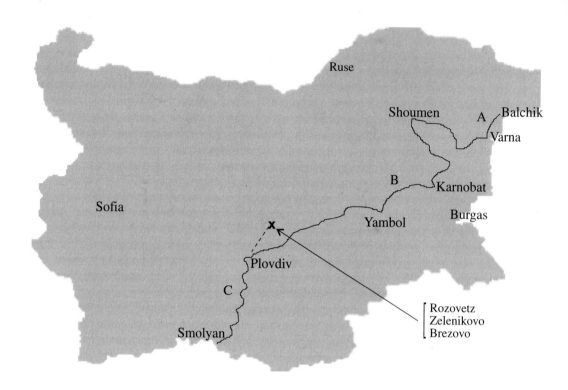

My escape route - 1951, from Balchik to the Greek border.

Our escape route – from Balchik, via Plovdiv, to the Greek border.

A. Balchik (Dobrudja) – Varna – By Foot
B. Varna – Shoumen – Yambol – Plovdiv – By Train
C. Plovdiv – Chepelare – Smolyan – Border – By Foot

Balchik – Varna – May 7, 45 klm. 6 Hours
Varna – Plovdiv – May 7-8. 18 Hours
Rodopi – „Goriani" – May 8 – June 19
Border – Paranesti (Greece), June 20-21